Handbook on Continuous Improvement Transformation

Aristide van Aartsengel • Selahattin Kurtoglu

Handbook on Continuous Improvement Transformation

The Lean Six Sigma Framework and Systematic Methodology for Implementation

Springer

Aristide van Aartsengel
Wijk Aan Zee
Netherlands

Selahattin Kurtoglu
Bochum
Germany

ISBN 978-3-642-35900-2   ISBN 978-3-642-35901-9 (eBook)
DOI 10.1007/978-3-642-35901-9
Springer Heidelberg New York Dordrecht London

Library of Congress Control Number: 2013933989

Printed on acid-free paper

Springer is a part of Springer Science+Business Media (www.springer.com)

Acknowledgments

Throughout the two books entitled: "A Guide to Continuous Improvement Transformation: Concepts, Processes, Implementation" and "Handbook on Continuous Improvement Transformation: The Lean Six Sigma Framework and Systematic Methodology for Implementation," we have referenced many illustrious practitioners to whom we are obviously indebted. The text of these volumes books has also been hugely improved through the friendly criticism and advice given by various extremely generous individuals.

We would like to express our gratitude to all those who taught us, who worked with us over the years, and who have helped us with this work or inspired new ideas. They are, literally, too numerous to mention. Many of the ideas and examples come from practice. We are therefore especially indebted to the many colleagues, managers, and CEOs who have allowed us to share their work on continuous improvement and project management.

We also wish to acknowledge dozens of people from our client organizations, practicing Kaizen in manufacturing plants, Business Process Management, Project Management, Lean Six Sigma, Lean Manufacturing, Total Quality Management (TQM), Total Quality Control (TQC), and Total Productive Maintenance (TPM), to whom we owe special thanks and who have shown the applicability of the ideas and methods described in this handbook.

We would also like to acknowledge all the client organizations over the years that have trusted our advice and provided us with the greatest laboratory there is—their organizations. Their willingness to test new hypotheses contributed greatly to the material. We extend a deep bow to IQPM Consulting for giving us such an interesting subject about which to learn.

Finally, our families deserve loving mention, and sincere thanks, for putting up with the hours of time spent hunched over our computers writing and revising the content of this book.

Wijk Aan Zee, Netherlands — Aristide van Aartsengel Ph.D.,
Bochum, Germany — Selahattin Kurtoglu Ph.D.,

Contents

1 Introduction . . . 1
1.1 The Purpose of This Book . . . 2
1.2 What Makes This Book Different . . . 4
1.3 How Is This Book Structured? . . . 5

2 Defining Lean Six Sigma . . . 7
2.1 Setting the Stage: Why Six of These Sigma? . . . 7
2.2 Standard Deviation, Quality and Cost . . . 11
2.3 Quality Related Costs Elements . . . 12
2.3.1 Prevention Costs . . . 13
2.3.2 Appraisal Costs . . . 13
2.3.3 Failure Costs . . . 14
2.3.4 The Cost of Quality . . . 15
2.4 Why Lean? . . . 17
2.4.1 Early Production Developments . . . 18
2.4.2 Scientific Management and Mass Production Developments . . . 18
2.4.3 Principles of Mass Production . . . 20
2.4.4 "Lean" or "Flexible" Production Method . . . 22
2.5 Conclusion . . . 24

3 Framework and Methodology . . . 25
3.1 Operational Definition of a Process . . . 25
3.2 Setting the Framework and Methodology . . . 27

4 "PDSA Initiate" Process Group . . . 31
4.1 Identify Customers and Stakeholders . . . 33
4.1.1 Develop a List of Customers and Stakeholders . . . 34
4.1.2 Analyze Stakeholders and Their Interests . . . 39
4.1.3 Record Stakeholders Information . . . 42
4.2 Develop Project Charter . . . 42
4.2.1 Project Purpose or Justification . . . 43
4.2.2 Project Success Criteria . . . 44
4.2.3 High-Level Description of the "Process to be Improved" . . . 44
4.2.4 Conclusion . . . 45

4.3 Develop Preliminary Project Scope Statement 49
4.3.1 Defining the Project Objectives 50
4.4 Perform Phase Review 51

5 "PDSA Plan" Process Group 53
5.1 The Purpose of Planning 53
5.2 The "PDSA Plan" Constituent Processes 54

6 Develop Project Management Plan 57
6.1 Elements of a "Process Improvement" Project Plan 58
6.1.1 Scope 58
6.1.2 Phases 59
6.1.3 Milestones 59
6.1.4 Activities 60
6.1.5 Tasks 60
6.1.6 Effort 60
6.1.7 Resources 61
6.1.8 Project Schedule 61
6.1.9 Project Risk 61
6.2 Collating the Materials 62

7 Develop Project Management Scope 65
7.1 Collect Requirements: V.O.B., V.O.C., & V.O.P. 67
7.2 Define Scope 69
7.3 Verify Scope 71
7.4 Control Scope 72
7.4.1 Scope Creep 72
7.4.2 Hope Creep 73
7.4.3 Effort Creep 73
7.4.4 Feature Creep 73

8 Collecting V.O.C. Requirements 75
8.1 Plan V.O.C. Capturing 75
8.1.1 Identify V.O.C. Data and Clarify Goals 77
8.1.2 Develop Operational Definitions and Procedures 77
8.1.3 Develop Sampling Strategy 89
8.1.4 Validate Data Collection System 92
8.2 Collect and Organize Data 125
8.2.1 Organize V.O.C. Data: Affinity Clustering 125
8.3 Analyze Data and Generate Customer Key Needs 128
8.4 Translate Customer Key Needs into CTXs 130
8.5 Set Specifications for CTXs 131
8.6 Conclusion 135

9 Create Work Breakdown Structure 137
9.1 Defining a Work Breakdown Structure 137
9.2 Developing a Work Breakdown Structure 138
9.3 Uses of a Work Breakdown Structure 141

10 Develop Time Management Plan 143
10.1 Define Activities 143
10.2 Assess Completeness of Activities 145
10.3 Sequence Activities 146
10.3.1 Network Diagram Formalism 147
10.3.2 Network Preparation 149
10.3.3 Constructing the Project Network Diagram 150
10.4 Estimate Activity Resources 151
10.5 Estimate Activity Durations 151

11 Develop Project Schedule Plan 155
11.1 Basic Approach to Scheduling 156
11.2 Update the Project Network Diagram 156
11.2.1 Showing Times on Arrow Networks 157
11.2.2 The Program Evaluation and Review Technique (PERT) 157
11.3 Develop Schedule Control Plan 174
11.3.1 Choose Control Subject 174
11.3.2 Establish Standard Performance 175
11.3.3 Plan and Collect Appropriate Data 175
11.3.4 Summarize Data and Establish Actual Performance . . . 175
11.3.5 Compare Actual Performance to Standard 176
11.3.6 Validate Control Subject 176
11.3.7 Take Action on Difference 176

12 Develop Resources Management Plan 179
12.1 Defining Resource Management 179
12.2 List the Resources to Be Consumed by the Project 180
12.2.1 Labor 180
12.2.2 Facilities 183
12.2.3 Equipment 184
12.2.4 Materials 184
12.3 Assign the Resources to Project Activities 185
12.4 Plan Project Team Development and Management 186

13 Develop Quality Management Plan 189
13.1 Develop Quality Plan 191
13.1.1 Collect Requirements: V.O.P. 192
13.1.2 Define Quality Plan 192
13.1.3 Verify Quality Plan 194
13.1.4 Control Quality Plan 195

13.2 Develop Quality Assurance Plan 195
13.2.1 Define the Quality Goals for the Processes 196
13.2.2 Identify All Relevant Organizational Process Assets . . . 196
13.2.3 Define Roles and Responsibilities of "Quality Assurance" Activities 197
13.2.4 Identify Tasks and Activities for "Quality Control" . . . 197
13.3 Develop Quality Control Plan 198
13.3.1 Choose Control Subject 198
13.3.2 Establish Standard of Performance 199
13.3.3 Plan and Collect Appropriate Data 200
13.3.4 Summarize Data and Establish Actual Performance . . . 200
13.3.5 Compare Actual Performance to Standards 200
13.3.6 Validate Control Subject 200
13.3.7 Take Action on the Difference 201
13.4 Conclusion 202

14 Collecting V.O.P. Requirements 203
14.1 Plan V.O.P. Data Capturing 203
14.1.1 Identify V.O.P. Data and Clarify Goals 204
14.1.2 Develop Operational Definitions and Procedures 205
14.1.3 Develop Sampling Strategy 210
14.1.4 Validate V.O.P. Data Collection System 211
14.2 Collect Data 211
14.3 Summarize Data & Display Patterns 211
14.3.1 Control Charts 213
14.3.2 Run Charts 221
14.3.3 Scatter Diagrams 223
14.3.4 Frequency Plots 224
14.3.5 Pareto Charts 227
14.4 Establish Process Performance 229
14.4.1 Process Yield: Rolled Throughput Yield 230
14.4.2 Process Defect Rate 231
14.4.3 Process Capability & Process Performance Indices . . . 233
14.5 Characterize Process & Revise Process Quality Targets 241
14.5.1 The Ideal State (No Failure) 242
14.5.2 The Threshold State (Process Outcome Failure) 243
14.5.3 The Brink of Failure (Process Failure) 243
14.5.4 The State of Total of Failure (Double Failure) 244
14.5.5 Summary of Process Characterization 245

15 Experimental Study: Design of Experiments 247
15.1 Designing and Conducting and Experimental Study 247
15.2 Basic Concepts 249
15.2.1 Replication 249
15.2.2 Extraneous Input Variables 249

15.2.3 Blocking (Planned Grouping) 249
15.2.4 Randomization 250
15.2.5 Randomized Block Design 251
15.2.6 Incomplete Block Designs 251
15.2.7 Balanced Incomplete Block Designs 252
15.2.8 Factorial Designs 252
15.2.9 2^k-Factorial Designs 253
15.2.10 Confounding 253

16 Develop Cost Management Plan 255
16.1 Plan Cost Data Collection 256
16.1.1 Cost Classifications for Assigning Costs 257
16.1.2 Cost Classifications for Predicting Cost Behavior 259
16.1.3 Cost Classifications for Management and Operations . . 264
16.1.4 Cost Classifications for Quality 265
16.1.5 Cost Classifications for Buying and Selling 268
16.1.6 Cost Classifications for Project Economics 269
16.1.7 Cost Classifications for Decision Making 270
16.2 Collect Costs Data 271
16.2.1 Personnel Costs 271
16.2.2 Operating and Maintenance (O&M) Costs 272
16.2.3 Capital Costs 273
16.2.4 Overhead Costs 273
16.2.5 Additional Costs 274
16.2.6 Why Estimate Costs 274
16.3 Allocate Costs to Activities 278
16.3.1 Make or Buy Analysis 279
16.3.2 Planned Value of Project Activities 284
16.4 Control Spending 287
16.4.1 Choose Control Subject 287
16.4.2 Establish Standard Performance 287
16.4.3 Plan and Collect Appropriate Data 287
16.4.4 Summarize Data and Establish Actual Performance . . . 289
16.4.5 Compare Actual Performance to Standard 296
16.4.6 Validate Control Subject 296
16.4.7 Take Action on Difference 297

17 Develop Procurement Management Plan 299
17.1 When to Develop a Procurement Plan? 300
17.2 Developing the Procurement Management Plan 300
17.2.1 Plan Procurement 302
17.2.2 Plan Contracting 312
17.2.3 Invite Tenders 337
17.2.4 Select Optimal Suppliers 340
17.2.5 Administer Contracts 351
17.2.6 Close Contracts 359

18 Develop Communication Management Plan 363
18.1 Project Communication 363
18.2 Project Communication Management 365
18.2.1 Communications Planning 366

19 Develop Risk Management Plan 381
19.1 Understanding the Nature of Risk 381
19.2 Characterizing Risk 383
19.3 Characterizing Project Risk Management 385
19.4 Develop Risk Management Planning 387
19.5 Identify Project Risks 390
19.5.1 Project Risks Classification 391
19.5.2 Risk Description 395
19.5.3 Project Risks Data Collection 396
19.6 Perform Risk Assessment 407
19.6.1 Likelihood of Occurrence of a Risk Event 407
19.6.2 Effect of Occurrence of a Risk Event: Risk Impact . . . 409
19.6.3 Risk Matrix: Importance or Ranking of Risks 410
19.6.4 Risk Prioritization 412
19.7 Develop Risk Response Planning 413
19.7.1 Planning Responses 414
19.7.2 Developing a Response Plan 419
19.8 Monitor and Control Risk 423
19.8.1 Choose Control Subject 423
19.8.2 Establish Standard Performance 423
19.8.3 Plan and Collect Appropriate Data 424
19.8.4 Summarize Data and Establish Actual Performance . . . 425
19.8.5 Compare Actual Performance to Standard 426
19.8.6 Validate Control Subject 426
19.8.7 Take Action on Difference 426

20 Conduct the Project Retrospective 429
20.1 Understanding the Reflection Process 429
20.2 When to Start the Reflection Process? 430
20.3 Layers of Reflection 431
20.4 Facilitating Learning and Continuous Innovation 432

21 Assess Overall Plan and Implementation 435
21.1 Perform Planning Phase Review 435
21.2 Identify and Document Lessons Learned 436

22 Conclusion to "PDSA Plan" 439

23 "PDSA Do" Process Group 443
23.1 The "PDSA Do" Constituent Processes 443

24 Build Deliverables . . . 447
24.1 Identify and Quantify Assignable Causes of Variations . . . 447
24.1.1 Process Behavior Charts . . . 448
24.1.2 Interviews . . . 449
24.1.3 Types of Interviews . . . 453
24.2 Conclusion . . . 454

25 Explore Cause-and-Effect Relationship . . . 455
25.1 Ishikawa's Cause-and-Effect Diagram . . . 455
25.1.1 Drawing Ishikawa's Cause-and-Effect Diagram . . . 457
25.2 Fault Tree Diagram (FTD) . . . 458
25.2.1 Drawing Fault Trees: Gates and Events . . . 459

26 Verify Identified Assignable Causes . . . 463
26.1 Plan Assignable Causes Data Capturing . . . 463
26.1.1 Identify Assignable Causes Data and Clarify Goals . . . 464
26.1.2 Develop Operational Definitions and Procedures . . . 464
26.1.3 Develop Sampling Strategy . . . 466
26.1.4 Validate Data Collection System . . . 467
26.2 Collect Cause-and-Effect Relationship Data . . . 467
26.2.1 Design and Conduct Experiments . . . 468
26.2.2 Design and Conduct Observational Studies . . . 468
26.3 Analyze Collected Data . . . 469
26.3.1 Summarize Data & Display Patterns . . . 470
26.3.2 Analyze Cause-and-Effect Relationship Data . . . 471

27 Analyze Process Steps and Tasks . . . 483
27.1 Identify Goal . . . 485
27.2 Explore Constraints . . . 486
27.3 Is Goal Carried Out to a Satisfactory Standard? . . . 486
27.4 Examine Operation . . . 488
27.4.1 Generate Hypotheses . . . 489
27.4.2 Examine Resources-Task Interaction . . . 497
27.4.3 Analyze Cognition Within the Discrete Element Context . . . 505
27.4.4 Estimate Cost-Benefits of Hypotheses: Value Added . . . 507
27.5 Examine Goals by Re-description . . . 508
27.6 Summarize Data & Display Value Stream Diagram . . . 509
27.6.1 Summarize Data . . . 509
27.6.2 Display Maps and Flowcharts Diagrams . . . 510

28 Generate Improvement Solutions . . . 519
28.1 Brainstorm Using Available Data Generated So Far . . . 519
28.2 Prioritize Potential Solutions . . . 522
28.3 Develop Prototype, Assess Risk & Pilot Solution(s) . . . 523
28.3.1 Develop Prototype . . . 524

28.3.2 Pilot Solution(s) 525
28.3.3 Assess and Reduce Risk 537
28.3.4 Conclude the Pilot 540
28.3.5 Develop Implementation Plan 540

29 Monitor and Control Execution 547
29.1 Perform Time Management 548
29.2 Perform Resource Management 550
29.3 Perform Quality Management 551
29.4 Perform Cost Management 552
29.5 Perform Procurement Management 554
29.6 Perform Communication Management 555
29.7 Perform Risk Management 556
29.8 Perform Deliverable Alteration Management 557
29.8.1 Submit Alteration Request 558
29.8.2 Review Alteration Request 559
29.8.3 Identify Alteration Feasibility 559
29.8.4 Approve Alteration Request 559
29.8.5 Implement Alteration Request 560
29.9 Conduct the Project Retrospective 560
29.10 Perform Phase Review 561
29.11 Identify and Document Lessons Learned 562

30 Conclusion to "PDSA Do" 565

31 "PDSA Study" Process Group 569
31.1 The "PDSA Study" Constituent Processes 569

32 Study Deliverables 573
32.1 Collect Retrospective Data: V.O.B., V.O.C., & V.O.P. 574
32.2 Summarize Overall Data and Display Patterns 575
32.3 Analyze Data and Validate Process Performance 575
32.4 Develop a Process Control Plan 578
32.5 Reinforce a Positive Context of Process Improvement 579
32.6 Continuously Monitor "Improved Process" and Context 583
32.6.1 Cumulative Sum (CUSUM) Control Charts 584
32.6.2 Exponentially Weighted Moving Average Control Charts 586
32.6.3 Continuously Monitor the People Aspect of the Context 587

33 Monitor and Control Study Execution 589
33.1 Perform Time Management 590
33.2 Perform Resource Management 590
33.3 Perform Quality Management 590
33.4 Perform Cost Management 592
33.5 Perform Procurement Management 592
33.6 Perform Communication Management 593

33.7 Perform Risk Management 594
33.8 Perform Deliverable Alteration Management 594
33.9 Perform Deliverable Acceptance Management 595
33.10 Conduct the Project Retrospective 598
33.11 Perform Phase Review 599
33.12 Identify and Document Lessons Learned 599

34 Conclusion to "PDSA Study" 603

35 "PDSA Act" Process Group 607
35.1 Implement "Improved Process" and Install All Deliverables 608
35.2 Complete Project Documentation 608
35.3 Reinforce Mechanisms and Build Capability 610
35.4 Create Standard Practices and Procedures 611
35.5 Release Resources and Adjourn Project Team 612
35.6 Settle Contractual Aspects and Final Accounting 613
35.7 Write Final Report 613
35.8 Conduct Post-implementation Review 615
35.9 Celebrate Success and Share the Wealth 617
35.9.1 Celebrate Success 617
35.9.2 Share the Wealth 619
35.10 Conclusion to "PDSA Act" Process Group 620

36 Conclusion 623
36.1 Data Collection System: The Fundamental Engine of "Process Improvement" 623
36.1.1 Predictive Data 624
36.1.2 Baseline Data 625
36.1.3 Formative Data 625
36.1.4 In-Process Data 627
36.1.5 Retrospective Data 627
36.2 Learning and Knowledge 629
36.3 Final Admonition 633

References 635

Index 639

List of Figures

Fig. 2.1	A plot of a normal distribution (or bell curve)	9
Fig. 2.2	A plot of a normal distribution with scrap and rework areas	12
Fig. 2.3	Quality costs categories and their relative magnitude	16
Fig. 3.1	Detailed PDSA cycle for improvement	28
Fig. 4.1	"PDSA Initiate" Process Group	32
Fig. 4.2	Influence/interest grid for stakeholder prioritization	41
Fig. 5.1	"PDSA Plan" Process Group	55
Fig. 7.1	Project scope management process	66
Fig. 8.1	The V.O.C. management process	76
Fig. 8.2	Sample template for V.O.C. data collection	79
Fig. 8.3	Sampling methods	91
Fig. 8.4	Variations in process outcome over time	93
Fig. 8.5	Bias and variability in shooting arrows at a target. Bias means the archer systematically misses in the same direction. Variability means that the arrows are scattered	95
Fig. 8.6	Sampling distribution for $\bar{Y}$	99
Fig. 8.7	Sampling distribution for $\bar{Y}$ and an observed value of $\bar{Y}$	100
Fig. 8.8	The intra-class correlation coefficient and the gauge R&R ratio	106
Fig. 8.9	Generic critical values of the chi-square distribution with n-1 degree of freedom	120
Fig. 8.10	Clustering customers' needs based on affinity	127
Fig. 8.11	Example of affinity clustering of the V.O.C.	128
Fig. 8.12	Kano model of customer key needs	130
Fig. 8.13	Sample CTX template	132
Fig. 8.14	Specification limits for a characteristic of the "process to be improved" outcomes	133
Fig. 9.1	Hierarchical visualization of the work breakdown structure	139
Fig. 10.1	Project time management process	144
Fig. 10.2	Representation of various activities and events	148
Fig. 11.1	Three different methods for showing times on arrow networks	158
Fig. 11.2	Example of PERT diagram	165
Fig. 11.3	The critical path on a PERT diagram	169
Fig. 11.4	The project schedule control process	174
Fig. 13.1	"Develop Quality Management Plan" process	190

Fig. 13.2 "Perform Planning of 'Quality Plan'" process 191
Fig. 13.3 The quality control process 199
Fig. 14.1 The V.O.P. management process 204
Fig. 14.2 Example of control chart 214
Fig. 14.3 Example of frequency (dots) plot 224
Fig. 14.4 Common shapes of frequency plots 226
Fig. 14.5 Example of Pareto chart 227
Fig. 14.6 Four examples of process capability 235
Fig. 14.7 Aligning V.O.C. (specifications) with V.O.P. 239
Fig. 14.8 Process outcomes versus process operation grid 241
Fig. 16.1 The cost management plan process 256
Fig. 16.2 The linearity assumption and the relevant range of variable cost 260
Fig. 16.3 The cost performance baseline chart 285
Fig. 16.4 The control spending process 288
Fig. 16.5 Earned value on the cost performance baseline chart 290
Fig. 16.6 Illustration of cost variance and schedule variance 292
Fig. 16.7 Cost performance baseline and earned schedule concept 293
Fig. 17.1 The project procurement management process 303
Fig. 17.2 The contract performance control process 354
Fig. 18.1 Influence/interest grid for stakeholder prioritization 372
Fig. 18.2 Example of A3 report template 378
Fig. 19.1 Example of likelihood, magnitude and subjective risk judgment 383
Fig. 19.2 The risk management process 386
Fig. 19.3 Example of project risks classification 393
Fig. 19.4 The risk response matrix 421
Fig. 19.5 The project risk monitoring and control process 424
Fig. 19.6 Example of cost effective risk control analysis 428
Fig. 22.1 Minimum activities of the "PDSA Plan" phase 441
Fig. 23.1 "PDSA Do" process group 444
Fig. 24.1 Effects of assignable causes in process outcome over time 448
Fig. 25.1 Ishikawa (fishbone) diagram 456
Fig. 25.2 Example of a fault-tree logic tree 460
Fig. 26.1 Scatter plot patterns 471
Fig. 27.1 The cycle of task analysis decisions 484
Fig. 27.2 A simple information-model of an operation. (**a**) Shows how the operation is represented using input, action and feedback. (**b**) Substitutes 'action' with planning for an action and executing the action 501
Fig. 27.3 Basic flowchart symbols 511
Fig. 28.1 Sequential building of knowledge with piloting through PDSA 532
Fig. 29.1 Deliverable alteration management process 558
Fig. 30.1 Minimum activities of the "PDSA Do" phase 567
Fig. 31.1 "PDSA Study" process group 570
Fig. 32.1 Example of control chart with baseline and retrospective data 576

Fig. 32.2 Example of reaction to "Process Improvement" transformation ... 582
Fig. 33.1 Deliverable acceptance management process ... 596
Fig. 34.1 Minimum activities of the "PDSA Study" phase ... 605
Fig. 35.1 Minimum activities of the "PDSA Act" phase ... 621
Fig. 36.1 Barriers to achieving business results from a project ... 626
Fig. 36.2 Data ability to influence project results ... 628

List of Tables

Table 2.1	12 digits Microsoft Excel calculations of p(z) and (1- p(z))	10
Table 4.1	Project customers list	36
Table 4.2	Project stakeholders list	37
Table 4.3	Project sponsors list	38
Table 4.4	The S.I.P.O.C.	45
Table 8.1	Four measurement scale levels	78
Table 8.2	Table of control limits constants for averages	111
Table 8.3	Table of constants for standard deviations	112
Table 8.4	Table constants for ranges	113
Table 8.5	Table constants for d_2^*	114
Table 8.6	One-way ANOVA table	125
Table 12.1	Labor listing	181
Table 12.2	Project manager responsibilities in the HRM practice areas	182
Table 12.3	Staff member responsibilities in the HRM practice areas	182
Table 12.4	Facilities listing	183
Table 12.5	Equipment listing	184
Table 12.6	Materials listing	185
Table 12.7	Resource schedule calendar	186
Table 14.1	Prioritization matrix template	208
Table 14.2	FMEA matrix template	210
Table 14.3	12 Digits microsoft excel calculations of process yield p(z) and process fall out (1−p(z))	232
Table 14.4	Plan of action for process capability	236
Table 14.5	Common process capability indices	237
Table 14.6	Common process performance indices	237
Table 14.7	Basic plan of action for process improvement	245
Table 16.1	Generic summary layout of a project costs	275
Table 17.1	Example of technical specifications	310
Table 18.1	The S.I.P.O.C.	364
Table 18.2	Stakeholders communications requirements	368
Table 18.3	Stakeholders communications schedule	369
Table 18.4	Stakeholders communications matrix	369
Table 19.1	Example of risk events and conditions internal to projects	388
Table 19.2	Generic risk management plan content	390

Table 19.3 Risk register 397
Table 19.4 Risk description 398
Table 19.5 Definition of risk impact scale for "Threat Events" on four project success criteria 410
Table 19.6 Definition of risk impact scale for "Opportunity Events" on four project success criteria 411
Table 19.7 Risk rating matrix 412
Table 21.1 Phase review form for the planning phase 437
Table 25.1 Classic fault tree diagram symbols 461
Table 28.1 Prioritization matrix template 522
Table 28.2 Deciding on the scale of the pilot 532
Table 29.1 Phase review form for the "PDSA Do" project phase 563
Table 32.1 Factors for cumulative sum control chart, $\alpha = 0.00135$ 585
Table 33.1 Phase review form for the "PDSA Study" project phase 600

Introduction 1

In today's hyper-competitive international marketplace, with severe economic turmoil, "Continuous Improvement" transformation is a condition for achieving and sustaining success. For an enterprise business not just to perform excellently, but to perform excellently consistently there must be improvement efforts in both the "Continuous Improvement" philosophy and break-through improvement methodology. Every enterprise business must have systematic methods for making smart decisions, attacking problems, improving its products (i.e. tangible products) and services (i.e. intangible products), repelling competitors, and keeping customers delighted. Anything less than a systematic, disciplined approach is leaving the enterprise business future in the hands of chance.

"Continuous Improvement" transformation may perhaps be the most misunderstood concept. In our first book entitled "A Guide to Continuous Improvement Transformation: Concepts, Processes, Implementation," we have examined the core of the art of "Continuous Improvement" transformation. We have delved into the key characteristics and constituents necessary to take the enterprise business to the next level to continue to exist in the long term; namely, the eight overarching determining factors of strategic management that matter the most. We have provided to our readers' enterprise business leaders—Project Managers, Green Belts, Black Belts, managers at all levels, and process improvement professionals—the necessary insight for the management thinking that must be put into practice on a daily basis. Hence, our first book answered the basic question to ask of management for a successful implementation of any improvement initiative: "Are we doing the right things?"

But even those who truly understand the essence of "Continuous Improvement" transformation and "do the right things" often struggle when it comes to execution. Applying the "Continuous Improvement" transformation philosophy at operation level on a daily basis is not as easy as it seems it should be.

A. van Aartsengel and S. Kurtoglu, *Handbook on Continuous Improvement Transformation*,
DOI 10.1007/978-3-642-35901-9_1,

1.1 The Purpose of This Book

The progressive realization of a "Continuous Improvement" transformation requires a framework and a systematic methodology for studying the constituent elements or processes associated with the determining factors of the system considered. It also requires a way of differentiating between the different types of variation present in those processes. In addition to the way of thinking described in our first book, and which must be put to practice on a daily basis, there are techniques to be learned.

The main focus of this second book is the framework and systematic methodology require for studying the constituent elements or processes and systems associated with the eight overarching determining factors of strategic management described in our first book. This second book establishes a sound basis for effective planning, scheduling, resourcing, decision making, management, and plan revision for the (production) line activities designed to support realization an enterprise business intended strategy. Hence, the rationale of this book is to provide an answer to the second basic question to ask of management for a successful implementation of any improvement initiative: "Are we doing things right?"

The (production) line activities designed to support realization of an enterprise business intended strategy include projects and operations activities that matter the most. They have fundamentally different objectives.

A project is a sequence of unique, complex, and connected activities having one goal or purpose and that must be completed by a specific time, within budget, and according to specification. It is a temporary effort undertaken to create a unique product, service, or result. The purpose of a project is to attain its objectives and then terminate. Within enterprise businesses, projects seldom exist in isolation. They originate as a result of alignment arising from the enterprise business intended strategy and business plans and, as such, exist alongside operations and within a portfolio of other projects. Projects are therefore utilized as a mean of achieving an enterprise business intended strategy. They conclude when their specific objectives have been attained. Operations activities are ongoing and repetitive efforts, the purposes of which are to sustain the enterprise business. When their objectives have been attained, operations adopt a new set of objectives and the work continues.

Although projects and operations sometimes overlap, both share the following characteristics: they are constrained by limited resources; they are selected following analyses of their added value in terms of costs and benefits to the enterprise business; they are performed by people; and they are planned, executed, and controlled. Another key characteristic that projects and operations also share is that they often use common series of sets of logically related discrete elements (tasks, actions, or steps) with well defined interfaces in order to achieve their objectives. These sets of logically related discrete elements (tasks, actions, or steps) are not goals in themselves within an enterprise business; they are mean to achieve operations and projects work. We define a process as:

> *A set of logically related discrete elements (tasks, actions, or steps) taken in order to achieve a particular end.*

In this definition, a discrete element, the performance of which is measurable, is meant to be the smallest identifiable and essential piece of activity that serves both as a unit of work and as a means of differentiating between the various aspects of a project or an operation work. Each discrete element is designed to create unique outcomes by ensuring proper control, acting on and adding value to the resources that support the work being completed.

From the perspective of this definition, a process acts on and adds value to the resources that support the activities being completed by a project or an operation work. Furthermore, each discrete element of a process has two aspects:

1. Its operational definition or specific technical content, which is addressed in a next chapter, and
2. Its context, which is represented by everything else that surrounds and affect the specific technical content.

A process is a set of logically related discrete element (tasks, actions, or steps) taken in order to achieve a particular end. But when most people think of process at work, it is much more than the operational definition or specific technical content of its discrete elements that they are reacting to: it is the patterns of interaction ensuing from the resulting specific technical content, plus the resulting context.

Thus a process is characterized by the patterns of interaction, coordination, communication, and decision making employees use to transform resources into products and services of greater worth. Processes include not just manufacturing processes, but those by which product development, procurement, market research, budgeting, employee development and compensation, and resource allocation are accomplished. Some processes are formal, in the sense that they are explicitly defined and documented. Others are informal: they are routines or ways of working that evolve over time. The former tend to be more visible, the latter less visible.

Processes are defined or evolve de facto to address specific tasks. This means that when employees use a process to execute the tasks for which it was designed, it is likely to perform efficiently. But when the same seemingly efficient process is used to tackle a very different task, it is likely to prove slow, bureaucratic, and inefficient. In contrast to the flexibility of resources, processes are inherently inflexible. In other words, a process that defines a capability in executing a certain task concurrently defines disabilities in executing other tasks. One of the dilemmas of management is that processes, by their very nature, are set up so that employees perform tasks in a consistent way, time after time. They are meant not to change or, if they must change, to change through tightly controlled procedures in order to avoid unproductive habits.

Improving processes—the subject of this book—offer an effective way to train enterprise business personnel to break unproductive habits and adopt the "Continuous Improvement" transformation philosophy while, at the same time, achieve breakthrough performance and unprecedented results. Through process improvement projects and operations activities, cross-functional teams learn how to make improvements in a methodological way. They learn how to apply specific improvement tools, establish relevant performance measures, and sustain their gains.

Most importantly, they learn how to work with one another to solve problems rapidly and in a highly effective way. After completion of a process improvement project or operation activity, these team members become ambassadors for change, spreading their learned behaviors across the enterprise business.

With each process improvement project or operation activity, the pool of ambassadors for change grows, fueling a cultural shift that begins to place "Continuous Improvement" transformation as the enterprise business' top priority and increasingly authorizes the employees themselves to design and implement operation improvements. After a series of many "process improvement" projects and operations activities that reach into various operating units, an enterprise business should be better positioned to begin a "Continuous Improvement" transformation. But while "process improvement" projects and operations activities provide the focus, structure, and skilled facilitation that enable "Continuous Improvement" maturity stage to take place within an enterprise business, the need for "process improvement" projects and operations activities never goes away.

1.2 What Makes This Book Different

With growing interest in Lean Six Sigma in the professional project management community and the commonality of activities presented in the many books, papers, training seminars and missives on improvement methodologies presented to management, comes the question: "How does 'Lean' and Six Sigma methodologies relate to the Project Management Body of Knowledge?"

The distinctive feature of this book that makes it different from the remaining literature is that it integrates the project management precepts covered in what has emerged as the world standard of project management knowledge—A Guide to the Project Management Body of Knowledge (Project Management Institute)—with the Plan-Do-Study-Act (PDSA) model for improvement and the Lean Six-Sigma concepts.

Unlike most of the project management methodologies available in the literature, this second book is not a guide to undertaking "process improvement" projects with useful tips, tools and techniques. It provides an entire methodology for undertaking "process improvement" projects or operations activities. It can be used by a student to learn how to complete a "process improvement" project from end-to-end, by a project manager to structure the way that a "process improvement" project should be undertaken and by a business owner to mandate the manner within which "process improvement" projects will be undertaken across the entire enterprise business.

The integration of the PDSA model for improvement, Lean Six Sigma methodology with the tools and techniques of project management in this book holds significant promise for enterprise businesses in need to get the most from their continuous improvement efforts. This integrated approach, which can be used for both transactional and manufacturing businesses, better define ways to accomplish cost reduction, process enhancement, faster implementation and new product or service development. Specifically, this framework:

1. Is useful as a roadmap for all size of projects: small, simple "process improvement" projects as well as large—system improvement projects;
2. Provides a framework for the application of improvement tools and methods;
3. Encourages planning to be based on good practices over a wide range of different projects and industries;
4. Is useful for both process and product improvement;
5. Can be used for the design of new processes and products;
6. Is applicable to all types of enterprise businesses;
7. Is applicable to all groups and levels in an enterprise business;
8. Facilitates the use of teamwork to make improvements;
9. Emphasizes and encourages the iterative learning process;
10. Allows project plans to adapt as learning occurs;
11. Offers a simple way to empower people in the enterprise business to take action.

It is a comprehensive framework that enterprise businesses or organizations can adopt, not a set of helpful hints for light reading. As such, it has been written in a clear, professional and formal manner.

1.3 How Is This Book Structured?

The structure of this book is reflected in the "Table of Contents." It consists of 36 chapters organized beyond the first two chapters into five parts associated with the PDSA model for improvement. We have followed the Project Management Body of Knowledge (PMBOK) standards advocated by the Project Management Institute (PMI) to foster consistency with the project management profession and to ensure comprehensiveness of the PDSA model.

Following an introductory chapter, the second chapter sets the stage by providing a short but comprehensive overview of what Lean Six-Sigma is and addresses the key characteristics of Six-Sigma. The material in Part I—Chaps. 3 and 4—focuses on defining the framework and "process improvement" project initiation. In Part II—Chaps. 5–22—we address the "process improvement" project planning. The chapters of this part are concerned with the essentials of planning a "process improvement" project in terms of activities, costs, and schedule. This second part also contains a discussion of phase-gate management retrospective and review. Part III of the text—Chaps. 23–30—then gets into actual "process improvement" project execution. In Part IV—Chaps. 31–34—we describe the project management processes necessary to sustain the deliverables built over the long term and to build new knowledge through learning from the deliverables built Part III. Part V (Chap. 35) discusses the steps required to ensure effective closeout of a "process improvement" project, by acting upon the built and studied deliverables based on what was learned from the previous project phase.

Defining Lean Six Sigma 2

The word sigma is the eighteenth letter of the Greek alphabet (Σ, σ), transliterated as 'S, s'. These symbols are used to denote a mathematical sum (Σ) and a standard deviation (σ); the term standard deviation was introduced to statistics by Karl Pearson (1894), which is a quantity calculated to indicate the extent of deviation for a group of elements as a whole from their expected central tendency. A low standard deviation indicates that the elements tend to be very close to their expected central tendency, whereas high standard deviation indicates that the elements are spread out over a large range from their expected central tendency.

2.1 Setting the Stage: Why Six of These Sigma?

Why six of these "sigma"?
What happened to the first five "sigma"?
What about those "sigma" coming after the sixth "sigma"?

When referring to the Greek letter, as a quantity calculated to indicate the extent of deviation for a group of elements as a whole from their expected central tendency, it is assumed that these elements are homogeneous and spread out in different spatial and temporal places following a regular and stable manner of performance (or occurrence). The stability of occurrence is a very important requirement as it allows the extraction of meaningful measures from these elements. Regardless of their origin and nature, these elements will display variations over time.

We can think of variation as change or slight difference in condition, amount, or level from the expected occurrence, typically within certain limits. Variation has two broad causes that have an impact on data collected from these elements: common (also called random, chance, or unknown) causes and assignable (also called special) causes.

Common causes of variation are inherent and an integral part in the business activities been considered. They can be though of as the "natural pulse of the business activities" and they are indicated by a stable, repeating pattern of variation.

A. van Aartsengel and S. Kurtoglu, *Handbook on Continuous Improvement Transformation*,
DOI 10.1007/978-3-642-35901-9_2, © Springer-Verlag Berlin Heidelberg 2013

Assignable causes of variation are those causes that are not intrinsically part of the business activities been considered but arise because of specific circumstances. When they occur, they signal a significant occurrence of change in the business activities and they lead to a statistically significant deviation from the norm. Assignable causes of variation are indicated by a disruption of the stable, repeating pattern of variation. They result in unpredictable performance of the business activities and must therefore be identified and systematically removed before taking any other steps to improve quality of the business activities considered.

In business applications, the elements considered are measurable features or measurable characteristics of business activities outcomes. Outcomes of business activities can be products, transactions, services delivered, sub-parts or particular features of these entities. In the remaining of this chapter, we will use the term "element" as a generic term to designate a measurable feature or a measurable characteristic of these entities. Here, the concept of outcome of business activities is multi-dimensional as it comprises a core benefit or service for which the customer has a need or want. It has a physical existence which is manifest in its price and quality, its performance, specification, design, reliability and longevity. It has a service that involves such things as its warranty, delivery, after-sales service and promotional support. And even beyond that, it has psychological characteristics such as the outcome image and brand and corporate images which are perceived by existing and potential customers.

Thus, by considering each element of a group as a balanced sum of a large enough number of unobserved random events acting additively and independently, each of which with finite mean and variance, the central limit theorem tells us that the occurrence pattern of the elements of the group will tend to follow a normal distribution in nature.

A normal distribution is a very important statistical data distribution pattern occurring in many natural phenomena, such as height, blood pressure, lengths of objects produced by machines, etc. The frequency of occurrence of certain data, when graphed as a histogram (data scores on the horizontal axis, amount of data or frequency on the vertical axis), creates a bell-shaped curve known as a normal curve, or normal distribution.

As illustrated in Fig. 2.1, normal distributions are symmetrical with a single central peak at the mean (μ, average) of the data. The shape of the curve is described as bell-shaped with the graph falling off evenly on either side of the mean. Fifty percent of the distribution lies to the left of the mean and fifty percent lies to the right of the mean. The spread of a normal distribution is controlled by the standard deviation, σ. The smaller the standard deviation, the more concentrated the data around the mean.

This tendency of elements to form a normal distribution is somewhat analogous to the tendency of water to run down a hill—it is simply the easiest and most natural way of going. In order to have water run down a hill, all we need is water and a hill. In order to have numerical values form a normal distribution, all we need is the summation (Σ)—the combined additive result—of a multiplicity of random coincidences. This simple but very important principle, upon which this whole handbook rest on, is embodied on the formal side of probability theory by central limit theorem.

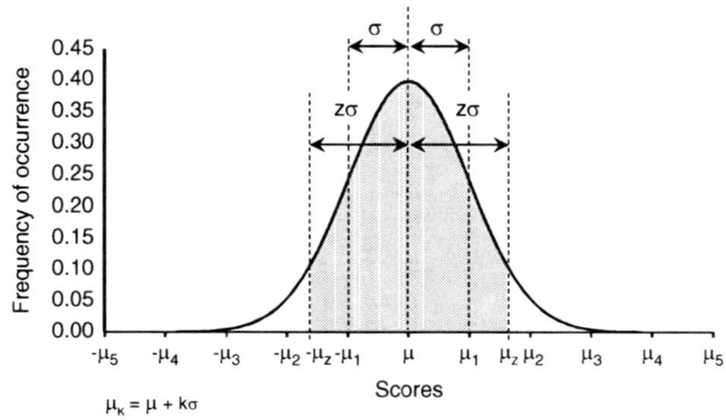

Fig. 2.1 A plot of a normal distribution (or bell curve)

For occurrence patterns normally distributed, the proportion $\rho(z)$ of elements falling within z standard deviations around the central tendency (i.e. the mean) is determined to be:

$$p(z) = \mathrm{erf}\left(z/\sqrt{2}\right)$$

Where erf is the error function.

For various values of z, the percentages of elements falling within and beyond z standard deviations of the central tendency are shown in Table 2.1.

During his work on "Economic Control of Quality of Manufactured Product" (Shewhart, Economic Control of Quality of Manufactured Product, 1931), Shewhart created the control chart with 3 standard deviations around the central tendency as a performance permissible limit of variations. Shewhart's use of 3-sigma limits, as opposed to any other multiple of sigma, did not stem from any specific mathematical computation. Rather, the choice of 3-sigma limits was seen to be an acceptable economic value, and it was also justified by "empirical evidence that it works." No calculations from the normal distribution, or any other distribution, were involved in the choice of the multiplier of 3. Certainly, Shewhart did then check that this multiplier turned out to be reasonable under the artificial conditions of a normal distribution—and plenty of other circumstances as well.

From Table 2.1, we can observe that a business application which operates at a performance permissible limit of variations of 3 standard deviations around its expected central tendency will result in 2.7 elements falling beyond 3 standard deviations from the expected central tendency, out of one thousand occurrences.

Similarly, a business application which operates at a performance permissible limit of variations of 4.5 standard deviations around its expected central tendency will result in 6.8 elements falling beyond 4.5 standard deviations from the expected central tendency, out of one million occurrences.

Furthermore, a business application which operates at a performance permissible limit of variations of 6 standard deviations around its expected central tendency will result in 1.97 elements falling out of 6 standard deviations from the expected central tendency, out of one milliard occurrences.

Table 2.1 12 digits Microsoft Excel calculations of p(z) and (1- p(z))

z	% falling within $z\sigma$	% falling beyond $z\sigma$	Amount falling beyond $z\sigma$ out of:		
			One thousand occurrences	**One million occurrences**	**One billion occurrences**
1.0	0.682689492137	0.317310507863	317.3	317310.5	317310507.9
1.5	0.866385597462	0.133614402538	133.6	133614.4	133614402.5
2.0	0.954499736104	0.045500263896	45.5	45500.3	45500263.9
2.5	0.987580669348	0.012419330652	12.4	12419.3	12419330.7
3.0	0.997300203937	0.002699796063	2.7	2699.8	2699796.1
3.5	0.999534741842	0.000465258158	0.5	465.3	465258.2
4.0	0.999936657516	0.000063342484	0.1	63.3	63342.5
4.5	0.999993204654	0.000006795346	0.0	6.8	6795.3
5.0	0.999999426697	0.000000573303	0.0	0.6	573.3
5.5	0.999999962021	0.000000037979	0.0	0.0	38.0
6.0	0.999999998027	0.000000001973	0.0	0.0	1.97
6.5	0.999999999920	0.000000000080	0.0	0.0	0.08
7.0	0.999999999997	0.000000000003	0.0	0.0	0.0

While a chosen performance permissible limit of variations around the expected central tendency might work well for certain business applications, it might not operate optimally or cost effectively for applications with a higher performance permissible limit of variations. A pacemaker business application might need higher standards, for example, whereas a direct mail advertising campaign might need lower standards. An automobile car factory business application might need higher standards, for example, whereas a hotel customer service might need lower standards.

In this book, the basis and justification for choosing 6 (as opposed to 3 or 4.5, for example) standard deviations as the permissible limit of variations around the expected central tendency for stable business applications is due to the fact that it results in all produced/occurred elements falling within 6 standard deviations from the expected central tendency, out of one million occurrences, while not more than of two elements are likely to fall beyond 6 standard deviations from the expected central tendency, out of one billion occurrences.[1]

[1] A popularly prearranged definition of a "six sigma" process, in the "six sigma" literature, is one in which there are about 3.4 defects per million opportunities, under a mythological assumption that an unpredictable process will not shift location more than ± 1.5 sigma. This assumption does not hold true in most high temperatures combustion applications where heat transfer by radiation is predominant. When a high temperature combustion process is operated unpredictably there is no limit on the size of the shifts that can occur.

2.2 Standard Deviation, Quality and Cost

In business applications which operate at a performance permissible limit of variations of z standard deviations around the expected central tendency, every element within those business applications is intended to add value to the enterprise (businesses & customers) as a whole. It has a set of requirements or descriptions of what an element needs to add value to the enterprise. When a particular element meets those requirements, it is said that it has achieved quality, provided that the requirements accurately describe what the businesses and the customers actually need. Those occurred elements falling beyond z standard deviations of the expected central tendency are often regarded as flaw, unacceptable, or in non conformance quality within the group considered. They will undergo more or less corrective actions: rework, scrapping (of whatever can not be reworked) and conformance use.

Within the enterprise as a whole, we can consider three views for describing the overall quality of an element.

First is the view of the business producing an element—the business is primarily concerned with the design, engineering, and activities involved in producing an element. Quality is then measured by the degree of conformance to predetermined specifications and standards, and deviations from these standards can lead to defects also referred to as non conformance quality, unacceptable, or poor quality and low reliability. Hence, efforts for quality improvement are aimed at eliminating defects (components and subsystems of elements that are out of conformance), minimizing the need for scrap and rework, and hence overall reduction in production costs.

Controlling and improving quality of business activities outcomes has become an important business strategy for many organizations; manufacturers, distributors, transportation companies, financial services organizations; health care providers, and government agencies. Quality is a competitive advantage. A business that can delight customers by improving and controlling quality can dominate its competitors.

Second is the view of the consumers or users of the produced element—to consumers, a high-quality element is one that well satisfies their preferences and expectations. This consideration can include a number of characteristics, some of which contribute little or nothing to the functionality of the element but are significant in providing customer satisfaction.

Quality has become one of the most important consumer decision factors in the selection among competing products and services. The phenomenon is widespread, regardless of whether the consumer is an individual, an industrial organization, a retail store, a bank or financial institution, or a military defense program. Consequently, understanding and improving quality are key factors leading to business success, growth, and enhanced competitiveness. There is a substantial return on investment from improved quality and from successfully employing quality as an integral part of overall business strategy.

A third view relating to quality of an element is to consider the element itself as a system and to incorporate those characteristics that pertain directly to the value it adds to the enterprise through its usage and functionality. This approach should include overlap of the businesses and customers views.

Fig. 2.2 A plot of a normal distribution with scrap and rework areas

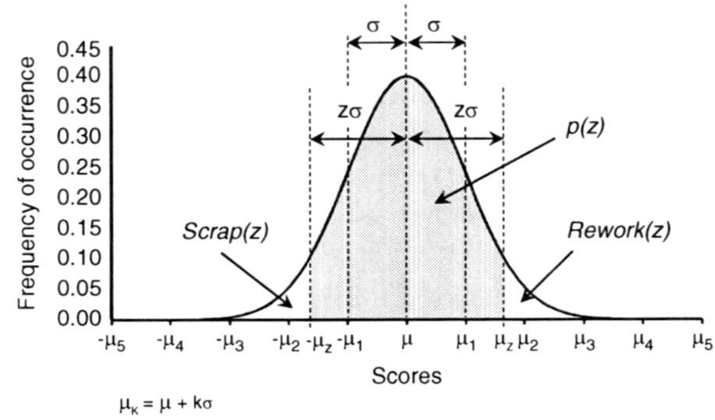

Thus, keeping to the minimum and almost to none those elements falling beyond z standard deviations of the expected central tendency is the key concern of businesses as these elements have nominal production costs associated to them and eventually excess costs associated to their corrective actions (rework, scrapping and conformance use).

For a given element, we can think of its associated excess cost as the cost incurred as a result of deviation of the element from the expected central tendency of the group as a whole. The excess cost, which is equal to the sum of an excess cost of production plus and excess cost of conformance use, is equal to zero at the expected central tendency for a group of elements as a whole.

We can also think of the cost of scrap as the cost of the raw materials plus the cost of all processing done to the element, including the cost of inspection and the cost of disposal of the element as well. In business practices, scrapping an element is done only when it is cheaper to scrap it than to use or keep it within the group. The expected cost of reworking an element is less than or equal to the cost of scrapping the element. In business applications which operate at a performance permissible limit of variations of z standard deviations, Fig. 2.2 shows the data score intervals of those elements in non conformance quality.

The total cost of corrective actions (rework, scrapping and conformance use) is known as the Cost of Quality (CoQ). It is a measure of the costs specifically associated with achievement or non achievement of an element quality—including all elements requirements established by the business and its contracts with its customers. Requirements include marketing specifications, end-product and process specifications, purchase orders, engineering drawings, company procedures, work instructions, professional or industry standards, governmental regulations, and any other documents or customer needs that can affect the definition of an element.

2.3 Quality Related Costs Elements

Over the last several decades, quality costs have been divided into several categories (Campanella, 1999). By increasing magnitude, these are: prevention, appraisal, and failure costs.

2.3.1 Prevention Costs

Prevention costs are the costs of all activities specifically designed to prevent poor quality in element. These costs can be divided into two categories: costs related to non-conforming elements and costs incurred because the business activities to produce them are themselves less than adequate.

There are those costs that may be regarded as an essential part of business activities, for example field testing, design proving, failure modes and effect analysis. These are really costs associated with performing good business practice; they would be incurred regardless of the failure and appraisal costs and are not to be considered in this definition of prevention costs.

Costs that are considered in the definition of prevention costs are those that must be incurred if the current cost of failure and appraisal is to be reduced. These represent an investment in the "Continuous Improvement" initiative and, if effective, should result in a significant reduction of the overall costs. Obviously, these costs are likely to be small otherwise the failures would not occur and relevant appraisal cost would not be necessary.

2.3.2 Appraisal Costs

These are costs associated with measuring, evaluating or auditing elements to assure conformance to quality standards and performance requirements. These costs can be divided into two categories: costs related to non-conforming elements and costs incurred because the business activities to produce them are themselves less than adequate.

There are those costs that must be incurred regardless of the likelihood of occurrence of the associated adverse risk event, because the consequences of such an event are severe and potentially life threatening. Such is the case for many of the controls and procedures at power stations. This form costs are not to be considered in this definition of appraisal costs. Because they will always be incurred regardless of the likelihood of occurrence of a threatening risk event.

Costs that are considered in the definition of appraisal costs are those that are related directly to the likelihood of occurrence of error or failure. In this case, the amount of appraisal costs increases as the likelihood of occurrence of error increases more or less in direct proportion and vice-versa. Business activities which are included embrace all the costs of: incoming and source inspection/test of purchased material; in-process and final inspection/test; product, process or service audits; calibration of measuring and test equipment; and associated supplies and materials; which are carried out for no other reason than that the related failure or non achievement of an element quality occurred.

2.3.3 Failure Costs

These are costs resulting from elements not conforming to requirements or customer/user needs. Failure costs are divided into internal and external failure categories.

2.3.3.1 Internal Failure Costs

These are failure costs occurring prior to delivery or shipment of an element to the customer. Internal failure costs can be many and varied. They include all costs and losses due to performing again what has already been done, or repairing or modifying the result of an activity, the cost of post mortems and all other consequential costs together with the waste of resources performing the business activities that need to be redone.

The consequential costs will include the effect on the balance sheet of excessive inventory and work-in-process (WIP) resulting from quality related deficiencies. In service industries, the equivalent problems do not show in inventory, but are hidden in direct costs. Most inventory and work-in-process, other than work actually being processed, can be regarded as Quality-Related costs. These include:

1. Reworking, redoing or repeating activities already performed because of inadequate performance at the first attempt. Costs of modification resulting from previous undetected design or planning weaknesses. These costs include the associated design or planning business activities, changes to tools and cost of retraining if procedures and methods are changed.
2. Retro design of a business activity element with a known design fault and all of the associated new features, fixtures and tools. Extra space in stores to accommodate replacement parts with different issue numbers. Revisions to parts lists, instruction manuals and the increased complexity of related service activities.
3. Increases to inventory and work-in-process due to disruptions to the smooth flow of work.
4. Modifications due to poor quality design.
5. Storage space

2.3.3.2 External Failure Costs

These are failure costs occurring after delivery or shipment of the product—and during or after furnishing of a service—to the customer. These costs can be further subdivided into residual and random categories.

The residual non conformances of produced elements to requirements or customer/user needs include the underlying costs of warranty calls, servicing, complaints, etc.. . .

Some of the more spectacular costs may be found in the random category which, if they occur, can produce catastrophic results. These will include product recall or product withdrawal. Enterprise businesses often spend fortunes on advertising how good their products or services are; then suddenly they are plunged without warning into huge expenditure telling the public that they have put their lives at risk. In many cases, this negative publicity is overwhelmed by media attention, which places the very survival of the enterprise business at stake.

Other external costs which can also be included in the records include:

1. Failed product (resp. service) launches which are due to deficiencies in the product (resp. service) and identified and exposed by its first customers. These costs are invariably incurred when an enterprise business is overzealous in its attempt to obtain prior franchise with an innovative new product (or service) and is a common problem. In these cases of failed product (resp. service) launches, the enterprise business tries to take shortcuts and fails to test and prove the product (or service) performance characteristics prior to launch. This results in the customer unwittingly being the first inspector of the product (or service).
2. Failure to meet either the emotional or specified needs of the customer: this is usually caused by poor voice of the customer capturing, poor market research and poor competitor-related information, inadequate and misdirected promotion, wrong launch time, short shelf-life in the case of chemical, food and pharmaceutical products, contamination, poor packaging and consequent adverse publicity.
3. Customer complaints and the recording and analysis of customer complaints, and the cost of running a customer service department (i.e. a euphemism for customer complaints department).
4. Excessive after-delivery, service or maintenance support. Excessive costs including storage, delivery and all related administration, particularly those that infer, conceal from or mislead the public.

The failure costs go far beyond the internal and external costs indicated above. They include the devastatingly demotivating impact on employees within an enterprise. Employees want to feel good about the quality of their work. But regrettably, some enterprise businesses make decisions, and design systems, that deprive employees of their right to pride in workmanship, a prerogative that Edwards Deming considered one of the keys to motivation in the workplace (Deming, The New Economics: For Industry, Government, Education, 1994; Deming, 1982).

2.3.4 The Cost of Quality

The Cost of Quality is the total of the costs of the above costs. As indicated already, it is a measure of the costs specifically associated with achievement or non achievement of an element quality—including all elements requirements established by the business and its contracts with its customers. It is not the cost of creating a quality element; it is the cost of NOT creating a quality element. It represents the difference between the actual cost of an element and what the reduced cost would be if it did not deviate from the central tendency within the group as a whole. It is the total of the costs incurred by:

1. Investing in the prevention of nonconformance to requirements.
2. Appraising an element for conformance to requirements.
3. Failing to meet requirements.

Fig. 2.3 Quality costs categories and their relative magnitude

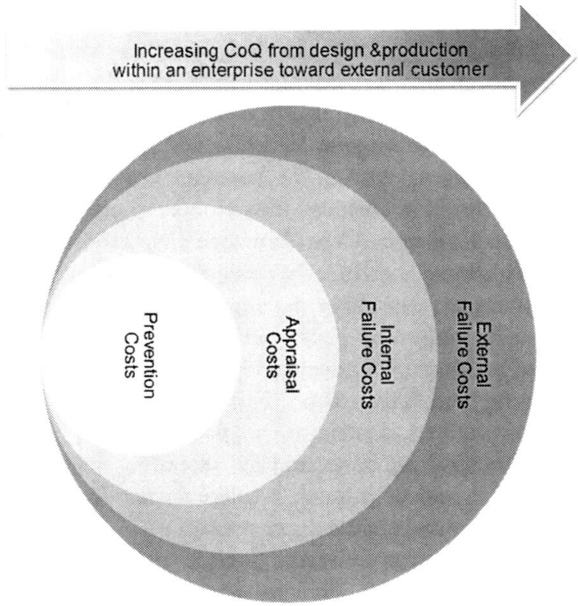

For a given element, the Cost of Quality increases as the element moves toward the consumer, as shown in Fig. 2.3.

We have mentioned in the previous section that a low standard deviation indicates that the occurred elements tend to be very close to their expected central tendency. When this happens, there will be few excess costs associated with the use of those occurred elements. Also, a high standard deviation indicates that the occurred elements are spread out over a large range from their expected central tendency. As an occurred element falls further and further away from the expected central tendency, the costs of keeping it within the group and using it will increase. Moreover, these costs are often extremely high and they increase up to the point where it will be cheaper to scrap or rework the unacceptable element than it will be to try to keep it within the group and use it further.

For business applications operating at 6 standard deviations as the permissible limit of variations, all occurred elements will certainly fall within 6 standard deviations from the expected central tendency, out of one million occurrences, while not more than of two elements are likely to fall beyond 6 standard deviations from the expected central tendency, out of one billion occurrences. Thus, fewer revenues are spent on rework and scrapping of elements in non conformance quality.

From this, it becomes self evident that focusing on minimizing (standard) deviations in key business applications or "making zero-defect products, profitably," hence minimizing excess costs and controlling quality in those applications, is the true goal behind the adoption of "Six standard deviations as the permissible limit of variations from the expected central tendency" (i.e. six-sigma) in enterprises.

2.4 Why Lean?

We have indicated in a previous section that, in business applications, the elements considered are measurable features or measurable characteristics of business activities outcomes. Furthermore, the concept of outcome of business activities is multi-dimensional:

1. It comprises a core benefit or service for which the customer has a need or want.
2. It has a physical existence which is manifest in its price and quality, its performance, specification, design, reliability and longevity.
3. It has a warranty, a delivery, an after-sales service and promotional support.
4. It has psychological characteristics such as the outcome image and brand and corporate images which are perceived by existing and potential customers.

An enterprise business needs activities that are worthy and can handle today's values and complexities accurately and efficiently; activities that are positioned for the future and therefore can move ahead with the business and not struggle along behind.

In this book, we shall think of a "Lean" business activity, which is often known simply as "Lean" or "Flexible," as a business activity that considers the expenditure of resources for any goal other than the creation of value in the element considered for the end customer to be wasteful. Eliminating waste is invariably the first and simplest way of improving the way things are done, in much the same manner as removing assignable causes of variation. We shall say that:

"A 'Lean' business activity is a business activity that is:

1. *Effective—Producing the desired outcome correctly the first time;*
2. *Efficient—Minimizing the resources used to produce the desired outcome in the shortest time;*
3. *Flexible or Adaptable—Being able to adapt to changing customers and to the circumstances surrounding the business and its market needs."*

The term "Lean" in the production context was first coined by John Krafcik in his 1988 article, "Triumph of the Lean Production System," based on his master's thesis at the MIT Sloan School of Management (Krafcik, 1988).

Toyota production line is the most often cited exemplar of "Lean" business activity where the insight is to continually improve both the efficiency and the effectiveness of work by eliminating unnecessary actions and activities (Koichi & Takahiro, 2009; Shingo & Dillon, 1989; Womack, Jones, & Roos, 2007; Ohno, 1988; Monden, 2011; Wang, 2010; Dennis, 2007). This insight descends from Taylor's 'scientific management' and much of the subsequent 'human relations' work that was focused on how to bring management and labor together in productive partnership from which both should gain (Taylor, 1911; Fayol, 1949).

2.4.1 Early Production Developments

Preceding 'scientific management,' the nineteenth-century factory production system was characterized by ad hoc organization, decentralized management, production organized on a craft basis with informal relations between employers and employees, and casually defined jobs and job assignments. Work was performed by highly skilled craftsmen who often prepared their basic raw materials, carried the product through each of the stages of manufacture, and ended with the finished product. These skilled craftsmen used with unsystematic workshop methods based on customary practice—the "rules of thumb" wielded by skilled craftsmen—as well as the "arbitrariness, greed, and lack of control."

Typically, the craftsman spent several years at apprenticeship learning each aspect of his trade; often he designed and made his own tools. He was identified with his product and his craft, enjoyed a close association with his customers, and had a clear understanding of his contribution and his position in society. While his product may be of extremely high quality, the uniqueness can be detrimental as seen in the case of early automobiles: No two products were exactly identical, and in many cases each product was intentionally made different from others. In craft production, all or most aspects of the work process are determined by the worker in accordance with the empirical lore that makes up craft principles.

By the end of the nineteenth century, however, increased competition, novel technologies, pressures from government and labor, and a growing consciousness of the potential of the factory had inspired a wide-ranging effort to improve organization and management.

2.4.2 Scientific Management and Mass Production Developments

The central figure in the movement to improve organization and management by the end of the nineteenth century was the American engineer, inventor, and management theorist Frederick W. Taylor. The events of Taylor's early years played a large and important part in these activities. Daniel Nelson, in "A Mental Revolution: Scientific Management since Taylor," chronicles the following (Nelson, 1992):

> *Born in 1856 into an aristocratic Philadelphia family, Taylor had the benefit of tutors and exclusive schools, extended travel, and associations with the Philadelphia elite. After attending Phillips Exeter Academy, he rejected a university education in favor of a traditional apprenticeship and an industrial career, which began in the machine shop of the Midvale Steel Company in 1878. ... Taylor left in 1893 to become a self-employed consultant.*
>
> *By that time he had taken important steps toward a new role. He had a substantial reputation as an inventor of industrial machinery and broad experience as an industrial manager. He had also undertaken several experiments that forced him to think more explicitly about organizations and people. One of these, an effort to compute operating times for machine tools with a stopwatch, would evolve into time and motion study, his signature contribution to industrial management.*

Taylor's groundwork was time and motion study which involved the detailed study of work and the assessment of what a normal competent worker would achieve working at normal speed for a given time.

In Taylor's view, the task of factory management was to determine the best way for the worker to perform the work, to provide the proper tools and training, and to provide incentives for good performance. After carefully studying the smallest parts of simple tasks, such as the shoveling of dry materials, Taylor was able to design methods and tools that permitted workers to produce significantly more with less physical effort. He broke each task down into its individual motions, analyzed these to determine which were essential. Later, by making detailed stopwatch measurements of the time required to perform each step of manufacture, Taylor brought a quantitative approach to the organization of production functions. With unnecessary motion eliminated, the worker, following a machinelike routine, became far more productive.

At the same time, Frank B. Gilbreth and his wife, Lillian M. Gilbreth, U.S. industrial engineers, began their pioneering studies of the movements by which people carry out tasks (Merrill, 1970). Using the then new technology of motion pictures, the Gilbreths analyzed the design of motion patterns and work areas with a view to achieving maximum economy of effort. The "time-and-motion" studies of Taylor and the Gilbreths provided important tools for the design of contemporary manufacturing systems.

Daniel Nelson further records that:

... Taylor had become associated with two enterprises that were reshaping the industrial environment. The first was the rapidly maturing engineering profession, whose advocates sought an identity based on rigorous formal education, frequent contact, mutually accepted standards of behavior, and social responsibility. In factories, mines, and railroad yards, they rejected the empiricism of the practitioner for scientific experimentation and analysis. They acknowledged the primacy of the profit motive, but they insisted that reason and truth were essential to continued financial success.

The second, closely related development was the systematic management movement, an effort among engineers and sympathizers to substitute administrative systems for the informal methods of industrial management that had evolved with the factory system. Systematic management was a rebellion against tradition, empiricism, and the assumption that common sense, personal relationships, and craft knowledge were sufficient to run a small factory. In the large, capital intensive, technologically advanced operations of the late nineteenth century, 'rule-of thumb' methods resulted in confusion and waste. The revisionists' answer was to replace traditional managers with engineers and to substitute managerial systems for guesswork and ad hoc evaluations.

By the time Taylor began his career as an engineer and manager, cost accounting systems, methods for planning and scheduling production and organizing materials, and incentive wage plans were staples of engineering publications and trade journals. Their objective was an unimpeded flow of materials and information. In human terms, proponents of systematic management sought to transfer power from the first-line supervisor to the plant manager and to force all employees to pay greater attention to the manager's goals. Most threatening, perhaps, they advocated decisions based on performance rather than on personal qualities and associations.

... By 1901 Taylor had fashioned scientific management from systematic management. As the events of Taylor's career make clear, the two approaches were intimately related.... His first report on his work, 'Shop Management' (1903), portrayed an integrated complex of systematic management methods, supplemented by refinements and additions like time

study. Between 1907 and 1909, with the aid of one of his shrewdest associates, Morris L. Cooke, he wrote a sequel to 'Shop Management' that ultimately became The Principles of Scientific Management (1911). Rather than discuss the specific methods he introduced in factories and shops, Taylor used colorful stories and language to illuminate 'principles' of management. To suggest the integrated character and broad applicability of scientific management, he equated it with a 'complete mental revolution'.

Though Taylor used the words "a complete mental revolution" to describe his contributions to factory or "shop" management, Morris L. Cooke, a friend and professional associate, and Louis Brandeis, a prominent attorney, deliberately chose the words "scientific management" to promote their contention that Taylor's methods were an alternative to railroad price increases in a rate case they were preparing for the Interstate Commerce Commission.

Taylor's 'scientific management' and much of the subsequent 'human relations' work clearly accommodated some subjective judgment. This was the basis of many and various incentive payment systems which were applied across much of manufacturing industry and which became the symbolic focus of industrial disputes through the first half of the twentieth century.

Taylor's 'scientific management' was concerned first and foremost with how a business could survive. Its aims were twofold. Firstly, to improve both the efficiency and the effectiveness of work by eliminating unnecessary actions and activities, improving methods and building in suitable relaxation breaks. Secondly, to share the resulting benefit between employer and employee and so remove the distrust between workers and management which had resulted in 'soldiering', a phenomenon of workers purposely operating well below their capacity, or slow working and restricting output intended by the workers to safeguard employments.

Much of the credit for bringing these early concepts of time and motion, and 'human relations' studies together in a coherent form, and creating the modern, integrated, mass production operation, belongs to the U.S. industrialist Henry Ford and his colleagues at the Ford Motor Company, where in 1913 a moving-belt conveyor was used in the assembly of flywheel magnetos. With it assembly time was cut from 18 min per magneto to 5 min. The approach was then applied to automobile body and motor assembly. The design of these production lines was highly analytical and sought the optimum division of tasks among work stations, optimum line speed, optimum work height, and careful synchronization of simultaneous operations.

The success of Ford's operation led to the adoption of mass production principles by industry in the United States and Europe. The methods made major contributions to the large growth in manufacturing productivity that has characterized the twentieth century and produced phenomenal increases in material wealth and improvements in living standards in the industrialized countries.

2.4.3 Principles of Mass Production

The efficiencies of mass production result from the careful, systematic application of the 'scientific management' ideas and concepts. The following summary lists the four basic principles of mass production:

1. *Division of Labor*—The careful division of the total production operation into specialized tasks comprising relatively simple, highly repetitive motion patterns and minimal handling or positioning of the work-piece. This permits the development of human motion patterns that are easily learned and rapidly performed with a minimum of unnecessary motion or mental readjustment.
2. *Standardization of tasks*—The simplification and standardization of component parts to permit large production runs of parts that are readily fitted to other parts without adjustment. The imposition of other standards (e.g., dimensional tolerances, parts location, material types, stock thickness, common fasteners, packaging material) on all parts of the product further increases the economies that can be achieved.
3. *Use of machinery and automation of work*—The development and use of specialized machines, materials, and processes. The selection of materials and development of tools and machines for each operation minimizes the amount of human effort required, maximizes the output per unit of capital investment, reduces the number of off-standard units produced, and reduces raw material costs.
4. *Systematic planning of work*—The systematic engineering and planning of the total production process permit the best balance between human effort and machinery, the most effective division of labor and specialization of skills, and the total integration of the production system to optimize productivity and minimize costs.

To achieve the maximum benefits that application of these principles can provide, careful, skilled industrial engineering and management are required. In a mass production factory, planning begins with the original design of the product; raw materials and component parts must be adaptable to production and handling by mass techniques. The entire production process is planned in detail, including the flows of materials and information throughout the process. Production volume must be carefully estimated because the selection of techniques depends upon the volume to be produced and anticipated short-term changes in demand. It must be large enough, first, to permit the task to be divided into its sub-elements and assigned to different individuals; second, to justify the substantial capital investment often required for specialized machines and processes; and third, to permit large production runs so that human effort and capital are efficiently employed.

The need for detailed advance planning extends beyond the production system itself. The large, continuous flow of product from the factory requires equally well-planned distribution and marketing operations to bring the product to the consumer. Advertising, market research, transportation problems, licensing, and tariffs must all be considered in establishing a mass production operation. Thus, mass production planning implies a complete system plan from raw material to consumer.

In addition to lowering cost, the application of the principles of mass production have led to major improvements in uniformity and quality. The large volume, standardized design, and standardized materials and processes facilitate statistical control and inspection techniques to monitor production and control quality. This leads to assurance that quality levels are achieved without incurring the large costs that would be necessary for detailed inspection of all products.

2.4.4 "Lean" or "Flexible" Production Method

A major problem of mass production based on continuous or assembly line processes is that the resulting system is inherently inflexible. Since maximum efficiency is desired, tools, machines, and work positions are often quite precisely adapted to details of the parts produced but not necessarily to the workers involved in the process. Changes in product design may render expensive tooling and machinery obsolete and make it difficult to reorganize the tasks of workers. One answer has been to design machinery with built-in flexibility; for relatively little extra cost, tooling can be changed to adapt the machine to accommodate design changes.

Similarly, a production line is usually designed to operate most efficiently at a specified rate. If the required production levels fall below that rate, operators and machines are being inefficiently used; and if the rate goes too high, operators must work overtime, machine maintenance cannot keep up, breakdowns occur, and the costs of production rise. Thus, it is extremely important to anticipate production demands accurately. Planning, an important function of management and engineering design, can alleviate the problems of increased demand by incorporating excess capacity in the facilities that would require the longest time to procure and install. Then, if production loads increase, it is easier to bring the entire system up to the new level. Similarly, if large fluctuations in demand cannot be avoided, flexibility to accommodate these changes economically must be planned into the system.

The ideas of Taylor's 'scientific management' and Henry Ford's operation spread wider than its origins in the study of work, from the "efficiency movement" of the 1920s, through the depression-era "rationalization" and wartime mobilization, up to postwar "productivity" drives and quality-control campaigns. Imported to Japan, these ideas were embraced—and ultimately transformed—in Japan's industrial workshops. Adaptation of Taylor's 'scientific management' and Henry Ford's operation to improve production as a response to specific demands of postwar Japanese automobile gave rise to innovation as Japanese managers sought a "revised" model that combined mechanistic efficiency with respect for the humanity of labor.

The Toyota production line paradigm evolved from the shops of Toyota Motor Company in the thirty years after World War II. The model pioneered by Toyota is an integrated system characterized by a flow of processing information backward from final assembly, "flexible" and multipurpose or multifunctional machinery and workers to make a wide range of product (i.e. automobile) components, tightly rationalized simplified and standardized tasks, low lead and setup times, and small-lots operations for manufacturing, in-house conveyance, and deliveries from subcontractors (Tsutsui, 2001; Cusumano, 1985). It remains, however, noticeably consistent with Taylor's 'scientific management' in general approach and adapting to customer demands. It constitutes a more rigorous and stringent application of Taylor's 'scientific management' principles than the standards that were applied at Henry Ford's factories.

Nowadays, "lean" production or Toyota production line is the envy of the world for its efficient and humane management practices. 'Humane', as Toyota understands it, means:

to eliminate from the work force worthless, unproductive persons who should not be there and to awaken in all the awareness that they can improve the work place through their own efforts and to foster a feeling of belongingness...

"Lean" production appears "human centered" only to the extend that tapping the skills and defining duties of individual workers would allow for further ascent of labor productivity. It has the objective of diluting individual worker skills. Despite the currently widespread preconception that the Toyota Production System is uniquely humane, flexible, and participative, the 'scientific management' ideals of control, discipline, and expertise are paramount in the "lean" approach to workplace labor relations. Rigid obedience, rather than nurturing inclusion, seems to define Toyota's shop-floor strategy. Supervisors must drill into the minds of workers that they must strictly abide by standard operations.

As Taiichi Ohno in an interview conducted by Koichi Shimokawa and Takahiro Fujimoto has indicated (Koichi & Takahiro, 2009):

The Toyota Production System is one and the same with Total Quality Control (TQC) and with its principle of zero defects. They are simply different names for the same basic approach.

It rests on two pillars. The first pillar is Sakichi Toyoda's Jidoka, the essence of which states that: *"Turning out defective work is not what we are here for."* The second pillar is Kiichiro Toyoda's just in time, the essence of which states that: *"Just make what is needed in time, but don't make too much..."*

The Toyota production line paradigm may have revolutionary implications, yet, to a large degree, the changes that Toyota made were "evolutionary" adaptations to the circumstances surrounding the company and its domestic market needs. Faced with a complex landscape of restrictions and opportunities—rapid growth in demand, low production volumes, highly diversified product lines, competitive pressure to reduce costs and improve quality, limited capital, and increasingly scarce labor—William Tsutsui (2001) has shown that the creators of the Toyota production line system finely modified Taylor's 'scientific management' methods and mind-sets to address urgent needs. It is an ingenious and practical rearrangement of the Taylor building blocks of the Ford approach, resulting in a new model—and seemingly non Ford model—of industrial production. The critical environment factor in this evolution was the lack of: sustained labor opposition, powerful craft unions and hostile workers.

Taken as a whole, the Toyota Production system—or "lean" production—can be seen as an innovative model of mass production achieved by mobilizing the 'scientific management' approaches and adapting them to the specific demands of postwar Japanese automobile manufacturing.

2.5 Conclusion

The precise and narrowly defined concept of "6 standard deviations as the permissible limit of variations around the expected central tendency" coupled with the insight of "Lean" business activities have grown over time and in media to represent a framework for quality improvement and control, the goal of which is to facilitate quality improvement efforts that will lead to operating cost reduction opportunities (Harry & Schroeder, 2006; Eckes, 2002; Pyzdek & Keller, 2009; Bertels & Strong, 2003; Pande, Neuman, & Cavanagh, 2000, 2001; Breyfogle, 2003; Webb & Gorman, 2006; Truscott, 2003; Summers, 2007; Perez-Wilson, 1999; Sodhi & Sodhi, 2008; Breyfogle, Cupello, & Meadows, 2001; Gupta, 2004; Przekop, 2005). This framework covers four perspectives—philosophy, economics, marketing, and operations management. Philosophy is focusing on definitional issues; economics is focusing on profit maximization and market equilibrium; marketing is focusing on the determinants of buying behavior and customer satisfaction; and operations management is focusing on engineering practices and manufacturing control.

To keep the balance of economic activity between services and manufacturing operations, we shall say that:

> *A 'Lean' Six Sigma business activity is a business activity operating at a performance permissible limit of variations of 6 standard deviations around its expected central tendency and that is:*
>
> 1. *Effective—Producing the desired outcome correctly the first time;*
> 2. *Efficient—Minimizing the resources used to produce the desired outcome in the shortest time;*
> 3. *Flexible or Adaptable—Being able to adapt to changing customers and to the circumstances surrounding the business and its market needs.*

Framework and Methodology

3

The progressive realization of the enterprise full potential by moving from its current maturity stage towards a higher (ultimately "Continuous Improvement") maturity stage, requires a framework and a systematic methodology for studying the constituent elements or processes and systems associated with the eight overarching determining factors. It also requires a way of differentiating between the different types of variation present in those processes and systems. In addition to the way of thinking described throughout the chapters of our first book, and which must be put to practice, there are techniques to be learned. In this chapter, we will describe the framework and systematic methodology for improving processes used within projects and operations work.

3.1 Operational Definition of a Process

The full history of the skills, knowledge and competencies used to improve specific technical activities or processes performed within projects and operations work is certainly not one that originated with quality professionals. It may have started with the building of the pyramids, which must clearly have involved some understanding of organization and execution of tasks among the Egyptians. It may have also started earlier back in the days of cave men and women who struggle together and allocate tasks with the common goal of survival. While this ancient history may be of some academic interest, it is now understood that in the twentieth century, this history really started with Shewhart's work on "Statistical Method from the Viewpoint of Quality Control" (Shewhart, 1939).

The fundamental concept, which underpinned much of the thoughts in the development of a framework and a methodology for improving the specific technical activities performed of within projects and operations work, was that of "operational definition" of work activities or their outcomes.

From the process perspective, we can think of an "operational definition" of a process as:

A. van Aartsengel and S. Kurtoglu, *Handbook on Continuous Improvement Transformation*,
DOI 10.1007/978-3-642-35901-9_3,

> *A repeatable demonstration of the process outcome(s), in terms of its set of logically related and validated discrete elements (tasks, actions, or steps) established in order to describe and define the process purpose and outcome(s). It is the "recipe" of the process outcome(s). It is a description of the logically related and validated discrete elements (tasks, actions, or steps) that were established in order to describe and define the process purpose and outcome(s), as applied to a specific situation to facilitate the collection of meaningful (standardized) data.*

With this perspective, the validation of the discrete elements, which constitute the process, is designed to distinguish them it from their background of empirical experience and testing, and not to define them in terms of some inherent or private essence.

W. Edwards Deming in "The New Economics" (Deming, 1994), delineates an "operational definition" as *"a procedure agreed upon for translation of concept into measurement of some kind."*

Hence, an "operational definition" specifically states how to measure the item being defined.

Thus, a process outcome can be defined in terms of how it is produced. For example, 100 °C may be crudely defined by describing the process of heating water until it is observed to boil.

There are three aspects that form the basis for an "operational definition" of a process. These are: the purpose, the methodology, and the performance measure.

1. *The purpose*—It provides a sense of direction and focus to the resources that support the activities being completed by the process within a project or an operation work. It also provides a sense of discovery and a sense of destiny. These add a social edge to the process and are an objective that the people resources perceive as being inherently valuable. Any situation in which the purpose remains unspecified will rapidly deteriorate into chaos. However, merely specifying the purpose is not enough.
2. *The methodology*—It relates to a constructive generic plan and guidelines for achieving the defined purpose. It may entail a description of generic discrete elements (tasks, actions, or steps) or, metaphorically, may be extended to explications of philosophically coherent concepts or theories as they relate to a particular project or operation work. Until the methodology has been established, the "purpose" aspect is merely nothing more than wishes and hopes.
3. *The performance measure*—As indicated in the previous chapter, the performance measure is a criterion of success stated in relation to the activities being completed by the process or in relation to its purpose. The goal of a "performance measure" is to enable improvement.

Walter Shewhart discussed these three aspects of "operational definition" in the context of making a product. He referred to them under the headings of (1) Specifications, (2) Production, and (3) Inspection. Edwards Deming talked about these three questions in terms of (1) having a Criterion, (2) having a Test Method for determining compliance to the criterion, and (3) having a Decision Rule for interpreting the results of the test.

3.2 Setting the Framework and Methodology

Regardless of the nomenclature, these three fundamental aspects of "operational definition" form the essence of both how to get things done and the systematic methodology for studying processes and systems. They were popularized by Shewhart and Deming, from the work of the philosopher C. I. Lewis, to provide the starting point for what grew into the Shewhart or PDSA model for improvement of work activities.

Improving the activities being completed by a specific process is a complex undertaking requiring a number of different technical skills, knowledge, tools and competencies. For this reason, we regard this undertaking as a project; namely, a "process improvement" project. The project approach has long been favored for undertakings such as product development or improvement that involve a significant expenditure of personnel, time, and resources, especially when they are considered essential to the well-being of the enterprise. The project approach has also been shown effective when applied to apparently lesser problems, especially when applied serially to accomplish incremental change and even to affect quality improvement breakthrough.

The management process outlined here is applicable generally to any "process improvement" project undertaking, whether to product or process development, or to product development in the service sector. In short, with some suitable modifications and fine tuning to fit the specific task at hand, the project approach works for just about any "process improvement" undertaking worth the effort. We will also refer to the specific process, the activities and outcome of which must be improved, as the "process to be improved" throughout the remaining of this book.

The primary attraction of the project concept as a management tool is its focus on results and the means to achieve those results. It is structured; there is a beginning, a middle and an end. When a project has been completed successfully, something happens; a new product, a new service, an improved process, comes into being where it did not exist before.

There certainly are plenty of choices for managing projects (PMI, A Guide to the Project Management Body of Knowledge (PMBOK Guide), 2010; Schmidt, 2009; Wysocki, Effective Project Management: Traditional, Agile, Extreme, 2011; Schwaber, 2004; Mantel, Meredith, Shafer, & Sutton, 2010; Kerzner, Project Management: A Systems Approach to Planning, Scheduling, and Controlling, 2009), and there is a vast amount of literature on project management (Verzuh, The Fast Forward MBA in Project Management, 2011; Verzuh, 2003; Westland, 2007; Bennett, 2003; Crawford, 2006; Richardson, 2010; Kerzner, 2004; Kerzner, 2010; Nicholas & Steyn, 2012; Schwalbe, 2010; Maylor, 2010; Wysocki, 2004; Ahuja, Dozzi, & AbouRizk, 1994; Torkzadeh & Gholamreza, 2008; Hill, 2009; Marco, 2011; Moore, 2002; Curlee & Gordon, 2010; Tonchia & Cozzi, 2008; Badiru, 1996; Morris, Pinto, & Söderlund, 2011; Lientz & Rea, 2002; Rosenau & Githens, 2005; Cockrell, 2001; Carmichael, 2000; Kliem & Anderson, 2003; Stasiowski & Burstein, 1994).

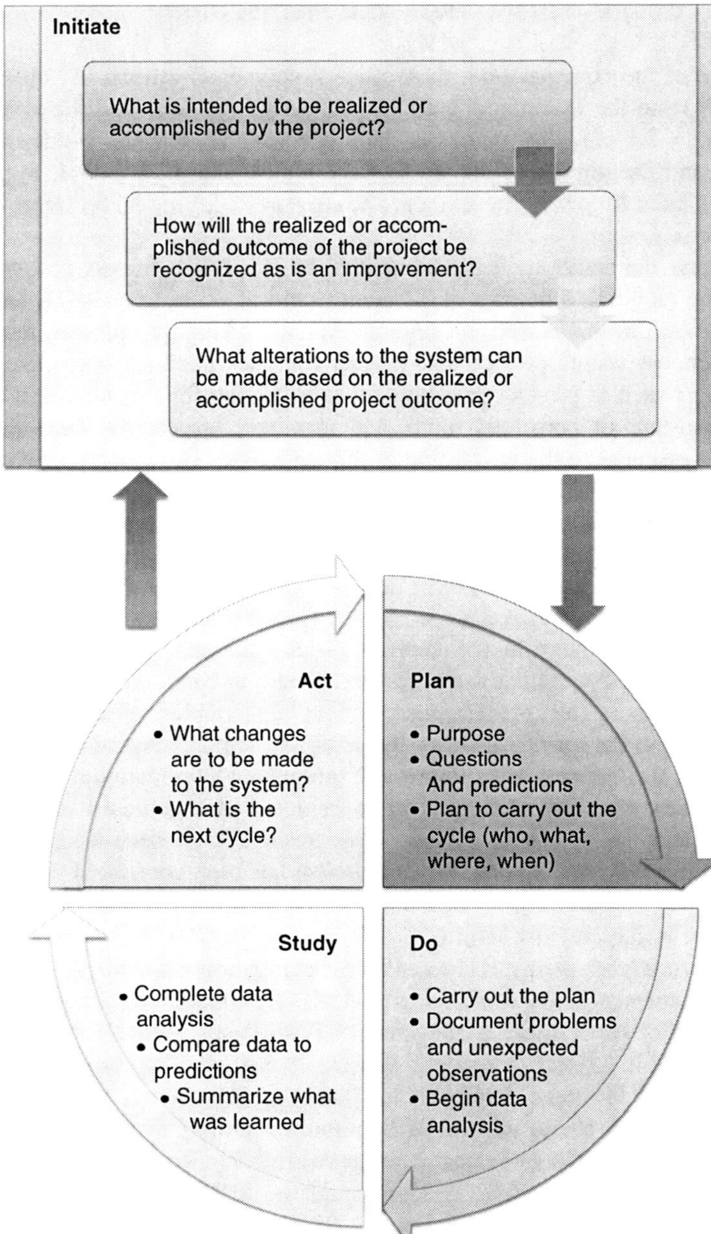

Fig. 3.1 Detailed PDSA cycle for improvement

The PDSA model for improvement is intended to drive all process improvement projects through its Plan—Do—Study—Act (PDSA) Cycle, illustrated in Fig. 3.1 adapted from (Langley et al., 2009), and by persistently asking a set of fundamental questions around the three aspects of an "operational definition." These fundamental questions, which form the basis and the preliminary step of the PDSA model, can be formulated as follow:

1. What is intended to be realized or accomplished by the "process improvement" project?
2. How will the realized or accomplished outcome of the "process improvement" project be recognized as is an improvement?
3. What alterations to the system affected by the "process to be improved" can be made based on the realized or accomplished outcome of the "process improvement" project?

From a project management perspective, the PDSA model is a framework for application of knowledge, skills, tools and techniques to "process improvement" project activities to meet the "process improvement" project requirements. This application of knowledge requires the effective management of appropriate project management processes

Since its founding in 1969, the Project Management Institute (PMI) has grown to be the organization of choice for project management professionalism. With more than 330,000 members worldwide in over 192 countries, and more than 400,000 people with the Project Management Professional (PMP) credential, the Institute is the largest and leading non-profit professional association in the area of project management, dedicated to the development of the project management profession. Its project management precepts covered in what has emerged as the world standard of project management knowledge—A Guide to the Project Management Body of Knowledge (PMI, A Guide to the Project Management Body of Knowledge (PMBOK Guide), 2010)—provides a generic structure and the required processes to successfully complete a generic project.

Consequently, the material contained in the following sections has been built on that generic structure to foster consistency with the project management profession and to ensure comprehensiveness of the PDSA model in compliance with the principles elucidated in the PMBOK Guide.

Using the PDSA model nomenclature, there are five key phases with corresponding process groups that govern the management of a "process improvement" project: initiating, planning, executing, studying, and acting upon the results:

1. "PDSA Initiate" Process Group
2. "PDSA Plan" Process Group
3. "PDSA Do" Process Group
4. "PDSA Study" Process Group
5. "PDSA Act" Process Group

These five Process Groups have clear dependencies and are performed in the same sequence on each "process improvement" project. They are independent of application areas or industry focus. Their constituent processes have a natural progression inherent in the work to be performed and they must be used in conjunction with a life cycle that covers the phases of the "process improvement" project. The actions taken during the course of one of these constituent project management processes typically affect that specific process and other related project management processes.

For example, a scope alteration typically affects the project cost, but may not affect the communication plan or the quality of the project outcome. These constituent project management process interactions often require tradeoffs among project requirements and objectives, and the specific performance tradeoffs will vary from project to project and enterprise business to enterprise business. A successful management of a "process improvement" project takes account of actively managing these interactions to meet sponsor, customer, and other stakeholder requirements. In some circumstances, a constituent project management process or set of processes will need to be iterated several times in order to achieve the required outcome.

"PDSA Initiate" Process Group

4

The "PDSA Initiate" phase is the first phase of the "process improvement" project management life cycle. It is the start of a process that takes the project brief, as developed, selected and prioritized, through to the delivery of the project's outcomes back into the business, as illustrated in Fig. 3.1.

The "PDSA Initiate" Process Group, illustrated in Fig. 4.1, consists of those project management processes performed to lay out the foundation for the "process improvement" project. These project management processes define the "process improvement" project by formulating preliminary answers to the three fundamental question of the PDSA model and by obtaining authorization to start the project. There are two activities involved in this groundwork:

1. The project manager must determine the purpose, goals, and constraints of the "process improvement" project. He or she must answer the three fundamental questions, which form the basis and the preliminary step of the PDSA model:
   - What is intended to be realized or accomplished by the "process improvement" project?
   - How will the realized or accomplished outcome of the "process improvement" project be recognized as is an improvement?
   - What alterations to the system affected by the "process to be improved" can be made based on the realized or accomplished outcome of the "process improvement" project?

   The answers become the foundation for making all project decisions because they describe the cost-schedule-quality equilibrium and keep the "process improvement" project aligned with the enterprise business intended strategy.
2. The project manager must establish basic project management controls. He or she must get agreement on which people and business functions or external organizations are involved in the project and what their roles will be. He or she also needs to clarify the chain of command, communication strategy, and project alteration control process. The documented acceptance of these decisions and strategies communicates expectations about the way the "process improvement"

A. van Aartsengel and S. Kurtoglu, *Handbook on Continuous Improvement Transformation*,
DOI 10.1007/978-3-642-35901-9_4, © Springer-Verlag Berlin Heidelberg 2013

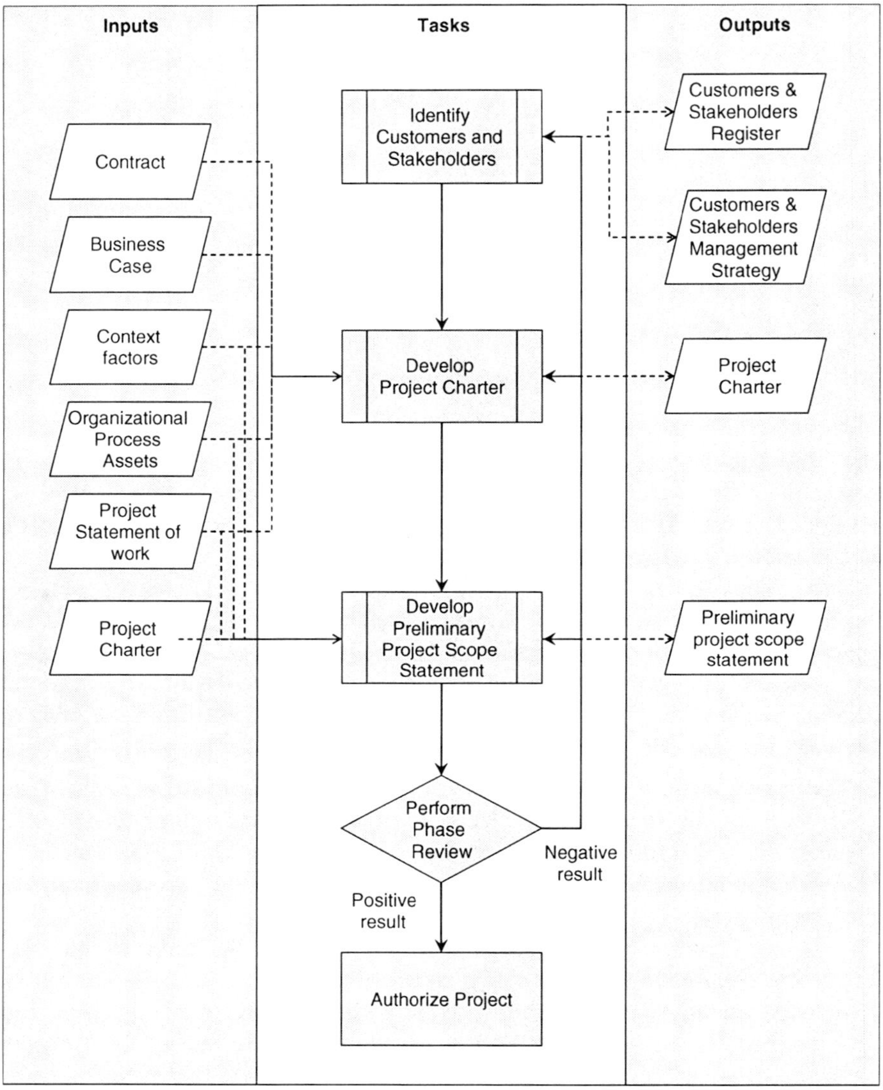

Fig. 4.1 "PDSA Initiate" Process Group

project will be managed. It also becomes an agreement to which the project manager can refer to keep everyone accountable to their responsibilities in the project.

The "PDSA Initiate" Process Group includes the following project management processes:

1. Identify Customers and Stakeholders
2. Develop Project Charter
3. Develop Preliminary Project Scope Statement

Within these constituent project management processes, the initial scope is defined and initial resources (including financial resources as specified in the strategic alignment plans) that the enterprise business is willing to invest are further refined and committed. Internal and external customers and stakeholders who will interact and influence the overall outcome of the project are identified. This information is captured in the project charter. From the strategic alignment, as described in our first book entitled "A Guide to Continuous Improvement Transformation: Concepts, Processes, Implementation," the relationship of the selected "process improvement" project to the enterprise business intended strategy identifies both:

1. The management responsibilities within the enterprise business; and,
2. The reasons why this specific "process improvement" project is the best alternative to close the strategic gap.
3. The project is officially authorized when the project charter gains approval.

The constituent project management processes of the "Initiate" Process Group, illustrated in Fig. 4.1, may be triggered and performed by organizational, program, or portfolio processes external to the actual "process improvement" project's scope of control. For example, prior to starting the "process improvement" project, the need for high-level requirements may be documented as part of a larger enterprise business initiative.

To help keep the project focus on the business need the project was undertaken to address, these constituent project management processes should be invoked and reviewed at start of each phase. Involving customers and stakeholders during this initial phase generally improves the probability of shared ownership, deliverables acceptance, and customer and other stakeholder satisfaction.

4.1 Identify Customers and Stakeholders

Identifying customers and stakeholders is a primary task because all the important decisions during the definition and planning stages of the "process improvement" project are made by these customers and stakeholders. These are the people who, under the guidance of the project manager, establish agreements on the goals and constraints of the project, construct the strategies and schedules, and approve the budget.

The "Identify Customers and Stakeholders" project management process prompts the project team to identify all people or functions within the enterprise business impacted by the "process improvement" project, and to document relevant information regarding interests, involvement, and impact on the project success. It links the project with the people or functions that will be affected by the project outcome to ensure a successful project completion. The people or functions that will be affected by the project outcomes fill a variety of roles, each important to the project's success or failure. The steps of the process for identifying these people or functions are:

1. Develop a list of customers and stakeholders;
2. Identify aspects of their relationship with the project; and
3. Categorize each identified customer and stakeholder.

Customers and stakeholder identification is necessary for managing their expectations and influence in relation to project requirements. This identification is an ongoing task. Throughout the initial stages of the "process improvement" project, the project manager must continue to clarify who the customers and stakeholders are and what roles they will play. Consequently, the outputs of this process—Customers and Stakeholder Register, and Customers and Stakeholders Management Strategy—should be revised during the subsequent phases of the project.

The following sections describe the roles of the customers and the five primary stakeholders and the impact each has on the success of the project. It is important to keep in mind that these are all roles. They can, therefore, be filled by one or more people, and an individual can play more than one role.

4.1.1 Develop a List of Customers and Stakeholders

When identifying affected stakeholders, a systematic approach often works well, starting with delineating the project's sphere of influence. Here, it is important to think beyond the obvious. Directly affected people or functions within the enterprise business are easy to identify, whereas indirectly affected people or functions—and, as a result, secondary stakeholders—are sometimes harder to identify. The project manager should think of as many ways as possible that "process improvement" project might bring benefits or problems to people not directly in its path. Given that, there are a number of ways to identify stakeholders. Often, the use of more than one technique will yield the best results.

1. With the brainstorming technique, you—as the project manager—should get together with people in the enterprise business, executives or their representatives, and others already involved in or informed about the "process improvement" project and start calling out categories and names. Part of the point of brainstorming is to come out with anything that comes to mind, even if it seems silly. On reflection, the silly ideas can turn out to be among the best, so be as far-ranging as you can. After 10 or 15 min, stop and discuss each suggestion, perhaps identifying each as a primary, secondary, and/or key customer or stakeholder.
2. Collect categories and names from people in the enterprise business, if they are not available to be part of a brainstorming session.
3. Consult with functions within the enterprise business that either are or have been involved in similar "process improvement" projects.
4. Get more ideas from stakeholders as you identify them.
5. If appropriate, advertise. Use some combination of the internal media—often free, through various community service arrangements within the enterprise business—community meetings, community and organizational newsletters, social media, targeted emails, etc... You may find people who consider themselves customers or stakeholders whom you have not thought about.

4.1.1.1 Customers

Whenever a "process improvement" project exists, the "process to be improved" owner or somebody else will be paying for it. And whoever pays usually gets the first and last word on the product description, budget, and the criteria by which success will be measured. Although other stakeholders may try to pinch in extra requirements, the final say on the product will come from the customer, because this customer is paying the bills.

Customers contribute funding and requirements regarding the "process to be improved." Determining who fills the role of customer can present real challenges to the project manager. In making this determination, the project manager must be guided by two basic questions: "Who is authorized to make decisions about the product?" and "Who will pay for this project?" Consequently, the project manager must distinguish between the people with final authority over product requirements, those who must be consulted as the requirements are developed, and those who simply need to be informed what the requirements are.

List the customers who intend to use the deliverables produced by the "process improvement" project. Customers for any "process improvement" project are either internal or external to the project.

1. *External Project Customers*—Consumers or users who pay for the project outcomes. By definition, external stakeholders are not part of the enterprise business that carries out the project work. Although they normally want the project to succeed, their stake is often focused more inwardly. This is true of most external stakeholders, except those for whom the project is been done (external customers). Most external groups—particularly those supplying goods and services—are inclined to take a parochial view of the project. This means that are often unreliable to put what is best for the project ahead of what is best for them. This may sound cynical, but it is reality. Projects that address the needs of external customers are typically characterized by contracts.
2. *Internal Project Customers*—individuals within the enterprise business who will use the deliverables or information produced at various stages of the project (internal to the project, not necessarily to the enterprise business). The internal customer often pays for the project and receives the benefits (business impact) and/or project deliverables.

The success of the "process improvement" project will be based primarily on whether or not the deliverables produced by the project match the requirements of the customers identified in Table 4.1.

4.1.1.2 Stakeholders

Develop a list of stakeholders for this project. A stakeholder is anyone who has a vested interest in the "process improvement" project. Stakeholders are individuals, functions and organizations who are actively involved in the project, or whose interest may be positively or negatively affected as a result of project execution or successful project completion.

Table 4.1 Project customers list

Customer	Representative
Customer group	Customer name and contact information
…	…

Stakeholders have a key role in defining the project success criteria and their interest and power should not be overlooked. Stakeholders must be identified, their level of interest and power to influence the success of the project analyzed, and plans devised for their management. Throughout the life of the "process improvement" project, stakeholders can be extremely helpful in solving personnel or performance problems.

"Making timely decisions based on the facts provided by the project team" is the other major responsibility of stakeholders. Identifying and characterizing the stakeholders who will make decisions can be delicate. We advise to start with the obvious ones:

1. Stakeholders whose operations will be affected by the outcome of the "process improvement" project;
2. Stakeholders representing other stakeholders, such as the customer;
3. The project sponsors to whom the project manager reports

For each of these stakeholders, remember to keep in mind why they will be interested in the "process improvement" project and which decisions they will influence. Having identified the obvious decision makers, the project manager should proceed to identify the less obvious ones, such as those with veto authority.

One way to characterize stakeholders is by their relationship to the "process improvement" project.

1. Primary stakeholders are those people or groups that stand to be directly affected, either positively or negatively, by the project or its outcomes.
2. Secondary stakeholders are those people or groups that are indirectly affected, either positively or negatively, by the project or its outcomes.
3. Key stakeholders, who might belong to either or neither of the first two groups, are those people or groups that can have a positive or negative effect on the project or its outcomes, or that are important within or to the project.

While an "interest" in an effort or organization could be just that—economically, intellectually, academically, philosophically, or politically motivated attention—stakeholders are generally said to have an "interest" in the project based on whether they can affect or be affected by it. The more they stand to benefit or lose by it, the stronger their interest is likely to be. The more heavily involved they are in the "process improvement" project, the stronger their interest as well. For instance, a financial controller within an enterprise business will have an interest in the cost implications of the project, and a CEO will have an interest in whether the "process improvement" project helps to achieve the vision of the enterprise business. Other

Table 4.2 Project stakeholders list

Stakeholder	Stakeholder interest
CEO	Alignment with enterprise business intended strategy
Financial controller	Alignment with enterprise business budget
Health and safety officer	Alignment with health and safety standards
Quality officer	Alignment with quality standards
Regulatory body	Compliance with legislations
Industry body	Compliance with codes of practice
…	…

examples of stakeholders include enterprise business executives, legislative bodies and regulatory bodies. Complete Table 4.2.

4.1.1.3 Sponsors

Project sponsors are individuals or groups that represent external project customers by advocating the "process improvement" project. They may be internal or external to the enterprise business, but they are committed to active involvement throughout the project lifecycle and with a very high stake in the project outcome. Sponsors ensure that the project remains a viable proposition and that the intended benefits are realized, resolving issues outside the control of the project manager. They are the individuals or groups with formal authority who are ultimately responsible for the project.

A "process improvement" project is intended to implement change that allows an enterprise business to fulfill its intended strategy. This emphasizes benefits realization, rather than delivery of deliverables. Consequently, the role of the sponsors is to direct the "process improvement" project with benefits in mind, as opposed to the project manager, who manages the project with delivery in mind. There are two basic aspects in understanding the importance of sponsors to the "process improvement" project that the project manager should be aware of. First, sponsors are ultimately responsible for the success of the project. The real, formal authority that comes from their title and position in the enterprise business endows them with this responsibility. Second, the sponsor's most important task is to help the project team be successful.

The best sponsors know that they are not sponsoring a project; they are sponsoring the project manager and the project team. The sponsors' task is to help these people be successful. Sponsors are the primary risk taker and owner of the "process improvement" project's business case. Ideally, there should be only one sponsor per "process improvement" project.

The project sponsors have relationships with all project stakeholders, but even more frequently with the project manager. Project sponsors perform different roles during the project lifecycle: sellers, coaches and mentors, filters, business judges, motivators, negotiators, protectors, and upper management links.

Table 4.3 Project sponsors list

Sponsor	Representative
Sponsor group	Sponsor name and contact information
...	...

1. *As sellers*: The project sponsors must be able to sell the project to project stakeholders. They believe in the project, speak positively about it, and can sell the benefits.
2. *As coaches and mentors*: Good project sponsors increase the level of confidence felt by the project manager. The project sponsors must have the ability to install a sense of confidence in the project and protect the project manager from loosing that confidence.
3. *As filters*: The project sponsors must be able to stimulate project leaders by allowing them to focus on the work at hand.

The project sponsors, to be listed in a table similar to Table 4.3, are the principal "owners" of the project. Their primary contribution to the "process improvement" project is their authority. Tangible ways sponsors lend their authority to projects include:

1. Defining the vision and high-level objectives for the project;
2. Approving the requirements, timetable, resources and budget;
3. Authorizing the provision of funds/resources (internal or external);
4. Approving the project plan and quality plan;
5. Ensuring that major business risks are identified and managed;
6. Approving any major changes in scope;
7. Receiving project review group minutes and taking action accordingly;
8. Resolving issues escalated by the project manager/project review group;
9. Ensuring the participation of all business resource, where required;
10. Providing final acceptance of the solution upon project completion;
11. Monitoring and maintaining the priority of the project relative to other projects

4.1.1.4 Project Review Group

The project review group may include both business and third-party representatives, and is put in place to ensure that the project progresses according to plan. Responsibilities of the project review group include:

1. Assisting the project sponsor with the definition of the project vision and objectives;
2. Undertaking quality reviews prior to the completion of each project milestone;
3. Ensuring that all business risks are identified and managed accordingly;
4. Ensuring conformance to the standards and processes identified in the quality plan;
5. Ensuring that appropriate client/vendor contractual documentation is established.

4.1.1.5 Project Manager

The project manager ensures that the daily activities undertaken on the "process improvement" project are in accordance with the approved project plans. The project manager is responsible for ensuring that the project produces the required deliverables on time, within budgeted cost and to the level of quality outlined within the quality plan. Responsibilities of the project manager include:

1. Documenting the detailed project plan and quality plan;
2. Ensuring that all required resources are assigned to the project and clearly tasked;
3. Managing assigned resources according to the defined scope of the project;
4. Implementing the project processes (time, cost, quality, alteration/change, risk, issue, procurement, communication, acceptance management);
5. Monitoring and reporting project performance (schedule, cost, quality and risk);
6. Ensuring compliance with the processes and standards outlined in the quality plan;
7. Adjusting the project plan to monitor and control the progress of the project;
8. Reporting and escalating project risks and issues;
9. Managing project interdependencies.

4.1.1.6 Project Team

Every "process improvement" project has a project team. The project team consists of every person who works on the project, including consultants, suppliers, utility companies, and resource agencies. Each team member is an internal customer for some deliverables and a supplier of other deliverables. Project team members are responsible for delivering products with the quality promised, in a timely and cost effective manner. Specific responsibilities include:

1. Completing tasks allocated by the project manager;
2. Reporting progress to the project manager on a frequent basis;
3. Maintaining documentation relating to the execution of allocated tasks;
4. Escalating risks and issues to be addressed by the project manager.

4.1.2 Analyze Stakeholders and Their Interests

Once the customers and stakeholders have been identified, the next task is to understand their interests. Some will have an investment in carrying the "process improvement" project forward, but others may be equally intent on preventing it from happening or making sure it's unsuccessful.

Stakeholder analysis (also called stakeholder mapping) helps decide which stakeholders might have the most influence over the success or failure of the "process improvement" project, which might be the most important supporters, and which might be the most important opponents. Once that information has been determined, plans for dealing with stakeholders with different interests and different levels of influence can be made.

Stakeholder interests may vary. Some stakeholders' interests may be best served by carrying the "process improvement" project forward, others' by stopping or weakening it. Even among stakeholders from the same group, there may be conflicting concerns. Some of the many ways that stakeholder interests may manifest themselves are as follow:

1. Potential beneficiaries may be wildly supportive of the "process improvement" project, seeing it as an opportunity or the pathway to a "better life"... or they may be ambivalent or resentful toward it. The "process improvement" project may be embarrassing to them or may seem burdensome. They may not understand it, or they may not see the benefit that will come from it. They may be afraid to try something new, on the assumption that they will fail, or will end up worse off than they are. They may be distrustful of any people or functions engaged in such a process improvement effort, and feel they are being looked down on.
2. Some stakeholders may have economic concerns that may also work in favor of a "process improvement" project.. Sometimes these concerns are merely selfish or greedy but in most cases, they are legitimate.
3. Business people may have concerns about the "process improvement" project. While it may be good for the enterprise business as a whole, it may actually hurt some business functions.
4. Organizations, agencies, and institutions may have a financial stake in the "process improvement" project because of funding concerns. Their ability to be funded for conducting activities related to the project may mean the difference between laying off and keeping staff members, or even between survival and closing the doors.
5. Legislators and policy makers may be concerned with public perceptions that they are wasting public money by funding a particular the "process improvement" project.
6. The work of staff members engaged in carrying out the "process improvement" project can be drastically changed by the necessity to learn new methods, increases in paperwork, or any number of other requirements. Depending on the situation, they may be more than willing to take on these responsibilities, may have ideas about how they can be made less burdensome, or may resent and dislike them.

Having identified all the stakeholders and recorded their concerns, the project manager has to respond to their concerns in some way—at least by acknowledging them, whether they can be satisfied or not—and the project manager must find a way to move forward with as much support from stakeholders as he/she can muster.

It is not practical, and usually not necessary, to engage with all stakeholder groups with the same level of intensity all of the time during the course of the project. Being strategic and clear as to whom you are engaging with and why, before jumping in, can help save both time and money. This requires prioritizing the stakeholders and, depending on who they are and what interests they might have, figuring out the most appropriate ways to engage them. Stakeholder analysis should assist in this prioritization by assessing the significance of the project to each

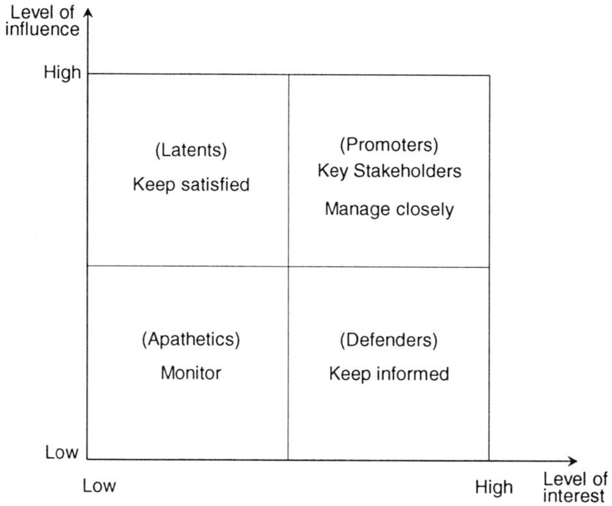

Fig. 4.2 Influence/interest grid for stakeholder prioritization

stakeholder group from their perspective, and vice versa. It is important to keep in mind that the situation might be dynamic and that both stakeholders and their interests might change over time, in terms of level of relevance to the project and the need to actively engage at various stages.

Stakeholder analysis (stakeholder mapping) is a way of determining who among stakeholders can have the most positive or negative influence on the "process improvement" project, who is likely to be most affected by the "process improvement" project, and how one should work with stakeholders with different levels of interest and influence.

Most methods of stakeholder analysis or mapping divide stakeholders into one of four groups, each occupying one space in a four-space grid—the influence versus interest grids—as illustrated in Fig. 4.2.

As shown in Fig. 4.2, low to high influence over the "process improvement" project runs along a line from the bottom to the top of the grid, and low to high interest in the "process improvement" project runs along a line from left to right. Both influence and interest can be either positive or negative, depending on the perspectives of the stakeholders in question. The lines describing them are continuous, meaning that people can have any degree of interest from none to as high as possible, including any of the points in between. The "key stakeholders" would generally appear in the upper right quadrant.

By mapping the sphere of influence of different stakeholders, the project manager can begin to identify distinct groups by impact area, and from this prioritize stakeholders for consultation. While priority should be given to stakeholders who are directly and adversely affected, drawing the line between who is affected and who is not can be challenging. For this reason, defining stakeholders too narrowly should also be avoided.

The purpose of this diagram is to help in understanding and determining what kind of influence each stakeholder has on the "process improvement" project and its potential success. That knowledge in turn can help deciding on how to manage stakeholders—how to marshal the help of those that support the "process improvement" project, how to involve those who could be helpful, and how to convert—or at least neutralize—those who may adversely affect the "process improvement" objectives.

An assumption that most proponents of this analysis technique seem to make is that the stakeholders most important to the success of your effort are in the upper right section of the grid, and those least important are in the lower left. The names in parentheses are another way to define the same stakeholder characteristics in terms of how they relate to the effort.

1. *Promoters* have both great interest in the "process improvement" project and the power to help make it successful (or to derail it).
2. *Defenders* have a vested interest and can voice their support in the "process improvement" project, but have little actual power to influence the "process improvement" project in any way.
3. *Latents* have no particular interest or involvement in the "process improvement" project, but have the power to influence it greatly if they become interested.
4. *Apathetics* have little interest and little power, and may not even know that the "process improvement" project exists.

Interest here means one or both of two things: (1) the stakeholder is interested intellectually or philosophically in the "process improvement" project; and/or (2) the stakeholder is affected by the "process improvement" project. The level of interest, in this second sense, corresponds to how great the effect is.

4.1.3 Record Stakeholders Information

The information issued from identification of customer and stakeholder and analysis of stakeholders' interest and influence is typically kept in a document called "Customer and Stakeholder Register." The Stakeholder Registry is a project management document that has the list of stakeholders and relevant information about them.

4.2 Develop Project Charter

This project management process is concerned with developing a document—the project charter—that formally authorizes the "process improvement" project.

A project charter announces that a new project has begun. The purpose of the charter is to demonstrate management support for the project and the project manager. It is a simple, powerful tool. It provides the project manager or the project team leader with the authority to apply organizational resources to the "process improvement" project activities. It documents the business needs, the project justification, the current understanding of the customers and the stakeholder's needs and expectations, the "process to be improved," or result that it is intended to satisfy, such as:

1. "Process improvement" project purpose or justification;
2. S.M.A.R.T[1] project goal;
3. Project success criteria;
4. High-level requirements;
5. High-level project description;
6. High-level description of the "process to be improved" (S.I.P.O.C.);
7. High-level characteristics of the "process to be improved";
8. Summary milestone schedule;
9. Summary budget;
10. Project approval requirements;
11. Assigned project manager, responsibility, and authority level; and
12. Name and responsibility of the person(s) authorizing the project charter.

4.2.1 Project Purpose or Justification

The purpose of the "process improvement" project is to provide an answer to the first fundamental question of the PDSA model: *"What is intended to be realized or accomplished by the "process improvement" project?"* What the project intends to do to address the problem or improvement opportunity identified. The purpose of the goal statement is to the enterprise business management to value the idea enough to read on. In other words, the management should think enough of the idea to conclude that it warrants further attention and consideration.

A project has one goal. The goal gives purpose and direction to the project. It defines the final deliverable or outcome of the project so that everyone understands what is to be accomplished in clear terms. The goal statement will be used as a continual point of reference for any questions that arise regarding scope or purpose.

The goal statement must not contain any language or terminology that might not be understandable to anyone having occasion to read it. It is written in the language of the business so that anyone who reads it will understand it without further explanation from the proposer.

Identifying the "process improvement" project success criteria relates to the second fundamental question of the PDSA model: *"How will the realized or accomplished outcome of the 'process improvement' project be recognized as is*

[1] Doran's S.M.A.R.T. characteristics provide the criteria for a goal statement (Doran, 1981):
Specific—Be specific in targeting an objective.
Measurable—Establish a measurable indicator(s) of progress.
Attainable—Make the goal attainable for completion.
Realistic—State what can realistically be done with available resources.
Time-related—State when the goal can be achieved: that is, duration.

an improvement?", "Why do we want to undertake this project?" The project success criteria are the measurable business values that will result from undertaking this project. Whatever criteria are used, they must answer the second fundamental question of the PDSA model.

4.2.2 Project Success Criteria

The success criteria form a statement of achievement. It is a statement of the business value to be achieved, and therefore, it provides a basis for the enterprise business management to authorize the project. It is essential that the success criteria be quantifiable and measurable, and if possible, expressed in terms of business value. Regardless of how the success criteria are defined, they all reduce to one of three types:

1. *Increased revenue*—As a part of the success criteria, that increase should be measured in hard money currency or as a percentage of a specific revenue number.
2. *Reduced costs*—Again, this criterion can be stated as a money currency amount or a percentage of some specific cost. Be careful here because quite often a cost reduction means staff reductions. Staff reductions do not mean the shifting of resources to other places in the enterprise business. Moving staff from one area to another is not a cost reduction.
3. *Improved product or service*—Here, the metric could be more difficult to define. It is usually expressed as percentage improvement in customer satisfaction or a reduction in the product defects or a reduction in the frequency or type of customer complaints.

In some cases, it will take some degree of creativity to identify the success criteria. For example, customer satisfaction may have to be measured by some pre- and post-surveys. In other cases, a surrogate might be acceptable if directly measuring the business value of the project is impossible.

The best choice for success criteria is to state clearly the bottom-line impact of the project on the enterprise business intended strategy. This is expressed in terms such as increase margins, higher net revenues, reduce turnaround time, improve productivity, and reduce cost of manufacture or sales, and so on.

4.2.3 High-Level Description of the "Process to be Improved"

A high-level description of the "process to be improved" is one of the most fundamental building blocks in a process improvement methodology. It provides a way to build the initial controlled and organized view of the "process to be improved" and sets the foundation for applying a process improvement methodology.

A very effective diagram often used for this purpose is the Suppliers-Inputs-Process-Outputs-Customers (S.I.P.O.C.) diagram, , illustrated in Table 4.4, which

Table 4.4 The S.I.P.O.C.

S Suppliers	I Inputs	P Process	O Outputs	C Customers
Suppliers are systems, people, organizations, or other sources of the materials, information, or other resources that are consumed or transformed in the process.	Inputs are materials, information, and other resources provided by the suppliers that are consumed and transformed in the process.	The process is a set of logically related discrete elements (tasks, actions, or steps) taken in order to achieve a particular end.	Outputs are the outcomes (products or services) produced by the process and used by the customers.	Customers are the persons, group of people, companies, systems, and downstream processes, recipients of the process outcomes.

maps the process at a high-level with four to seven steps, with clearly defined process boundaries (start and end points) so that everyone involved understands the limits of the analysis. Then working from the right letter of its acronym, it identifies the customers, the outcomes or the process, the inputs to the process and the suppliers.

The S.I.P.O.C. diagram is built from the inside-out, starting at the center with a four to seven steps high-level map of the "process to be improved," followed by an identification of the process outcomes and the associated customers, and finishing with identification of the inputs to the process and the sources of these inputs; i.e., their suppliers.

4.2.4 Conclusion

The project charter links the project to ongoing work within the enterprise business and authorizes the project. The "Develop Project Charter" process is also used to validate or refine the assumptions and decisions made during its previous iteration. As indicated by the Project Management Body of Knowledge guidelines, (PMI, 2004, 2010), key inputs to a successful development of the project charter include:

1. The project statement of work,
2. The business case, and
3. The Enterprise organizational process assets.

4.2.4.1 Project Statement of Work

The statement of work (SOW) is a narrative description of the work required for the project. The complexity of the statement of work is determined by the desires of enterprise business management, the customers, and the stakeholders.

For projects internal to the enterprise business, the statement of work is prepared by the project office with input from the stakeholders. The reason for this is that stakeholders tend to write in terms that only the stakeholders understand their meaning. Since the project office is usually composed of personnel with writing skills, it is only fitting that the project office prepare the statement of work and submit it to the stakeholders for verification and approval.

For projects external to the enterprise business, as in competitive bidding, the contractor may have to prepare the statement of work for the customer because the customer may not have a team of people trained in statement of work preparation. In this case, as before, the contractor would submit the statement of work to the customer for approval. It is also quite common for the project manager to rewrite a customer's statement of work so that the contractor's managers can price out the effort.

In a competitive bidding environment, the reader should be aware of the fact that there are two statement of work—the statement of work used in the proposal and a contract statement of work (CSOW). There might also be a proposal work breakdown structure (WBS) and a contract work breakdown structure (CWBS). Special care must be taken by contract and negotiation teams that all discrepancies between the SOW/WBS and CSOW/CWBS are discovered, or additional costs may be incurred.

A good (or winning) proposal is no guarantee that the customer or contractor understands the statement of work. For large "process improvement" projects, fact-finding must be carried out before final negotiations because it is essential that both the customer and the contractor understand and agree on the statement of work, what work is required, what work is proposed, the factual basis for the costs, and other related elements. In addition, it is imperative that there be agreement between the final CSOW and CWBS.

The statement of work includes:

1. *The purpose statement or business need.* The enterprise business need may be based on a market demand, technological advance, required training, legal requirement, or governmental standard. "What is intended to be realized or accomplished by the 'process improvement' project?" This is the question that the purpose statement attempts to answer. "Why?" is always a useful question, particularly when significant amounts of time and money are involved.
2. *Process scope description.* It documents the process characteristics of the "process to be improved" that the project will be undertaken to improve. The description should also document the relationship between the "process to be improved" and the business need that the project will address. The process scope statement puts some boundaries on the "process to be improved." Scope creep is one of the most common project afflictions. It means adding work, little by little, until all original cost and schedule estimates are completely unachievable. The process scope statement should describe the major activities of the "process improvement" project in such a way that it will be absolutely clear if extra work is added later on.
3. *Strategic plan.* All projects should support the enterprise business intended strategic goals. As illustrated in the previous chapter, the intended strategic plan of the performing enterprise business should be considered as a factor when making project selection decisions and prioritization.

As project manager, you need to write out the statement of work and then present it to the stakeholders. Even though you may not know all the answers, it is easier for a group to work with an existing document than to formulate it by committee. The stakeholders will have sufficient opportunities to give their input and make changes once the SOW is presented to them.

4.2.4.2 Business Case

Identifying a business problem or opportunity to be addressed is the basis to initiating a project. A business case is created to define the problem or opportunity in detail and identify a preferred solution for implementation. The business case or similar document provides the necessary information from a business standpoint to determine whether or not the "process improvement" project is worth investing in. Typically, at a high level, the business need, which will be addressed by the "process improvement" project and the cost benefit analysis, are contained in the business case to justify the project. The requesting enterprise business function or customer, either internal customer or external customer in the case of external projects, may write the business case. The business case is created as a result of one or more of the following:

1. *Market demand:* e.g., an automobile plant factory authorizing a project to improve the process development process to allow building more fuel-efficient cars in response to gasoline shortages;
2. *Business need:* e.g., a quality training company authorizing a project to improve the quality process course to allow a course expansion to increase its revenues;
3. *Customer request:* e.g., a turbo-machinery plant authorizing a project to improve the turbine development process to allow compatibility with plane turbine engines;
4. *Technological advance*: e.g., an electronics firm authorizing a new project to develop a faster, cheaper, and smaller laptop after advances in computer memory and electronics technology;
5. *Legal requirement*: e.g., a paint manufacturer authorizing a project to establish guidelines for handling toxic materials; or
6. *Social need*: e.g., a nongovernmental organization in a developing country authorizing a project to improve the water distribution process to provide potable water systems, latrines, and sanitation education to communities.

As "process improvement" projects are often carried out in multi-phases within enterprise business procedures, the business case is referred to throughout the project to determine whether the costs, benefits, risks and issues align with those originally documented. It should be periodically reviewed to ensure that the "process improvement" project is on track to deliver the business benefits. In the early stages of the project life-cycle, periodic review of the business case by the sponsoring function also helps to confirm that the project is still required. At the end of the project, a post-implementation review (PIR) will be undertaken to determine whether the "process improvement" project delivered the business benefits outlined in the business case. As such, the success of the "process improvement" project is measured against the ability of the project to meet the criteria outlined in the business case.

4.2.4.3 Enterprise Organizational Process Assets

The fourth edition PMBOK®, (PMI, A Guide to the Project Management Body of Knowledge (PMBOK Guide), 2010), defines organizational process assets as:

> *"Any or all process related assets, from any or all of the organizations involved in the project that can be used to influence the project's success."*

Examples of organizational process assets include: plans, procedures, lessons learned, historical information, schedules, risk data and earned value data. The key concept is that these are assets a project manager may use for the benefit of the project. They work together with Enterprise Environmental Factors (EEF's) to bring those things outside the project team into focus.

In regards to the process of effectively managing of projects, it is important that the project team compiles and creates an essential and complete listing of organizational process assets. This listing of organizational process assets details a thorough listing of any and all organizational process-oriented assets, from the entirety of the enterprise business functions that are involved in any and all elements of the "process improvement" project, particularly and of significance referring to any elements as such that can influence to success of the project one way or another. The types of assets that can fall under this categorical heading include any formal or informal plans that have been derived, as well as any project related policies and procedures. Also of importance for inclusion in the compilation of the enterprise business process assets list is the outline and delineation of the enterprise business knowledge base and such as any lessons that have been learned and a historical information recording.

Almost, every enterprise business keeps a database of all the information pertaining to the enterprise business. So organizational process assets may include but not limited to all the documents, templates, policies, procedures, plans, guidelines, lesson learned, historical data and information, earned value, estimating, risk etc. These assets are typically grouped into two broad categories—Processes and Procedures, and the Corporate Knowledge Base (PMI, 2004, 2010):

1. Enterprise business processes and procedures for conducting work:
   - Enterprise business standard processes, such as standards, policies: e.g., safety and health policy, and project management policy.
   - Standard product and project life cycles, and quality policies and procedures: e.g., process audits, improvement targets, checklists, and standardized process definitions for use in the organization.
   - Standardized guidelines, work instructions, proposal evaluation criteria, and performance measurement criteria.
   - Templates: e.g., risk templates, work breakdown structure templates, and project schedule network diagram templates.
   - Guidelines and criteria for tailoring the enterprise business set of standard processes to satisfy the specific needs of the project.
   - Enterprise business communication requirements: e.g., specific communication technology available, allowed communication media, record retention, and security requirements.

- Project closure guidelines or requirements: e.g., final project audits, project evaluations, product validations, and acceptance criteria.
- Financial controls procedures: e.g., time reporting, required expenditure and disbursement reviews, accounting codes, and standard contract provisions.
- Issue and defect management procedures defining issue and defect controls, issue and defect identification and resolution, and action item tracking.
- Change control procedures, including the steps by which official company standards, policies, plans, and procedures—or any project documents—will be modified, and how any changes will be approved and validated.
- Risk control procedures, including risk categories, probability definition and impact, and probability and impact matrix.
- Procedures for approving and issuing work authorizations.

2. Enterprise business corporate knowledge base for storing and retrieving information:
   - Process measurement database used to collect and make available measurement data on processes and products.
   - Project files: e.g., scope, cost, schedule, and quality baselines, performance measurement baselines, project calendars, project schedule network diagrams, risk registers, planned response actions, and defined risk impact.
   - Historical information and lessons learned knowledge base: e.g., project records and documents, all project closure information and documentation, information about both the results of previous project selection decisions and previous project performance information, and information from the risk management effort.
   - Issue and defect management database containing issue and defect status, control information, issue and defect resolution, and action item results.
   - Configuration management knowledge base containing the versions and baselines of all official company standards, policies, procedures, and any project documents.
   - Financial database containing information such as labor hours, incurred costs, budgets, and any project cost overruns.

4.3 Develop Preliminary Project Scope Statement

This is the project management process necessary for producing a preliminary high-level definition of the "process improvement" project using the project charter with other inputs. The scope statement is a short description of the project scope. It is used as the basis for future project decisions and the criteria used to determine if major phases of the project and the project as a whole have been completed successfully. The scope statement forms the basis for an agreement between the project team and the project customer by identifying:

1. The justification for the project
2. Project objectives
3. Major project deliverables
4. Success criteria

This project management process addresses and documents the "process improvement" project and deliverables requirements, the boundaries of the project, the methods of acceptance, and a high-level scope control. It also validates or refines the scope of each "process improvement" project phase. A project scope statement includes (PMI, 2004, 2010):

1. Project and "process to be improved" objectives
2. "Process to be improved" requirements and characteristics
3. "Process to be improved" acceptance criteria
4. Project boundaries
5. Project requirements and deliverables
6. Project constraints
7. Project assumptions and constraints
8. Initial project organization
9. Initial defined risks
10. Schedule milestones
11. Initial work breakdown structure
12. Order of magnitude cost estimate
13. Project configuration management requirements
14. Approval requirements

The preliminary scope statement is a pre-version of the project scope statement on a higher level. It expresses how the project manager understands the project charter. The preliminary project scope statement paints a very broad description of what the "process improvement" project will create for the enterprise business and the project stakeholders. It is often an initial companion piece to the project charter and is expected to be altered and refined in detail once the project moves into the planning process group.

The project scope statement content will vary depending upon the application area and complexity of the "process improvement" project and can include some or all of the components identified above. During subsequent phases of the "process improvement" project, the "Develop Preliminary Project Scope Statement" process validates and refines, if required, the project scope defined for that phase.

4.3.1 Defining the Project Objectives

An objective statement is a more detailed version of the goal statement. The purpose of objectives statements is to clarify the exact boundaries of the goal statement and define the boundaries or the scope of the "process improvement" project. In fact, the objective statements written for a specific goal statement are nothing more than a decomposition of the goal statement into a set of necessary and sufficient objective statements. That is, every objective must be accomplished in order to reach the goal, and no objective is superfluous. An objective statement should contain four parts:

1. *An outcome*—A statement of what is to be accomplished.
2. *A time frame*—The expected completion date.

3. *A performance measure*—Metrics that will measure success.
4. *An action*—How the objective will be met.

A good exercise to test the validity of the objective statements is to ask if it is clear what is in and what is not in the project. Statements of objectives should specify a future state, rather than be activity based. We like to think of them as statements that clarify the goal by providing details about the goal. Think of them as sub-goals and you will not be far off the mark.

It is also important to keep in mind that these are the current objective statements. They may be altered during the course of planning the "process improvement" project. This will happen as the details of the project work are defined. When interfacing with customers and stakeholders, we all have the tendency to put more on our plates than we need. The result is to include project activities and tasks that extend beyond the boundaries defined in the project charter. When this occurs, the project manager should stop the planning session and ask whether the activity is outside the scope of the project, and if so, whether the scope should be altered to include the new activity or delete the new activity from the project plan.

4.4 Perform Phase Review

At the end of the initiation phase, a phase review must be performed. This is a checkpoint to ensure that the project has achieved its stated objectives by providing answers to the three fundamental questions, which form the basis and the preliminary step of the PDSA model:

1. What is intended to be realized or accomplished by the "process improvement" project?
2. How will the realized or accomplished outcome of the "process improvement" project be recognized as is an improvement?
3. What alterations to the system affected by the "process to be improved" can be made based on the realized or accomplished outcome of the "process improvement" project?

A phase review form should be completed to formally request approval to proceed to the next phase of a project. It should be completed by the project manager and approved by the project sponsor. To obtain approval, the project manager will usually present the current status of the project to the project board for consideration. The project board (chaired by the project sponsor) may decide to cancel the project, undertake further work within the existing project phase or grant approval to begin the next phase of the project.

5 "PDSA Plan" Process Group

In the PDSA model, planning the "process improvement" project is indispensable. A credible and robust plan is one of the foundation stones of effective "process improvement" project management. Not only is it a roadmap to how the work will be performed, but it is also a tool for decision making. It suggests alternative approaches, schedules, and resource requirements from which the project manager can select the best alternative. Planning the "process improvement" project is as much an art, which requires prior experience and common sense, as it is an exact science.

5.1 The Purpose of Planning

Too many "process improvement" projects start life doomed to failure. Poorly defined business requirements and unrealistic delivery deadlines are all too common. Your "process improvement" project planning phase should get more attention than any other aspect of the project management—and justifiably so.

It is hard to imagine how a "process improvement" project could be successful without some planning. In addition to being important, a "process improvement" project planning is also an enormous subject that consists of two components.

1. The first is almost strategic; it consists of understanding some of the principles and philosophies of "process improvement" planning.
2. The second component of "process improvement" project planning is tactical, operational and almost mechanical; it consists of the step-by-step process of creating a detailed "process improvement" project plan, collecting factual data, using estimates, of characteristic value (traits, behaviors, qualities, figures or parameter) of the target population, such as average value or standard deviation, as raw material and making inference is to drawing conclusions from factual evidence.

A. van Aartsengel and S. Kurtoglu, *Handbook on Continuous Improvement Transformation*,
DOI 10.1007/978-3-642-35901-9_5,

Each "process improvement" project poses new questions regarding what, how, by whom, in what order, for how much, and by when, and the purpose of planning is to provide answers these questions and to help:

1. To form a view of what tasks there are in the "process improvement" project and thus how long it will take, and from this be able to derive what resources will be required.
2. To explain to senior managers and other stakeholders how the "process improvement" project will be delivered.
3. To enable people involved in the "process improvement" project to be allocated to work and for them to understand how their work fits within the project.

Project planning puts together the details of how to meet the project's goals, given the constraints. Common estimating and scheduling techniques will lay out just how much work the "process improvement" project entails, who will do the work, when it will be accomplished, and how much it will cost. Along the way, risk management activities will identify the areas of greatest uncertainty and create strategies to manage them. The detailed strategy laid out in the plan becomes a reality check for the cost-schedule-quality equilibrium developed during project definition.

In planning a "process improvement" project, it is essential that the customers and stakeholders have a good understanding of its underlying objectives and the most important requirements. If the customers and stakeholders are not clear about what they want to achieve, the project manager's task of putting together a plan is made more difficult. It also raises serious questions about the wisdom of proceeding any further. It is important for you, as project manager, and your customers to achieve a shared understanding of the 'Big Picture'. It is not unusual to find that customers have developed a detailed view of what they want without giving a great deal of thought to what they would like the "process improvement" project to achieve.

5.2 The "PDSA Plan" Constituent Processes

The "PDSA Plan" Process Group encompasses the processes needed to establish requirements for customer, staff, business, and management. It helps gather information from many sources with each having varying levels of completeness and confidence.

The PDSA model, the "PDSA Plan" Process Group, illustrated in Fig. 5.1, which reflects a structure that mirrors the perspective of the Project Management Institute's PMBOK Guide, consists of those project management processes performed to: establish the total scope of the effort, define and refine the objectives, and develop the course of action required to attain those objectives.

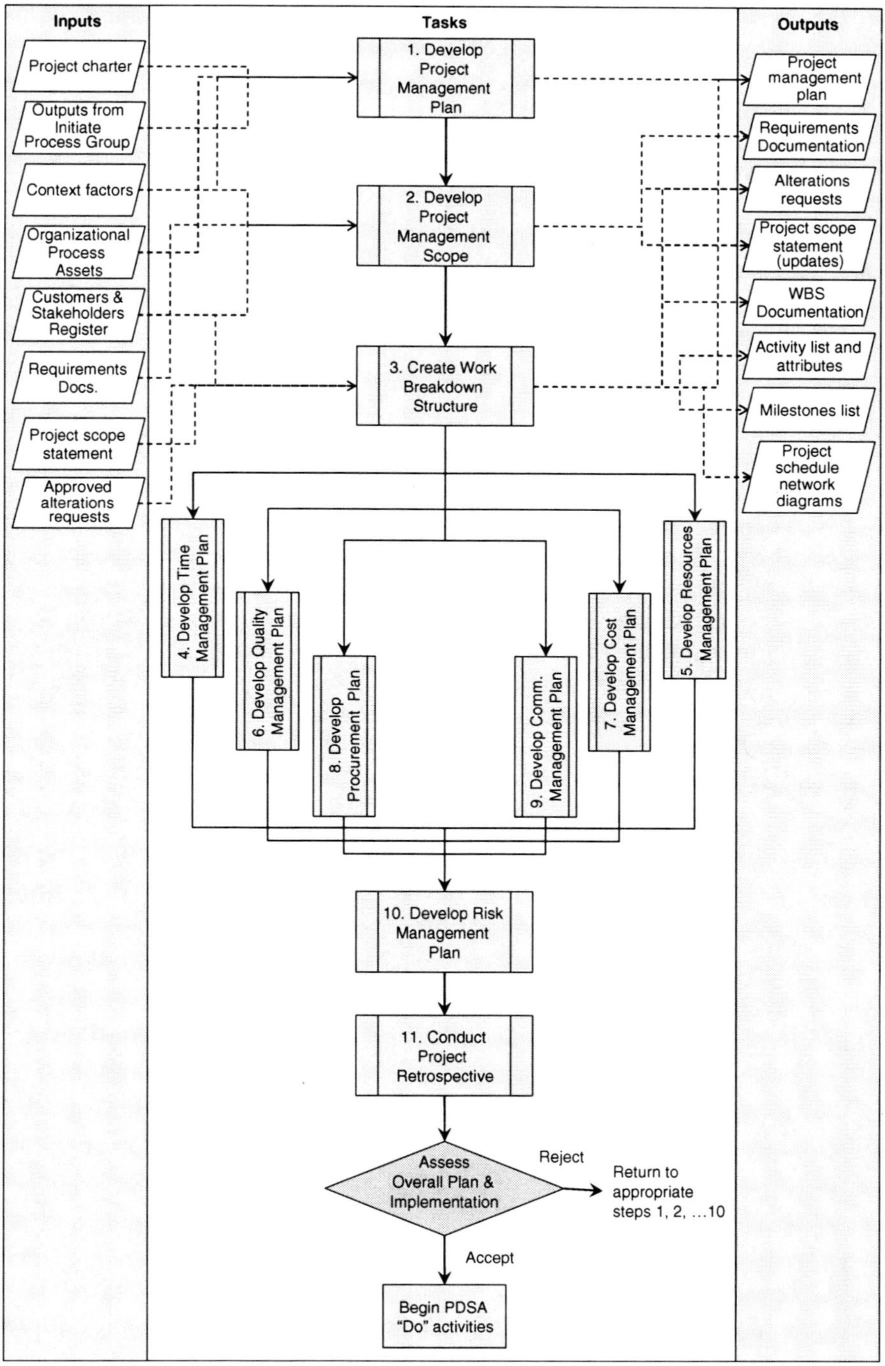

Fig. 5.1 "PDSA Plan" Process Group

The constituent project management processes used during the capturing of the project scope, illustrated in Fig. 5.1, include the following:

1. Develop Project Management Plan
2. Develop Project Management Scope
3. Create Work Breakdown Structure
4. Develop Time Management Plan
5. Develop Resource Management Plan
6. Develop Quality Management Plan
7. Develop Cost Management Plan
8. Develop Procurement Management Plan
9. Develop Communication Management Plan
10. Develop Risk Management Plan
11. Conduct Project Retrospective
12. Asses Overall Project Plan

Interactions among the project management processes within the "PDSA Plan" Process Group depend on the nature of the "process improvement" project. For example, for some "process improvement" projects there will be little or no identifiable risk until after significant planning has been carried out. At that time, the project team might recognize that the cost and schedule targets are overly aggressive, thus involving considerably more risk than previously understood. The results of the iterations are documented as updates to the project management or project documents.

Develop Project Management Plan

6

This is the project management process for documenting the actions necessary to define, prepare, integrate, and coordinate all subsidiary plans into one reference document: The "process improvement" project plan.

The "process improvement" project plan captures what you have been asked to do and how you, as project manager or as process improvement team leader, intend to deliver. It documents all the key points relating to a "process improvement" project ranging from its objectives and deliverables right through to the key milestones and resource requirements. A good "process improvement" project plan is one of the foundation stones for any "process improvement" project and should inspire confidence in all concerned. It cannot be discarded by the project team.

There are three key benefits to developing a credible and robust "process improvement" project plan. These include:

1. *Reducing uncertainty*—Even though one would never expect the "process improvement" project work to occur exactly as planned, planning the improvement intervention work allows the project team to consider the likely outcomes and to put the necessary corrective measures in place.
2. *Increasing understanding and learning*—The mere act of planning gives the project team a better understanding of the goals and objectives of the "process improvement" project. It increases the learning capability through the series of data collected to understand the extent of the improvement intervention, to diagnose problems and begin addressing them.
3. *Improve efficiency*—Once the project team has defined the "process improvement" project plan and the necessary resources to carry out the plan, it can schedule the work to take advantage of resource availability. It also can schedule work in parallel; that is, team members can perform tasks concurrently, rather than in series. By performing tasks concurrently, the project team can shorten the total duration of the project. It can maximize the use of resources and complete the project work in less time than by taking other approaches.

A. van Aartsengel and S. Kurtoglu, *Handbook on Continuous Improvement Transformation*,
DOI 10.1007/978-3-642-35901-9_6, © Springer-Verlag Berlin Heidelberg 2013

6.1 Elements of a "Process Improvement" Project Plan

The "process improvement" project plan is the primary source of information for how the project will be planned, executed, monitored and controlled, and closed. It can be either summary level or detailed, and it documents the suite of planning documents outputs of the constituent processes of the "PDSA Plan" Process Group listed in the previous chapter.

The "process improvement" project plan can assume many different shapes, sizes, and forms. In the project management profession, some people equate the plan with the schedule, but as we will see, in a "process improvement" project there is much more to a project plan than just a schedule. Project plans are often considered to consist of three fundamental "dimensions":

1. Scope: what is to be done?
2. Cost: how much money that will be spent and how it's budgeted over time;
3. Time: how long it will take to execute work—individually and as a total project;

The elements required in the "process improvement" project plan fall into one of the following nine categories.

6.1.1 Scope

The first step towards creating a project plan is to reconfirm the project scope statement, as preliminary defined in the "PDSA Initiate" project phase. The term scope actually has two meanings that are quite different in concept. It is important to understand what each meaning represents and how they are applied in discussions you may have during the development of the "process improvement" project plan.

Project scope is a term that is most closely associated with the mission, goals, and objectives of the "process improvement" project. It may be thought of as the overall size of the project or a high-level description of what the project will tackle. For example, building and installing a few automobile high-bay storage racks has a much smaller project scope than installing a computer-controlled storage and retrieval system.

Scope of work refers to all of the individual elements of the improvement intervention work (taken collectively) that must be performed to accomplish the project objectives. The efforts represented by all of the items that appear on the schedule or in the activities listing constitute the scope of work.

This first step in the planning process also consists of identifying exactly the scope of work. In this stage, you identify major elements of work and then break them down systematically into smaller and smaller pieces, until each piece becomes a comfortable size to estimate, execute, and monitor.

6.1.2 Phases

The second step towards creating a project plan is to list and describe the major phases within the "process improvement" project. A phase is a set of activities to be undertaken to deliver a substantial portion of an overall "process improvement" intervention. It includes the following:

1. A list of the constituent processes of the "PDSA Plan" Process Group selected by the project management team.
2. The level of implementation of each selected constituent process.
3. The descriptions of the tools and techniques to be used for accomplishing those constituent processes of the "PDSA Plan" Process Group.
4. How the selected constituent processes will be used to manage the specific "process improvement" project, including the dependencies and interactions among those constituent processes, and the essential inputs and outputs.
5. How work will be executed to build the deliverables and accomplish the "process improvement" project objectives.

6.1.3 Milestones

The third step towards creating a project plan is to list and describe the key project milestones. Throughout the entire life cycle of a "process improvement" project, there are going to be a number of natural time gradients that exists, and are more than likely to be defined by the eventual and natural ebbing and flowing of the improvement intervention workload and the momentum of the team's performance and the business of the schedule. However, it is typically helpful in attempting to assure that the improvement intervention is moving effectively as well as to allow points in time for the project team to perform retrospective; that is, to pause and look back on what has been accomplished to date as well as what may need to be changed in the future for the project team leader with or without the help of the project team to establish a number or series of project milestones.

These milestones can occur at any point throughout the "process improvement" project and specifically refer to any significant or substantive point, time, or event in the life cycle of the project. These milestones typically refer to points at which large schedule events or series of events have been completed, and a new phase ort phases are set to begin. They are governed by "deliverables," which provide the evidence that would indicate successful completion of a milestone.

A deliverable is an input/output term that refers specifically to the unique and individual products, elements, results, or items that are produced for delivery at the conclusion of a specific milestone, or at the conclusion of the project as a whole. Deliverables can come in a number of different variations. They can be any agreed tangible item that will define the completion of a phase of work and presented at a milestone. These may be:

1. Written report
2. Prototype
3. A model
4. Alternative documentation
5. Plan drawings
6. Designs

Deliverables towards the end of a project life are typically referred to as external deliverables, and these typically require the review and/or approval of the customer or financially responsible party.

6.1.4 Activities

The fourth step towards creating a project plan is to list and describe the key activities in the "process improvement" project. An activity is a set of tasks that are required to be undertaken to complete a portion of a "process improvement" project. Typically, key activities will include:

1. Defining the improvement intervention
2. Collecting factual evidence
3. Estimating and making inference on the collected evidence
4. Analyzing the evidence as well as the "process to be improved"
5. Devising solutions to the "process to be improved" underperformance
6. Selecting appropriate solutions and assessing impact on the business
7. Building knowledge and transformational learning about the solutions
8. Acting upon the built knowledge and system affected by the solutions

6.1.5 Tasks

The fifth step towards creating a project plan is to list and describe all key tasks required to undertake each activity in the "process improvement" project. A task is an item of work to be completed within a project activity.

6.1.6 Effort

The sixth step towards creating a project plan is to quantify the likely 'effort' required to complete each task listed above. The principal output of this portion of the planning process is a task-based timeline that the project team will use as a map for executing the work and that you, as project manager or project team leader will use as a guide for verifying that work is getting done on time.

6.1.7 Resources

The seventh step towards creating a project plan is to identify the critical resources required to complete each task listed above. As mentioned already in a previous chapter, the critical resources are assets such as the people, technology, products, facilities, equipment, channels, and brand required to deliver the value proposition to the targeted customer. The focus here is on the critical elements that create value for the customer and the enterprise business, and the way those elements interact. Every enterprise business also has generic resources that do not create competitive differentiation.

6.1.8 Project Schedule

The eighth step towards creating a project plan is to create a detailed project schedule, listing each of the phases, activities and tasks required to complete the project. The project schedule is a fairly broad and all encompassing concept that while seemingly easy to grasp, must truly be mastered in order for all members of the project staff to effectively manage the project in a capable manner from start to finish.

The project schedule typically will include all elements of the project from the pre-planning stages of the project through all ongoing project processes that may take place during the active project period, to any and all project related process that may occur at the conclusion and or closing stages of the project. The project schedule, as a project related input/.output mechanism, typically keeps careful track of any and all planned dates for the performance of particular schedule activities, as well as any predetermined dates that are expected to be met and followed up on in regards to the implementation of any particular project milestones.

6.1.9 Project Risk

The eighth step towards creating a project plan is to list the reasonably frequent risks that strike projects similar to the one being undertaken—late subcontractor deliveries, bad weather, unreasonable deadlines, equipment failure, complex coordination problems, and similar happenings. The argument is made that crises cannot be predicted. The only uncertainty, however, is which of the crises will occur and when. In most "process improvement" projects there are times when dependence on a subcontractor, material or machine availability is critical to progress on the project. Plans to deal with such potential crises should be a standard part of the "process improvement" project plan. It is well to remember that no amount of current planning can solve current crisis—but preplanning may prevent or soften the impact of some.

6.2 Collating the Materials

Collating into one document all of the materials listed in the section above creates your "process improvement" project plan document. This document forms the basis upon which the project is measured and it will be referred to throughout the project life cycle. It is a dynamic document and the project team should expect it to be altered during the project life cycle. It is continuously modified and refined in terms of content, structure, and level of detail.

As the project definition becomes more refined, work is broken down into ever-increasing levels of detail, assumptions are verified or refuted, and actual results are achieved, the project plan must keep pace. From this information, the "PDSA Plan" Process Group develops the "process improvement" project management plan and a suite of planning documents which help guide the project team through the remaining phases of the project. The multi-dimensional nature of project management creates repeated loops for additional analysis.

As more project information or characteristics are gathered and understood, follow-on actions may are also developed. Significant alterations occurring throughout the "process improvement" project life cycle may trigger a need to revisit one or more of the project management planning document and, possibly. In project management practice, this progressive detailing of the project management plan is often called "rolling wave planning," indicating that planning and documentation are iterative and ongoing processes.

The project management plan and a suite of planning documents which help guide the project team through the remaining phases of the "process improvement" project will explore all aspects of the scope, time, costs, quality, communication, risk, and procurements. Update arising from approved alterations during the "process improvement" project may significantly impact parts of the project management plan and the "process improvement" project documents. Updates to these documents should provide a greater precision with respect to schedule, costs, and resource requirements to meet the defined project scope.

While planning and developing the project plan and project documents for the "process improvement" project, within the enterprise business procedures, the project team should encourage involvement from all appropriate customers and stakeholders. Quite often, these customers and stakeholders possess the skills and knowledge that can be leveraged in developing the project plan and any subsidiary plans. The project team must continuously create and re-enforce a positive context in which customers and stakeholders can contribute appropriately.

Creating a credible, accurate and robust "process improvement" project plan by collating materials issued by each "PDSA Plan" constituent process management process requires a significant amount of effort and the input of many people. Enterprise businesses vary considerably in their general approach to project planning. The specific procedures that your enterprise business prescribes reflect its philosophy toward planning and control of projects. If your enterprise business management tends to be extremely action-oriented or to not believe in the value of planning, it is likely that your planning procedures will be minimal.

In such environment, projects may be hastily initiated and a significant amount of upfront planning is done without much thought or without properly considering alternatives or risks. Conversely, if your enterprise business management has a bias toward certainty or control, that is likely to be reflected in the development and use of rigorous planning procedures.

Develop Project Management Scope

7

This chapter is concerned with the project management process required to ensure that the project includes all the work required, and only the work required, to complete the "process improvement" project successfully. In the project context, and in accordance with the Project Management Body of Knowledge indication, the term scope can refer to:

1. *Project Scope*—The work that needs to be accomplished to deliver an improved process with the specified features and functions.
2. *Process scope*—The features and functions that characterize the actual "process to be improved." The process scope can remain constant while project scope expands.

The scope is a statement that defines the boundaries of the project. It tells not only what will be done but also what will not be done. In the information systems industry, the scope is often referred to as a functional specification. In the engineering profession, it is generally called a statement of work.

The scope may also be referred to as a document of understanding, a scoping statement, a project initiation document, and a project request form. Whatever its name, this document is the foundation for all project work to follow. It is critical that the scope be correct. So many times we have seen "process improvement" projects get off to a terrible start simply because there never was a clear understanding of exactly what was to be done. If you do not know what you are going to do and where you are going, how will you know when and if you ever get there?

Managing a "process improvement" project scope is primarily concerned with defining and controlling what is and what is not included in the project. The constituent project management processes used during the development of the project scope, illustrated in Fig. 7.1, which reflects a structure that mirrors the

A. van Aartsengel and S. Kurtoglu, *Handbook on Continuous Improvement Transformation*,
DOI 10.1007/978-3-642-35901-9_7, © Springer-Verlag Berlin Heidelberg 2013

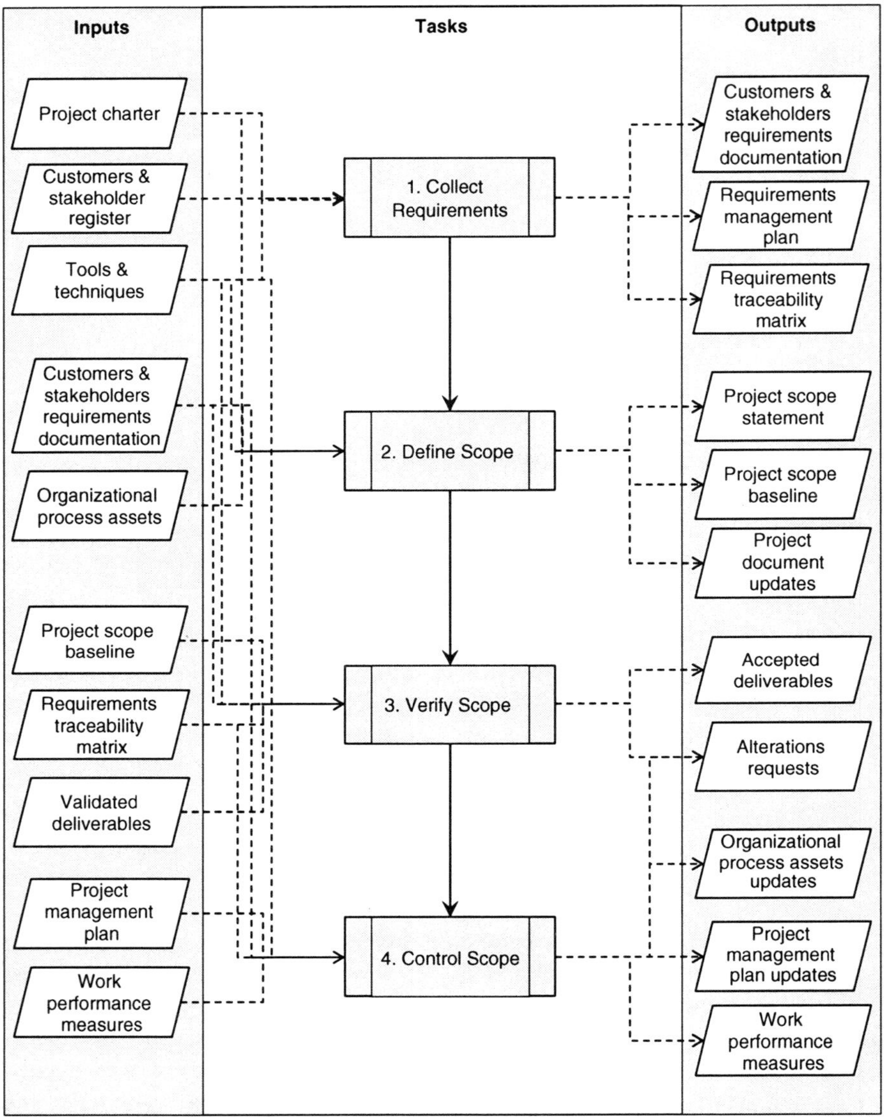

Fig. 7.1 Project scope management process

perspective of the Project Management Institute's PMBOK Guide, include the following:

1. Collect Requirements
2. Define Scope
3. Verify Scope
4. Control Scope

These four constituent processes interact with each other and with the project management processes in the PDSA "Process Groups." As the PMBOK Guide remind us, each aspect of executing any of these constituent processes can involve effort from one or more persons, based on the needs of the project. Each aspect occurs at least once in every "process improvement" project and occurs in one or more project phases.

The project management constituent processes utilized to manage project scope as well as the supporting tools and techniques, vary by application area and are defined as part of the project life cycle. The approved detailed project scope statement is the scope baseline for the "process improvement" project. This baseline scope should be monitored, verified and controlled throughout the lifecycle of the project.

Performance completion of the "process improvement" project scope is measured against the project management plan, while performance completion of the process scope is measured against the requirements of the actual "process to be improved."

7.1 Collect Requirements: V.O.B., V.O.C., & V.O.P.

The first step in developing the project management scope is to "Collect Requirements." It is an important concept in "process improvement" project as it describes conditions that must be met in order to produce a satisfactory deliverable. It relates to defining and documenting the "process improvement" project and "process to be improved" features and functions needed to fulfill the business needs and expectations (Voice of the Business—V.O.B.), the customers and the stakeholder's needs and expectations (Voice of the Customer—V.O.C.), and the "process to be improved" needs and expectations (Voice of the Process—V.O.P.) from the quality perspective. The project's success is directly influenced by the care taken in capturing and managing these requirements.

The most strident needs and expectations are those related to the customers. Indeed, the bottom line for every enterprise business is the value of its products and services in the eyes of potential customers. Without continuing enthusiasm from customers, business sustainability may not last. It is customers' opinions that will determine the value of the process outcomes. Customers' opinions of the value of the process outcomes determine the "customer value" of these outcomes. The customer value of a process outcome consists of key factors that determine how well customers will appreciate this outcome. For a given process, the customer value may change over time, and a new process outcome that better fits the changing customer value could be a breakthrough product or service.

Nominally, the customer value of a process outcome can be defined as the difference between the perceived benefit (i.e., benefits) and the perceived cost (i.e. liabilities) (Sherden, 1994; Gale, 1994):

$$customer\ value = Benefits - Liabilities$$

The benefits include the following categories:

1. Functional benefits
   - Process outcome functions, functional performance levels
   - Economic benefits, revenues (for investment services)
   - Reliability and durability
2. Psychological benefits
   - Prestige and emotional factors, such as reputation.
   - Perceived dependability (for example, people prefer known products and services rather than an unknown ones).
   - Social and ethical reasons (for example, environmentally friendly products and services).
   - Psychological awe (many first-in-market products and services not only provide unique functions, but also give customers a tremendous delight).
3. Service and convenience benefits
   - Availability (how easy is it to access the process outcome?)
   - Service (how easy is it to get service in case of process outcome problems or failure?)

The liabilities include the following:

1. Economic liabilities
   - Price
   - Acquisition cost (such as transportation and shipping costs, time and effort spent to obtain the process outcome).
   - Usage cost (additional cost to use the process outcome in addition to the purchasing price, such as installation).
   - Maintenance costs
   - Ownership costs
   - Disposal costs
2. Psychological liabilities
   - Uncertainty about the dependability of the process outcome
   - Self-esteem liability of using an unknown process outcome
   - Psychological liability of poor performance of the process outcome
3. Service and convenience liability
   - Liability due to lack of service
   - Liability due to poor service
   - Liability due to poor availability (such as delivery time, distance to shop)

For each particular "process to be improved" outcome, the profile of benefits and liabilities will be very different, and customers and stakeholders will give the benefits and liabilities different relative importance. Consequently, stakeholder risk tolerances must also be captured accurately.

Stakeholder risk tolerances are a vital input because different members of the customer, project, and management teams may have different perspectives on what constitutes liabilities and "acceptable" risk. This is rarely preordained or predetermined. The data collection team must plan to capture this information by vigorously pursuing the key stakeholders to identify what they are and are not willing to accept. This extends beyond simple thresholds for cost and schedule. Some stakeholders have passionate perspectives on project visibility. Some want to ensure the "process improvement" project is regularly in the public eye and consistently in the best possible light. Others, by contrast, want to ensure that project publicity is kept to an absolute minimum and consider any public exposure "bad exposure." Thresholds can be established for a variety of issues, ranging from satisfaction survey responses to team attrition to technology exposure. Failure to develop an acute awareness of the stakeholder's tolerances may lead to unidentified risks or improperly assigned impact levels during the planning phase of the project.

Collecting the Voice of the Customer is a practice used in process improvement undertakings to capture the customer's requirements, expectations, and entitlements that will be flowed into the "process to be improved." However, the Voice of the Customer is not only voice in a "process improvement" project. Two additional voices must be considered: the Voice of the Process (V.O.P.) and the Voice of the Business (V.O.B.).

Voice of the Business (V.O.B.)—This is the voice of profit and return on investment. Every "process improvement" project has to enable the enterprise business sustainability and meet the needs of the employees and shareholders.

Voice of the Process (V.O.P.)—The "process to be improved" must meet the requirements of the customers and stakeholders, and the ability of this process to meet these requirements is called Voice of the Process. It is a construct for examining what the "process to be improved" is telling about its inputs and outputs and the resources required to transform the inputs into outputs. We will elaborate further on the Voice of the Process in the "Develop Quality Management Plan" step of the "PDSA Plan" Process Group. This is the subject of Chap. 25.

Voice of the Customer (V.O.C.)—This is the voice calling back at the "process to be improved" from beyond its outcomes that offer compensation in return for satisfaction of the customers and stakeholders needs and wants. This voice represents the stated and unstated needs, wants, and desires of the customers and stakeholders, generally referred to as the customers and stakeholders' requirements. This is the subject of a next chapter.

7.2 Define Scope

The second step in developing the project management scope is "Define Scope." It relates to developing a detailed description of the extent of work and effort of the "process improvement" project and the "process to be improved." The preparation of a detailed project scope statement is critical to project success and builds upon the major deliverables, assumptions, and constraints that are documented during project initiation.

During planning, the preliminary project scope statement is refined and described with greater specificity as more information about the project is known. Existing risks, assumptions, and constraints are analyzed for completeness; and additional risks, assumptions, and constraints are added as necessary. Key tools and techniques used in defining the scope include but are not limited to:

1. Expert judgment
2. Process analysis
3. Alternative identification
4. Facilitated workshop

Expert Judgment—Expert judgment is often used to analyze the information needed to develop the project scope statement. Such judgment and expertise is applied to any technical details. Such expertise is provided by any group or individual with specialized knowledge or training, and is available from many sources, including:

1. Other functions or business units within the enterprise;
2. Consultants;
3. Stakeholders, customers, and sponsors;
4. Professional and technical associations;
5. Industry groups; and
6. Subject matter experts.

Process Analysis—Each application area has one or more generally accepted methods for translating high-level process descriptions into tangible deliverables. Process analysis includes techniques such as process breakdown, systems analysis, systems engineering, value engineering, value analysis, and functional analysis.

Alternatives Identification—Identifying alternatives is a technique used to generate different approaches to execute and perform the work of the project. A variety of general management techniques is often used here, the most common of which are brainstorming and lateral thinking.

The key outcome of preparation of a detailed project scope is the project scope statement. It describes, in detail, the project's deliverables and the work required to create those deliverables. The project scope statement also provides a common understanding of the project scope among project stakeholders. It may contain explicit scope exclusions that can assist in managing customers and stakeholder expectations. It enables the project team to perform more detailed planning, guides the project team's work during execution, and provides the baseline for evaluating whether requests for changes or additional work are contained within or outside the project's boundaries.

The degree and level of detail to which the project scope statement defines the work that will be performed and the work that is excluded can determine how well the project team can control the overall project scope. The detailed project scope statement includes, either directly, or by reference to other documents, the following:

1. *Process scope description*—It progressively elaborates the characteristics of the "process to be improved" that are described in the project charter.
2. *Project deliverables*—Deliverables include both the outputs that comprise the "process to be improved" of the project, as well as ancillary results, such as project management reports and documentation. The deliverables may be described at a summary level or in great detail.
3. *Project boundaries*—It generally identifies what is included within the project. It describes well the importance of the project knowing what the project has agreed to do, and, perhaps of greater importance, knowing what the project has not agreed to do. It must state explicitly what is out of scope for the "process improvement" project; and whether a stakeholder should assume that a particular stated outcome requirement could be included in the project when, in actuality, it is not.
4. *Process acceptance criteria*—It defines the performance measure for accepting completed and improved process.
5. *Project constraints*—It lists and describes the specific project constraints associated with the project scope that limits the team's options. For example, a predefined budget or any imposed dates (schedule milestones) that are issued by the customer. When a "process improvement" project is performed under contract, contractual provisions will generally be constraints. Information on constraints may be listed in the project scope statement or in a separate record.
6. *Project assumptions*—It lists and describes the specific project assumptions associated with the project scope and the potential impact of those assumptions if they prove to be false. Project teams frequently identify, document, and validate assumptions as part of their planning process. Information on assumptions may be listed in the project scope statement or in a separate record.

7.3 Verify Scope

The third step in developing the project management scope is "Verify Scope." It relates to formalizing acceptance of the completed project deliverables. As the "Project Management Body of Knowledge" in its guidelines indicates, verifying the project scope includes reviewing deliverables to ensure that each is completed satisfactorily. If the project is terminated early, the project scope verification process should establish and document the level and extent of completion.

Scope verification is performed through inspection. Inspection comprises activities such as measuring, examining, and verifying to determine whether work and deliverables meet requirements and "process to be improved" acceptance criteria. Inspections are sometimes called reviews, process reviews, audits, and walkthroughs. In some application areas, these different terms have narrow and specific meanings.

Scope verification differs from quality control in that scope verification is primarily concerned with acceptance of the deliverables, while quality control is primarily concerned with correctness of the deliverables and meeting the quality requirements specified for the deliverables. Quality control is generally performed before scope verification, but these two processes can be performed in parallel.

The "Verify Scope" project management process also documents those completed deliverables that have been formally accepted. Those completed deliverables that have not been formally accepted are documented, along with the reasons for non-acceptance. Scope verification includes supporting documentation received from the customer or sponsor and acknowledging formal stakeholder acceptance of the project's deliverables, as well as alterations requests.

7.4 Control Scope

The last step in developing the project management scope is "Control Scope." It relates to monitoring the status of the project and "process to be improved" scope and controlling alterations. Controlling the project scope ensures that all requested alterations and recommended corrective actions are taken into account and processed. Project scope control is also used to manage the actual alterations when they occur and is integrated with the other control processes. Uncontrolled alterations are often referred to as project scope creep, hope creep, effort creep, or feature creep.

7.4.1 Scope Creep

It is the term that has come to mean any alteration in the project that was not in the original plan. Alteration is constant. To expect otherwise is simply unrealistic. Alterations occur for several reasons that have nothing to do with the ability or foresight of the customer, the stakeholder, or the project manager. Market conditions are dynamic. The competition can introduce or announce an upcoming new version of its product. The enterprise business management might decide that getting to the market before the competition is necessary. The task of the project manager is to figure out how these alterations can be accommodated. Regardless of how the scope alteration occurs, it is the task of the project manager to figure out how, if at all, the alteration can be accommodated.

7.4.2 Hope Creep

It is the result of a project team member's getting behind schedule, reporting that he or she is on schedule, but hoping to get back on schedule by the next report date. Hope creep is a real problem for the project manager. There will be several activity managers within a "process improvement" project, team members who manage a hunk of work. They do not want to report bad news to the team leader, so they are prone to say that their work is proceeding according to schedule when, in fact, it is not. It is their hope that they will catch up by the next report period, so they mislead the project manager or project team leader into thinking that they are on schedule. The activity managers hope that they will catch up by completing some work ahead of schedule to make up the slippage. The project manager must be able to verify the accuracy of the status reports received from the team members. This does not mean that the project manager has to check into the details of every status report. Random checks can be used effectively.

7.4.3 Effort Creep

It is the result of a team member's working but not making progress proportionate to the work expended. Most of us have worked on projects that always seem to be 95 % complete no matter how much effort is expended to complete it. Each week the project status report records progress, but the amount of work remaining doesn't seem to decrease proportionately. Other than random checks, the only effective thing that the project manager can do is to increase the frequency of status reporting by those team members who seem to suffer from effort creep.

7.4.4 Feature Creep

Closely related to scope creep is feature creep. Feature creep results when team members arbitrarily add features and functions to the deliverable that they think customers and stakeholders would want to have. The problem is that the customer did not specify the feature, probably for good reason. If the team member has strong feelings about the need for this new feature, formal alteration management procedures can be employed.

Alteration is inevitable, thereby mandating some type of alteration control process. Scope controlling is performed through variance analysis and re-planning techniques. With the variance analysis technique, project performance measurements are used to assess the magnitude of variation to the original scope baseline. Here, important aspects of project scope control include determining the cause of variance relative to the scope baseline and deciding whether corrective action is required. With the re-planning technique, approved alteration requests affecting the project scope can require modifications to the project scope statement, and the customers and stakeholders requirements documentation. These approved alteration requests can cause updates to components of the project management plan.

The results of project scope control include establishing work performance measurements and updating "organizational process assets." These results can generate alteration requests, which are processed for review. Alteration requests can include preventive or corrective actions or defect repairs.

8 Collecting V.O.C. Requirements

Collecting the customers and stakeholders' requirements is as much about defining and managing customers' and stakeholders' expectations as any other key project deliverables and will be the very foundation of completing to "process improvement" project. It is also about focusing the improvement effort by gathering information on the current situation. Its purpose is to build, as precisely as possible, a factual understanding of existing "process to be improved" conditions and problems or causes of underperformance. Cost, schedule, and quality planning are all built upon these requirements.

In other words, the purpose of collecting the customers and stakeholders' requirements is to get sufficient and accurate information to complete improvement of the "process to be improved." Most importantly, the purpose is to get accurate and sufficient data to derive complete functional requirements for the "process to be improved" outcomes from the V.O.C. data. This clear purpose will determine what kind of V.O.C. data should be collected.

The constituent project management processes used during the capturing of the project scope, illustrated in Fig. 8.1, include the following:

1. Plan V.O.C. Capturing
2. Collect and Organize Data
3. Analyze Data and Generate Customer Key Needs
4. Translate Key Needs into CTXs
5. Set Specifications for CTXs

8.1 Plan V.O.C. Capturing

The first step in collecting the customers and stakeholders' requirements is "Plan V.O.C. Capturing." This is the project management process for documenting the actions necessary to define, prepare, integrate, and coordinate all subsidiary V.O.C. capturing actions into one document. It builds on the customer and a stakeholder

A. van Aartsengel and S. Kurtoglu, *Handbook on Continuous Improvement Transformation*, DOI 10.1007/978-3-642-35901-9_8,

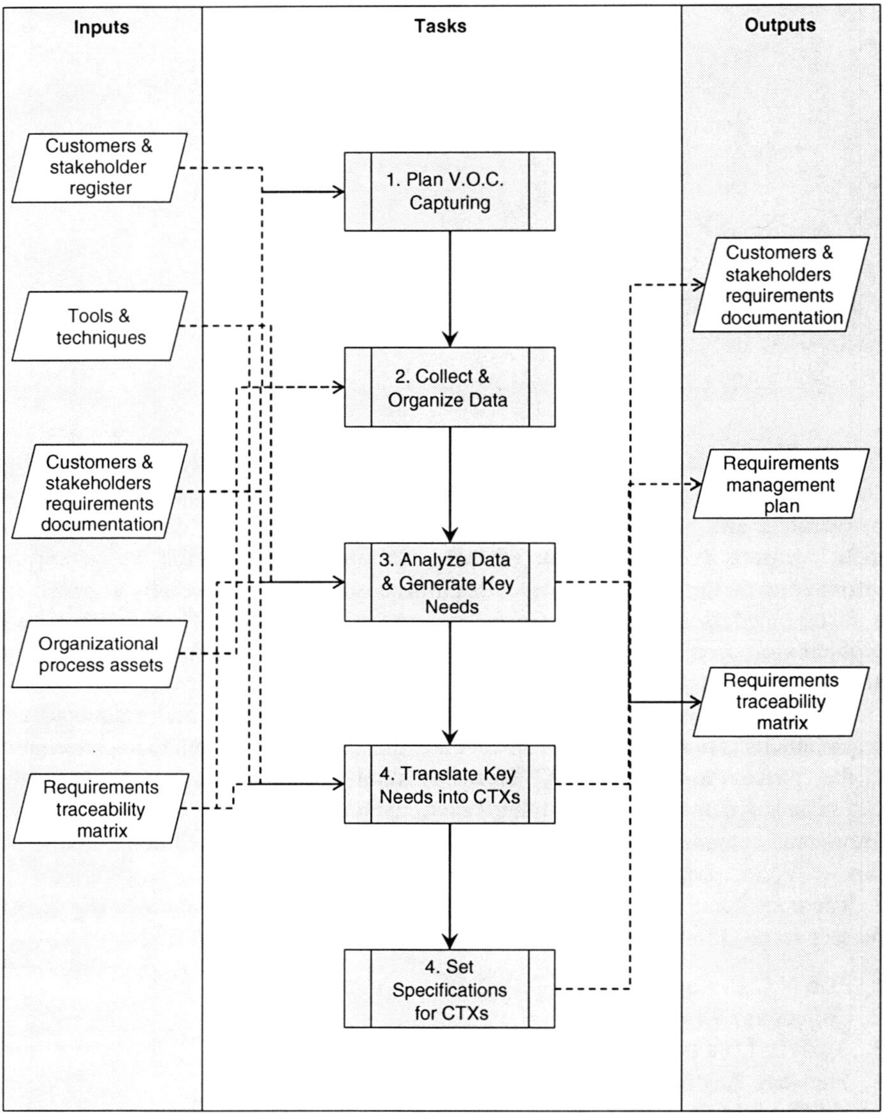

Fig. 8.1 The V.O.C. management process

register and it clarifies what is needed to be known from the customers and stakeholders to improve the "process to be improved" to their satisfaction.

Planning for V.O.C. data collection includes, but is not limited to the following steps:

1. Identify V.O.C. data and clarify goals
2. Develop operational definitions and procedures
3. Develop sampling strategy
4. Validate data collection system

8.1.1 Identify V.O.C. Data and Clarify Goals

The first step in planning for V.O.C. data collection is to identify the V.O.C. data and clarify goals. The purpose here is to ensure that the V.O.C. data, which the project team collects, will provide the answers needed to carry on the "process improvement" project successfully. Knowing what type of data the project team will be dealing with also tells which tool should be used to capture it. The right V.O.C. data should:

1. Describe the issue or problem that the "process to be improved" is facing;
2. Describe related conditions that might provide clues about causes of underperformance of the "process to be improved";
3. Lead to analysis in ways that answer the project team questions.

Desired V.O.C. data characteristics are: sufficient, relevant, representative, and contextual. In general, there are two types of data: qualitative and quantitative data.

Qualitative V.O.C. data are obtained from those customer needs in which items are described in terms of words and narratives statements. They can be grouped by highlighting key words, extracting themes, and elaborating concepts.

Quantitative V.O.C. data are obtained from those customer requirements in which items are described in terms of measurable quantity and in which a range numerical values are used without implying that a particular numerical value refers to a particular distinct category.

However, data originally obtained as qualitative information about individual items may give rise to quantitative data if they are summarized by means of counts; and conversely, data that are originally quantitative are sometimes grouped into categories to become qualitative data. When a given data set is numerical in nature, it is necessary to carefully distinguish the actual nature of the customer requirement being quantified.

Another important characteristic of quantitative data is its measurement scale. Table 8.1 describes four measurement scales with the latter ones being more useful for statistical analysis.

One of the most important things that the project team should do in planning for V.O.C. data collection is to draw and label the graph that will communicate the findings before the actual V.O.C. data collection begins. This directs the project team to exactly what V.O.C. data is needed. Moreover, it raises questions that the project team might not have though of, which it can add to the planning. This will prevent having to return for V.O.C. data that the project team had not though of.

8.1.2 Develop Operational Definitions and Procedures

An operational definition for a V.O.C. data is a description of term as applied to a specific situation of the "process improvement" project to facilitate the collection of meaningful (standardized) V.O.C. data.

When collecting V.O.C. data it is important to define terms very clearly in order to assure all those collecting and analyzing the data have the same understanding. Any V.O.C. data for which an "operational definition" has not been defined often lead to inconsistencies and erroneous results.

Table 8.1 Four measurement scale levels

Data Type	Scale	Description	Example
Attribute	Nominal	Data consists of names or categories only. No ordering scheme is possible.	A parking lot has cars of the following colors: Red 5 White 4 Blue 7 Black 6
	Ordinal (Ranking)	Data is arranged in some order but differences between values cannot be determined or are meaningless.	A survey question: Individual commitment is required for realizing alignment: 1. Strongly disagree 2. Disagree 3. Neither agree nor disagree 4. Agree 5. Strongly agree The difference between Strongly agree (5) and Agree (4) does not have the same meaning as the difference between Disagree (2) and Strongly disagree (1).
Variable	Interval	Data is arranged in order and differences can be found. However, there is no inherent starting point and ratios are meaningless.	The temperature of three heated metal pieces is 300°C, 600°C, and 900°C respectively. Three times 300°C is not the same as 900°C in temperature measurement.
	Ratio	An extension of the interval scale that includes an inherent zero starting point. Both the difference between values and the ratios are meaningful.	Product A costs 900,- monetary units; product B costs 500,-monetary units.

It is easy to assume that those collecting the data understand what and how to complete the task. However, people have different opinions and views, and these will affect the V.O.C. data collection. Therefore, operational definitions should be very precise and be written to avoid possible variation in interpretations and to ensure consistent data collection.

The procedures associated with an operational definition for a V.O.C. defines exactly how the project team will proceed to collect and record the V.O.C. data. Figure 8.2 shows a sample template for V.O.C. data collection. During this planning step, the following must also be considered by the project team:

1. Importance of the Voice of the Customer (V.O.C.) data;
2. Accuracy of V.O.C. data;
3. Completeness of V.O.C. data capturing.

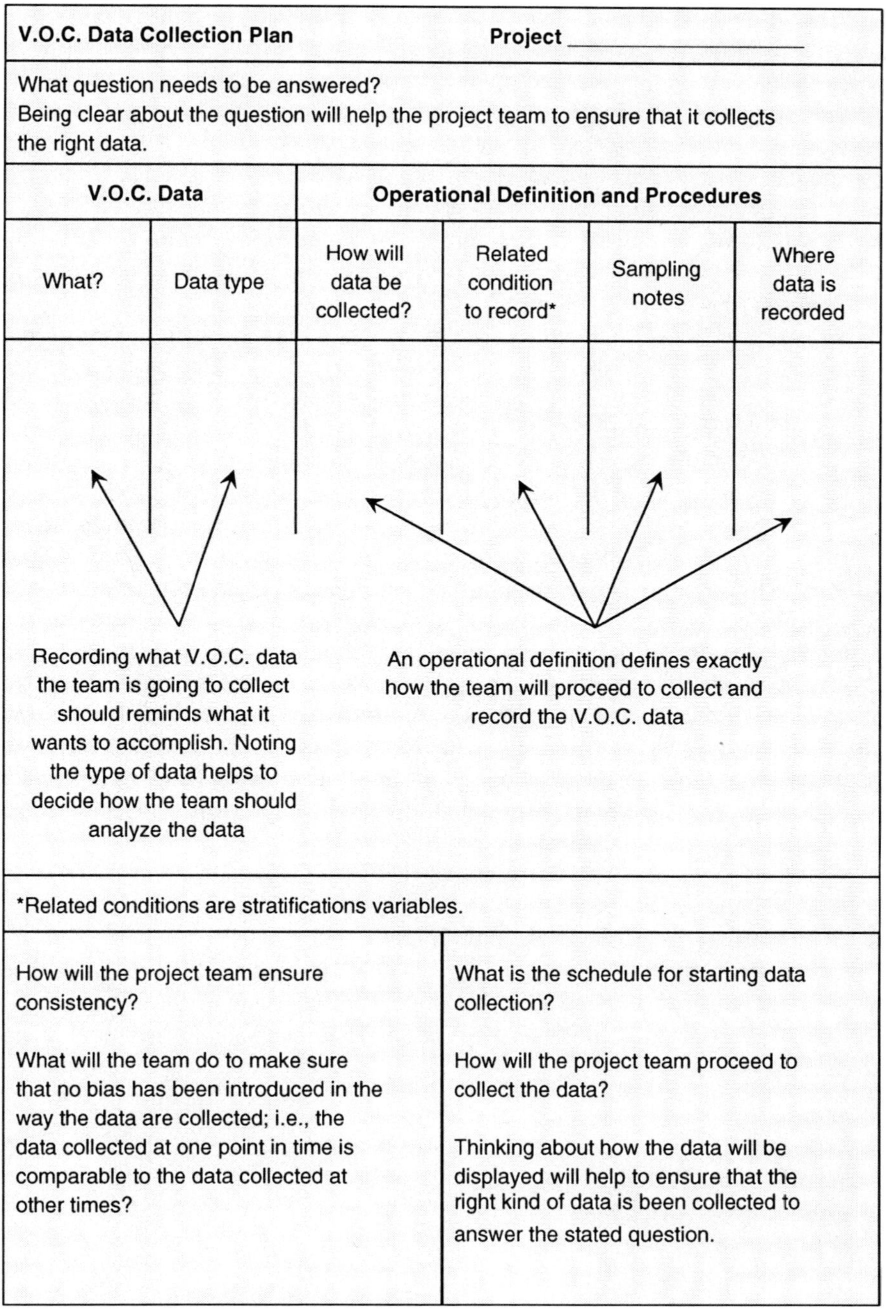

Fig. 8.2 Sample template for V.O.C. data collection

Importance of the Voice of the Customer (V.O.C.) data—Information on the customer value of the "process to be improved" outcomes can only come from the voice of the customer. In detailed product (resp. Service) improvement and development, the V.O.C. is also the source of information on the product (resp. Service) development process; enough information must be gathered so that the mappings in the product (resp. Service) improvement and development process can be carried out flawlessly. Therefore, operational definitions and procedures for the V.O.C. needs should be developed as these are the source of information for both the strategically important customer value proposition, and all the building blocks in the product (resp. Service) design.

Accuracy of V.O.C. data—Because of the importance of the voice of the customer, the V.O.C. data must be captured accurately. Only with accurate V.O.C. data can an accurate customer value proposition and "process to be improved" outcome performance specification be developed. There are many methods for capturing the voice of the customer, but arbitrary use of these methods will not lead to capturing the voice of the customer accurately.

Completeness of V.O.C. data capturing—Not only accurate capturing of the voice of the customer must be performed, but a sufficient amount of V.O.C. information needed to define customer value. Sometimes, the project team will have to collect enough V.O.C. information to figure out ways to redesign the customer value proposition to create breakthrough improvements on the "process to be improved." The voice of the customer is the source of information for process improvement, and sufficient V.O.C. data must be captured to drive the process improvement development. Specifically, sufficient V.O.C. information must be gathered to drive the transformations from customer attributes to functional requirements, from functional requirements to design parameters, and from design parameters to process variables.

8.1.2.1 Sources for Collecting V.O.C. Data

There are two sources from which the V.O.C. data can be collected: reactive sources and proactive sources. Reactive sources include such things as customers' complaints, service calls, and warranty claims. Proactive sources include, but are not limited to:

1. Interviews
2. Focus groups
3. Facilitated workshops
4. Group creativity techniques
5. Group decision making techniques
6. Questionnaires and surveys
7. Case studies
8. Observation
9. Prototypes and experiments

Interviews

An interview is a formal or informal approach to discover information from customers and stakeholders by talking to them directly. It is typically performed by asking prepared and spontaneous questions and recording the responses. Interviews are often conducted "one-on-one," but may involve multiple interviewers and/or multiple interviewees. Interviewing experienced project participants, stakeholders, and subject matter experts can aid in identifying and defining the features and functions of the desired project deliverables. The interview technique is relatively simple. Basically, it consists of identifying appropriate project participants and then methodically questioning them about the features and functions of the desired deliverables as related to the project. The technique can be used with individuals on a one-to-one basis or with groups of experts.

When conducted properly, interviews provide very reliable qualitative information. Transforming qualitative information into quantitative distributions or other measures depends on the skill of the interviewer. Moreover, the technique is not without problems. Those problems include:

1. Wrong project participants identified
2. Poor quality information obtained
3. Participants unwillingness to share information
4. Changing opinions
5. Conflicting judgments

Focus Groups

As one of the most popular tools for gathering information in today's market place, focus groups require understanding of purpose and good grounding in the technique to be effective. Thomas Greenbaum (1998) provides excellent information on conducting effective focus groups.

A focus group is a V.O.C. data collection tool in which a small group of people (typically eight to ten prequalified customers and stakeholders and subject matter experts) engages in a roundtable discussion about their expectations and attitudes about a proposed "process to be improved" outcome (i.e., product, service or result) in an informal setting. The focus group discussion is typically directed by a moderator who guides the discussion in order to obtain the group's opinions about or reactions to specific products or marketing-oriented issues, known as test concepts. While focus groups can provide the project team with a great deal of helpful information, their use as a research tool is limited in that it is difficult to measure the results objectively. In addition, the cost and logistical complexity of focus group research is frequently cited as a deterrent, especially for small size enterprise businesses. Nonetheless, many small businesses find focus groups to be useful means of staying close to consumers and their ever-changing attitudes and feelings. By providing qualitative information from well-defined target audiences, focus groups can aid businesses in decision making and in the development of marketing strategies and promotional campaigns.

Traditionally, focus groups have been used by makers of consumer products to gather qualitative data from target groups of consumers. They are often used in the new product development process, for example, to test consumer reaction to new product concepts and prototypes. Focus groups are also used to test marketing programs, as they can provide an indication of how consumers will react to specific advertising messages and other types of marketing communications. In this way, focus groups can help advertising and promotion managers position a particular product, service, or institution with respect to their target audience. Reactions to new types of product packaging can also be determined. In addition, many companies have used focus groups as a tool to learn more about consumer habits, product usage, and service expectations.

A key factor in determining the success of focus groups is the composition of the group in terms of the participants' age, gender, and product usage. Focus group participants are generally selected on the basis of their use, knowledge, attitudes, or feelings about the products, services, or other test concepts that are the subject of the focus group. In selecting participants, the objective is to find individuals who can knowledgeably discuss the topics at hand and provide quality output that meets the specified research objectives.

The most common method of selecting participants for focus groups is from databases that contains demographic, psychographic, and lifestyle information about a large number of customers. Such databases are available from a variety of commercial vendors. A list of desired characteristics is drawn up and matched with the database to select participants for focus groups. These characteristics may include purchase behavior, attitudes, and demographic data such as age and gender. The goal is to select participants who would likely be in the target audience for the products, services, or concepts being tested.

There is no absolute ideal in terms of the number of participants, although eight to ten participants is the norm. Different moderators are comfortable with different sizes of focus groups, but most consultants encourage companies to utilize groups in the eight-ten person range. Supporters of this size contend that these groups are large enough to provide a nice range of perspective and make it difficult for one or two individuals to dominate the discussion (moderators should guard against such developments). Groups that include more than ten participants, however, are usually more difficult for moderators to control. Group interaction is also more difficult, and moderators have a harder time stimulating discussion. In addition, it is often more difficult for a moderator to spend time following up on the insights voiced by one individual when there are a dozen or more participants.

Moderators play an important role in determining the success of focus groups. Well-trained moderators can provide a great deal of added value in terms of their past experience, skills, and techniques. On the other hand, poorly trained moderators are likely to fail to generate quality output from their focus groups. In addition to professional, full-time focus group moderators, other types of individuals who often serve as moderators include professional researchers, academicians, marketing consultants, psychologists or psychiatrists, and company representatives.

Focus group moderators serve as discussion leaders. They try to stimulate discussion while saying as little as possible. They are not interviewers. They usually work from a guide that provides them with an outlined plan of how the discussion should flow. The guide includes topics to be covered together with probing questions that can be used to stimulate further discussion. Moderators try to include everyone in the discussion. They allocate available time to make sure the required topics are covered. When the discussion digresses, it is up to the moderator to refocus the group on the topic at hand.

When setting up a focus group session, it is important to give careful consideration to the physical setting where it will take place. The location should be one that encourages relaxed participation and informal, spontaneous comments. The focus group facility must be of adequate size and have comfortable seating for all of the participants. Living room and conference room settings both provide good locations for focus groups, but public places—such as restaurants and auditoriums—are generally regarded as too distracting for gaining optimal results. In selecting a focus group site it is also important to make it geographically convenient for the participants. Locations that are hard to find or located in out of the way places may cause delays and scheduling problems. Finally, sites should be determined with an eye toward the schedules and locations of managers and executives who should be in attendance.

Once the facility, moderator, and participants have been selected, typical focus group sessions begin with an introduction. During the introductory part of the session the moderator welcomes the participants, informs them of what will take place during the session, and generally sets the stage for the discussion to follow. Prior to the main discussion there is usually a warm-up phase. The warm-up is designed to make the participants feel at ease. During the warm-up participants generally introduce themselves to the group. General topic discussions, usually related to the specific topics that will be covered later, also form part of the warm-up stage. These general discussions help participants focus their attention. They also provide the moderator with some insight into the different participants.

Gradually the moderator moves the level of discussion from general topics to more specific ones. The moderator may present different concepts for discussion. These include the test concepts for which the group was convened. The moderator may choose to use props to focus the group's attention. Typical props include product samples, actual or concept ads, concept statements that participants read together, photographs, and television commercials.

Once all of the test concepts have been discussed and evaluated by the group, the moderator moves the discussion into a wrap-up phase. During this phase the best concepts are identified and their strengths and weaknesses discussed. Participants may be asked to write down their reactions to what they have seen and discussed. During this final phase, any outstanding issues that were omitted are covered. When all of the substantive discussions have been completed, the moderator closes the session by thanking the participants and giving them any final instructions. Participants should leave with a positive feeling about the experience and the company, if the company that arranged the focus group has been identified. After the participants have left, it is standard practice for the moderator and the client company observers to have a post-group discussion.

Following the conclusion of the focus group or series of focus group sessions, the moderator may prepare a report for the client company. The report generally provides a written summary of the results of the session or sessions as interpreted by the moderator. Focus group reports may be summary in nature or more detailed. In some cases the client company may not require a written report.

Facilitated Workshops

Requirements workshops are focused sessions that bring key cross-functional customers and stakeholders together to define requirements. Workshops are considered a primary technique for quickly defining cross-functional requirements and reconciling stakeholder differences. Because of their interactive group nature, well-facilitated sessions can build trust, foster relationships and consensus, and improve communication among the participants. Another benefit of this technique is that issues can be discovered, and resolved more quickly than in individual sessions.

For example, facilitated workshops called Joint Application Development (or Design) (J.A.D.) sessions are used in the software development industry. These facilitated sessions focus on bringing users and the development team together to improve the software development process.

In the manufacturing industry, Quality Function Deployment (QFD) described in the previous chapter is an example of another facilitated workshop technique that helps determine critical characteristics for new product development.

Group Creativity Techniques

Several group activities can be organized to identify project and "process to be improved" requirements. Some of the group creativity techniques that can be used are:

Brainstorming

With the brainstorming approach, the project team, through ideas generation under the leadership of a facilitator and judgment of multidisciplinary experts, identifies a comprehensive list of V.O.C. data for the current project. A brainstorm is more than a basic core dump of information. It is the expression of ideas that then feeds other ideas and concepts in a cascade of data. It encourages team members to build on one another's concepts and perceptions. It circumvents conventions by encouraging the free flow of information.

The brainstorming technique is a facilitated sharing of information, without criticism, on a topic of the facilitator's choosing. It brings forth information from participants without evaluation, drawing out as many answers as possible and documenting them. There are no limits to the information flow or direction during a brainstorming session. Brainstorming is designed to encourage thinking outside of conventional boundaries so as to generate new insights and possibilities.

The Delphi Technique

With the Delphi technique, the project team, through consensus from opinions of a selected group of industry experts, identifies a comprehensive list of V.O.C. data for the current project. The technique helps reduce bias in the data and keeps any one person from having undue influence on the outcome.

Although people with experience of particular subject matter are a key resource for expert interviews, they are not always readily available for such interviews and, in many instances, prefer not to make the time to participate in the data gathering process. The Delphi technique works to address that situation by affording an alternative means of educing information from experts in a fashion that neither pressures them nor forces to leave the comfort of their own environs.

The Delphi technique has the advantage of drawing information directly from experts without impinging on their busy schedules. It also allows for directed follow-up from the experts after their peers have been consulted.

The Delphi technique (created by the Rand Corporation in the 1960s) derives its name from the oracle at Delphi. In Greek mythology, the oracle god Apollo foretold the future through a priestess who, after being question, channeled all knowledge from the gods, which an interpreter then catalogued and translated. In the modem world, the project manager or facilitator takes on the role of the interpreter, translating the insights of experts into common terms and allowing for his or her review assessment. The cycle of question, response, and reiteration is repeated several times to ensure that the highest quality of information is extracted from the experts.

This technique is recommended when the project's experts cannot coordinate their schedules or when geographic distance separates them. The technique is also appropriate when bringing experts together to a common venue may generate excess friction.

The inputs for the Delphi technique are questions or questionnaires. The questionnaire addresses the features and functions of the desired project deliverables, allowing for progressive refinement of the answers provided until general consensus is achieved. The questionnaire should allow for sufficient focus on the areas of concern without directing the experts to specific responses.

Outputs from the process are progressively detailed because all iterations should draw the experts involved closer to consensus. The initial responses to the questionnaire will generally reflect the most intense biases of the experts. Through the iterations, the facilitator will attempt to define common ground within their responses, refining the responses until consensus is achieved.

The Delphi technique relies heavily on the facilitator's ability both to generate the original questions to submit to the experts and to distill the information from the experts as it is received. The process is simple but is potentially time-consuming. Its steps are as follow:

1. *Identify experts and ensure their participation.* The experts need not be individuals who have already done the work or dealt with the features and functions of the desired project deliverables under consideration, but they should be individuals who are attuned to the enterprise business. Experts can be defined

as anyone who has an informed stake in the project and its processes. Commitments for participation should come from the experts, their direct superiors, or both.
2. *Create the Delphi instrument.* Questions asked under the Delphi technique must be sufficiently specific to draw out information of value but also sufficiently general to allow for creative interpretation. Because the voice of the customer is inherently an inexact science, attempts to generate excessive precision may lead to false assumptions. The Delphi questions should avoid cultural and organizational bias and should not be directive, unless there is a need to evaluate the features and functions of the desired project deliverables issues in a niche rather than across the entire project spectrum.
3. *Have the experts respond to the instrument.* Classically, this is done remotely, allowing the experts sufficient time to reflect over their responses. However, some enterprise businesses have supported encouraging questionnaire completion en masse during meetings to expedite the process. No matter the approach, the idea is to pursue all the key insights of the experts. The approach (e-mail, social networks, and meetings) for gathering the experts' observations will largely determine the timing for the process as a whole.
4. *Review and restate the responses.* The facilitator will carefully review the responses, attempting to identify common areas, issues, and concerns. These will be documented and returned to the experts for their assessment and review. Again, this may happen by mail or in a meeting, although the classic approach is to conduct the Delphi method remotely.
5. *Gather the experts' opinions and repeat.* The process is repeated as many times as the facilitator deems appropriate in order to draw out the responses necessary to move forward. Three process cycles are considered a minimum to allow for thoughtful review and reassessment.
6. *Distribute and apply the data.* Once sufficient cycles have been completed, the facilitator should issue the final version of the documentation and explain how, when, and where it will be applied. This is important so that the experts can observe how their contributions will serve the project's needs.

Nominal Group Technique

The nominal group technique enhances brainstorming with a voting process used to rank the most useful ideas for further brainstorming or for prioritization.

Idea/Mind Mapping

Ideas created through individual brainstorming are consolidated into a single map to reflect commonality and differences in understanding, and generate new ideas.

Group Decision Making Techniques

Group decision making is an assessment process of multiple alternatives with an expected outcome in the form of future actions resolution. These can be used to generate, classify and prioritize requirements for the project. There are multiple methods of reaching a group decision, for example:

1. Unanimity: Everyone agrees on a single course of action.
2. Majority: Support from more than 50 % of the members of the group.
3. Consensus: The majority defines a course of action, but the minority agrees to accept it.
4. Plurality: The largest block in a group decides even if a majority is not achieved.
5. Dictatorship: One individual makes the decision for the group.

Almost any of the decision methods described above can be applied to the group techniques potentially used in the requirements gathering process.

Questionnaires and Surveys

Questionnaires and surveys are synonymous terms for the same thing. They are written sets of questions designed to collect verbal data. Asking questions is widely accepted as a cost-efficient (and sometimes the only) way, of gathering information about past behavior and experiences, private actions and motives, and beliefs, values and attitudes (i.e. subjective variables that cannot be measured directly). The use of verbal data has been made the keystone of collecting V.O.C. data. It helps teams understand what customers want. Surveys will differ depending on whether they will be administered through telephone, mail, or electronic means. A good survey may be time-consuming to construct, but a bad survey will only provide misleading data. The following are points to remember when creating a survey:

1. Write questions that are short, simple and clear
2. If it's possible to misunderstand, it will be!
3. Avoid leading questions
4. Phrase sensitive questions carefully
5. Use close-ended rather than open-ended
6. Keep alternatives short
7. Options should be mutually exclusive and exhaustive
8. Have a cover letter or introduction email or script
9. Keep the survey short
10. Start with interesting questions
11. Sensitive questions should be later in survey (such as age or income range)
12. Order questions carefully

Questionnaires/surveys are most appropriate with broad audiences, when quick turnaround is needed, and where statistical analysis is appropriate.

Case Studies

These are in-depth research documents that examine a specific real-life situation or imagined scenario business. They allow the retention of the holistic characteristics of real-life business events while investigating empirical business events. Case studies can serve several purposes: some generate theories, some are simply description of cases, and others are more analytical in nature and display cross-case comparisons. A case study is done about some feature of the "process to be improved" or something we do not sufficiently understand and want to from the customer perspective. Conducting the case study provides a picture of the customer perspective to help inform the "process to be improved" outcomes or to see unexpected details of the "process to be improved" outcomes. It is useful for exploring or investigating the customer experiences with the "process to be improved" outcomes.

Observation

One would think that observation would be the most natural, un-obstrusive way to gather V.O.C. data on a first-hand basis. It is unclear as to how much unavoidable effect the presence of an observer may have on the V.O.C. data. An observer cannot but affect and be affected by the setting of the V.O.C. data collection, and this interaction may lead to a distortion of the customer real perception of the "process to be improved" outcome characteristic being observed. This is consistent with systems thinking, which explains how an open system (e.g. a social system) is affected by changes in its environment. Customers may adapt their responses to what they think the observer collecting V.O.C. data wants to see, and even to how the observer reacts to the customers' actions (e.g. when the observer takes notes). Despite those potential problems, observation has the advantages of providing real-time and contextual data.

An Observation provides a direct way of viewing customers in their environment and how they use or consume the "process to be improved" outcome. It is particularly helpful for detailed processes when the people that use the outcome of the "process to be improved" have difficulty or are reluctant to articulate their requirements. Observation, also called "task shadowing" is usually done externally by the observer viewing the user perform his or her task. It can also be done by a "participant observer," who actually performs a process or procedure to experience how it is done to uncover hidden requirements. Watching customers use an existing product, reacting to an existing service, or perform a task for which the product or service is intended can reveal important details about customers needs. Observation may be completely passive, without any interaction with the customer, or may involve working side by side with a customer, allowing members of the "process improvement" project to develop first hand experience using the product or service.

Prototypes

Prototyping is a method of obtaining early feedback on requirements by providing a working model of the expected "process to be improved" outcome before actually building it. Since prototypes are tangible, it allows customers and stakeholders to experiment with a model of their final process outcome more quickly and less expensively than only discussing abstract representations of their requirements. Prototypes support the concept of progressive elaboration because they are used in iterative cycles of mock-up creation, user experimentation, feedback generation, and prototype revision. When enough feedback cycles have been performed, the requirements obtained from the prototype are sufficiently complete to move to a design or build phase.

8.1.3 Develop Sampling Strategy

There is no single voice of the customer. Even for the same "process to be improved" outcome, there are diverse voices from customers about what they think this outcome should be like.

For example, in the automobile market sector, some customers would prefer larger sizes and more power cars, while some customers would prefer low gas mileage, low cost, easy maintenance cars, and so on. There are several major segments in this automobile market, and within a market segment, the customers' opinions are similar; however, there are significant differences among different automobile market segments. In this case, the "process improvement" project team may want to capture the voice of the customer from all market segments and develop a portfolio of products to satisfy different needs. In order to do this, the project team should plan to target a representative set of customers from each automobile market segment to get complete information on customers' needs.

For many "process to be improved" outcomes, there might be multiple voices from each outcome. Considering again an example, in the automobile market sector; when buying a car, a husband and wife may have different opinions. When making a purchasing decision about commercial equipment in an enterprise business, the voice of the direct user, the voice of the purchasing department, and the voice of support personnel might be very different. In this case, all kinds of voices should be considered in the planning and the corresponding "process to be improved" outcome should take into account all these voices.

For a given "process to be improved" outcome, the "process improvement" project team also needs to capture the voice of the following types of customers:

1. *Current customers.* These are customers who are receiving or buying the "process to be improved" outcomes; the project team should plan to know what they want in order to keep them.
2. *Competitors' customers.* These are customers who need this kind of "process to be improved" outcomes, but they are not receiving or buying these outcomes, so the project team should plan to determine what they want in order to improve these outcomes to capture more market share.
3. *Potential customers.* These customers are not receiving or buying the "process to be improved" outcomes those from competitors. They are not customers of the current line of business of the current industry; the project team should plan to know why they are not and then integrate their V.O.C. data into project plans, and hopefully they can be customers in the future.
4. *Lead customers.* These are either current customers or competitors' customers. Lead customers are those customers who are the most advanced users of the "process to be improved" outcomes, customers who are pushing these outcomes to their limits, or customers who are adapting an existing outcome to new uses. The voice of lead customers is important to catch the future trend and develop a new generation of products or services.

The totality of all customers about which the voices should be collected can be relatively large and it might not be possible, nor it is necessary, to collect information from the total population of customers considered. It is incumbent on the project team to clearly define the target population. There are no strict rules to follow, and the project team must rely on logic and judgment. The population is defined in keeping with the questions to be answered and the objectives of capturing the V.O.C.

Sometimes, the entire population will be sufficiently small, and the project team can include the entire population in the study. Collecting the V.O.C. data in this case is called a "census V.O.C. data collection" because data is gathered on every customer of the target population.

Usually, the target population is too large for the project team to attempt to survey all of its customers. A small, but carefully chosen sample can be used to represent the population. The sample should reflects the characteristics of the population from which it is drawn and the goal in choosing a sample is to have a picture of the population, disturbed as little as possible by the act of gathering information.

Sampling methods are classified as either probability or non-probability, as shown in Fig. 8.3.

1. *In probability sampling*, each member of the population has a known non-zero probability of being selected. Probability methods include random sampling, systematic sampling, and stratified sampling.
2. *In non-probability sampling*, members are selected from the population in some nonrandom manner. These include convenience sampling, judgment sampling, quota sampling, and snowball sampling.

The advantage of probability sampling is that the sampling error can be calculated. The sampling error is the degree to which a sample might differ from the population. When inferring to the population, results are reported plus or minus the sampling error. In non-probability sampling, the degree to which the sample differs from the population remains unknown.

Random sampling is the purest form of probability sampling. Each member of the population has an equal and known chance of being selected. The major benefit of random sampling is that any differences between the sample and the population from which the sample was selected will not be systematic. Although randomly selected samples may differ from the larger population in important ways (especially if the sample is small), these differences are due to chance rather than to a systematic bias in the selection process.

Systematic sampling is often used instead of random sampling. It is also called an Nth name selection technique. After the required sample size has been calculated, every Nth record is selected from a list of population members. As long as the list does not contain any hidden order, this sampling method is as good as the random sampling method. Its only advantage over the random sampling technique is simplicity. Systematic sampling is frequently used to select a specified number of records from a computer file.

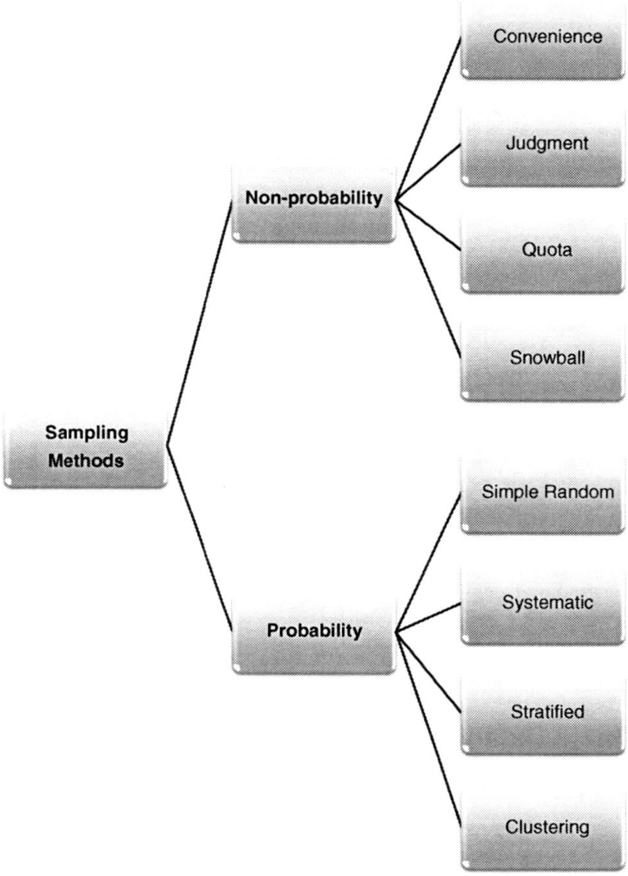

Fig. 8.3 Sampling methods

Stratified sampling is a commonly used probability method that is superior to random sampling because it reduces sampling error. A stratum is a subset of the population that shares at least one common characteristic. Examples of stratums might be males and females, or managers and non-managers. The project team first identifies the relevant stratums and their actual representation in the population. Random sampling is then used to select a sufficient number of subjects from each stratum. "Sufficient" refers to a sample size large enough for the project team to be reasonably confident that the stratum represents the population. Stratified sampling is often used when one or more of the stratums in the population have a low incidence relative to the other stratums.

Convenience sampling is used in exploratory study where the project team is interested in getting an inexpensive approximation of the truth. In convenience sampling, the project team generally selects customers on the basis of proximity, ease of access, and willingness to participate (i.e., convenience). This non-probability method is often used during preliminary study efforts to get a gross estimate of the results, without incurring the cost or time required to select a random sample.

Judgment sampling is a common non-probability method. The project team selects the sample based on judgment. This is usually and extension of convenience sampling. For example, a project team may decide to draw the entire sample from one "representative" market segment, even though the population includes all market segments. When using this method, the project team must be confident that the chosen sample is truly representative of the entire population.

Quota sampling is the non-probability equivalent of stratified sampling. Like stratified sampling, the project team first identifies the stratums and their proportions as they are represented in the population. Then convenience or judgment sampling is used to select the required number of subjects from each stratum. This differs from stratified sampling, where the stratums are filled by random sampling.

Snowball sampling is a special non-probability method used when the desired sample characteristic is rare. It may be extremely difficult or cost prohibitive to locate respondent customers in these situations. Snowball sampling relies on referrals from initial subjects to generate additional subjects. While this technique can dramatically lower search costs, it comes at the expense of introducing bias because the technique itself reduces the likelihood that the sample will represent a good cross section from the population.

8.1.4 Validate Data Collection System

The "data collection system" consists of data obtained from the sample, appraisers or people executing the data collection tasks, operational definitions and procedures followed to collect the data, and data collection instruments. The events associated with any one of these constituents are not conveyed to the other constituents; that is, the constituents of a "data collection system" are statistically independent.

The sample selected from the target population is one of all possible samples. Furthermore, any V.O.C. data collected from the sample is based on the sample characteristics. It may or may not be close to the true characteristic value (traits, behaviors, qualities, figures or parameter) of the target population. The difference between the V.O.C. data collected from the sample and the true characteristic value (traits, behaviors, qualities, figures or parameter) of the target population is called sampling error.

8.1.4.1 Understanding the Nature of Variation

It is important to note that the collected V.O.C. data which can be seen as outcomes of the process or act of collecting data using the "data collection system" will display variations over time. The goal of validating the "data collection system" is to minimize controllable factors that could exaggerate the amount of variation in such collected data. To achieve this, the project team must understand the nature of variation.

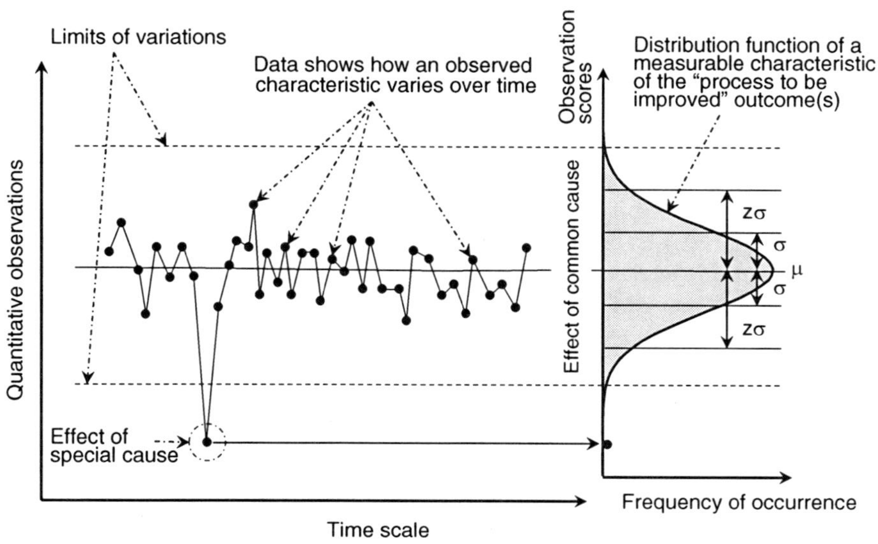

Fig. 8.4 Variations in process outcome over time

We can think of variation as change or slight difference in condition, amount, or level from the expected occurrence, typically within certain limits, as shown in Fig. 8.4. Variation has two broad causes that have an impact on data collected: common (also called random, chance, or unknown) causes and special (also called assignable) causes.

Common causes of variation are inherent and an integral part in the process been considered. They can be though of as the "natural pulse of the process been considered" and they are indicated by a stable, repeating pattern of variation.

A process behavior chart, illustrated in Fig. 8.4, is the unique operational definition of an assignable cause. Assignable causes of variation are those causes that are not intrinsically part of the process been considered but arise because of specific circumstances. When they occur, they signal a significant occurrence of change in the process and they lead to a statistically significant deviation from the norm. Assignable causes of variation are indicated by a disruption of the stable, repeating pattern of variation. They result in unpredictable process performance and must therefore be identified and systematically removed before taking other steps to improve quality of the system been considered.

A process that has only common causes, affecting its outcomes is referred to as be a stable process, or one that is in a state of statistical control. In a stable process, the causal system of variation remains essentially constant over time. This does not mean that there is no variation in the outcomes of the process, or that the variation is small, or that outcomes meet the specified requirements. It implies only that the variation is predictable within statistically established limits of variation. In practice, this means that improvement can be achieved only through a fundamental change to the process.

A process whose outcomes are affected by both common and assignable causes of variation is referred to as an unstable process. An unstable process is not necessarily one with large variations. Rather, the magnitude of variations from one period to the next is unpredictable. If assignable causes can be identified and removed, the process becomes stable; its performance becomes predictable. In practical terms, this implies that the system can be put back to an original level of performance by identifying the assignable causes and taking appropriate action. Once a change is made, continuing to plot data over time and observe the patterns helps to determine whether the change has eliminated the assignable cause.

Thus, quantifying the amount of variation in a process been considered is a critical step towards improvement. Understanding the difference between the two types of variations helps the project team decide what kinds of actions are most likely to lead to lasting improvement. It also helps to target the improvement efforts correctly and thereby avoid wasted resources.

The purpose of validating the "data collection system" is to ensure less bias and less variability by answering the question: "How much of the variation occurring in the 'data collection system' is due to the data collected from the sample?"

Bias refers to the difference between the data collected from the sample and the true characteristic value (traits, behaviors, qualities, figures or parameter) from the target population. Bias is consistent, repeated deviation of the data collected from the sample from the population parameter in the same direction when we take many samples.

Variability refers to the variation observed when the same data is collected using the "data collection system" repeatedly. Variability describes how "spread out" the values of the data collected are when taken many samples. It is made up of two components: repeatability and reproducibility. Repeatability is the part of variation in the collected data that occurs when you repeat collecting data under the same circumstances and with the same means and procedures. Large variability means that the result of sampling is not repeatable. Reproducibility is the part of variation in the collected data that occurs when you repeat collecting data under different circumstances and with different means, instruments and procedures.

To illustrate this graphically, we can think of the true value of the population parameter as the bull's eye on a target, and of the collected data from the sample as an arrow fired at the bull's eye.

Bias and variability describe what happens when an archer fires many arrows at the target. Bias means that the aim is off, and the arrows land consistently off the bull's eye in the same direction. The sample value do not center about the population value. Large variability means that repeated shots are widely scattered on the target. Repeated samples do not give similar results but differ widely among themselves. Figure 8.5 shows this target illustration of the bias and variability.

Notice that small precision (repeated shots are close together) can accompany large accuracy (the arrows are consistently away from the bull's eye in one direction). And small accuracy (the arrows center on the bull's eye) can accompany large precision (repeated shots are widely scattered). A good sampling scheme, like a good archer, must have both small accuracy and small precision.

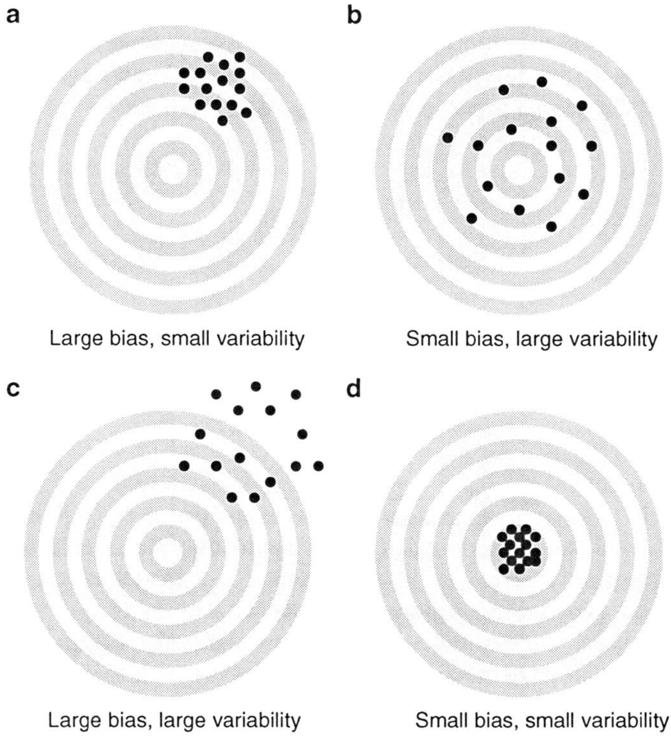

Fig. 8.5 Bias and variability in shooting arrows at a target. Bias means the archer systematically misses in the same direction. Variability means that the arrows are scattered

To manage bias and variability, the project team should consider using random sampling to reduce bias and also using a large enough sample to reduce variability. Simple random sampling produces unbiased estimates and the values of a characteristic describing the sample computed from a simple random sample neither consistently overestimate nor consistently underestimate the value of the population parameter.

8.1.4.2 Statistical Inference: Determining the Sample Size

The project team will use the data collected from samples to calculate estimates of characteristic value (traits, behaviors, qualities, figures or parameter) of the target population, such as average value or standard deviation. These descriptive statistics (i.e. estimates of characteristic) do nothing more than provide information about this specific sample from which the project team collected data. However, to make sound based decisions from the V.O.C. collected data, it is essential to infer or reach important conclusions about how well the statistics of selected samples generalize to the larger population.

Statistical inference is the process of using sample data to infer the distribution that generated the data. A typical statistical inference question is: "Given a sample of size n following a distribution $F, Y_1, \ldots, Y_n \sim F$, how do we infer F?" In "process improvement" business applications, we often want to infer only some feature of F such as its mean μ, or its standard deviation σ^2, using knowledge of the sample mean, $\bar{Y}$, and the sample standard deviation s^2, defined to be:

$$\bar{Y} = \frac{1}{n} \sum_{i=1}^{n} Y_i$$

$$\sigma^2 = \frac{1}{n} \sum_{i=1}^{n} (Y_i - \mu)^2$$

$$s^2 = \frac{1}{n-1} \sum_{i=1}^{n} (Y_i - \bar{Y})^2$$

To infer is to draw a conclusion from evidence. Statistical inference draws a conclusion about the characteristics of the population the sample is alleged to represent. Drawing conclusions in mathematics is a matter of starting from a hypothesis and using logical argument to prove without doubt that the conclusion follows. Statistical conclusions, however, are uncertain because they drawn conclusions about a population on the basis of data about a sample. So statistical inference has to state conclusions and also say how uncertain the conclusions are.

Before calculating estimates of parameters of the large enough total population of customers considered from data sample obtained using the "data collection system" and decide whether results are statistically significant, the project team should establish a standard, or benchmark. This is done through statistical inference by developing a hypothesis and establishing a criterion that will be used when deciding whether to retain or reject the established hypothesis.

The primary hypothesis of interest is the *null hypothesis*, H_0. As the name implies, the null hypothesis always suggests that there will be an absence of effect. For example, the null hypothesis could suggest that for a selected true characteristic value (traits, behaviors, qualities, figures or parameter) of the target population, if we were to randomly select a sample from that population, then the sample mean of the V.O.C. data collected will not be different from the mean of the target total population of customers considered. Notice that the null hypothesis always refers to an absence of effect in the population.

It might be known that there is a chance that the selected sample would have a different mean than the population considered, but the best guess is that the sample would have the same mean as the population considered. Therefore, the null hypothesis would be that the mean of the selected true characteristic value (traits, behaviors, qualities, figures or parameter) of the target population and the sample mean would not differ from each other (i.e., an absence of effect). We could write this hypothesis symbolically as follows:

$$H_0 : \mu = \bar{Y}$$

Where μ represents the characteristic (traits, behaviors, qualities, figures or parameter) population mean and $\bar{Y}$ represents the sample mean.

At this point of the hypothesis building process, the sample has not yet been selected and its mean has not yet been calculated. This entire hypothesis building process occurs a priori (i.e., before a test of statistical significance is conducted). Of course, where there is one hypothesis (the null), it is always possible to have alternative hypotheses. One alternative to the null hypothesis is the opposite hypothesis. Whereas the null hypothesis is that the sample and population means will equal each other, an alternative hypothesis could be that they will not equal each other. This alternative hypothesis (H_A or H_1) would be written symbolically as:

$$H_A : \mu \neq \bar{Y}$$

Where μ represents the characteristic (traits, behaviors, qualities, figures or parameter) population mean and $\bar{Y}$ represents the sample mean.

The alternative hypothesis indicated above does not include any speculation about whether the sample mean will be larger or smaller than the population mean, only that the two differ. This is known as a two-tailed alternative hypothesis. A different alternative hypothesis can also be used. For example, an alternative hypothesis could state that the sample mean would be larger than the population mean because historical data on the population mean is available. When the alternative hypothesis is directional (i.e., includes speculation about which value will be larger), this is known as one-tailed alternative hypothesis. We could write this one-tailed alternative hypothesis symbolically as follows:

$$H_A : \mu < \bar{Y}$$

Where μ represents the characteristic (traits, behaviors, qualities, figures or parameter) population mean and $\bar{X}$ represents the sample mean.

Let's suppose that the project team is using the two-tailed hypothesis and that the characteristic (traits, behaviors, qualities, figures or parameter) population mean and the sample mean are different from each other, with no direction of difference specified. At this point in the process, the null and alternative hypotheses have been established.

The project team may assume that all it needs to do is randomly select and collect a large enough sample of V.O.C. data, find their average, and see if it is different from or equal to the historical data on the population mean or to the predefined population mean. But, alas, it is not quite that simple. Suppose that the project team gets the sample and find the average of the characteristic considered is slightly higher than the population mean. Technically, that is different from the population mean, but is it different enough to be considered meaningful? Whenever a sample is selected at random from a population, there is always a chance that it will differ slightly from the population.

Although the best guess is that the sample mean of the characteristic considered will be the same as the population mean, it would be almost impossible for the sample to look exactly like the population. So the key question becomes this: "How different does the sample mean have to be from the population mean before the difference can be considered meaningful, or significant?" If the sample mean is just a little different from the population mean, the project team can shrug it off and say: "Well, the difference is probably just due to random sampling error, or chance." But how different do the sample and population means should to be before it can be concluded that the difference is probably not due to chance? That's where the significance level, or alpha level, or Type I error, comes into play.

Before the project team can conclude that the differences between the sample descriptive statistic of a true characteristic of a population and the population parameter are probably not just due to random sampling error, the project team have to decide how unlikely the chances are of getting a difference between the statistic of a true characteristic of a population and the population parameter just by chance if the null hypothesis is true. In other words, before the project team can reject the null hypothesis, it want to be reasonably sure that any difference between the sample statistic and the population parameter is not just due to random sampling error, or chance.

In most business applications, the convention is to set that level of significance at $\alpha = 0.05$ or $\alpha = 0.10$, partly because of tradition and partly because these levels represent (to some people) a reasonable level of certainty. The level of significance at $\alpha = 0.05$ (or $\alpha = 0.10$) level translates into a long-run chance of 1 in 20 (or 1 in 10) of not covering the population parameter. This seems reasonable and is comprehensible, whereas 1 chance in 1,000 or 1 in 10,000 is too small. In other words, it is generally agreed that if the probability of getting a difference between the sample statistic and the population parameter by chance is less than 5 % (or 10 %), then the null hypothesis can be rejected and it can be concluded that the differences between the statistic and the parameter are probably not due to chance. The agreed-upon probability of 0.05 or 0.10 (symbolized as $\alpha = 0.05$ or $\alpha = 0.10$) represents the Type I error rate that the project team is willing to accept before conducting statistical analysis on the data collection system.

Given that samples generally do not precisely represent the populations from which they are drawn, some difference between the sample statistic and the population parameter should be expected simply due to the luck of the draw, or random sampling error. If the project team reach into the total population of customers and pull out another random sample, the project team will probably get slightly different descriptive statistics (i.e. estimates of characteristic) again. Thus, some of the difference between a sample descriptive statistic, like the mean, and a population true characteristic (traits, behaviors, qualities, figures or parameter) will always be due to the random sampling error.

When considering a descriptive statistic like the mean of a characteristic, the sampling distribution of the mean is a normal distribution. Consequently, a random sampling method will produce many sample means that are close to the value of the population mean and fewer that are further away from the population mean.

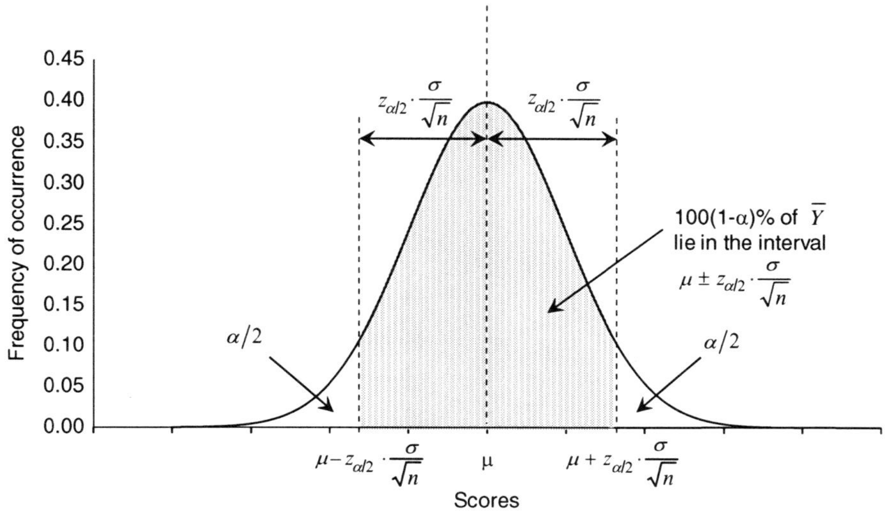

Fig. 8.6 Sampling distribution for $\bar{Y}$

The further the mean of the sample is from the population mean, the less likely it is to occur by chance, or random sampling error.

Furthermore, the Central Limit Theorem for the sample mean indicates that for large sample size n (crudely, $n \geq 30$), $\bar{Y}$ will be approximately normally distributed, with a mean μ and a standard error $\sigma/\sqrt{n}$. Then from the knowledge of the Empirical Rule and areas under a normal curve, it is known that the interval $\mu \pm z_{\alpha/2} \cdot \sigma/\sqrt{n}$ includes $100(1-\alpha)\%$ of those averages $\bar{Y}$ in repeated sampling, as illustrated in Fig. 8.6.

The quantity $z_{\alpha/2}$ is a value of the normal distribution score z having a tail area of $\alpha/2$ to its right. In other words, at a distance of $z_{\alpha/2}$ standard deviations to the right of the population mean μ, there is an area of $\alpha/2$ under the normal curve.

From Fig. 8.6 we can observe that for a considered characteristic, the sample mean $\bar{Y}$ may not be very close to the population mean μ, the quantity it is supposed to estimate. Thus, when the value of $\bar{Y}$ is reported, the project team should also provide an indication of how accurately $\bar{Y}$ estimates μ. This is accomplished by considering an interval of possible values for μ in place of using just a single value $\bar{Y}$.

Consider the interval $\bar{Y} \pm z_{\alpha/2} \cdot \sigma/\sqrt{n}$. As illustrated in Fig. 8.7, whenever a sample mean $\bar{Y}$ falls in the interval $\mu \pm z_{\alpha/2} \cdot \sigma/\sqrt{n}$, the sample interval $\bar{Y} \pm z_{\alpha/2} \cdot \sigma/\sqrt{n}$ will contain the population mean, μ. The probability of falling in the interval is $1-\alpha$, so the project team can state that $\bar{Y} \pm z_{\alpha/2} \cdot \sigma/\sqrt{n}$ is an interval estimate of μ with level of confidence $1-\alpha$. Thus, for a specified value of $1-\alpha$, a $100(1-\alpha)\%$ confidence interval for the population mean, μ is given as: $\bar{Y} \pm z_{\alpha/2} \cdot \sigma/\sqrt{n}$.

Because the level of confidence is $100(1-\alpha)\%$, it is expected that, in a large collection of includes $100(1-\alpha)\%$ confidence intervals, approximately $\alpha\%$ of the intervals would fail to include the population mean μ. Thus, in 100 intervals

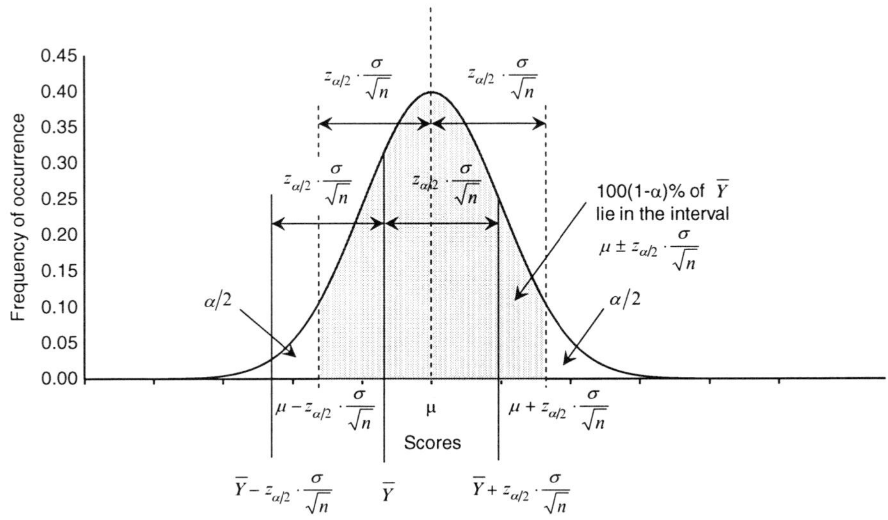

Fig. 8.7 Sampling distribution for $\bar{Y}$ and an observed value of $\bar{Y}$

the project team should expect approximately $\alpha\%$ intervals to not contain the population mean μ. It is crucial to understand that even when data are properly collected, a number of the data collected will yield results that in some sense are in error. This occurs when the project team collects only a small amount of data or selects only a small subset of the population.

The width, $2z_{\alpha/2} \cdot \sigma/\sqrt{n}$, of the confidence interval and the confidence coefficient $1 - \alpha$ measure the goodness of the inference on the population mean. For a given value of the confidence coefficient, the smaller the width of the interval, the more precise the inference. As indicated already, in most business applications, the convention is to set confidence coefficient $1 - \alpha$ to express how much assurance the data collection team places in whether the interval estimate encompasses the population parameter of interest. For a fixed sample size, increasing the level of confidence will result in an interval of greater width. Thus, the data collection team will generally express a desired level of confidence and specify the desired width of the interval.

In most situations when the population mean is unknown, the population standard deviation σ will also be unknown. Hence, it will be necessary to estimate both μ and σ from the data. However, for all practical purposes, if the sample size is relatively large (30 or more is the standard rule of thumb), the project team can estimate the population standard deviation σ with the sample standard deviation s in the confidence interval formula. Because the population standard deviation σ is estimated by the sample standard deviation s, the actual standard error of the mean $\sigma/\sqrt{n}$, is naturally estimated by $s/\sqrt{n}$. This estimation introduces another source of random error (s will vary randomly, from sample to sample, about σ) and, strictly speaking, invalidates the level of confidence for the interval estimate of μ. Fortunately, the formula is still a very good approximation for large sample sizes.

Once the data collection team has expressed a desired level of confidence and has specified the desired width of the interval, it must determine the number of observations/data to include in the sample; that it, it must determine the adequate sample size. The implications of determining the sample size are clear. Data collection costs money. If the sample size is too large, time and talent are wasted. Conversely, it is wasteful if the sample size is too small, because inadequate information has been collected for the time and effort expended. Also, it may be impossible to increase the sample size at a later time. How large is large enough or how many customers should the project team collect data from depends on the complexity of the "process to be improved" outcomes, diversity of the market, product or service use, and the sophistication of customers. When choosing a sample size, the project team must consider the following issues:

1. What population parameters do the team wants to estimate?
2. What is the cost of sampling (importance of information)?
3. How much is already known from the target population?
4. What is known about the spread (variability) of the population?
5. How hard is it to collect the identified V.O.C. data?
6. How precise does the project team want the final estimates to be?

Cost of sampling—The cost of sampling issue helps the project team to determine how precise estimates of the target customer population should be. If the decisions that will be made from the sampling activity are very valuable, then the project team should consider taking low risks and hence use larger sample sizes.

Availability of prior information—If the "process to be improved" has been studied before, then the project team can use that prior information to reduce sample sizes. This can be done by using prior estimates of characteristics of the target customer population and by stratifying the population to reduce variation within selected samples.

Inherent variability –The variance of an estimate of a characteristic of the target customer population is proportional to the inherent variability of the population divided by the sample size. Namely,

$$\text{Variance of an estimate} \approx \text{population variance/sample size}$$

This means that if the variability of the population is large, then the project team must consider taking several samples. Conversely, a small population variance indicates that few samples should suffice.

Level of confidence of the final estimates—The goal of the project team should be to reach levels of confidence higher than 90 % in capturing customer needs.

There are two key aspects to be considered in determining the appropriate sample size for estimating the population mean μ using a confidence interval: The tolerance error and the level of confidence.

1. *Tolerable error*—the tolerable error establishes the desired width of the interval. The tolerable error depends heavily on the context of the problem, and only someone who is familiar with the situation or the business application considered can make a reasonable judgment about its magnitude.

2. *Level of confidence*—It serves as benchmark for helping to decide whether to reject or retain our null hypothesis. If the probability value (which is obtained after calculating the statistic) is smaller than the specified alpha level, the project team should reject the null hypothesis. When the null hypothesis is rejected, the project team is concluding that the difference between the sample statistic and the population parameter is probably not due to chance, or random sampling error. However, this conclusion is reached, there is always a chance that the decision will be wrong, having made a Type I error. One goal of performing statistical inference and hypothesis testing is to avoid making such errors. Thus, to be on the extra safe side the project team may want to select a more conservative alpha level, such as 0.01, and say that unless the probability value is smaller than 0.01, the null hypothesis will be retained.

In selecting tolerance and level of confidence specifications for a given characteristic, the data collection team needs to consider that if the confidence interval of the population mean μ is too wide, then the estimation of the population mean μ will be imprecise and not very informative. Similarly, a very low level of confidence (say 50 %) will yield a confidence interval that very likely will be in error; that is, fail to contain the population mean μ. However, to obtain a confidence interval having a narrow width and a high level of confidence may require a large value for the sample size and hence be unreasonable in terms of cost and/or time.

For a given tolerable error W, which is the width of the confidence interval, once the $100(1-\alpha)\%$ confidence level is specified and an estimate of σ supplied, the required sample size n for a confidence interval of the for $\bar{Y} \pm W/2$, can be calculated using the following relation:

$$n = \left(2 \cdot z_{\alpha/2} \cdot \frac{\sigma}{W}\right)^2$$

Determining a sample size to estimate the population mean μ requires knowledge of the population variance σ^2 (or standard deviation σ). The data collection team can obtain an approximate sample size by estimating σ^2, using one of the two methods indicated below and then substitute the estimated value of σ^2 in the sample-size equation to determine an approximate sample size n.

1. Employ information from a prior collected data to calculate a sample variance σ^2. This value is used to approximate σ^2.
2. Use information on the range of the observations to obtain an estimate of σ.

8.1.4.3 Analyzing Quantitative Variations in the "Data Collection System"

The objective here is to answer the question about "How much of the variation occurring in the V.O.C. data is due to the data collection system?" For this purpose, the project team can make use of audits, qualitative and quantitative gage studies.

1. *Audit data collection system* studies—An audit data collection system study is a data collection system study where the data collected is compared to a known

correct standard. Any differences between the two reflect variation in the data collection system.

2. *Qualitative data collection system* studies—A qualitative data collection system study is a data collection system study where the accuracy, repeatability, and reproducibility of customer words and narratives statements are assessed. Customer words and narratives statements are usually the result of human judgment; "which category does this item belong to?" is often the question to be answered. When categorizing items (good/bad, yes/no, type of call, etc...) the project team needs a high degree of agreement on which way an item should be categorized.

   When the project team collects qualitative data, it is important to determine whether the team's ability to place items into correct categories is consistent and reliable. The risk of poor qualitative data collection is to make a decision that is not consistent with reality in twofolds: the project team may falsely accept bad customer words and narratives statements or it may falsely reject good customer words and narratives statements.
3. *Quantitative data collection system* studies—A quantitative data collection system study is a data collection system study where the variation in the system from multiple samples is analyzed to determine how much of it comes from difference in appraisers or people executing the data collection tasks, operational definitions and procedures followed to collect the data, or the samples themselves. The common tools and techniques used for quantitative data collection system study are (1) gauge repeatability and reproducibility, and (2) analysis of variance gauge repeatability and reproducibility.

The remaining of this section provides the necessary knowledge to perform a quantitative data collection study.

As we indicated previously, the "data collection system" consists of data collected from the sample, appraisers or people executing the data collection tasks, operational definitions and procedures followed to collect the data, and data collection instruments. Furthermore, the events associated with any one of these constituents are not conveyed to the other constituents; that is, the constituents of a "data collection system" are statistically independent.

By considering events associated with each of these constituents of a "data collection system" as a balanced sum of a large enough number of unobserved random events acting additively and independently, each of which with finite mean and variance, the central limit theorem tells us that the occurrence pattern of these events will tend to follow a normal distribution in nature.

Consequently, the total variation associated with the "data collection system" is the sum of the variation inherent in the collected data from the sample plus the variation inherent in the treatment (appraisers or people executing the data collection tasks and operational definitions and procedures, and data collection instruments) followed to collect the data. This is summarized by the following equation of variance:

$$\sigma^2_{Total} = \sigma^2_{Collected\ data} + \sigma^2_{Treatment}$$

Clearly, when the variation inherent in the treatment followed to collect the data is small relative to the variation inherent in the collected data from the sample, the effect of the treatment will be small, the total variation associated with the "data collection system" will be quite similar to the variation inherent in the collected data from the sample, and the "data collection system" will be said to be good.

As the variation inherent in the treatment increases in size relative to the variation inherent in the collected data from the sample, the effect of the treatment will become a more prominent part of the "data collection system," and the collected data will contain more and more noise. This is the characteristic of an ineffective "data collection system," where the contribution of the variation inherent in the collected data from the sample will be small compared with the total variation associated with the "data collection system."

The traditional way of quantifying this relationship was introduced in 1921 by Sir Ronald Fisher and is defined to be the ratio of the variation inherent in the collected data from the sample to the total variation associated with the "data collection system":

$$\eta^2 = \frac{\sigma^2_{Collected\ data}}{\sigma^2_{Total}}$$

η^2 (eta-squared), also called "effect size," is referred to as the intra-class correlation coefficient. It correctly describes that proportion of the total variation in the "data collection system" that can actually be attributed to the values of the collected data from the sample. It defines the strength of the relationship between the contribution of the variation inherent in the collected data from the sample and the total variation associated with the "data collection system." Eta-squared is a biased estimator of the variance explained by the model in the population. As the sample size gets larger the amount of bias in this interclass correlation coefficient gets smaller.

The amount by which variations coming from the collected data from the sample are attenuated by the effects of the treatment may be defined as:

$$Variations\ attenuation = 1 - \eta$$

The complement to the intra-class correlation coefficient, $1 - \eta^2$, characterizes that proportion of the total variation in the "data collection system" that must be attributed to the error in the treatment:

$$1 - \eta^2 = \frac{\sigma^2_{Treatment}}{\sigma^2_{Total}}$$

The intra-class correlation coefficient and its complement are the proper measures to use in characterizing common variations.

Comparing the contribution of the variation inherent in the collected data from the sample against the total variation associated with the "data collection system" is not difficult mathematically. The difficult part is to obtain a good estimate of the variation inherent using the collected data from the sample. Thus, estimates obtained from the ratio should be used in practice.

Estimates of this intra-class correlation coefficient typically conveys the estimated magnitude of the relationship between the contribution of the variation inherent in the collected data from the sample and the total variation associated with the "data collection system" without making any statement about whether the apparent relationship in the data reflects a true relationship in the population.

While the intra-class correlation coefficient is bounded by 0 and 1, its estimate from the complement ratio above can occasionally turn out to be less than zero. When this happens it is simply an indication that the variation inherent in the collected data from the sample is so small that it has been overwhelmed by the uncertainty in the estimates of both the total variation associated with the "data collection system" and the variation inherent in the treatment (appraisers or people executing the data collection tasks and operational definitions and procedures) followed to collect the data (Wheeler, Good Data, Bad Data, and Process Behavior Charts, 2003).

Rather than using an estimate of the intra-class correlation coefficient, the industrial technique known as a "gauge R&R study" uses the an estimate of the ratio of the standard deviation inherent to the treatment (appraisers or people executing the data collection tasks and operational definitions and procedures) followed to collect the data, to the standard deviation associated with the total variation of the "data collection system." Namely,

$$Gauge\ R\&R\ ratio = \sqrt{1 - \eta^2} = \frac{\sigma_{Treatment}}{\sigma_{Total}}$$

Gauge R&R ratio is completely different from the intra-class correlation coefficient. It characterizes the strength of variations coming from the "data collection system." It is a representation of how the "data collection system" variations are attenuated before they show up in the collected data from the sample.

The complement of the Gauge R&R ratio defines the amount by which variations in the treatment are attenuated before they show up in the "data collection system" results. Similarly, the complement of the square root of the intra-class correlation coefficient defines the amount by which variations in the collected data from the sample are attenuated before they show up in the "data collection system" results.

Figure 8.8 shows how estimates of Gauge R&R ratio and estimates of intra-class correlation are related. The upper scale shows values of the Gauge R&R ratio and the four categories into which all data collection systems can be sorted (Wheeler, Good Data, Bad Data, and Process Behavior Charts, 2003; Wheeler, An Honest Gauge R&R Study, 2009a). The lower scale shows the values for an estimate of the intra-class correlation.

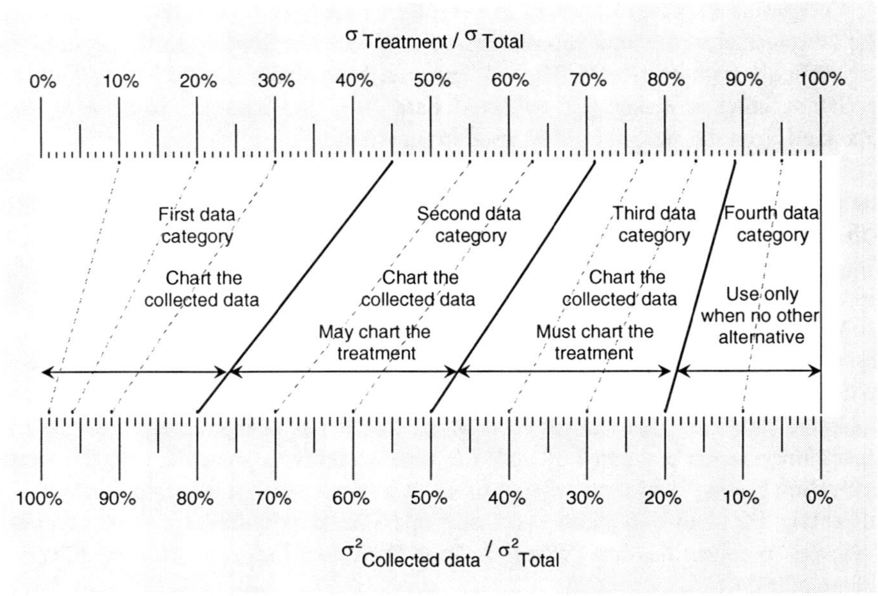

Fig. 8.8 The intra-class correlation coefficient and the gauge R&R ratio

The left side of Fig. 8.8 represents a data collection system with no treatment error—the values are determined by the collected data alone. The left side represents the limiting condition where the total variation associated with the "data collection system" is equal to the variation inherent in the collected data from the sample.

The right side of this figure represents a random number generator which produces values that are 100 % treatment error—pure noise containing no information about the collected data from the sample. The right side also represents the limiting condition where the total variation associated with the "data collection system" is equal to the variation inherent in the treatment (appraisers or people executing the data collection tasks and operational definitions and procedures, and data collection instruments) followed to collect the data.

Between these two extremes, the nonlinear relationship between the gauge R&R ratio and the intra-class correlation coefficient is shown by the diagonal lines which connect corresponding values on these two scales. In accordance with Wheeler's findings (Wheeler, An Honest Gauge R&R Study, 2009), when the Gauge R&R ratio is 10 %, any variations coming from the collected data from the sample will show up at 99.5 % of full strength in the "data collection system" results, while variations coming from the treatment will show up at 10 % of full strength in the "data collection system" results. Likewise, when the Gauge R&R ratio is 30 %, any variations coming from the collected data from the sample will show up at 95.4 % of full strength in the "data collection system" results, while variations from the treatment will only show up at 30 % of full strength in the "data collection system" results.

In evaluating a "data collection system," we recommend that the project team uses Wheeler's four categories of data classification shown in Fig. 8.8.

First Category of Data

The first data category will have an intra-class correlation coefficient between 1.00 and 0.80. Data in this category will have only slight attenuation for variations coming from the selected sample (less than 10 %) while variations from the treatment are greatly attenuated (more than 55 %). Therefore, whenever an estimate of the intra-class correlation coefficient exceeds 80 % any variations in the "data collection system" should be interpreted as coming from the collected data from the sample.

Second Category of Data

The second category of data will have an intra-class correlation coefficient between 0.80 and 0.50. In this region variations coming from the collected data from the sample will be attenuated between 10 and 30 %, while variations coming from the treatment will be attenuated from 55 to 30 %. This means that the "data collection system" will still be more sensitive to variations in the collected data from the sample than to variations in the treatment.

The second category of data still provides a very high likelihood of detecting variations in the collected data from the sample. Moreover, since variations in the collected data from the sample are less attenuated than variations coming from the treatment, it is more likely that any variations in the "data collection system" results come from the collected data from the sample.

However, with the second category of data the possibility of variations from the treatment showing up on the "data collection system" results cannot be ruled out. If the "data collection system" is reasonably predictable, then the process behavior chart for the collected data from the sample may be all that the project team may need to use to validate the "data collection system." But if there are doubts about the predictably of the "data collection system," then the project team may choose to maintain a process behavior chart for the treatment in addition to the process behavior chart for the collected data from the sample.

Third Category of Data

The third category of data will have an intra-class correlation coefficient between 0.50 and 0.20. Variations from the collected data from the sample will be attenuated between 30 and 55 %, while variations from the treatment will only be attenuated from 30 to 10 %. This means that the "data collection system" will be more sensitive to a shift in the treatment than to changes in the collected data from the sample.

The third category of data still detects variations in the collected data from the sample, but only when the possibility that a variation has occurred in the treatment can be ruled out. This means that in order to use data from the third category, the project team will have to maintain, on a concurrent basis, a process behavior chart that monitors variations in the treatment. Variations on both charts are interpreted as a change in the treatment.

Fourth Category of Data

The fourth category of data will have an intra-class correlation coefficient below 0.20. Here any variations coming from the collected data from the sample will be very highly attenuated (more than 55 %), while variations coming from the treatment will come through at nearly full strength in the "data collection system." Data from this category should only be used when no other alternative data are available as they very little useful information about the collected data from the sample itself.

8.1.4.4 Wheeler's Gauge R&R Process Guidelines

The "data collection system" can be further characterized by computing estimates of the Gauge R&R parameters to determine how much variations in the "data collection system" results comes from differences in the treatment (appraisers or people executing the data collection tasks and operational definitions and procedures) or from the data provided by the customers.

To do this, the project team can proceed by collecting preliminary data using m appraisers (people executing the data collection tasks), each collecting data from a sample of p customers, $n = 1, 2, 3, \ldots$ *or* 25 times each. The collected data must be balanced in the sense that each appraiser must collect data from each customer the same number of times. Let denote by Y_{ij}^{l} the $i-th$ value of the data collected from the $j-th$ customer by the $l-th$ appraiser.

Compute ranges—Arrange the $[n\ m\ p]$ data into $k = m^{*}p$ subgroups of size n, and compute the range for each of these k subgroups, as well as the Average Range for the k subgroups of size n:

$$R_j^l = max\left(\left|Y_{ij}^l\right|\right) - min\left(\left|Y_{ij}^l\right|\right), \quad i = 1, \ldots, n$$

$$\bar{R} = \sum_{l=1}^{m} \sum_{j=1}^{p} R_j^l$$

Compute the upper range limit—Use the average range for the k subgroups of size n, $\bar{R}$, to estimate the upper range limit. If any of the subgroup ranges exceed this upper limit the project team needs to find out why.

$$Upper\ Range\ Limit = D_4 \bar{R}$$

Compute the repeatability variance component—Use the average range from to estimate the repeatability variance component. This estimate is also called the repeatability or the equipment variation.

$$Repeatability\ Variance\ Component = \left(\frac{\bar{R}}{d_2}\right)^2$$

Compute the reproducibility variance component—Use the range of the m appraisers' averages to estimate the reproducibility variance component. This estimate is also called the reproducibility or the appraiser's variation.

$$l - th\ Appraiser\ average = \bar{Y}^l = \frac{1}{n^* p} \sum_{j=1}^{p} \sum_{i=1}^{n} Y_{ij}^l$$

$$Range\ of\ appraisers\ averages = R_Y = max\left(\left|\bar{Y}^l\right|\right) - min\left(\left|\bar{Y}^l\right|\right),\ l = 1, .., m$$

$$Reproducibility\ Variance\ Component = \left(\frac{R_Y}{d_2^*}\right)^2 - \frac{m}{n^* m^* p}\left(\frac{\bar{R}}{d_2}\right)^2$$

Compute the combined R&R variance component—Add the estimates above to get an estimated Combined R&R Variance Component, which is an estimate of the variance in the treatment:

$$R\&R\ Variance\ Component = \left(\frac{R_Y}{d_2^*}\right)^2 + \left(1 - \frac{m}{n^* m^* p}\right)\left(\frac{\bar{R}}{d_2}\right)^2$$

$$R\&R\ Variance\ Component = Estimate\ of\ \sigma^2_{Treatment}$$

The bias correction factor for ranges used here is the bias correction factor for estimating variances which is commonly known as d_2^*.

Compute the variance component of collected data from the sample—Use the range of the p customers' averages to estimate the collected data from the sample variance component:

$$j - th\ Customer\ average = \bar{Y}_j = \frac{1}{p^* n} \sum_{l=1}^{p} \sum_{i=1}^{n} Y_{ij}^l$$

$$Range\ of\ customers\ averages = R_Y = \max\left(\left|\bar{Y}_j\right|\right) - \min\left(\left|\bar{Y}_j\right|\right), j = 1, .., p$$

$$Collected\ data\ Variance\ Component = \left(\frac{R_{\bullet Y}}{d_2^*}\right)^2$$

$$Collected\ data\ Variance\ Component = Estimate\ of\ \sigma^2_{Collected\ data}$$

Compute the total variance of the "data collection system"—Add the estimates above to get the estimated Total Variance:

$$Estimate\ of\ Total\ Variance = \left(\frac{R_{\bullet Y}}{d_2^*}\right)^2 + \left(\frac{R_Y}{d_2^*}\right)^2 + \left(1 - \frac{m}{n^*m^*p}\right)\left(\frac{\bar{R}}{d_2}\right)^2$$

The proportion of the total variance of the "data collection system" results that is consumed by Repeatability is:

$$Repeatability\ proportion = \frac{\left(\frac{\bar{R}}{d_2}\right)^2}{\left(\frac{R_{\bullet Y}}{d_2^*}\right)^2 + \left(\frac{R_Y}{d_2^*}\right)^2 + \left(1 - \frac{m}{n^*m^*p}\right)\left(\frac{\bar{R}}{d_2}\right)^2}$$

The proportion of the total variance of the "data collection system" results that is consumed by Reproducibility is:

$$Repeatability\ proportion = \frac{\left(\frac{R_Y}{d_2^*}\right)^2 - \frac{m}{n^*m^*p}\left(\frac{\bar{R}}{d_2}\right)^2}{\left(\frac{R_{\bullet Y}}{d_2^*}\right)^2 + \left(\frac{R_Y}{d_2^*}\right)^2 + \left(1 - \frac{m}{n^*m^*p}\right)\left(\frac{\bar{R}}{d_2}\right)^2}$$

The proportion of the total variance of the "data collection system" results that is consumed by the combined Repeatability and Reproducibility is:

$$R\&R\ proportion = \frac{\left(\frac{R_Y}{d_2^*}\right)^2 + \left(1 - \frac{m}{n^*m^*p}\right)\left(\frac{\bar{R}}{d_2}\right)^2}{\left(\frac{R_{\bullet Y}}{d_2^*}\right)^2 + \left(\frac{R_Y}{d_2^*}\right)^2 + \left(1 - \frac{m}{n^*m^*p}\right)\left(\frac{\bar{R}}{d_2}\right)^2}$$

The proportion of the total variance of the "data collection system" results that is consumed by the variation from the collected data from the sample is an estimate of the intra-class correlation coefficient:

$$Estimate\ of\ \eta^2 = \frac{\left(\frac{R_{\bullet Y}}{d_2^*}\right)^2}{\left(\frac{R_{\bullet Y}}{d_2^*}\right)^2 + \left(\frac{R_Y}{d_2^*}\right)^2 + \left(1 - \frac{m}{n^*m^*p}\right)\left(\frac{\bar{R}}{d_2}\right)^2}$$

Use the estimated repeatability variance component to estimate the probable error of a single measurement:

$$Probable\ Error = 0.675 * \frac{\bar{R}}{d_2}$$

Table 8.2 Table of control limits constants for averages

Number of observations in subgroup n	Control limits		
	$\pm A\sigma$	$\pm A_2\bar{R}$	$\pm A_3\bar{S}$
	A	A_2	A_3
2	2.121	1.880	2.659
3	1.732	1.023	1.954
4	1.500	0.729	1.628
5	1.342	0.577	1.427
6	1.225	0.483	1.287
7	1.134	0.419	1.182
8	1.061	0.373	1.099
9	1.000	0.337	1.032
10	0.949	0.308	0.975
11	0.905	0.285	0.927
12	0.866	0.266	0.886
13	0.832	0.249	0.850
14	0.802	0.235	0.817
15	0.775	0.223	0.789
16	0.750	0.212	0.763
17	0.728	0.203	0.739
18	0.707	0.194	0.718
19	0.688	0.187	0.698
20	0.671	0.180	0.680
21	0.655	0.173	0.663
22	0.640	0.167	0.647
23	0.626	0.162	0.633
24	0.612	0.157	0.619
25	0.600	0.153	0.606

The constants in the relations above and are given in Tables 8.2, 8.3, 8.4, and 8.5 for sub-groups containing a relatively small number of collected data ($n \leq 25$). For subgroups containing more than 25 data ($n > 25$), the following definitions and approximations provides appropriate values for these constants:

Table 8.3 Table of constants for standard deviations

Number of observations in subgroup n	Constants for Center Line		Constants for Control Limits			
	c_4	$1/c_4$	B_3	B_4	B_5	B_6
2	0.7979	1.2533	0	3.267	0	2.606
3	0.8862	1.1284	0	2.568	0	2.276
4	0.9213	1.0854	0	2.266	0	2.088
5	0.9400	1.0638	0	2.089	0	1.964
6	0.9515	1.0510	0.030	1.970	0.029	1.874
7	0.9594	1.0423	0.118	1.882	0.113	1.806
8	0.9650	1.0363	0.185	1.815	0.179	1.751
9	0.9693	1.0317	0.239	1.761	0.232	1.707
10	0.9727	1.0281	0.284	1.716	0.276	1.669
11	0.9754	1.0252	0.321	1.679	0.313	1.637
12	0.9776	1.0229	0.354	1.646	0.346	1.610
13	0.9794	1.0210	0.382	1.618	0.374	1.585
14	0.9810	1.0194	0.406	1.594	0.399	1.563
15	0.9823	1.0180	0.428	1.572	0.421	1.544
16	0.9835	1.0168	0.448	1.552	0.440	1.526
17	0.9845	1.0157	0.466	1.534	0.458	1.511
18	0.9854	1.0148	0.482	1.518	0.475	1.496
19	0.9862	1.0140	0.497	1.503	0.490	1.483
20	0.9869	1.0133	0.510	1.490	0.504	1.470
21	0.9876	1.0126	0.523	1.477	0.516	1.459
22	0.9882	1.0119	0.534	1.466	0.528	1.448
23	0.9887	1.0114	0.545	1.455	0.539	1.438
24	0.9892	1.0109	0.555	1.445	0.549	1.429
25	0.9896	1.0105	0.565	1.435	0.559	1.420

$$c_4 \cong \frac{4(n-1)}{4n-3}$$

$$A = \frac{3}{\sqrt{n}}, A_3 = \frac{3}{c_4\sqrt{n}},$$

$$B_3 = 1 - \frac{3}{c_4\sqrt{2(n-1)}}, \qquad B_4 = 1 + \frac{3}{c_4\sqrt{2(n-1)}}$$

$$B_4 = c_4 - \frac{3}{\sqrt{2(n-1)}}, \qquad B_4 = c_4 + \frac{3}{\sqrt{2(n-1)}}$$

Table 8.4 Table constants for ranges

n	Constants for Center Line		Constants for Control Limits				
	d_2	$1/d_2$	d_3	$D_1 = d_2 - 3d_3$	$D_2 = d_2 + 3d_3$	D_3	D_4
2	1.128	0.8865	0.853	0	3.686	0	3.267
3	1.693	0.5907	0.888	0	4.358	0	2.574
4	2.059	0.4857	0.880	0	4.698	0	2.282
5	2.326	0.4299	0.864	0	4.918	0	2.114
6	2.534	0.3946	0.848	0	5.078	0	2.004
7	2.704	0.3698	0.833	0.204	5.204	0.076	1.924
8	2.847	0.3512	0.820	0.388	5.306	0.136	1.864
9	2.970	0.3367	0.808	0.547	5.393	0.184	1.816
10	3.078	0.3249	0.797	0.687	5.469	0.223	1.777
11	3.173	0.3152	0.787	0.811	5.535	0.256	1.744
12	3.258	0.3069	0.778	0.922	5.594	0.283	1.717
13	3.336	0.2998	0.770	1.025	5.647	0.307	1.693
14	3.407	0.2935	0.763	1.118	5.696	0.328	1.672
15	3.472	0.2880	0.756	1.203	5.741	0.347	1.653
16	3.532	0.2831	0.750	1.282	5.782	0.363	1.637
17	3.588	0.2787	0.744	1.356	5.820	0.378	1.622
18	3.640	0.2747	0.739	1.424	5.856	0.391	1.608
19	3.689	0.2711	0.734	1.487	5.891	0.03	1.597
20	3.735	0.2677	0.729	1.549	5.921	0.415	1.585
21	3.778	0.2647	0.724	1.605	5.951	0.425	1.575
22	3.819	0.2618	0.720	1.659	5.979	0.434	1.566
23	3.858	0.2592	0.716	1.710	6.006	0.443	1.557
24	3.895	0.2567	0.712	1.759	6.031	0.451	1.548
25	3.931	0.2544	0.708	1.806	6.056	0.459	1.541

8.1.4.5 Statistical Inference: Comparing Two Population Central Values

The inference that we have made in a previous sub-section above has concerned a parameter from a single target population. Quite often the project data collection team will be faced with an inference involving a preliminary data collection using m appraisers (people executing the data collection tasks), each collecting data from a sample of p customers, $n = 1, 2, 3, \ldots$ *or* 25 times each from different target customer populations or from stratums of a target customer population. For a specified characteristic of the target population, the project data collection team

Table 8.5 Table constants for d_2^*

Number of customers × Number of appraisers	Repeat readings taken													
	2	3	4	5	6	7	8	9	10	11	12	13	14	15
1	1.41	1.91	2.24	2.48	2.67	2.83	2.96	3.08	3.18	3.27	3.35	3.42	3.49	3.55
2	1.28	1.81	2.15	2.40	2.60	2.77	2.91	3.02	3.13	3.22	3.30	3.38	3.45	3.51
3	1.23	1.77	2.12	2.38	2.58	2.75	2.89	3.01	3.11	3.21	3.29	3.37	3.43	3.50
4	1.21	1.75	2.11	2.37	2.57	2.74	2.88	3.00	3.10	3.20	3.28	3.36	3.43	3.49
5	1.19	1.74	2.10	2.36	2.56	2.73	2.87	2.99	3.10	3.19	3.28	3.35	3.42	3.49
6	1.18	1.73	2.09	2.35	2.56	2.73	2.87	2.99	3.10	3.19	3.27	3.35	3.42	3.49
7	1.17	1.73	2.09	2.35	2.55	2.72	2.87	2.99	3.10	3.19	3.27	3.35	3.42	3.48
8	1.17	1.72	2.08	2.35	2.55	2.72	2.87	2.98	3.09	3.19	3.27	3.35	3.42	3.48
9	1.16	1.72	2.08	2.34	2.55	2.72	2.86	2.98	3.09	3.18	3.27	3.35	3.42	3.48
10	1.16	1.72	2.08	2.34	2.55	2.72	2.86	2.98	3.09	3.18	3.27	3.34	3.42	3.48
11	1.16	1.71	2.08	2.34	2.55	2.72	2.86	2.98	3.09	3.18	3.27	3.34	3.41	3.48
12	1.15	1.71	2.07	2.34	2.55	2.72	2.85	2.98	3.09	3.18	3.27	3.34	3.41	3.48
13	1.15	1.71	2.07	2.34	2.55	2.71	2.85	2.98	3.09	3.18	3.27	3.34	3.41	3.48
14	1.15	1.71	2.07	2.34	2.54	2.71	2.85	2.98	3.08	3.18	3.27	3.34	3.41	3.48
15	1.15	1.71	2.07	2.34	2.54	2.71	2.85	2.98	3.08	3.18	3.26	3.34	3.41	3.48
>15	1.13	1.69	2.06	2.33	2.53	2.70	2.85	2.97	3.08	3.17	3.26	3.34	3.41	3.47

might wish to compare the mean for two different populations or population stratums.

Having quantified the sensitivity of the "data collection system" to variations in the collected data from the sample and to variations in the treatment, the project team can further use several hypothesis tests to compare differences among parameters of different target populations and their statistical significances. Two hypothesis tests that are often used in business applications are the variance ratio F-Test and Student's t-Test, the latter named after Gosset, the statistician at the Dublin brewery of Arthur Guinness who published his work under the pseudonym 'Student'. Both tests are used for comparing two independent samples. When there are more than two sources of variability to be compared Student's t-Test is not relevant. Instead, the technique of analysis of variance (ANOVA) is appropriate.

In many sampling situations, the project team will select independent random samples from two target populations to compare the populations' parameters. The statistics (calculated estimates of the populations parameters) used to make these inferences will, in many cases, be the difference between the corresponding sample statistics.

Suppose that the project data collection team collects independent random samples of n_1 data with sample mean $\overline{Y^1}$ from one population and n_2 data with sample mean $\overline{Y^2}$ from a second population. The project data collection team will use the difference between the sample means, $\overline{Y^1} - \overline{Y^2}$, to make an inference about the difference between the population means, $\mu_1 - \mu_2$.

As indicated already, when considering a descriptive statistic like the mean of a characteristic, the sampling distribution of the mean is a normal distribution. Consequently, a random sampling method will produce many sample means that are close to the value of the population mean and fewer that are further away from the population mean. The further the mean of the sample is from the population mean, the less likely it is to occur by chance, or random sampling error.

Furthermore, the Central Limit Theorem for the difference between the sample means, $\overline{Y^1} - \overline{Y^2}$, indicates that for large sample sizes n_1 and n_2 (crudely, $n_1, n_2 \geq 30$), $\overline{Y^1} - \overline{Y^2}$ will be approximately normally distributed, with a mean $\mu_1 - \mu_2$ and a standard error $\sqrt{\sigma_1/n_1 + \sigma_2/n_2}$.

The sampling distribution for the difference between two sample means, $\overline{Y^1} - \overline{Y^2}$, can be used to answer the same types of questions answered about the inference made in a previous sub-section above concerning a parameter from a single target population.

Often in situations where the project data collection team is making inferences about the difference between two sample means, $\mu_1 - \mu_2$, based on random samples independently selected from two populations, three case should be distinguished:

1. Both population distributions have equal variance, hence $\sigma_1 = \sigma_2$.
2. Both sample sizes n_1 and n_2 are large enough (crudely, $n_1, n_2 \geq 30$).
3. The sample sizes n_1 or n_2 are not large enough.

Let assume that the population distribution have equal variance, but different means μ_1 and μ_2. The project data collection team should summarize the data into the statistics: sample means $\overline{Y^1}$ and $\overline{Y^2}$, and sample standard deviations s_1 and s_2. Then, compare the two populations by constructing appropriate graphs, confidence intervals for the mean $\mu_1 - \mu_2$, and tests of hypotheses concerning the difference mean $\mu_1 - \mu_2$.

A logical point estimate for the difference in the two population means is the sample difference $\overline{Y^1} - \overline{Y^2}$. The standard error for the difference in sample means is more complicated than for a single sample mean, but the confidence interval has the same form:

$$point\ estimate \pm t_{\alpha/2} \cdot (standard\ error)$$

A general confidence interval for $\mu_1 - \mu_2$ with confidence level of $100(1-\alpha)\%$ is given as:

$$\left(\overline{Y^1} - \overline{Y^2}\right) \pm t_{\alpha/2} \cdot s_p \cdot \sqrt{\frac{1}{n_1} + \frac{1}{n_2}}$$

Where

$$s_p = \sqrt{\frac{(n_1 - 1)s_1^2 + (n_2 - 1)s_2^2}{n_1 + n_2 - 2}}$$

The sampling distribution of $\overline{Y^1} - \overline{Y^2}$ is a normal distribution, with standard deviation:

$$\sigma_{\overline{Y^1} - \overline{Y^2}} = \sigma\sqrt{\frac{1}{n_1} + \frac{1}{n_2}}$$

If the common value of the populations standard deviation σ, was known, then the project data collection team would use the percentile $z_{\alpha/2}$ in the relation that defines the confidence interval for $\mu_1 - \mu_2$. Because the common value of the populations' standard deviation σ is unknown in most cases, its value must be estimated. This estimate is denoted by s_p and is formed by combining (pooling) the two independent estimates of the populations' standard deviation σ, s_1, and s_2. In fact, s_p^2 is a weighted average of the sample variances s_1^2 and s_2^2. The project data collection team has to estimate the standard deviation of the point estimate of $\mu_1 - \mu_2$, so it must use the percentile from Student's t-distribution $t_{\alpha/2}$ in place of the normal percentile, $z_{\alpha/2}$. The degrees of freedom for the Student's t-percentile are $n_1 + n_2 - 2$, because there is a total of $n_1 + n_2$ data values and two parameters μ_1 and μ_2 that must be estimated prior to estimating the standard deviation σ.

The results above assume that the two populations from which the sample data are collected have a common variance σ^2. If the confidence interval presented were valid only when this assumption was met exactly, the estimation procedure would be of limited use. Fortunately, the confidence coefficient remains relatively stable if the sample sizes are approximately equal. For those situations in which this condition does not hold, the project data collection team should use alternative procedures outlined below.

The project data collection team can also test a hypothesis about the difference between two population means. It, we might, for example, specify that the null hypothesis would be that the difference mean $\mu_1 - \mu_2$ of the two target populations is statistically insignificant to some fixed value D_0 (i.e., an absence of effect). We could write this hypothesis symbolically as follows:

$$H_0 : \mu_1 - \mu_2 = D_0$$

Of course, where there is one hypothesis (the null), it is always possible to have alternative hypotheses. One alternative to the null hypothesis is the opposite hypothesis. Whereas the null hypothesis is that the difference mean $\mu_1 - \mu_2$ of the two target population is statistically significant to some fixed value D_0. This alternative hypothesis (H_A) would be written symbolically as:

$$H_A : \mu_1 - \mu_2 \neq D_0$$

Using Student's t-distribution defined to be:

$$t = \frac{\overline{Y^1} - \overline{Y^2} - D_0}{s_p \sqrt{\frac{1}{n_1} + \frac{1}{n_2}}}$$

The project data collection team should reject the null hypothesis if the calculated absolute value of t is greater or equal than the percentile $t_{\alpha/2}$; that is, $|t| \geq t_{\alpha/2}$.

If the sample sizes n_1 and n_2 are not large enough ($n_1 < 30$, *or* $n_2 < 30$), the project data collection team can still use the estimate above from the family of Student's t-distributions.

In the situation in which the sample variances (s_1^2 and s_2^2) suggest unequal population variances, $\sigma_1^2 \neq \sigma_2^2$, percentage points of a t distribution with modified degrees of freedom, known as Satterthwaite's approximation or separate-variance t-Test, can be used to set the rejection region for t. This approximate t test is summarized as:

$$t_{Satterhwaite} = \frac{\overline{Y^1} - \overline{Y^2} - D_0}{\sqrt{\frac{s_1^2}{n_1} + \frac{s_2^2}{n_2}}}$$

With the degrees of freedom for Satterthwaite's approximation of Student's t-percentile defined to be round down nearest integer to:

$$\frac{(n_1 - 1)(n_2 - 1)}{(1 - c)^2(n_1 - 1) + c^2(n_2 - 1)}$$

Where

$$c = \frac{1}{1 + \left(\frac{s_2 n_1}{s_1 n_2}\right)^2}$$

8.1.4.6 Statistical Inference on a Single Population Variance

In "process improvement" business applications, the variability of a population's variance is as important as the population mean. It might be known that there is a chance that the selected sample would have a different variance than the population considered, but the best guess is that the sample would have the same variance as the population considered. Therefore, the null hypothesis would be that the variance of the selected true characteristic value (traits, behaviors, qualities, figures or parameter) of the target population and the sample variance would not differ from each other (i.e., an absence of effect). We could write this hypothesis symbolically as follows:

$$H_0 : \sigma^2 = s^2$$

Where σ^2 represents the characteristic (traits, behaviors, qualities, figures or parameter) population variance and s^2 represents the sample variance.

As indicated already, where there is one hypothesis (the null), it is always possible to have alternative hypotheses. One alternative to the null hypothesis is the opposite hypothesis. Whereas the null hypothesis is that the sample and population variance will equal each other, an alternative hypothesis could be that they will not equal each other. This alternative hypothesis (H_A) would be written symbolically as:

$$H_A : \sigma^2 \neq s^2$$

Where σ^2 represents the characteristic (traits, behaviors, qualities, figures or parameter) population variance and s^2 represents the sample variance.

The alternative hypothesis indicated above does not include any speculation about whether the sample mean will be larger or smaller than the population variance, only that the two differ. This is also known as a two-tailed alternative hypothesis. A different alternative hypothesis can also be used. For example, an alternative hypothesis could state that the sample variance would be larger than the population variance because historical data on the population variance is available. When the alternative hypothesis is directional (i.e., includes speculation about which value will be larger), this is known as one-tailed alternative hypothesis. We could write this one-tailed alternative hypothesis symbolically as follows:

$$H_A : \sigma^2 < s^2$$

Where σ^2 represents the characteristic (traits, behaviors, qualities, figures or parameter) population variance and s^2 represents the sample variance.

Let's suppose that the project team is using the two-tailed hypothesis and that the characteristic (traits, behaviors, qualities, figures or parameter) population variance and the sample variance are different from each other, with no direction of difference specified. At this point in the process, the null and alternative hypotheses have been established.

The project team may assume that all it needs to do is randomly select and collect a large enough sample of V.O.C. data, find their variance, and see if it is different from or equal to the historical data on the population variance or to the predefined population variance. But, as with inference on the population mean, it is not quite that simple. Suppose that the project team gets the sample and find the variance of the characteristic considered is slightly higher than the population variance. Technically, that is different from the population variance, but is it different enough to be considered meaningful? Whenever a sample is selected at random from a population, there is always a chance that it will differ slightly from the population.

Although the best guess is that the sample variance of the characteristic considered will be the same as the population variance, it would be almost impossible for the sample to look exactly like the population. So the key question still remains this: "How different does the sample variance have to be from the population variance before the difference can be considered meaningful, or significant?" If the sample variance is just a little different from the population variance, the project team can wave it off and say: "Well, the difference is probably just due to random sampling error, or chance." But how different do the sample and population variance should to be before it can be concluded that the difference is probably not due to chance?

Before the project team can conclude that the differences between the sample descriptive statistic of a true characteristic of a population and the population parameter are probably not just due to random sampling error, the project team have to decide how unlikely the chances are of getting a difference between the statistic of a true characteristic of a population and the population parameter just by chance if the null hypothesis is true. In other words, before the project team can reject the null hypothesis, it want to be reasonably sure that any difference between the sample statistic and the population parameter is not just due to random sampling error, or chance. For the population variance, this is done by considering the statistic $(n-1)s^2/\sigma^2$ from repeated samples of size n from a normal population whose variance is σ^2. The statistic $(n-1)s^2/\sigma^2$ follows a chi-square distribution with $n-1$ degree of freedom, as illustrated in Fig. 8.9.

Because the chi-square distribution is not symmetrical, the confidence intervals based on this distribution do not have the usual form, estimate error, as we saw for the population mean and the normal distribution. The $100(1-\alpha)\%$ confidence interval for the population variance σ^2 is obtained by dividing the estimator of σ^2, s^2, by the lower and upper $\alpha/2$ percentiles at degree of freedom, $\chi^2_{1-\alpha/2}(n-1)$ and $\chi^2_{\alpha/2}(n-1)$, as follows:

$$\frac{(n-1)s^2}{X^2_{\alpha/2}(n-1)} < \sigma^2 < \frac{(n-1)s^2}{X^2_{1-\alpha/2}(n-1)}$$

The confidence interval for σ is found by taking square roots throughout.

In addition to estimating a population variance, the project data collection team can construct a statistical test of the null hypothesis that the characteristic (traits, behaviors, qualities, figures or parameter) population variance σ^2 equals a specified value, σ_0^2.

$$H_0 : \sigma^2 = \sigma_0^2$$

With the alternative hypothesis (H_A) written symbolically as:

$$H_A : \sigma^2 \neq \sigma_0^2$$

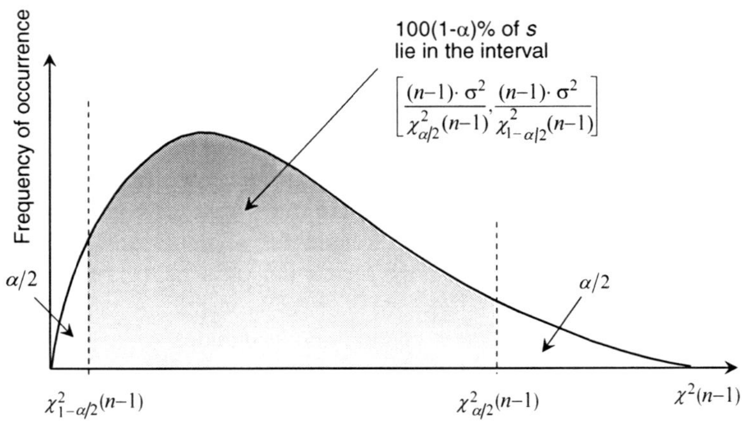

Fig. 8.9 Generic critical values of the chi-square distributionwith n-1 degree of freedom

Using the quantity defined to be:

$$X^2 = \frac{(n-1)s^2}{\sigma_0^2}$$

The project data collection team should reject the null hypothesis if the calculated value of χ^2 is not in the confidence interval for the population variance σ^2.

The inference method described above about the target population variance σ^2 is based on the condition that the random sample is selected from a population having a normal distribution similar to the requirements for using Student's t-distribution based inference procedures. However, when the sample size is moderate to large ($n \geq 30$), Student's t-distribution based procedures can be used to make inferences on the population mean even when the normality condition does not hold, because for moderate to large sample sizes the Central Limit Theorem provides that the sampling distribution of the sample mean is approximately normal.

Unfortunately, the same type of result does not hold for the chi-square based procedures for making inferences about the target population variance σ^2; that is, if the population distribution of the characteristic considered by the project data collection team is distinctly non-normal, then these procedures for are not appropriate even if the sample size is large. Population non-normality, in the form of skewness or heavy tails, can have serious effects on the nominal significance and confidence probabilities for the variance σ^2. If a normal probability plot of the sample data shows substantial skewness or a substantial number of outliers, the project data collection team should not apply the chi-square-based inference procedures described above. There are some alternative approaches that involve computationally elaborate inference procedures. One such procedure is the bootstrap.

Bootstrapping is a technique that provides a simple and practical way to estimate the uncertainty in sample statistics like the sample variance. The project data collection team can use bootstrap techniques to estimate the sampling distribution of sample variance. The estimated sampling distribution is then manipulated to produce confidence intervals for and rejection regions for tests of hypotheses about the target population variance σ^2. Information about bootstrapping can be found in the books by Efron and Tibshirani (1993) and by Manly (1998).

8.1.4.7 Statistical Inference: Comparing Two Population Variances

In many business applications in which two processes or two suppliers of a product or service are been compared, the project data collection team needs to compare the standard deviations of the target populations associated with process measurements. Another major application of a test for the equality of two population variances is for evaluating the validity of the equal variance condition (that is, $\sigma_1^2 = \sigma_2^2$) for a two-sample Student's t-Test.

In the previous sub-sections, the tests of hypotheses concerned either population means or a shift parameter. For both types of parameters, it was important to provide an estimate of the effect size along with the conclusion of the test of hypotheses. In the case of testing population means, the effect size was in terms of the difference in the two means, $\mu_1 - \mu_2$. When comparing population variances, the appropriate measure is the ratio of the population variances, σ_1^2/σ_2^2. Thus, the project data collection team needs to formulate a confidence interval for the ratio σ_1^2/σ_2^2. The results developed below require that the two population distributions both have normal distributions.

Assuming that the project data collection team is interested in comparing the variance of the first population, σ_1^2, to the variance of the second population σ_2^2, when random samples of sizes n_1 and n_2 have been independently collected from two normally distributed populations characteristic, a $100(1-\alpha)\%$ confidence interval for the ratio σ_1^2/σ_2^2 is given to be:

$$\frac{s_1^2}{s_2^2}F(1-\alpha/2,\ n_2-1, n_1-1) \le \frac{\sigma_1^2}{\sigma_2^2} \le \frac{s_1^2}{s_2^2}F(\alpha/2, n_2-1, n_1-1)$$

Where $F(\alpha, v_2, v_1)$ is the α percentile of Fisher's distribution with degrees of freedom v_2 and v_1.

The following ratio is known to possess a probability distribution in repeated sampling referred to as a Fisher distribution.

$$f = \frac{s_1^2/\sigma_1^2}{s_2^2/\sigma_2^2} = \frac{s_1^2/s_2^2}{\sigma_1^2/\sigma_2^2}$$

A statistical test comparing the variance of the first population, σ_1^2, to the variance of the second population σ_2^2, utilizes the test statistic defined to be:

$$f = \frac{s_1^2}{s_2^2}$$

This statistic follows from the null hypothesis written symbolically as:

$$H_0 : \sigma_1^2 = \sigma_2^2$$

With the alternative hypothesis (H_A) written symbolically as:

$$H_A : \sigma_1^2 \neq \sigma_2^2$$

With these, the project data collection team should reject the null hypothesis if the calculated value of f is less or equal than the lower percentile value of Fisher's distribution, $F(1 - \alpha/2, n_2 - 1, n_1 - 1)$, or it should reject the null hypothesis if the calculated value of F is greater or equal than the upper percentile value of Fisher's distribution, $F(\alpha/2, n_2 - 1, n_1 - 1)$. The percentiles of Fisher's distribution follow the relation:

$$F(1 - \alpha, v_1, v_2) = \frac{1}{F(\alpha, v_2, v_1)}$$

Note that the degrees of freedom have been reversed for the upper percentile on the right-hand side of the equation above.

8.1.4.8 Inferences About More Than Two Population Central Values: ANOVA

In many practical/scientific settings, the number of populations for which the project data collection team might want to make comparisons will be higher than two. When there are more than two sources of variability to be compared Student's t-Test is not relevant. Instead, the technique of analysis of variance (ANOVA) is appropriate.

Consider a completely randomized design in which the project data collection team collects preliminary data on a selected characteristic from p different target customer populations or from stratums of a target customer population. Let assume that the population distribution have equal variance, but different means. If Y_{ij} denotes the $j-th$ collected data in a sample of size n_i from the $i-th$ target population, the project data collection team could display the sample data for this completely randomized design as a matrix form $\{Y_{ij}, i = 1, \cdots, p; j = 1, \cdots, n_i\}$. The project data collection team should summarize the data into the statistics:

1. Sample means $\overline{Y^l}$ of the sample of size n_i from the $i-th$ target population;
2. Overall average of all samples mean, $\bar{Y} = \left(\sum_{i=1}^{p} \overline{Y^i}\right)/n$; with $n = \sum_{i=1}^{p} n_i$ been the total sample size.

The project data collection team can subsequently measure the variability of the n sample collected data Y_{ij} about the overall mean using the total sum of squares (TSS) defined to be:

$$TSS = \sum_{i=1}^{p} \sum_{j=1}^{n_i} (Y_{ij} - \bar{Y})^2 = (n-1)s^2$$

The double summation in TSS means that the project data collection team must sum the squared deviations for all rows and columns of the one-way classification. It is possible to partition the total sum of squares as follows:

$$TSS = \sum_{i=1}^{p} \sum_{j=1}^{n_i} (Y_{ij} - \bar{Y})^2 = \sum_{i=1}^{p} \sum_{j=1}^{n_i} \left(Y_{ij} - \overline{Y^i}\right)^2 + \sum_{i=1}^{p} \left(\overline{Y^i} - \bar{Y}\right)^2 n_i$$

The first quantity on the right side of the equation measures the variability of an observation Y_{ij} about its sample mean $\overline{Y^i}$. Thus,

$$SSW = \sum_{i=1}^{p} \sum_{j=1}^{n_i} \left(Y_{ij} - \overline{Y^i}\right)^2 = \sum_{i=1}^{p} (n_i - 1)s_i^2 = (n-p)s_w^2$$

SSW is a measure of the within-sample variability. SSW is referred to as the within sample sum of squares and is used to compute s_w^2.

The second expression in the total sum of squares equation measures the variability of the sample means about the overall mean $\bar{Y}$. This quantity, which measures the variability between (or among) the sample means, is referred to as the sum of squares between samples (SSB) and is used to compute s_B^2.

$$SSB = \sum_{i=1}^{p} \left(\overline{Y^i} - \bar{Y}\right)^2 n_i = (p-1)s_B^2$$

Although the formulas for TSS, SSW, and SSB are easily interpreted, they are not easy to use for manual calculations. Instead, we recommend using a computer software program.

An analysis of variance for a completely randomized design with p populations has the following null and alternative hypotheses:

$$H_0 : \mu_i = \mu_k;\ i, k = 1, \cdots, p$$

With the alternative hypothesis (H_A) written symbolically as:

H_A: At least one of the t population means differs from the rest.

The sum of squares divided by its degrees of freedom is often referred to as a mean square. In much the same vein, s_B^2 is often referred to as the mean square between samples and s_w^2 is referred to as the mean square within samples. These quantities are mean squares because they both are averages of squared deviations. There are only $n-p$ linearly independent deviations $Y_{ij} - \overline{Y^i}$ in SSW because $\sum_{j=1}^{n_i}\left(Y_{ij} - \overline{Y^i}\right) = 0$ for each of the p samples. Hence, SSW is divided by $n-p$ and not n. Similarly, there are only $p-1$ linearly independent deviations $\overline{Y^i} - \bar{Y}$ in SSB, because $\sum_{i=1}^{p}\left(\overline{Y^i} - \bar{Y}\right)n_i = 0$. Hence, SSB is divided by $p-1$.

Using the quantity defined to be:

$$f = \frac{s_B^2}{s_w^2}$$

The project data collection team should reject the null hypothesis if the calculated value of f exceeds the tabulated percentile value of Fisher's distribution, $F(\alpha, p-1, n-p)$, or it should reject the null hypothesis if the calculated value of F is greater or equal than the upper percentile value of Fisher's distribution, $F(\alpha/2, n_2 - 1, n_1 - 1)$. The percentiles of Fisher's distribution follow the relation:

These results are often summarized in an analysis of variance table. The format of an ANOVA table is shown in Table 8.6. The ANOVA table lists the sources of variability in the first column. The second column lists the sums of squares associated with each source of variability. The total sum of squares, TSS, can be partitioned into two parts, therefore SSB and SSW must add up to TSS in the ANOVA table. The third column of the table gives the degrees of freedom associated with the sources of variability. Again, the project data collection team has a check; $(p-1)+(n-p)$ must add up to $n-1$. The mean squares are found in the fourth column of Table 8.6, and the F-Test for the equality of the p population means is given in the fifth column.

To summarize, the purpose of a one-way ANOVA is to divide up the variance in some dependent variable into two components: the variance attributable to between-group differences, and the variance attributable to within-group differences, also known as error.

When the project data collection team selects a sample from a population and calculates the mean for that sample on some variable, that sample mean is the best predictor of the population mean. In other words, if the mean of the population in not known, the best guess about what the population mean is would have to come from the mean of a sample drawn randomly from that population. Any scores in the sample that differ from the sample mean are believed to include what statisticians call error. The variation that is fond among the scores in a sample is not just considered error. In fact, it is thought to represent a specific kind of error: random error.

When the project data collection team selects a sample at random from a population, it is expected that the data of that sample will not all have identical

Table 8.6 One-way ANOVA table

Source	Sum of squares	Degrees of freedom	Mean Square	F-Test
Between samples	SSB	$p-1$	$s_B^2 = \frac{SSB}{p-1}$	$\frac{s_B^2}{s_w^2}$
Within samples	SSW	$n-p$	$s_w^2 = \frac{SSW}{n-p}$	
Total	TSS	$n-1$		

scores on the variable of interest. That is, it is expected that there will be some variability in the scores of the data of the sample. That is just what happens when sample data is collected randomly from a population. Therefore, the variation in scores that occurs among the data of the sample is just considered random error.

The question that the project data collection team addresses using ANOVA is this: "Is the average amount of difference, or variation, between the scores of data of different samples large or small compared to the average amount of variation within each sample, otherwise known as random error?" To answer this question, the project data collection team has to determine three quantities. First, it has to calculate the average amount of variation within each of our samples. Second, it has to find the average amount of variation between the groups. Once the project data collection team has found these two statistics, it must find their ratio by dividing the mean square between by the mean square error. This ratio provides the F value, from which a family of F distributions can be inspected to see if the differences between the groups are statistically significant.

8.2 Collect and Organize Data

Once the customer and stakeholder value proposition and plan for collecting customer and stakeholder requirements are established, the next step is to begin the data collection, analyze reactive data from the customers and stakeholder's needs, and then fill the gaps with proactive approaches.

Once data about customer needs has been gathered, it must be organized. The mass of interview notes, requirements documents, market research, and customer data needs to be distilled into a handful of statements that express key customer needs. Affinity clustering is useful tools to achieve this goal.

8.2.1 Organize V.O.C. Data: Affinity Clustering

Affinity clustering is a useful tool that organizes language data into related groups. It is also called the KJ Method, named after the Japanese anthropologist, Jiro Kawakita (Kawakita, A Scientific Exploration of Intellect, 1977; Kawakita, The KJ Method:

Chaos Speaks for Itself, 1986), who developed a method of establishing an orderly system from chaotic information. The goals of an affinity clustering are to:

1. Stress creative and intuitive thinking;
2. Help to identify patterns in large amount of data collected;
3. Allow the project members to gather large amounts of language data;
4. Help to organize customers' needs, ideas, issues, and opinions.

The affinity clustering is a four-step process for organizing and summarizing a large amount of data (needs, ideas, issues, solutions, problems) into logical categories so that it is possible to understand the essence of a problem or solution.

When using an affinity clustering, the "process improvement" project team should write all relevant facts and information on individual cards, which are then collate, shuffle, spread out, and read carefully. The cards are shuffle because they could originate from several sources. The shuffled cards should then be reviewed, classified, and sorted on the similarity, affinity, and characteristics of the needs. The four steps of an affinity clustering process can be summarized as:

1. Collect data and prepare cards
2. Sort needs/ideas into related clusters
3. Create header/title cards for each cluster
4. Write reports and perform further analysis

8.2.1.1 Collecting Data and Preparing Cards First

The project team should prepare some index cards or sticky notes for use. Then write each idea from the "words and notes" raw data on one of the cards or notes, one idea for each card.

Sorting Needs/Ideas into Related Groups

The project team will then proceed to sort the cards into groups using the following process:

1. Place all the cards in one pile.
2. Draw one card at a time from this pile, and examine its content. If the customer need/idea from this card is related to the customer need/idea from the card that was drawn before, place them together; these two cards belong to a cluster.
3. Keep drawing the cards one at a time. Examine the card to see if the customer need/idea from the card is similar to any existing cluster; if the answer is yes, place this card into that cluster. If the answer is no, this card can start a new cluster. Keep doing this step until all the cards are drawn and all of the ideas are sorted into clusters.
4. After all cards are drawn, re-examine all the clusters. Some cards can be moved around so that each cluster customer needs/ideas look more coherent. If a customer need or an idea seems equally applicable to two clusters, create a duplicate of that card and place one in each cluster.
5. It is possible for one card to stand alone and form a cluster, as illustrated in Fig. 8.10.

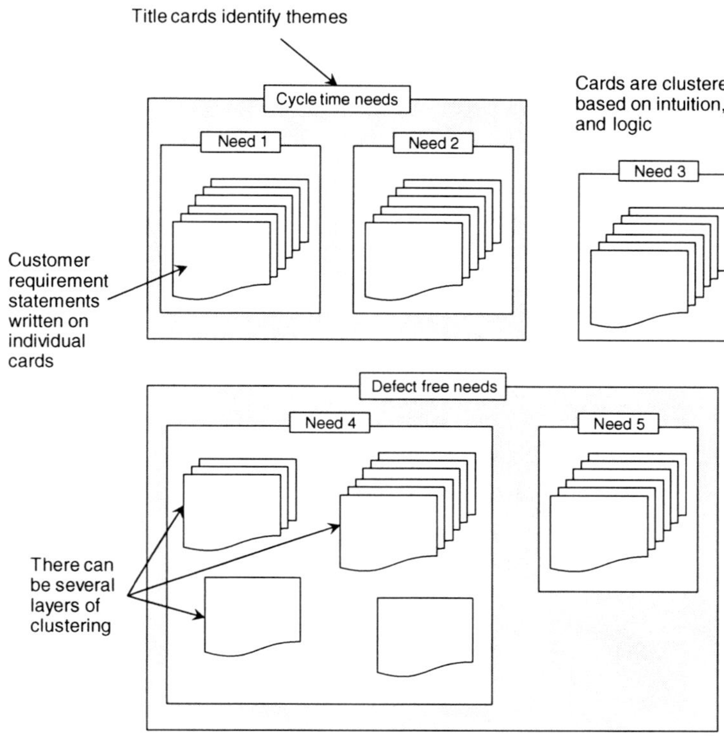

Fig. 8.10 Clustering customers' needs based on affinity

6. The ideal clustering result should have the following features:
   - The ideas within a cluster should be closely related;
   - There should be significant differences between clusters.

Creating Header/Title Cards for Each Cluster

Create header cards for each cluster. A header is a title that captures the essential theme and link among the customer needs and ideas contained in a cluster of cards. Figure 8.11 shows an example of need or idea clusters and headers.

Here are some characteristics of header or title cards:

1. The header or title should be the best word or phrase that describes the meaning of each cluster. The meaning of the header should stand alone and be clear to outside readers without reading the contents of the cards in the cluster.
2. During the process of creating headers or titles, it is possible to cluster anew customer needs or ideas so that the headers have clear and better meanings. It is also possible that a large cluster will be subdivided into several small clusters under different headers.
3. It may take several iterations to finalize the header process in order to best capture the meaning of each cluster.
4. Clarify and finalize headers through consensus.
5. It is possible that hierarchical clusters or multilevel clusters will be adopted.

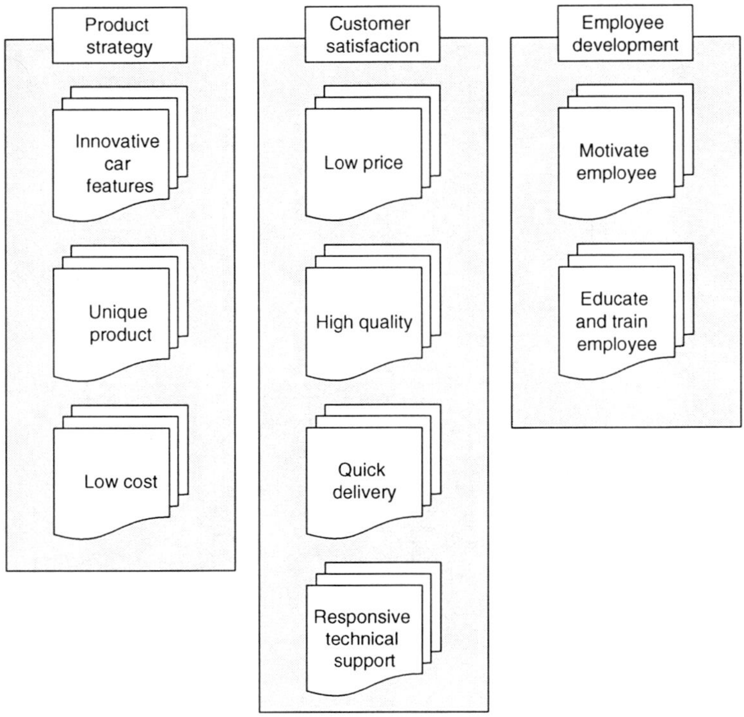

Fig. 8.11 Example of affinity clustering of the V.O.C.

Writing reports and performing further analysis—An affinity clustering can help to reveal hidden clusters and structures in a large amount of fragmented notes and words, so that the "process improvement" project team can see how everything fits together. Once this initial clustering of customer needs has been performed, the project team will need to issue a written report (dictionary) outlining the meaning of these clusters.

8.3 Analyze Data and Generate Customer Key Needs

The mass of interview notes, requirements documents, market research, and customer data has been distilled and clustered into a handful of statements that express key customer needs using an affinity clustering. While an affinity clustering is used to organize language data into related groups, it does not provide a means to gain further insight into the data. The Kano model provides a means to achieve this purpose. It is a useful tool in gaining a thorough understanding of a customer's needs.

From the collected and organized customers and stakeholders requirements it is very important that the project team gains further insight into the data and sorts "wants" from "needs." Indeed, the root cause of many problems that arise in the course of a "process improvement" project originates from a disconnection between what the customers and stakeholders say they want and what they really need.

The disconnection may arise because the customers are swept up in euphoria over a new process outcome feature and are so captivated with what they see from other similar products, for example, that they have convinced themselves that they have to have it in the process outcome without any further thought of exactly what it is they really need.

The disconnection can also arise because the customers and stakeholders apparently do not really know what they need. If there is any reason to believe that what the customers and stakeholders say they want is different from what they need, then project team has the responsibility of sifting and sorting "wants" from "needs." It would be a mistake to proceed without having the assurance that "wants" and "needs" are in alignment.

From the quality perspective, the V.O.C. activities resulting from collecting and organizing customers and stakeholders requirements provide an array of customer attributes. These attributes are "spoken" by the customer and are often called "performance quality." For example, gas mileage of an automobile is a performance quality; it is also "the more, the better." However, more requirements have to be addressed than just those directly spoken by the customer. These are known as "unspoken" attributes.

Unspoken attributes are the basic quality features that the customer automatically assumes will be in the "process to be improved" outcomes. Such attributes are implied in the functional requirements of the design of the "process to be improved" outcomes or assumed from historical experience. For example, customers automatically expect their lawnmower to cut grass to the specified level, but they wouldn't discuss it on a survey unless they had trouble with one in the past. Unspoken attributes have a "peculiar" property—they don't increase customer satisfaction, but if they are not delivered, they have a strong negative effect on customer satisfaction.

Noriaki Kano summarized these finding into a model, illustrated in Fig. 8.12, and which involves two dimensions:

1. Achievement (the horizontal axis) which runs from the "process to be improved" did not achieve expectations at all to the supplier did achieve expectations very well.
2. Satisfaction (the vertical axis) that goes from total customer dissatisfaction with the "process to be improved" outcomes (i.e. product or service) to total customer satisfaction with the "process to be improved" outcomes.

The Kano model isolates and identifies three key levels of customer expectations: that is, what it takes to positively impact customer satisfaction. Figure 8.12 portrays the three key levels of customer needs: must be, more is better, and delighters.

Must be, threshold or basic attributes—These needs are expected by the customer as they are crucial parts of product or service improvement. They are basically the features that the "process to be improved" outcomes must have in order to meet customer demands. If they are unfulfilled, the customer will be dissatisfied, but even if they are completely fulfilled the customer would not be particularly satisfied. These attributes are either there or not. They are generally

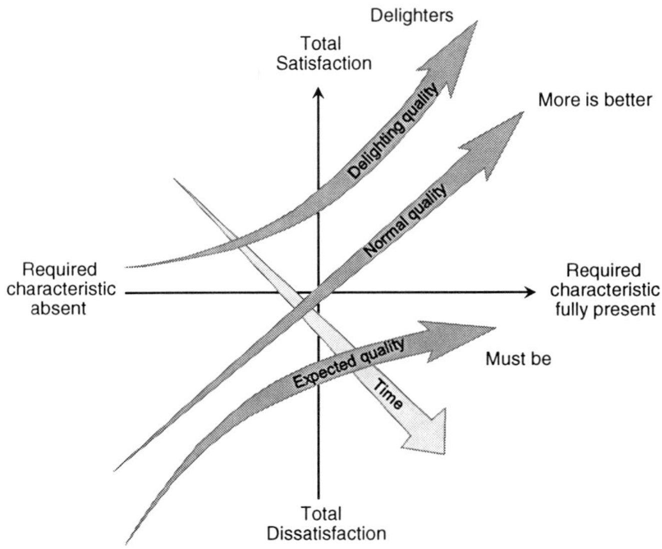

Fig. 8.12 Kano model of customer key needs

taken for granted unless they are absent. The car safety in the automotive industry is an example of such attribute. Another example of a threshold attribute would be a steering wheel in a vehicle. The vehicle is no good if it is not able to be steered. If these attributes are overlooked, the outcomes are completely incomplete. This is the first and most important characteristic of the Kano model.

More is better or performance attributes—These needs have a linear effect on customer satisfaction. The more these needs are met, the more satisfied these customers are. For example, cheep airline tickets. Customers generally discuss or bring up issues related to the "More is better" characteristics.

Delighters or excitement attributes—These needs do not cause dissatisfaction when not present, but satisfy customer in a nonlinear fashion when they are present. Delighters are generally not mentioned, since the customers are not dissatisfied with their presence. For example, in the automotive industry, van owners were delighted by the second van side door and by baby-seat anchor bolts.

8.4 Translate Customer Key Needs into CTXs

The ultimate goal of the "process improvement" project is to design and develop a product or service that customers really want. That is why it is important to spend so much effort capturing the V.O.C. and identifying key needs. However, we cannot design a good product simply by using the V.O.C. key needs because the key V.O.C. needs do not give employees working on the "process to be improved" enough information to set technical specifications.

The identified V.O.C. key needs must to be developed into clear, specific, quantitative requirements in order to be really helpful in the development of the "process to be improved" outcomes. These quantitative requirements are called Critical-to-X characteristics (CTXs), where X represents Quality, Cost, or Schedule on the "process to be improved." CTXs are the measurable product or service characteristics that the customer considers important, and whose performance standards or specification limits must be met to satisfy customer requirements. They usually have four components: characteristic, measure, target, and specification limits.

A CTS or Critical to Schedule characteristic has a major influence on the capacity of the "process to be improved" to deliver its outcomes on time.

A CTC or Critical to Cost characteristic has a major influence on the cost of producing the "process to be improved" outcomes. It often involves increasing the production capacity without increasing the resources. Factors Critical to Cost include parameters that impact work in progress, finished goods inventory, overhead, delivery, material and labor, even when the costs can be passed on to the customer.

A CTQ or Critical to Quality characteristic has a major influence on the suitability for use of the product or service produced by the "process to be improved." A Critical to Quality characteristic conveys quality requirements and it aligns upgrading and creative efforts with the customer key needs. It represents the expectation of a customer from the "process to be improved" outcomes.

Critical to Quality (CTQ) factors are most familiar to operational personnel since they directly impact the functional requirements specified by the customers. For example, "light in weight" may be one of the customers' key requirements for a power saw, but the statement "light in weight" is not a CTQ, because it does not give either a performance measure or specification limits. The statement "the weight of power saw should be no more than 3 kg" is a CTQ because weight is a key performance factor that is important to customers, and this statement gives a very specific performance specification.

A typical tool useful in translating customer key needs into quantified CTXs requirements for the outcomes of the "process to be improved" is the CTX tree. Figure 8.13 shows a sample CTX tree template. With each customer key need, there are a number of drivers that can be developed. Each driver will be decomposed into CTXs.

8.5 Set Specifications for CTXs

After translating customer key needs into quantified CTXs requirements for the outcomes of the "process to be improved," the project team should set target specifications. These preliminary specifications represent the hope and aspirations of the project team. During planning, the preliminary specifications are refined and described with greater specificity as more information about the project is known.

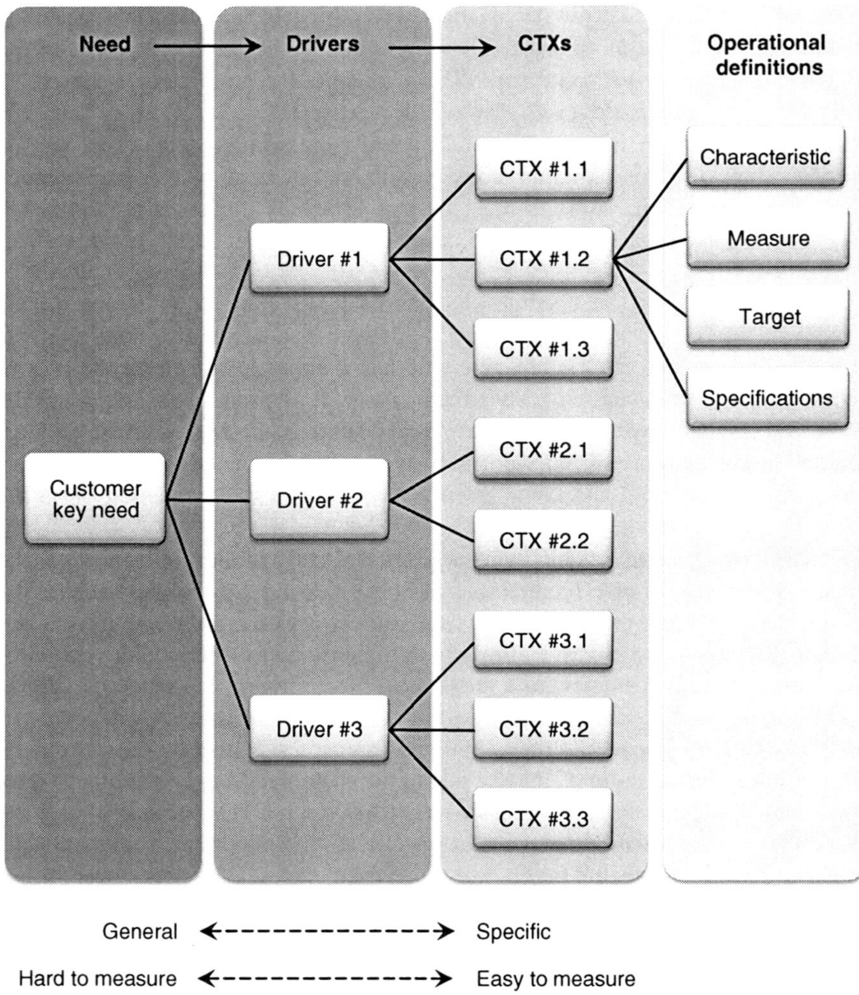

Fig. 8.13 Sample CTX template

Customers and stakeholders' key needs and their associated CTXs are conditions or capabilities that must be met or possessed by the "process to be improved." Base limits on CTXs measures set performance values on the "process to be improved" outcomes beyond which the customer and stakeholder satisfaction starts to fall off appreciably. A specification for the "process to be improved" outcomes is that value of the considered characteristic that separates acceptable from non acceptable performance of the "process to be improved" outcomes. It spells out in precise and measurable details what performance level the "process to be improved" outcomes must meet.

Two types of values are often used as specifications: an ideal value and a marginally accepted value. The ideal value is the one for which the resulting

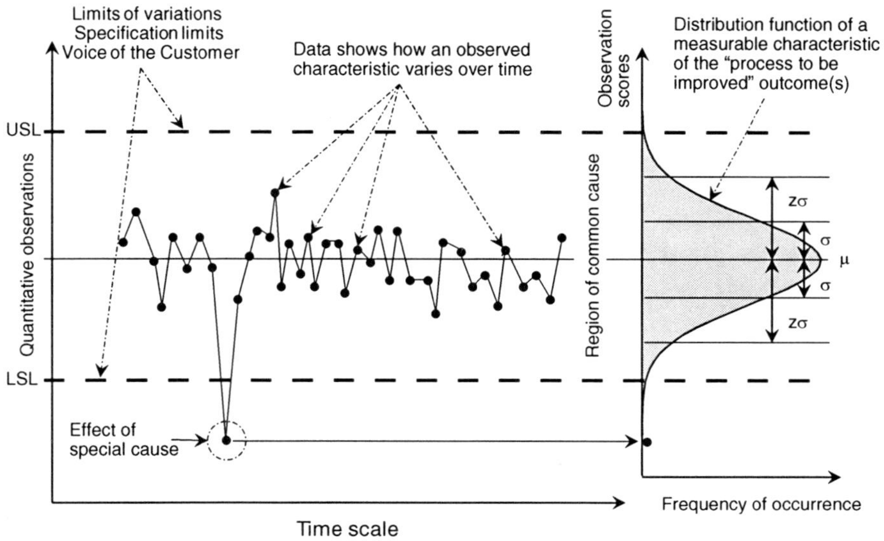

Fig. 8.14 Specification limits for a characteristic of the "process to be improved" outcomes

performance of the "process to be improved" outcomes will be at its highest level, hence resulting in the highest customer satisfaction. The marginally accepted value is the one for which the resulting performance of the "process to be improved" outcomes would just barely make these outcomes acceptable to the customers. Both of these two values are useful in guiding subsequent stages of improvement of the "process to be improved." There are five ways of expressing these two specification values:

1. *At least X—Lower Specification Limit (LSL):* These specifications, illustrated in Fig. 8.14, establish targets for the lower bound on a measure of the considered characteristic, but higher are still better.
2. *At most X—Upper Specification Limit (USL):* These specifications establish targets for the upper bound on a measure of the considered characteristic, with smaller values being better.
3. *Between X and Y*—These specifications establish both upper and lower bounds for the value of a measure of the considered characteristic.
4. *Exactly X*—These specifications establish a target of a particular value of a measure of the considered characteristic, with any deviation degrading performance of the "process to be improved" outcomes. These types of specifications are to be avoided if possible because they substantially constrain the improvement efforts. Often, upon reconsideration, the team will realize that what initially appears as an "exactly X" specification can be expressed as a "between X and Y" specification.
5. *A set of discrete values*—Some measures of the considered characteristic will have values corresponding to several discrete choices.

The desirable range of values for one measure of the considered characteristic may depend on another. In situations where the project team feels that this level of complexity is warranted, such specifications can easily be included, although we recommend that this level of complexity not be introduced until the final phase of the specifications process. However, the project team must ensure that each specification is:

1. *Reasonable*—The specification is based on a realistic assessment of the customer's actual key needs and relate directly to the performance of a characteristic.
2. *Understandable*—The specification is clearly stated and defined so that there can be no argument about its interpretation.
3. *Measurable*—The characteristic performance can be measured against the specification to avoid debate with the customer as to whether the specification has been met or not.
4. *Believable*—The project team has a buy-in into the specification and will strive to meet it.
5. *Attainable*—The level and range of the specification can be reached.

Using these five different types of expressions for values of the measures of the considered characteristic, the project team should set the preliminary specifications at this step of the planning. This is done by simply proceeding down the list of identified CTXs characteristic measures and assigning both the marginally acceptable and ideal target values for each measure. These decisions can be facilitated by a measure-based competitive benchmarking. To set the target values for CTXs, the team has many considerations, including the capability of competing products or services available at the time, competitors' future product or service capabilities (if these are predictable), and the mission statement of the "process improvement" project.

Because most of the values are expressed in terms of bounds (upper or lower or both), the project team is establishing the boundaries of the acceptable space for the "process to be improved" outcomes. These outcomes will hopefully meet some of the ideal values specified but they can be acceptable to customers even if they exhibit one or more marginally acceptable characteristics.

These specifications are preliminary because until a concept for improving the "process to be improved" is chosen and some of the design/development details for improvement are worked out; namely, the quality management plan, the risk management plan and the costs management plan, many of the exact tradeoffs are uncertain. Once a concept for improving the "process to be improved" is chosen and quality management plan, risk management plan and costs management plan have been worked out, the specifications must be refined and finalized, making tradeoffs where necessary.

Finalizing the specifications is difficult because of tradeoffs—inverse relationships between two specifications that are inherent in the selected concept for improving the "process to be improved." Tradeoffs frequently occur between different technical performance measures on selected characteristics and almost always occur between technical performance measures and cost. The difficult part of refining the specifications is choosing how such tradeoffs will be resolved.

Finalizing specifications can be accomplished in a group session in which feasible combinations of values are determined through the use of design/development details and then the quality, risk and cost implications are explored. In an iterative fashion, the project team will converge on the specifications which will most favorably position the "process to be improved" outcomes to best satisfy the customer needs, also ensuring adequate profits.

8.6 Conclusion

Collecting the customers and stakeholders' requirements is as much about defining and managing customers' and stakeholders' expectations as any other key project deliverables and will be the very foundation of completing to "process improvement" project. It is also about focusing the improvement effort by gathering information on the current situation. Its purpose is to build, as precisely as possible, a factual understanding of existing "process to be improved" conditions and problems or causes of underperformance. Cost, schedule, and quality planning are all built upon these requirements. The key outputs in collecting the requirement include:

1. Customers and stakeholders requirements documentation
2. Requirement management plan
3. Requirements traceability matrix

Customers and Stakeholders Requirements Documentation—This document describes how individual requirements meet the business need for the project. Requirements may start out at a high-level and become progressively more detailed as more is known. Before setting the base line, requirements must be unambiguous (measurable and testable), traceable, complete, and consistent.The format of a customer and stakeholder requirements document may range from a simple document listing all the requirements categorized by customer and stakeholder and priority, to more elaborate forms containing executive summary, detailed descriptions, and attachments. Components of customer and stakeholder requirements documentation can include but are not limited to:

1. Business problem to be solved or opportunity to be seized, describing the limitations of the current situation and why the "process improvement" project has been undertaken;
2. Business and "process improvement" project objectives for traceability;
3. Functional requirements, describing business process, information, and interaction with the "process to be improved" outcome, as appropriate which can be documented textually in a requirements list, in models, or both;
4. Non-functional requirements, such as level of service, performance, security, compliance, supportability, retention/purge, etc...:
5. Quality requirements;
6. Business rules stating the guiding principles of the organization;
7. Impacts to other organizational areas, such as call center, sales force, technology groups;

8. Impacts to other entities inside or outside the performing organization;
9. Support and training requirements; and
10. Requirements assumptions, specifications and constraints.

Requirements Management Plan—The requirements management plan documents how requirements will be analyzed, documented, and managed throughout the project life cycle. The project manager must choose the most effective phase-to-phase relationship for the project and document this approach in the requirements management plan. Many of the requirements management plan components will be based on that relationship. Components of the requirements management plan can include but are not limited to:

1. How requirements activities will be planned, tracked, and reported;
2. Configuration management activities such as how changes to the product, service or result requirements will be initiated, how impacts will be analyzed, how they will be traced, tracked, and reported, as well as the authorization levels required to approve these changes;
3. Requirements prioritization process;
4. Product metrics that will be used and the rationale for using them; and
5. Traceability structure, that is, which requirements attributes will be captured on the traceability matrix and to which other project documents requirements will be traced.

Requirements Traceability Matrix—The requirements traceability matrix is a table that links requirements to their origin and traces them throughout the project life cycle. The implementation of a requirements traceability matrix helps to ensure that each requirement adds business value by linking it to the business and project objectives. It provides a means to track requirements throughout the project life cycle, helping to ensure that requirements approved in the customers and stakeholders requirements documentation are delivered at the end of the project. Finally, it provides a structure for managing alterations to the "process to be improved" scope. The requirements traceability matrix includes but is not limited to tracing:

1. Requirements to business problems, opportunities, goals, objectives;
2. Requirements to project objectives;
3. Requirements to project scope;
4. Requirements to "process to be improved" design;
5. Requirements to "process to be improved" development;
6. Requirements to test strategy and test scenarios; and
7. High-level requirements to more detailed requirements.

Attributes associated with each requirement can be recorded in the requirements traceability matrix. These attributes help to define key information about the requirement. Typical attributes used in the requirements traceability matrix may include: a unique identifier, a textual description of the requirement, the rationale for inclusion, owner, requirement, source, priority, version, current status (such as active, cancelled, deferred, added, approved) and date completed.

Additional attributes to ensure that the requirement has met customers' and stakeholders' satisfaction may include stability, complexity, and acceptance criteria.

9 Create Work Breakdown Structure

Identifying and breaking down the work to be done is the logical starting point in the entire planning process described in a previous chapter. This chapter is concerned with the project management process required to subdivide project deliverables and project work into smaller, more manageable components.

The work breakdown structure is a method of portrayal of a project, exploding it in a level-by-level fashion, down to the degree of detail needed for effective planning and controlling, this without considering the order of events. The work breakdown structure is to the project manager what the organization chart is to the enterprise business executive: it defines the project manager's universe. Its purpose is to define discrete quantities of work so that:

1. They can be uniquely identified for what they are;
2. They can be seen for their contribution to the project;
3. They can be monitored and controlled from the perspective of time, cost and content;
4. Responsibility for achievement and performance can be allocated; and
5. Meaningful historic data can be recorded at the end of the project.

The work breakdown structure is used as input for every process of creating the project schedule and project budget. In that sense, it is the foundation of the schedule and the budget. If the foundation is weak, the schedule and budget will never be strong and it is hard to recover from a weak work breakdown structure.

9.1 Defining a Work Breakdown Structure

Creating an appropriate work breakdown structure is an essential step in handling any complex project. The work breakdown structure is a deliverable-oriented hierarchical decomposition of the work to be executed by the project team, to accomplish the project objectives and create the required deliverables, with each descending level of the work breakdown structure representing an increasingly detailed definition of the project work. It does not specify the order in which the

A. van Aartsengel and S. Kurtoglu, *Handbook on Continuous Improvement Transformation*,
DOI 10.1007/978-3-642-35901-9_9,

decomposed work will be carried out. However, it organizes and defines the total scope of the "process improvement" project, including all deliverable end items and the major functional tasks that must be performed, and represents the work specified in the current approved project scope statement.

The project work breakdown structure is developed through the process of decomposition of the project goal. This decomposition relates to defining a tree structure, which shows a subdivision of effort required to achieve the project goal. It is developed by starting with the end project goal and successively subdividing it into manageable components in terms of size, duration, and responsibility (e.g., systems, subsystems, components, tasks, subtasks, and work packages) which include all steps necessary to achieve the project goal. Here, a work package is a complete description of how the tasks that make up an activity will actually be done. It includes a description of the what, who, when, and how of the work.

9.2 Developing a Work Breakdown Structure

A hierarchical visualization of the work breakdown tree structure is shown in Fig. 8.1. As indicated in the previous section, the work breakdown tree structure is created through a decomposition technique. Decomposition is the subdivision of project deliverables into smaller, more manageable components until the work and deliverables are defined to the work package level. The work package level is the lowest level in the work breakdown structure, and is the point at which the cost and schedule for the work can be reliably estimated. The level of detail for work packages will vary with the size and complexity of the "process improvement" project.

As illustrated in Fig. 9.1, the project goal statement is defined as a Level 0 activity in the work breakdown structure. The next level, Level 1, is a decomposition of the Level 0 activity into a set of activities defined as Level 1 activities. These Level 1 activities are major portions of work. They provide a reflection of the management approach, the major chunks of effort, and the critical subprojects. When the work associated with each Level 1 activity is complete, the Level 0 activity is complete. For this example, that means that the project is complete.

The subordinate Level 2 and lower levels simply reflect a further decomposition of defined work into progressively smaller segments. Level 1 is likely the most critical subdivision for any project because it reflects the management approach.

As a general rule, when an activity at Level n is decomposed into a set of activities at Level $n + 1$ and the work associated with those activities is complete, the activity at Level n, from which they were defined, is complete.

Decomposition of the upper level work breakdown structure components requires subdividing the work for each of the deliverables or subprojects into its fundamental components, where the work breakdown structure components represent verifiable products, services, or results. Verifying the correctness of the decomposition requires determining that the lower-level work breakdown structure components are those that are necessary and sufficient for completion of the corresponding

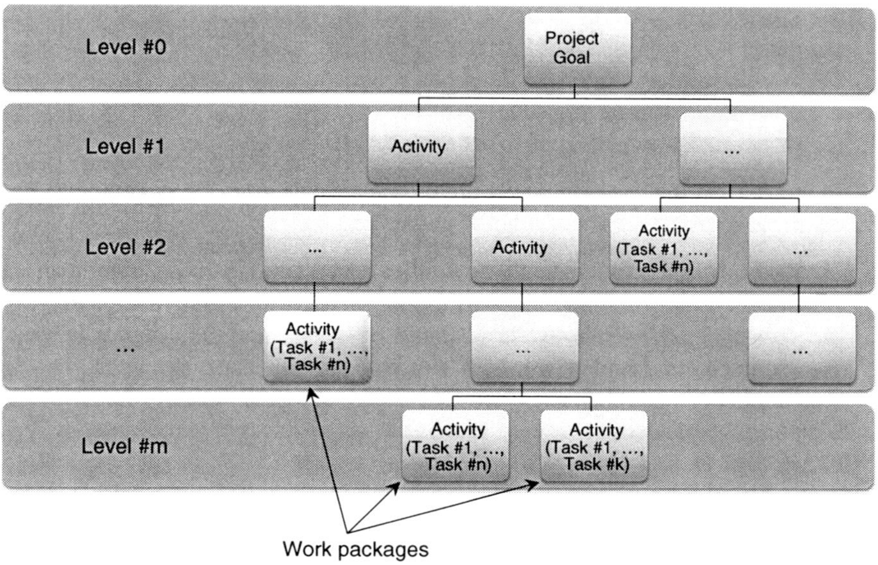

Fig. 9.1 Hierarchical visualization of the work breakdown structure

higher level deliverables. Here, the completion criteria answer these two critical questions about each work package: (1) "What does it mean to be complete with this activity?" and (2) "How will we know it was done correctly?"

> As the PMBOK Guide indicates (PMI, 2004), *"different deliverables can have different levels of decomposition and to arrive at a work package, the work for some deliverables needs to be decomposed only to the next level, while others need more levels of decomposition. As the work is decomposed to lower levels of detail, the ability to plan, manage, and control the work is enhanced. However, excessive decomposition can lead to non-productive management effort, inefficient use of resources, and decreased efficiency in performing the work."*

Decomposition of the total project work into work packages involves the following activities:

1. Identifying and analyzing the deliverables and related work;
2. Structuring and organizing the work breakdown structure;
3. Decomposing the upper work breakdown structure levels into lower level detailed components;
4. Developing and assigning identification codes to the work breakdown structure components; and
5. Verifying that the degree of decomposition of the work is necessary and sufficient.

There is not just one unique work breakdown structure for each project. A variety of different work breakdown structure structures may be generated for the same project, each being preferable under different conditions. However, the following are key points to remember when structuring a work breakdown structure:

1. The work breakdown structure represents work content and not an execution sequence.
2. The work breakdown structure should be generic in nature so that it may be used in the future for similar projects.
3. The work breakdown structure is not a product structure tree, or bill of materials.

To ensure that the work packages are of manageable size, the project team can follow these common "rules of thumb" guidelines:

1. *The 8/80 rule.* No activity should be smaller than 8 labor hours or larger than 80. This translates into keeping the work packages between 1 and 10 days long.
2. *The reporting period rule.* No task should be longer than the distance between two status points. In other words, if you hold weekly status meetings, then no task should be longer than one week. This rule is especially useful when it is time to report schedule status, because you, as project manager or team leader, will no longer have to hear about activity statuses that are 25, 40, or 68 % complete. If you have followed a weekly reporting rule, tasks will be reported as either complete (100 %), started (50 %), or not started (0 %). No task should be at 50 % for two consecutive status meetings.
3. *The "if it's useful" rule.* As the project team considers whether to break activities down further, it should remember that there are three reasons to do so:
   - The activity is easier to estimate. Smaller activities tend to have less uncertainty, leading to more accurate estimates.
   - The activity is easier to assign. Large activities assigned to many people lose accountability. Breaking down the activity can help to clarify who is responsible. Another potential benefit is that having smaller activities assigned to fewer people can give you greater flexibility in scheduling the activity and the resource.
   - The activity is easier to track. The same logic applies as in the reporting period rule. Because smaller activities create more tangible status points, you will have more accurate progress reports.
4. If breaking down an activity in a certain way is not useful—that is, if it does not make it easier to estimate, assign, or track—then do not break it down!

By essentially forcing the project team to define its necessary work activities into progressively greater detail with the use of a work breakdown structure, the total scope of the project will then take form. The work breakdown structure in total will define what is inside and what is outside of any given project.

When properly developed, a work breakdown structure allows the project team to identify every single element of work (task) required to complete the project. Once this has done this, the project team is now able to move rapidly forward in the planning process. For each of those tasks, the project team now needs to consider important characteristics to be used as input for future planning steps. These include:

1. *Time:* The number of days (weeks?) that will be spent working on the activity.
2. *Cost:* How much will be spent on labor and materials;
3. *Scope:* The work that will be done, how it will be done, and what will be produced.
4. *Responsibility:* The person accountable for its successful completion.

5. *Resources:* Supporting labor, materials, or supplies needed.
6. *Quality:* How well the work should be done; how well any outputs should perform.
7. *Dependencies:* Dependencies are logical relationships between tasks which influence the way that a project will be undertaken. Dependencies may be internal to the project (between project activities) or external to the project (between a project activity and a business activity). Overall, there are four types of dependency:
   - Finish-to-start (the item this activity depends on must finish before this activity can start);
   - Finish-to-finish (the item this activity depends on must finish before this activity can finish);
   - Start-to-start (the item this activity depends on must start before this activity can start);
   - Start-to-finish (the item this activity depends on must start before this activity can finish).

The work breakdown structure is finalized by establishing at the lowest level of each work breakdown structure element, a management control point called a "control account."

The control account is a critical point for performance measurement to take place, for this is where the integration of scope, schedule, and resources will take place and where the project will measure its performance throughout the duration of the project and compare it to its earned value. Each control account may include one or more work packages, but each of the work packages may be associated with only one control account.

9.3 Uses of a Work Breakdown Structure

The work breakdown structure clarifies and provides necessary details for a number of project management activities. It has four uses:

1. Thought Process Tool;
2. Architectural Design Tool;
3. Planning Tool; and
4. Project Status Reporting Tool.

Thought Process Tool—As a thought process tool, it helps the project manager and the project team to visualize exactly how the work of the project can be defined and managed effectively. It would not be unusual to consider alternative ways of decomposing the work until an alternative is found with which the project manager is comfortable.

Architectural Design Tool—As architectural design tool, it provides is a picture of the work of the project and how the items of work are related to one another. It must make sense. In that context, it is a design tool.

Planning Tool—As a planning tool, the work breakdown tree structure also reflects the project scope evolution as it becomes more detailed until the work package level is reached. Planning using the work breakdown tree structure is known as "Rolling Wave Planning." It is a form of progressive elaboration planning where the work to be accomplished in the near term is planned in detail at a low level of the work breakdown tree structure. Future work is planned for work breakdown tree structure components that are at a relatively high level of the tree. The work to be performed within another one or two reporting periods in the near future is planned in detail as work is being completed during the current period. Therefore, activities can exist at various levels of detail in the project's life cycle. During early strategic planning, when information is less defined, activities may be kept at the milestone level.

Thus, as a planning tool, in the planning phase, the work breakdown structure gives the project team a detailed representation of the project as a collection of activities that must be completed in order for the project to be completed. It is at the lowest activity level of work breakdown structure that we will estimate effort, elapsed time, and resource requirements; build a schedule of when the work will be completed; and estimate deliverable dates and project completion.

Project Status Reporting Tool—As a project status reporting tool, it is used as a structure for reporting project status. The project activities are consolidated (that is, rolled up) from the bottom as lower-level activities are completed. As work is completed, activities will be completed. Completion of lower-level activities causes higher-level activities to be partially complete. Some of these higher-level activities may represent significant progress whose completion will be milestone events in the course of the project. Thus, the work breakdown structure defines milestone events that can be reported to senior management and to the customer.

10 Develop Time Management Plan

This chapter is concerned with the project management process required to implement conscious control over the amount of time spent on specific activities, especially to increase efficiency and productivity.

Time management may be aided by a range of skills, tools, and techniques used to manage time when accomplishing specific activity tasks. This set encompasses a wide scope of actions, and these include analysis of time spent, monitoring, scheduling, and prioritizing.

The constituent project management processes used during the development of the project scope, illustrated in Fig. 10.1, include the following:

1. Define Activities
2. Assess Completeness of Activities
3. Sequence Activities
4. Estimate Activity Resources
5. Estimate Activity Durations
6. Develop Project Schedule
7. Develop Schedule Control Plan

These seven constituent processes interact with each other and with the project management processes in the PDSA "Process Groups." Each aspect of executing any of these can involve effort from one or more persons, based on the needs of the project. Each aspect occurs at least once in every "process improvement" project and occurs in one or more project phases.

10.1 Define Activities

The first step in developing the project time management plan is "Define Activities." It relates to identifying the specific actions to be performed to produce the project deliverables.

A. van Aartsengel and S. Kurtoglu, *Handbook on Continuous Improvement Transformation*, DOI 10.1007/978-3-642-35901-9_10,

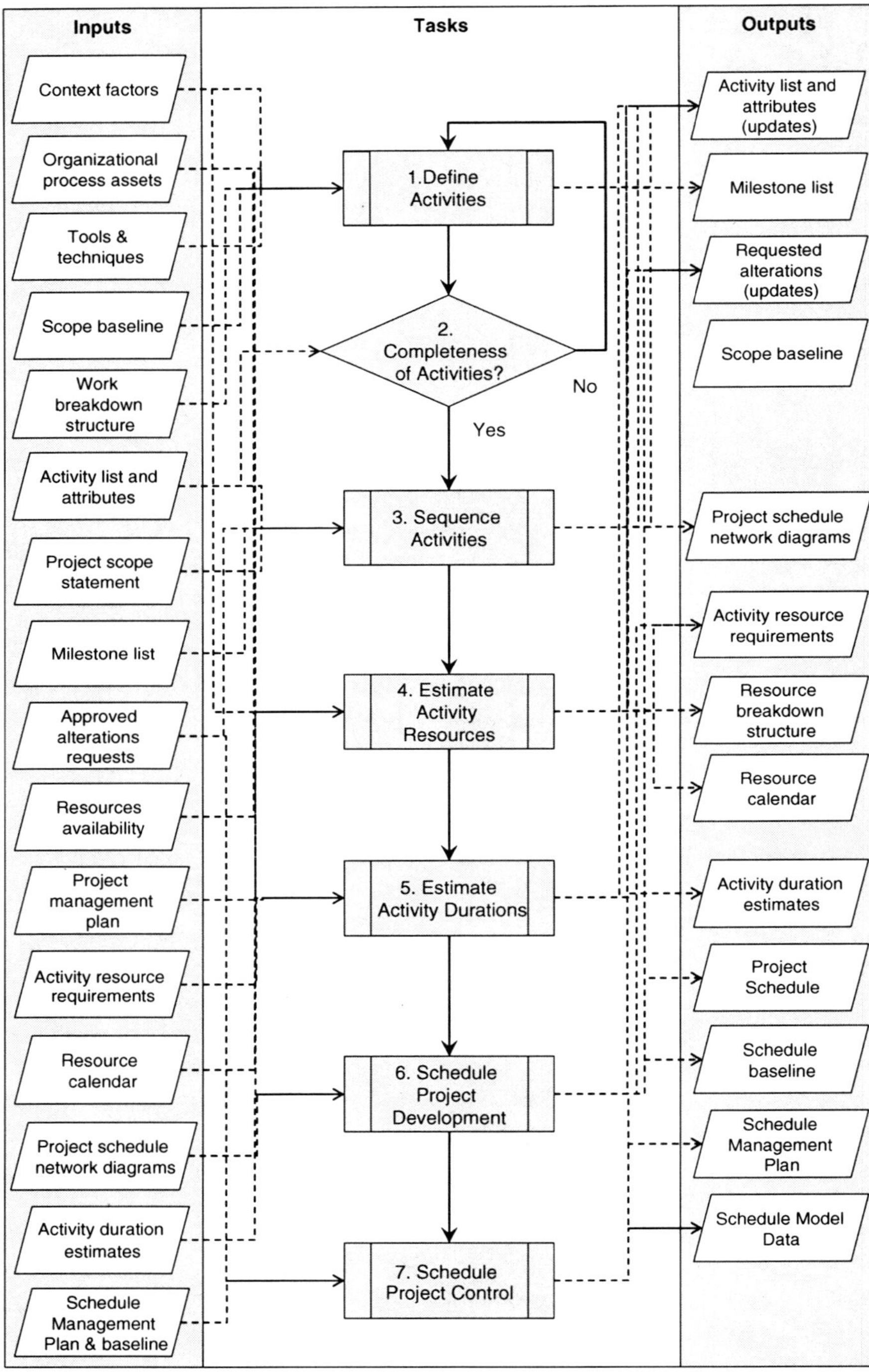

Fig. 10.1 Project time management process

The work breakdown structure identifies the work package as the deliverables at the lowest level in the work breakdown tree structure. These project work packages are typically decomposed into smaller components called activities to provide a basis for estimating, scheduling, executing, and monitoring and controlling the project work. Implicit in this process is defining and planning the schedule activities such that the project objectives will be met. This first step, then defines the final outputs as scheduled activities rather than work package deliverables.

The activity list can be developed either sequentially or concurrently, with the work breakdown tree structure in the scope baseline being used as basis for development of the final activity list. The final activity list is a comprehensive list which includes all scheduled activities that are planned to be performed on the project. The activity list includes the activity identifier and a scope of work description for each scheduled activity in sufficient detail to ensure that project team members understand what work is required to be completed.

The activity attributes (identifier, name, description, predecessor, successor, leads, lags, resources requirements, estimates of duration, date constraints, assumptions, person responsible, etc.. . .) extend the activity by identifying the multiple components associated with each activity.

10.2 Assess Completeness of Activities

The second step in developing the project time management plan is "Assess Completeness of Activities." To be considered complete, each activity at the lowest levels of decomposition must possess specific characteristics that allow meeting the planning and scheduling needs. Six criteria for assessing this completion has been introduced by Robert Wysocki (Wysocki, Effective Project Management: Traditional, Agile, Extreme, 2011). These are:

1. Activity status/completion is measurable;
2. Activity has start/end events are clearly defined;
3. Activity has a deliverable;
4. Activity time/cost is easily estimated;
5. Activity duration is within acceptable limits;
6. Work assignments are independent.

Activity has start/end events are clearly defined—Each activity should have a clearly defined start and end event. Once the start event has occurred, work can begin on the activity. The deliverable is most likely the end event that signals work is closed on the activity. For example, using the systems documentation example, the start event might be notification to the team member who will manage the creation of the systems documentation that the final acceptance tests of the system are complete. The end event would be notification to the project manager that the customer has approved the system documentation.

Activity has a deliverable—The result of completing the work that makes up the activity is the production of a deliverable. The deliverable is a visible sign that the activity is complete. This sign could be an approving manager's signature, a physical product or document, the authorization to proceed to the next activity, or some other sign of completion.

Activity time/cost is easily estimated—Each activity should have an estimated time and cost of completion. Being able to do this at the lowest level of decomposition in the work breakdown structure allows the aggregation to higher levels and estimates the total project cost and the completion date. By successively decomposing activities to finer levels of granularity, you are likely to encounter primitive activities that you have performed before. This experience at lower levels of definition gives you a stronger base on which to estimate activity cost and duration for similar activities.

Activity duration is within acceptable limits—While there is no fixed rule for the duration of an activity, we recommend that activities have duration of less than two calendar weeks. This seems to be a common practice in many organizations. Even for long projects where contractors may be responsible for major pieces of work, they will generate plans that decompose their work to activities having this activity duration. There will be exceptions when the activity defines process work, such as will occur in many manufacturing situations. There will be exceptions, especially for those activities whose work is repetitive and simple.

Activities are independent—It is important that each activity be independent. Once work has begun on the activity, it can continue reasonably without interruption and without the need of additional input or information until the activity is complete. The work effort could be contiguous, but it can be scheduled otherwise for a variety of reasons. You can choose to schedule it in parts because of resource availability, but you could have scheduled it as one continuous stream of work. Related to activity independence is the temptation to micromanage an activity. Best practices suggest that you manage an individual's work down to units of one week.

If an activity does not possess these six characteristics, it must be decomposed into small parts which satisfy these criteria. As soon as an activity possesses the six characteristics, there is no need to further decompose it. As soon as every activity in the work breakdown structure possesses these six characteristics, the work breakdown structure is defined as complete.

10.3 Sequence Activities

The third step in developing the project time management plan is "Sequence Activities." It relates to identifying and documenting relationships among activities. Sequencing is performed using network diagrams. These are schematic displays of the project's schedule activities and the logical relationships among them, also referred to as dependencies.

10.3.1 Network Diagram Formalism

The Network is a graphical diagram representing all dependencies and interrelationships amongst various activities of a project. It is also known as arrow diagram. The network is formed with the help of arrows and circles, which give technological relationships to the activities involved. The head of an arrow indicates the direction of progress in the project. Preparation of network needs thorough understanding of the network logic and basic elements of the network, which are activities, events, technological relationships, dummy activities, path, etc.. . .

In a network diagram, an activity is represented by an arrow (→). This is am element of a project having definite end. Thus, an activity is defined as a recognizable part of a work in the project that consumes time and resources for its completion. The description of the activity, if mentioned, is written below the arrow and the time required to perform the activity is written above the arrow. The tail end of an arrow shows the starting point of an activity, whereas the head shows the end or completion of the activity; that is, the starting and the end points of an activity are described by a preceding (tail) as well as the succeeding (head) events. Activities originating from a certain event cannot start until the activities terminating at the same event have been completed. Furthermore, each activity must have a definite start and a definite end.

In a network diagram, the arrow is not a vector quantity and, thus, need not be drawn on scale. It may be straight, long, short or bent, but not broken. The same activity cannot be represented by two arrows. The arrows, indicating the activities, move in one unique direction only; that is, from left to right and not vice-versa.

An activity should be independent; that is, each arrow is used to represent exactly one and only one operation (activity) of a project, indicating which activity to precede, follow, or to take place simultaneously. However, a number of arrows may be used to represent different parts of the same operation.

A project is made up of a number of activities, identified through the work breakdown structure, which are interrelated. Three possible relationships are:

1. *Concurrent relationship*—Activities that run parallel are known as "concurrent" activities.
2. *Succeeding relationship*—Activities that depend upon completion of others activities are known as "succeeding" activities.
3. *Preceding relationship*—Activities on which succeeding activities depend upon completion of others activities are known as "preceding" activities.

An activity that shows only dependence, but neither does not consume time or resources, is known as "dummy" activity. Such activities are used purely for convenience in drawing networks and are indicated by dotted line. Since dummies

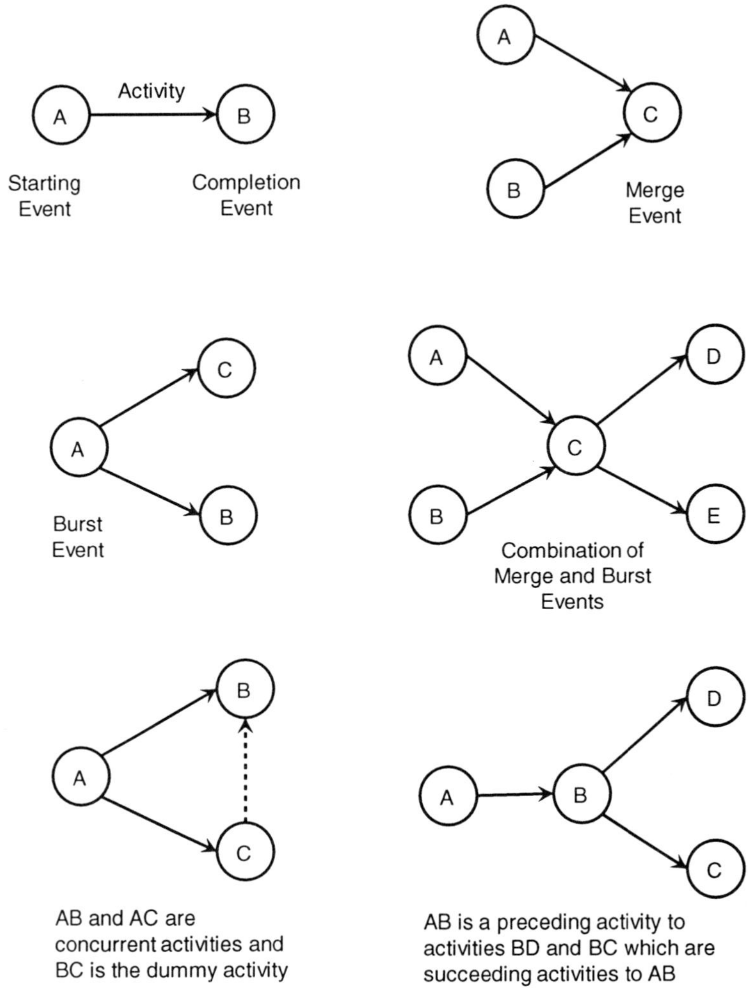

Fig. 10.2 Representation of various activities and events

are as important as zeros in arithmetic, they should be used to serve the following purposes:

1. Establish and maintain realistic and perfect logical relationships between one activity and other.
2. To maintain the uniqueness in the numbering system as every activity may have a distinct set of events (numbers or letters) by which the activity can be identified.
3. To show the relationship between events; that is, when an activity has to be completed before another can start.

In a network diagram, an event is defined as an accomplishment occurring at an instantaneous point of time; that is, a point in time and not a passage of time, but consuming no time or resources by itself. In order words, an event represents a point in time that signifies the completion of some activities and the start of some new ones. An event is represented by a circle, rectangle, or square, etc.. . ., as shown in Fig. 10.2. Thus, each arrow, representing the activity, must be bounded (circle, square, etc.. . .).

The start and end event, however, are reduced to a common event representing the completion of the first activity and the start of the next. Often, a single event may represent the joint start of more than one activity (known as burst event) or the joint completion of more than one activity (known as merge event) or both (known as merge and burst event). A rectangle represents a milestone; that is, an important event in the project. A hexagon is an indication of interface events; that is, events common to two or more than two networks.

10.3.2 Network Preparation

The rules for constructing a network diagram can be summarized as follows:

1. Each activity is represented by only one arrow in the network; that is, no single activity can be represented twice in the network or an event cannot occur twice. Thus, a path of activities cannot form a loop that returns to any previously accomplished; that is, no event can depend for its completion upon completion of a succeeding event. However, a case of one activity, being broken down into segments, each segment representing a separate task, should be differentiated.
2. No two activities can be identified by the same head and tail events. This type of situation may arise due to two or more activities being performed concurrently. In such situation, a dummy activity is introduced.
3. No event can occur until each activity preceding it has been performed to completion or has been finished.
4. An activity, succeeding an event, cannot be started until that even has occurred.
5. Each activity must terminate in an event.
6. Time flows from left to right.
7. Each activity on the network should be completed to reach the end objective.
8. All individual tasks of a project should be visualized very clearly as to show on the network. In order to develop a network with correct precedence relationships, the following three questions should be answered with respect to each activity (arrow):
   - What activities should precede (completed) the one being started; that, before this activity can start?
   - What activities should follow this activity?
   - What activities can proceed concurrently?

10.3.3 Constructing the Project Network Diagram

The activities and the activity duration are the basic building blocks needed to construct the project network diagram. This graphic picture provides the project team with two additional pieces of critical schedule information about the project:

1. The earliest time at which work can begin on every activity that makes up the project.
2. The earliest expected completion date of the project.

A graphical picture of the project is produced by spelling out the rules for constructing a network diagram summarized in the previous section to capture the precedence or parallel relationships among the project activities. It represents all activities and events in their logical sequence. The resulting network diagram is logically sequenced to be read from left to right. Every event in the network, except the start event and the end event, must have at least one event that comes before it (its immediate predecessor) and one event that comes after it (its immediate successor). The following are steps for the process of constructing a project network diagram:

1. *For each work package activity in the work breakdown structure, determine the logical relationships (also called precedence relationships) with other activities.* That is, determine which activities depend on other activities. Some dependencies are mandatory, being inherent in the nature of the work. Other dependencies are discretionary, as defined by the project team. These are preferred dependencies based on "best practices." Here, the project team should remember that an activity may depend on more than one other activity.
2. *Arrange the activities into logical sequences or paths.* Place activities that are not physically or logically dependent on each other in separate paths. Each activity in a given path must be dependent on the activity that immediately precedes it. In other words, an activity cannot begin until its preceding activities have been completed.
3. *Review each path to ensure that it makes sense.* The activities in a given path build on each other. All paths come together at the end of the project. No activity can lead to a dead end. If the project team discovers that it has overlooked an activity that should be part of the project, it must go back and add it to the work breakdown structure.

In constructing the project network diagram, the project team should keep in mind that not every single item on the work breakdown structure needs to be on the network diagram; only activities with dependence or relationship need to be shown on the network diagram. The network diagram represents activities that must be performed to complete the project. Every activity on the network diagram must be completed in order for the project to finish.

10.4 Estimate Activity Resources

The fourth step in developing the project time management plan is "Estimate Activity Resources." It involves determining the type and quantities of assets, such as people, material, equipment, physical facilities, inventories, and supplies which have limited availabilities, can be scheduled, or can be leased from an outside party, and which are required to perform each scheduled activity. Some resources are fixed; others are variable only in the long term. In any case, they are central to the scheduling of project activities and the orderly completion of the project. For "process improvement" in systems development projects, people are the major resource. Another valuable resource for systems projects is the availability of computer processing time (mostly for testing purposes), which can present significant problems to the project manager with regard to project scheduling.

The tools and techniques used to estimate the activity resource requirements are based on expert judgment, alternatives analysis, enterprise business recorded estimating data, or bottom up estimating. These requirements can then be aggregated to determine the estimated resources for each work package. The amount of detail and the level of specificity of the resource requirement descriptions can vary by application area. The resource requirements documentation for each scheduled activity can include the basis of estimate for each resource, as well as the assumptions that were made in determining which types of resources are applied, their availability, and what quantity are used.

10.5 Estimate Activity Durations

The fifth step in developing the project time management plan is "Estimate Activity Durations." Estimating activity durations requires that each project activity be examined to determine how much time is needed to complete it.

The project manager could think of the duration of an activity as the elapsed time expressed in convenient units of time such as hours, days, months etc... The duration of an activity is influenced by the amount of resources scheduled to work on it. The work effort is labor required to complete an activity. That labor can be consecutive or nonconsecutive hours.

The duration of an activity may vary randomly. Because the factors that will be operative when work is in progress on an activity cannot be known, how long it will take to complete the activity cannot be know exactly. There will, of course, be varying estimates with varying precision for each activity.

Consequently, one of the project manager or project leader goals in estimating activity duration is to define the activity to a level of granularity so that the activity duration estimates have a narrow variance; that is, the estimate is as good as it can be used at the planning stages of the "process improvement" project. As "process improvement" project work is carried out, the project manager or project leader will be able to improve the earlier estimates of activities scheduled later in the project.

The estimated time assigned to activities should be realistic rather than desirable; that is, it should be acceptable to the project team members responsible to carry out the project tasks. The time required for the performance of an activity under normal availability of resources depends upon the size and nature of the "process improvement" project. Its estimation takes into account the following assumptions:

1. The activity is sufficiently well defined.
2. The estimate of activity duration is independent of any influence from other activities.
3. Resources required to carry out the activity are available.
4. The estimate of activity duration includes only normal delays and interruption due to breakdowns, absenteeism, etc.. . . The effects of unforeseen events such as hazard, strikes, etc.. . ., are not considered.
5. The time estimate is single time for the activities of repetitive nature, with the assumption that the project manager has a fair idea of the activity durations from their earlier experience.
6. For new activities, it may be difficult to establish one time estimate with reasonable accuracy and, thus, multiple, time estimates, based on PERT concepts of Optimistic, Pessimistic, and Most Likely Time, described in the next section, should be used.

The activity time estimates should be carried out by the persons most familiar with, or actually responsible for the performance of the activity, and/or who have a sound knowledge of the process involved in the completion of the project. There are several factors that can affect the actual activity duration:

1. Varying skill levels
2. Unexpected events
3. Efficiency of work time
4. Mistakes and misunderstandings
5. Common cause variation

Varying skill levels—A higher- or lower-skilled person assigned to the activity, may affect the actual duration to vary from planned duration. These varying skill levels can be both a help and a hindrance to completing the activity work.

Unexpected events—Random acts of nature, vendor delays, incorrect shipments of materials, traffic jams, power failures, and sabotage are but a few of the possibilities.

Efficiency of work time—Every time a worker is interrupted, it takes more time to get up to the level of productivity prior to the time of the interruption. It is difficult to control the frequency or time of interruptions, although their occurrences are highly probable.

Mistakes and misunderstandings—Despite all of efforts to be complete and clearance in describing the work to be performed, mistakes and misunderstanding are facts of live that may occur a few times. This will take its toll in rework or scrapping semi-completed work.

Common causes of variation—Apart from all of these factors that can influence activity duration, the reality is that durations will vary for no reason other than the statistical variation that arises because the duration is in fact a random variable.

Several tools and techniques can be used to determine estimates of activity durations, four of which are suitable for initial planning. These are:

1. Similarity technique
2. Historical data technique
3. Expert judgment technique
4. Delphi technique

Similarity technique—The similarity technique extrapolates data from similar activities successfully completed in other projects to determine estimates for the present activity duration. In most cases, using the approximations from those activities provides estimates that are good enough.

Historical data technique—Every good "process improvement" project should contain a project notebook that records the estimated and actual activity duration. This historical record can be used on other projects. The recorded data becomes enterprise business knowledge base for estimating activity durations. This technique differs from the previous technique in that it uses a record, rather than depending on memory.

An enterprise business can built an extensive database of activity duration history that records not only estimated and actual duration but also the characteristics of activities, the skill set of the people working on these activities, and other activity attributes found useful. When an activity duration estimate is needed, a query to the database with appropriate attributes and, with some rather sophisticated regression models, could provide an estimate the activity duration.

Expert judgment—Activity duration can be estimated directly by experts with relevant experience in performing similar activities. Such experts should be identified by the project manager and invited to consider all aspects of the project activity, suggesting possible duration estimates based on their previous experience and areas of expertise. The experts' bias should be taken into account in this process.

Applying the Delphi Technique—The Delphi Technique, described in a previous section, is an essential technique that refers to an information gathering technique in which the opinions of those whose opinions are most valuable, traditionally industry experts, is solicited, with the ultimate hope and goal of attaining a consensus.

The Delphi Technique is a group technique that extracts and summarizes the knowledge of the group to arrive at an estimate. Typically, the polling of these industry experts is done on an anonymous basis, in hopes of attaining opinions that are unfettered by fears or a possibility of identification.

Using the Delphi Technique, after the group is briefed on the project and the nature of the activities, each individual in the group is asked, which is typically, but not always, presented to the expert by a third-party facilitator, to make his or her best guess of the activity durations. The responses from all experts are typically combined in the form of an overall summary, which is then provided to the experts for a review and for the opportunity to make further comments. This process typically results in consensus within a number of rounds, and this technique typically helps minimize bias, and minimizes the possibility that any one person can have too much influence on the outcomes. Even though the technique seems rather simplistic, it has been shown to be effective in the absence of expert advice.

11 Develop Project Schedule Plan

This chapter is concerned with the sixth step in developing the project time management plan: "Develop Project Schedule." It is the project management process necessary for analyzing activity sequences, durations, resource requirements, and schedule constraints to create the project schedule.

A schedule is the conversion of a project action plan into an operating timetable. As such, it serves as the basis for monitoring and controlling project activity and, taken together with the plan and budget, is probably the major tool for the management of projects.

In a project environment, the scheduling function is more important than it would be in an ongoing operation because projects lack the continuity of day-to-day operations and often present much more complex problems of coordination. The "Develop Project Schedule" process can require that duration estimates and resource estimates are reviewed and revised to create an approved project schedule that can serve as a baseline against which progress can be tracked. Indeed, project scheduling is so important that a detailed schedule is sometimes a customer-specified requirement. This process continues throughout the project as work progresses, the project management plan changes, and anticipated risk events occur or disappear as new risks are identified.

Not all project activities need to be scheduled at the same level of detail. In fact, there may be several schedules (e.g. in a production industry environment, the master schedule, the development and testing schedule, the assembly schedule). These schedules are typically based on the previously determined action plan and/or work breakdown structure, and it is good practice to create a schedule for each major task level in the work breakdown structure that will cover the work packages. It is rarely necessary, however, to list all work packages. One can focus mainly on those that need to be monitored for maintaining adequate control over the project. Such packages are usually difficult, expensive, or have a relatively short time frame for their accomplishment.

A. van Aartsengel and S. Kurtoglu, *Handbook on Continuous Improvement Transformation*,
DOI 10.1007/978-3-642-35901-9_11,

11.1 Basic Approach to Scheduling

The basic approach of all scheduling techniques is to form a network of activity and event relationships that graphically portrays the sequential relations between the tasks in a project. Tasks that must precede or follow other tasks are then clearly identified, in time as well as function. Such a network, previously described on sequencing activities, is a powerful tool for planning and controlling a project, and has the following benefits:

1. It is a consistent framework for planning, scheduling, monitoring, and controlling the project.
2. It illustrates the interdependence of all tasks, work packages, and work elements.
3. It denotes the times when specific individuals and resources must be available for work on a given task.
4. It aids in ensuring that the proper communications take place between departments and functions.
5. It determines an expected project completion date.
6. It identifies so-called critical activities that, if delayed, will delay the project completion time.
7. It also identifies activities with slack that can be delayed for specified periods without penalty, or from which resources may be temporarily borrowed without harm.
8. It determines the dates on which tasks may be started or must be started if the project is to stay on schedule.
9. It illustrates which tasks must be coordinated to avoid resource or timing conflicts.
10. It also illustrates which tasks may be run, or must be run, in parallel to achieve the predetermined project completion date.
11. It relieves some interpersonal conflict by clearly showing task dependencies.
12. It may, depending on the information used, allow an estimate of the probability of project completion by various dates, or the date corresponding to a particular a priori probability.

11.2 Update the Project Network Diagram

At this point in the project management life cycle, the project team has identified the set of activities in the project as output from the work breakdown structure building exercise; it has also constructed a preliminary project network diagram as output from the activity sequencing exercise. The next task for the project team is to update the project network diagram with activity durations, or schedule the project activities, and to subsequently analyze the network diagram.

11.2.1 Showing Times on Arrow Networks

The times written on arrow networks usually refer to the events rather than directly to the activities. Project managers, however, need to know the times when each activity should start and finish. Although these times are easily derived from an arrow network, they cannot easily be shown owing to lack of space on the arrow diagram. This is demonstrated in Fig. 11.1, using a fragment from a larger network.

Version 1 in Fig. 11.1 shows an arrow network notation according to the early British Standard BS 4335:1987. This notation, although favored by some writers, is not well suited to freehand sketching (the principal remaining role for arrow networks) because:

1. Relatively large diameter event circles are required, which reduce the amount of network detail that can be drawn on a sheet of paper.
2. Each event must be drawn very carefully, taking time that is not usually available in a brainstorming planning session.

Version 2 in Fig. 11.1 is a form of notation that allows rapid freehand sketching and economy of space on a sheet or roll of paper.

Now consider activity A-B in Fig. 11.1 (using either version 1 or 2). The time analysis data for this activity are not all immediately apparent from the network. Certainly its earliest possible start is 30 units of time, the earliest completion time for event 25 units of time. The latest permissible start for this activity is the latest permissible time for event B minus the activity duration, which is 90 minus 10, giving day 80 (not the 75 units of time shown as the latest permissible completion for event A). This arises because other activities entering and leaving events A and B affect the times for those events independently of activity A-B.

The "missing" time analysis data for any activity can be added if desired (and if space and drafting time allow) using version 3.

The most common approaches to project scheduling are described using network techniques such as the Program Evaluation and Review Technique (PERT) and the Gantt Charts. The following sub-section outlines the PERT approach to project scheduling.

11.2.2 The Program Evaluation and Review Technique (PERT)

The Program Evaluation and Review Technique (PERT), originally developed by the U.S. Navy in cooperation with Booz-Allen Hamilton and the Lockheed Corporation for the Polaris missile/submarine project in 1958, is considered a project management classic. Its objectives included managing project schedule by establishing the shortest development schedule, monitoring project progress, and funding or applying necessary resources to maintain the schedule. Despite its age (relative to other project risk techniques), PERT has worn the test of time well.

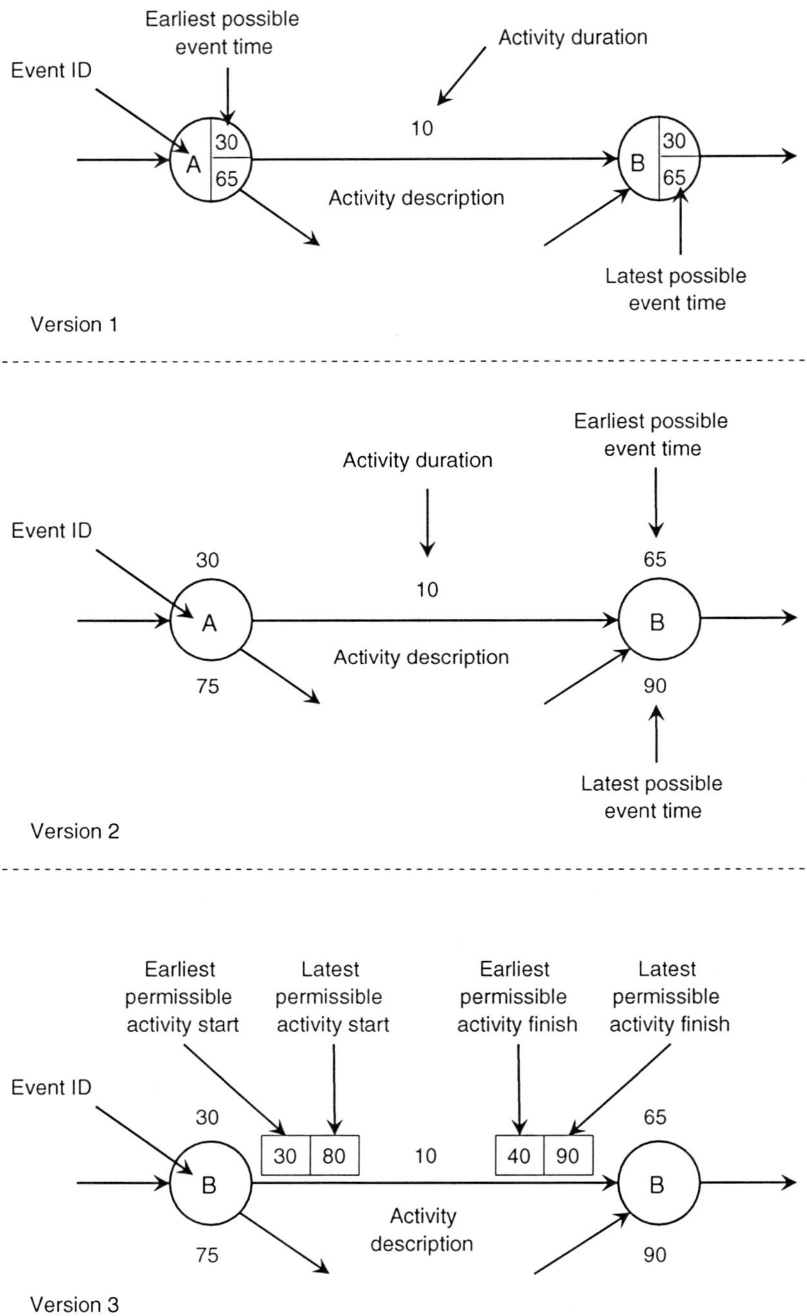

Fig. 11.1 Three different methods for showing times on arrow networks

11.2.2.1 Analyzing the Project Network Diagram

The following are steps of the calculations algorithm to be used for analyzing the project network diagram.

1. Determine the amount of time likely to be consumed by each activity.
2. Make a forward pass through the diagram, calculating for each event (node) the earliest time (T_E) at which all of the activities entering the node will be completed. To find T_E, look at all of the activities which enter a node. The earliest time T_E is the latest of the arrival times of entering arcs, namely:

$$T_E(Node) = \max_{Arc \in \{entering\ arcs\}} \{T_E(N.a.t.o.a) + Arc\ duration\}$$

   Where $N.a.t.o.a$ is the node at tail of arc. By definition, T_E of the starting node is zero.
3. Make a backward pass through the diagram, calculating for each event (node) the latest allowable event time (T_L) at which the outflow activities can begin without causing a late arrival at the next node for one of those activities. To find T_L, look at all of the activities which exit a node. The latest time T_L is the earliest of the leaving times for the exiting arcs, namely:

$$T_L(Node) = \min_{Arc \in \{arcs\}} \{T_E(N.a.h.o.a) - Arc\ duration\}$$

   Where $N.a.h.o.a$ is the node at head of arc. By definition, T_L of the ending arc is equal to its T_E.
4. Calculate the node slack time (S_N) for each node (event). This is the amount of time by which an event could be adjusted later than its earliest time T_E without causing problems downstream.

$$S_N(Node) = T_L(Node) - T_E(Node)$$

5. Calculate the total arc slack time (S_A) for each arc (activity). This is the amount of time by which an activity could be adjusted later than the earliest time T_E of the node at its tail without causing problems later.

$$S_A(Arc) = T_L(N.a.h.o.a) - T_E(N.a.t.o.a) - Arc\ duration$$

6. Calculate the critical and sub-critical paths. The critical path connects the nodes at which $S_N(Node) = 0$ via the arcs at which $S_N(Node) = 0$. It should be no

surprise that the critical path connects the nodes and arcs which have no slack. If there is slack, then the activity does not need to be carried out on time, which is exactly the opposite definition of the critical path!

7. Revise the project network diagram to meet the project objectives in terms of time-cost trade-off.

Step 1

The purpose of the first step of this calculation algorithm is to determine the amount of time likely to be consumed by each activity, if reasonable estimates have not yet been established. These activities duration estimates are of great importance as they constitute the basis for the subsequent analysis of the project network. The successful completion of the project on schedule is very much dependent on this basic data, from which the activity time estimates are established.

New "process improvement" projects, not attempted earlier, are associated with an element of uncertainty of the time required for their accomplishment due to the absence of historic data or previous experience. This uncertainty for time estimates would result in an incorrect estimation of the duration of the project as a whole. Such an uncertainty factor can be dealt with using a statistical method. When it is difficult to estimate the elapse time for an activity precisely, its likelihood of achievement is expressed in three time estimates rather than a positive assurance. These are:

1. *Optimistic Time* (O): It is the minimum possible time required to accomplish an activity under ideal conditions; i.e., assuming everything proceeds better than is normally expected
2. *Pessimistic Time* (P): It is the maximum possible time required to accomplish a task, assuming everything goes wrong (but excluding major catastrophes).
3. *Most Likely Time* (M): It is the best estimate of the time required to accomplish a task, assuming everything proceeds as normal. Even if the work is repeated under identical conditions, the time required would be almost the same.

The range, specified by the optimistic and pessimistic estimates (O and P, respectively), should essentially represent every possible estimate of the duration of the activity. The most likely estimate need not coincide with the midpoint $(O + P)/2$, and may occur to its left or right. Because of these properties, it is intuitively justified to assume that the duration of each activity may follow a β-distribution with its unimodel point occurring at M and its end points at O and P. Therefore, the expressions for the expected time and variance can be developed as following:

1. *Expected Time*: the best estimate of the time required to accomplish a task, accounting for the fact that things don't always proceed as normal (the implication being that the expected time is the average time the task would require if the task were repeated on a number of occasions over an extended period of time).

$$Expected\ Time = \frac{O + 4M + P}{6}$$

This expected time is more close to M and it is the value that is used in all the project network calculations.

The range (O, P) is assumed to enclose about 6 standard deviations of the distribution, since nearly 90 % or more of any probability density function lies within 3 standard deviations. Thus the variance associated with the expected time is given as:

$$Variance = \left(\frac{P - O}{6}\right)^2$$

As indicated already, the duration of an activity may vary randomly. Because the factors that will be operative when work is in progress on an activity cannot be known, how long it will take to complete the activity cannot be know exactly. There will, of course, be varying estimates with varying precision for each activity. There will likely be uncertainty and risk. It is, therefore, natural for the project manager to know the risks involved ad extent of uncertainty associated with the project.

Based on the spread of $P - O$, the activities may be seen as deterministic if $P - O$ spread is small, and the activities may be seen as variable if $P - O$ spread is fairly large. As discussed above, three-time estimates $O, P,$ *and* M are used for variable activities.

For the calculation of the probability of project completion in a given time, the following points should be kept in mind:

In the majority of situations, the data on the probability of occurrence of an activity against its durations will conform to a β-distribution. Using this distribution, the probability of completing an activity within its expected duration is 50 %.

The expected time of the activity is located at one-third of the distance between M midrange $(O + P)/2$ away from the most likely time.

The variance $((P - O)/6)^2$ helps in determining the probability of achieving the target completion date of the project or any stage of the project.

Although the expected time of each activity independently has a β-distribution, yet the completion time T of the project has a normal distribution with mean equal to the sum of expected times of activities along the project critical path and with variance equal to the sum of variance associated with expected times of activities.

Step 2

The purpose of the second step of the calculation algorithm is to establish the date at which each activity should start and end to maintain:

1. The earliest starting date, which implies the earliest time of an activity of the tail event.
2. The latest starting date, which is the latest finishing time minus the activity duration.

The earliest expected time of an event is the earliest time by which the event can be completed. Since an event cannot occur earlier than this point of time, it is termed as Earliest Time Event. Once the elapse time for each activity is worked out on the project network, the Earliest Expected Time for each event (i.e., the earliest time when an event may be expected to occur) is calculated by forward pass computation from the beginning event by summing up the activities times through the longer path on the project network. When two or more activities constrain a single event, the expected time is calculated along each path and the longest time is chosen as the expected time of the given event.

Initial project event can be started arbitrarily with an occurrence time of zero. Based on zero starting time, the earliest expected time for the next event is computed by adding duration of the activity path leading to this event. In other words, head and tail event times are treated as the boundaries between which the activities can move. But when two or more activities are joining a particular event, then the expected time is calculated by taking the higher or largest of all the time values for the merging activities leading to an event.

Step 3

The purpose of the third step of the calculation algorithm is to establish the date at which each activity should finish and end to maintain:

1. The earliest finishing date, which equals the earliest event time of tail event plus the duration of the activity emanating from the tail activity.
2. The latest finishing date, which is the latest event time of the head event.

The latest allowable event time is the latest time when an event can occur without creating any expected delay in the completion of the end event. The latest allowable event time for the end event is set equal to the expected time of the end event or equal to any pre-determined, specified, or directed time. The significance of the latest allowable event time lies in the fact that if any of the events is delayed beyond the permissible latest allowable event time values, it is bound to affect the project completion in the desired period. Thus, the latest allowable event time acts as a warning signal to achieve the event as expected. Otherwise, it would lead to slippages in terms of time, cost, as well as performance. The latest allowable event time for any given event is calculated by backward pass computations from the end event.

Generally, value of end event expected time is set equal to the latest allowable event time. The values of the latest allowable event time are worked out by subtraction process; that is, subtract from the latest allowable event time of each immediately following event the elapse time leading to that event from its successor event; that is, the latest allowable event time of an event is equal to the latest allowable event time of its successor event minus the duration of activity joining the two events. If two alternatives paths back to the event, different results would be obtained. The smallest out of the different results' values is chosen. This value also represents the longest backward time path from the end event.

Step 4

The purpose of the fourth step of the calculation algorithm is to establish the slack or float at which an event can be delayed beyond its expected time without affecting the latest allowable event time of the final event. The float or slack is a measure of the excess time and resources available to complete a task. It is the amount of time that a project task can be delayed without causing a delay in any subsequent tasks (free float) or the whole project (total float). The slack of an event may be zero, positive or negative.

Zero slack means that exactly enough time has been allowed for the activity and spare time is not available; that is, the activity work would be on time.

A positive slack means that there is enough time needed to finish the activity work. If the slack of the end event is positive; that is, the directed time is later than the computed expected time for the end event, the project would be ahead of schedule. A relatively large positive slack identifies the network path which will allow the reduction of resources in its share without causing any delay in the completion of the project as a whole. These spare resources can be transferred from such paths to other paths requiring resources. This results in reduction of the total duration of the overall project.

A negative slack means that sufficient time has not been allowed to accomplish an event and it indicates "apparent trouble." Where negative slack occurs, attention should be focused to these areas, most warranting the action to reduce the time required to complete the activity work.

Following the determination of the critical path, the slacks for the noncritical activities must be computed. Naturally, a critical activity must have zero slack. In fact, this is the main reason why it is critical.

Step 5

The purpose of the fifth step of the calculation algorithm is to establish the total slack time (S_A) for each activity. This is the amount of time by which an activity can be delayed without affecting the project schedule. It represents the maximum leeway available to an activity when all preceding activities occur at the earliest possible time and all succeeding activities occur at the latest possible time.

Step 6

The purpose of the sixth step of the calculation algorithm is to establish the critical path of the project network in order to estimate the total project duration and to assign starting and finishing times to all activities involved in the project. The critical path connects the events at which the slack is zero via the activities at total slack is zero.

The application of the PERT should ultimately yield a schedule that specifies the start and completion dates of each activity. The project network diagram represents the first step towards achieving that goal. Due to inter relation among the various activities, the determination of the start and completion time requires special calculations. These calculations are performed directly on the project network diagram using simple arithmetic. The end result is to classify the activities of the project as critical or non-critical. An activity is said to be critical if a delay in its start will cause a delay in the completion date of the entire project. A critical activity has total float equal to zero. An activity with zero float is not necessarily on the critical path since its path may not be the longest. A non-critical activity is such that the time between its earliest start and its latest completion dates, as allowed by the project plan, is longer than its actual duration. In this case, the non-critical activity is said to have slack or float time.

Floats or slacks suggest the leeway available and its knowledge provides the flexibility from the scheduling perspective. The project network, therefore, may contain a path (or paths) which have no leeway or have zero slack. Such a path (or paths) is (resp. are) called critical path(s) which have no leeway or have zero slack. The critical path has the least algebraic slack and it determines the minimum or earliest time required for completion of the overall project as the sequence of the activities on this path imparts the most rigorous time constrain on attainment of the end event. It represents the longest possible continuous pathway taken from the initial event to the terminal event. It determines the total calendar time required for the project; and, therefore, any time delays along the critical path will delay the reaching of the terminal event by at least the same amount. If any time is to be saved on the overall project, it should be saved on the activities, which fall on the critical path.

Since all the project network paths, other than the critical path, are shorter, they have some amount of slack or free time. The identification of the critical path from the slack or non-critical paths would indicate possibilities of diverging resources from activities on non-critical paths to activities on the critical path; thereby, leading to reduction in the total project duration. As such, the project can be brought to completion within a desired schedule of completion by the application of a given measure of resources.

Step 7

The purpose of the seventh step of the calculation algorithm is to allow updating of the project plan. A project may not follow exactly the time schedule developed for it when it is actually executed. There are abound to be unexpected delays and difficulties in terms of delay in supply of materials, non-availability of some machines and/or breakdown of machines, non-availability of skilled manpower, natural calamity, force majeure event, etc.... In such case, it may be necessary to review the progress of the project network planning and scheduling. Such review of the progress helps in taking stock of the progress that has been made and making necessary change in the initial schedule in terms of time and resources required by incomplete activities in the project. Updating the project network plan can be done in two ways:

1. Use the revised time estimates of incomplete activities and calculate from the initial event the earliest and latest completion time of each event in the usual manner to establish the project completion time.
2. Change the completed work to zero duration and represent all the activities already finished by an arrow called elapsed time arrow.

11.2.2.2 Example of Application of PERT Calculations Algorithm

Consider the assembly of a product using a just-in-time manufacturing strategy, as shown in the PERT network diagram in Fig. 11.2, and assume that certain steps in

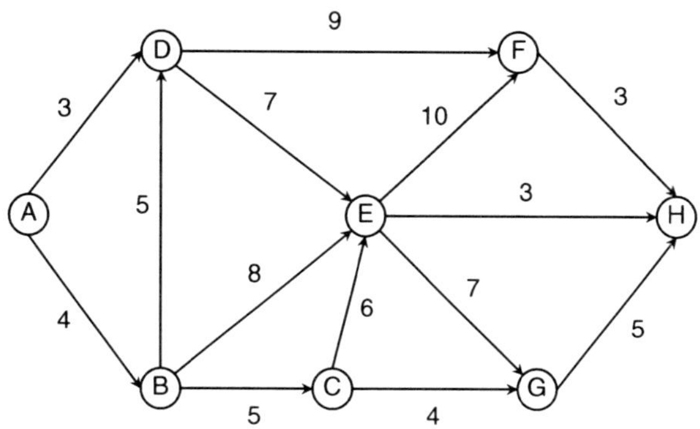

Fig. 11.2 Example of PERT diagram

the assembly line are to install the product's core component, install the body shell, and install the wheels with arbitrary interstitial steps. Perhaps the assembly of the core component (Activity A-B) happens at the same time as the pre-assembly of the roof trusses (Activity A-D). However, finalizing of the body shell (Activity D-E), cannot begin until both A-D and B-D (assembly of the product frame), are done. Of course B-D cannot start until the core component has been assembled (Activity A-B). All of this precedence and parallelism information is neatly captured in the PERT diagram.

The arrows or directed arcs in Fig. 11.2 are labeled with numbers, which show the amount of time (in consistent units of time) that each activity is expected to take. Let's find the critical path for the PERT diagram. Note that there is a predefined order in which the earliest time T_E can be calculated. For example, the earliest time $T_E(D)$ of node D cannot be found until the earliest time $T_E(B)$ of node B is known.

The starting node in Fig. 11.2 is node A, and by definition the earliest time $T_E(A)$ of the starting node is 0. To calculate earliest time $T_E(Node)$ at a node, we need to know the earliest time $T_E(N.a.t.o.a)$ of the node at the tail of every entering arc, therefore we can next only calculate the earliest time $T_E(B)$ of node B. This is simple since there is only one inflow arc, from node A. Thus:

$$T_E(B) = T_E(A) + duration\ of\ (Activity\ A - B)$$
$$T_E(B) = 0 + 4 = 4$$

The complete set of earliest time calculations follows:

Node A:

$$T_E(A) = Starting\ node = 0$$

Node B:

$$T_E(B) = T_E(A) + duration\ of\ (Activity\ A - B)$$
$$T_E(B) = 0 + 4 = 4$$

Node D:

$$T_E(D) = max\begin{Bmatrix} T_E(A) + duration\ of\ (Activity\ A - D), \\ T_E(B) + duration\ of\ (Activity\ B - D) \end{Bmatrix}$$
$$T_E(D) = max\begin{Bmatrix} 0 + 4, \\ 4 + 5 \end{Bmatrix} = 9$$

Node C:

$$T_E(C) = T_E(B) + duration\ of\ (Activity\ B - C)$$
$$T_E(C) = 4 + 5 = 9$$

Node E:

$$T_E(E) = max\begin{Bmatrix} T_E(D) + duration\ of\ (Activity\ D - E), \\ T_E(B) + duration\ of\ (Activity\ B - E), \\ T_E(C) + duration\ of\ (Activity\ C - E) \end{Bmatrix}$$
$$T_E(E) = max\begin{Bmatrix} 9 + 7, \\ 4 + 8, \\ 9 + 6 \end{Bmatrix} = 16$$

Node F:

$$T_E(F) = max\begin{Bmatrix} T_E(D) + duration\ of\ (Activity\ D - F), \\ T_E(E) + duration\ of\ (Activity\ E - F) \end{Bmatrix}$$
$$T_E(F) = max\begin{Bmatrix} 9 + 9, \\ 16 + 10 \end{Bmatrix} = 26$$

Node G:

$$T_E(G) = max\begin{Bmatrix} T_E(E) + duration\ of\ (Activity\ E - G), \\ T_E(C) + duration\ of\ (Activity\ C - G) \end{Bmatrix}$$
$$T_E(G) = max\begin{Bmatrix} 16 + 7, \\ 9 + 4 \end{Bmatrix} = 23$$

Node H:

$$T_E(H) = max\begin{Bmatrix} T_E(F) + duration\ of\ (Activity\ F - H), \\ T_E(E) + duration\ of\ (Activity\ E - H), \\ T_E(G) + duration\ of\ (Activity\ G - H) \end{Bmatrix}$$

$$T_E(H) = max\begin{Bmatrix} 26 + 3, \\ 16 + 3, \\ 23 + 5 \end{Bmatrix} = 29$$

The shortest time within which the project can be completed is now known. It is the same as the earliest time of the ending node, node H, i.e. 29 units of time. But we still need to complete the remaining 4 steps of the algorithm to positively identify the critical path.

The backwards pass in the second step of the PERL calculations algorithm begins with the ending node H. By definition, the latest allowable event time $T_L(H)$ of the ending node is equal to its earliest time $T_E(H)$; that is, $T_L(H) = T_E(H) = 29$. The complete set of latest allowable event time calculations follows:

Node H:

$$T_L(H) = Ending\ node = 29$$

Node F:

$$T_L(F) = T_L(H) - duration\ of\ (Activity\ F - H)$$
$$T_L(F) = 29 - 3 = 26$$

Node G:

$$T_L(G) = T_L(H) - duration\ of\ (Activity\ G - H)$$
$$T_L(G) = 29 - 5 = 24$$

Node E:

$$T_L(E) = min\begin{Bmatrix} T_L(F) - duration\ of\ (Activity\ E - F), \\ T_L(H) - duration\ of\ (Activity\ E - H), \\ T_L(G) - duration\ of\ (Activity\ E - G) \end{Bmatrix}$$

$$T_L(E) = min\begin{Bmatrix} 26 - 10, \\ 29 - 3, \\ 24 - 7 \end{Bmatrix} = 16$$

Node C:

$$T_L(C) = min\left\{ \begin{array}{l} T_L(E) - duration\ of\ (Activity\ C - E), \\ T_L(G) - duration\ of\ (Activity\ C - G) \end{array} \right\}$$

$$T_L(C) = min\left\{ \begin{array}{c} 16 - 6, \\ 24 - 4 \end{array} \right\} = 10$$

Node D:

$$T_L(D) = min\left\{ \begin{array}{l} T_L(F) - duration\ of\ (Activity\ D - F), \\ T_L(E) - duration\ of\ (Activity\ D - E) \end{array} \right\}$$

$$T_L(D) = min\left\{ \begin{array}{c} 26 - 9, \\ 16 - 7 \end{array} \right\} = 9$$

Node B:

$$T_L(B) = min\left\{ \begin{array}{l} T_L(D) - duration\ of\ (Activity\ B - D), \\ T_L(E) - duration\ of\ (Activity\ B - E), \\ T_L(C) - duration\ of\ (Activity\ B - C) \end{array} \right\}$$

$$T_L(B) = min\left\{ \begin{array}{c} 9 - 5, \\ 16 - 8, \\ 10 - 5 \end{array} \right\} = 4$$

Node A:

$$T_L(A) = min\left\{ \begin{array}{l} T_L(D) - duration\ of\ (Activity\ A - D), \\ T_L(B) - duration\ of\ (Activity\ A - B) \end{array} \right\}$$

$$T_L(A) = min\left\{ \begin{array}{c} 9 - 3, \\ 4 - 4 \end{array} \right\} = 0$$

By performing step 4 of the PERL calculations algorithm, the nodes slack time for each node event are found to be:

Node	A	B	C	D	E	F	G	H
$S_N(Node)$	0	0	1	0	0	0	1	0

In much the same vein, by performing step 5 of the PERL calculations algorithm, the arcs slack time for each arc activity are found to be:

Arc	AB	AD	BC	BD	BE	CE	CG	DE	DF	EF	EG	EH	FH	GH
$S_A(Arc)$	0	6	1	0	4	1	11	0	8	0	1	10	0	1

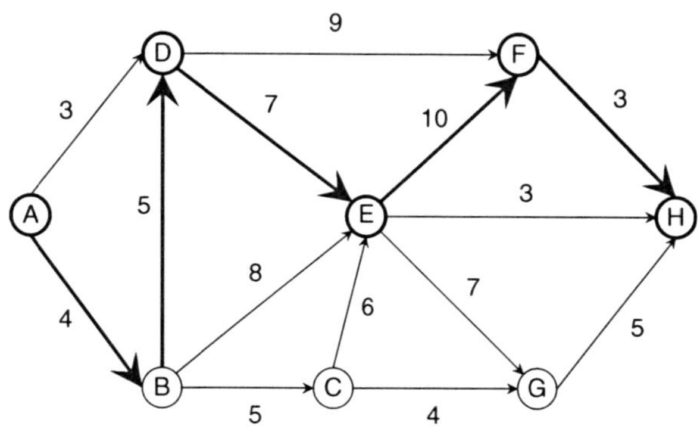

Fig. 11.3 The critical path on a PERT diagram

Finally, by performing step 6 of the PERL calculations algorithm, we find the critical path by linking the nodes having no slack via the arcs having no slack. Figure 11.3 shows the critical path for the PERT diagram example considered above.

The nodes and arcs having no slack are shown in boldface. Notice that the critical path through the PERT diagram is actually the longest path through the network. If you only needed the critical path and its length, it is easy to convert Dijkstra's shortest route algorithm to a longest route algorithm to find it, as would be done in operational research applications.

Sometimes a situation arises in which one activity must precede two different events. How can this happen when a single arc can terminate only at a single event node? The solution lies in the use of dummy arcs which have duration equal to zero.

11.2.2.3 Classification of a Project Network Scheduling

Having created the initial project network diagram, one of the following two situations will be present:

1. The initial project completion date meets the requested completion date. Usually this is not the case, but it does sometimes happen.
2. The more likely situation is that the initial project completion date is later than the requested completion date. In other words, the project manager has to find a way to compress some time out of the project schedule.

The project team eventually needs to address two considerations: the project completion date and the resource availability under the revised project schedule. In other word, the project team needs to classify the project as either time constrained or resource constrained. Project managers need to consult their priority matrix to determine which case fits their "process improvement" project. One simple test to determine if the project is time or resource constrained is to ask the question: "If the critical path is delayed, will resources be added to get back on schedule?" If the answer is yes, the project is assumed to be time constrained; otherwise the project is assumed to be resource constrained.

A time-constrained project is one that must be completed by an imposed date. If required, resources can be added to ensure the project is completed by a specific date. Although time is the critical factor, resource usage should be no more than is necessary and sufficient.

A resource-constrained project is one that assumes the level of resources available cannot be exceeded. If the resources are inadequate, it will be acceptable to delay the project, but as little as possible.

In scheduling terms, time constrained means time (project duration) is fixed and resources are flexible, while resource constrained means resources are fixed and time is flexible.

11.2.2.4 Project Completion Date: Compressing Network Schedule

Almost without exception, the initial project calculations will result in a project completion date beyond the required completion date. The project manager has fewer options for accelerating project completion when additional resources are either not available or the budget is severely constrained. This is especially true once the schedule has been established.

To address this problem, the project team should analyze the network diagram to identify areas where it can compress the project duration. The project team should look for pairs of activities that allow a conversion of activities that are currently worked on in series into more parallel patterns of work. Work on the successor activity might begin once the predecessor activity has reached a certain stage of completion. In many cases, some of the deliverables from the predecessor can be made available to the successor so that work might begin on it.

Sometimes it is possible to rearrange the logic of the project network schedule so that critical activities are done in parallel (concurrently) rather than sequentially. This alternative is a good one if the project situation is right. When this alternative is given serious attention, it is amazing to observe how creative project team members can be in finding ways to restructure sequential activities in parallel. One of the most common methods for restructuring activities in the production sector is to change a finish-to-start relationship to a start-to-start relationship. For example, instead of waiting for the final design to be approved, manufacturing engineers can begin building the production line as soon as key specifications have been established. Changing activities from sequential to parallel usually requires closer coordination among those responsible for the activities affected but can produce tremendous time savings.

A project network schedule may assume adequate resources and show activities occurring in parallel. However, parallel activities hold potential for resource conflicts. The caution, however, is that project risk increases because the project team would has created a potential rework situation if changes are made in the predecessor after work has started on the successor. Schedule compressions affect only the timeframe in which work will be done; they do not reduce the amount of work to be done. In other words, schedule compression shortens the project schedule without changing the project scope, to meet schedule constraints, imposed dates, or other schedule objectives. The result is the need for more coordination and communication, especially between the activities affected by the dependency changes.

First, the project team needs to identify approaches for locating potential relationships changes. It should focus its attention on the critical path activities because these are the activities that determine the completion date of the project, the very thing the project team wants to impact. One might be tempted to look at critical path activities that come early in the life of the project, thinking that a jump on the scheduling problem can be obtained, but this usually is not a good approach for the following reason: At the early stages of a project, the project team is little more than a group of people who have not worked together before. Because the project team is going to make relationships changes on the project network events and activities, the project team is going to introduce risk into the project, which it is not ready to assume in the early stage of the project. Thus, the project manager should give members some time to function as a real team before intentionally increasing the risk they will have to contend with. That means the project manager should look downstream on the critical path for those compression opportunities.

A second factor to consider is to focus on activities that can be partitioned. An activity that can be partitioned is one whose work can be assigned to more than one individual working in parallel. If an activity can be partitioned, it is a candidate for consideration. The project team could be able to partition it so that when some of it is finished, they can begin working on successor activities that depend on the part that is complete. Once the project team has identified a candidate set of activities that can be partitioned, it needs to assess the extent to which the schedule might be compressed by starting the activity's successor activity earlier. There is not much to gain by considering activities with short duration times.

11.2.2.5 Time-Constrained Projects: Smoothing Resource Demand

The interrelationships and interactions among time and resource constraints are complex for even small project networks. Some effort to examine these interactions before the project begins frequently uncovers surprising problems. Project managers who do not consider resource availability in moderately complex projects usually learn of the problem when it is too late to correct. A deficit of resources can significantly alter project dependency relationships, completion dates, and project costs. Project managers must be careful to schedule resources to ensure availability in the right quantities and at the right time. Fortunately, there are computer software programs that can identify resource problems during the early project planning phase when corrective changes can be considered. These programs only require activity resource needs and availability information to schedule resources.

Scheduling time-constrained projects focuses on resource utilization. When demand for a specific resource type is erratic, it is difficult to manage, and utilization may be very poor. Practitioners have attacked the utilization problem using resource leveling techniques that balance demand for a resource.

When looking at the resource availability under the revised project schedule, it is assumed that the resources are not available according to the project schedule. In this situation, the project manager has to revert to the original project definition, budget, time, and resource allocations to resolve the scheduling problem, which may require additional time, budget, and resource allocation in order to comply with the requested deliverables and deliverable schedule.

Resource leveling is part of the broader topic of resource management. This is an area that has always created problems for enterprise businesses. Some of the situations that enterprise businesses have to deal with include, but are not limited to:

1. Committing human resources to more than they can reasonably handle in the given timeframe, reasoning that they will find a way to get the work done.
2. Changing project priorities and not considering the impact on existing resource schedules.
3. The absence of a resource management function that can measure and monitor the capacity of the resource pool and the extent to which it is already committed to projects.
4. Employee turnover that is not reflected in the resource schedule

When a "process improvement" project is declared time constrained, the goal will be to reduce the peak requirement for the resource and thereby increase the utilization of the resource. A quick examination of the earliest start times of the project network schedule should suggest activities that have slack that can be used to reduce the peak requirement for the resource. Resource leveling is a process that the project manager follows to schedule how each resource is allocated to activities in order to accomplish the work within the scheduled start and finish dates of the activity. Its purpose is to delay non-critical activities by using positive slack to reduce peak demand and fill in the valleys for the resources. The resource schedule needs to be leveled for two reasons.

1. To ensure that no resource is over-allocated. That is, the project manager does not schedule a resource to more than 100 % of its available time.
2. The project manager wants the number of resources (human resource, in most cases) to follow a logical pattern throughout the life of the project. You would not want the number of people working on the project to fluctuate wildly from day to day or from week to week. That would impose too many management and coordination problems. Resource leveling avoids this by ensuring that the number of resources working on a project at any time is fairly constant. The ideal project would have the number of people resources relatively level over the planning phases, building gradually to a maximum during the project work phases, and decreasing through the closing phases. Such increases and decreases are manageable and expected in the life of a well-planned project.

Two approaches are often used to level project resources:

1. *Utilizing Available Slack*—Slack was defined in a sub-section above as the amount of time by which an event or an activity could be adjusted later than its earliest time without causing problems downstream; i.e., without causing a delay in the completion of the project. It can be used to alleviate the over-allocation of resources. With this approach, one or more of the project events or activities are postponed to a date that is later than their early start date but no later than their late finish date. In other words, the events and activities are rescheduled but remain within their $T_L(Event) - T_E(Event)$ window.

2. *Smoothing*—Occasionally, limited overtime is required to accomplish the work within the scheduled start and finish dates of the activity. Overtime can help alleviate some resource over-allocation because it allows more work to be done within the same scheduled start and finish dates. This is called "smoothing," and through its use, the project manager can eliminate resource over-allocations, which appear as peak requirement for the resource in the resource loading graphs. In effect, what we is done is to move some of the work from normal workdays to days that otherwise are not available for work. To the person doing the work, it is overtime.

The downside of leveling is a loss of flexibility that occurs from reducing slack. The risk of activities delaying the project also increases because slack reduction can create more critical activities and/or near-critical activities. Pushing leveling too far for a perfectly level resource profile is risky. Every activity then becomes critical.

11.2.2.6 Output of the "Develop Project Schedule" Process

The key outcomes of this sixth step of developing the project time management plan are: the project schedule, the schedule model data, the schedule baseline, and the requested alterations.

The Project Schedule—It includes at least a planned start date and planned finish date for each scheduled activity. If resource planning is done at an early stage, then the project schedule would remain preliminary until resource assignments have been confirmed, and scheduled start dates and finish dates are established. As the Project Management Body of Knowledge indicates, this process usually happens no later than completion of the project management plan. A project target schedule may also be developed with defined target start dates and target finish dates for each scheduled activity. The project schedule can be presented in summary form (schedule network diagrams, bar charts, or milestones charts), sometimes referred to as the master schedule or milestone schedule, or presented in detail.

The Schedule Model Data—It represents supporting data for the project schedule and it includes at least the schedule milestones, schedule activities, activity attributes and documentation of all identified assumptions and constraints. The amount of additional data varies by application area. Information frequently supplied as supporting detail includes, but is not limited to:

1. Resource requirements by time period, often in the form of a resource histogram.
2. Alternative schedules, such as best-case or worst-case, not resource leveled, resource leveled, with or without imposed dates.
3. Schedule contingency reserves.

Schedule Baseline—A schedule baseline is a specific version of the project schedule developed from the schedule network analysis of the schedule model. It is accepted and approved by the project team as the schedule baseline with baseline start dates and baseline finish dates for the accomplishment of the project activities. It sets a benchmark against which schedule performance is measured and forecasts projected.

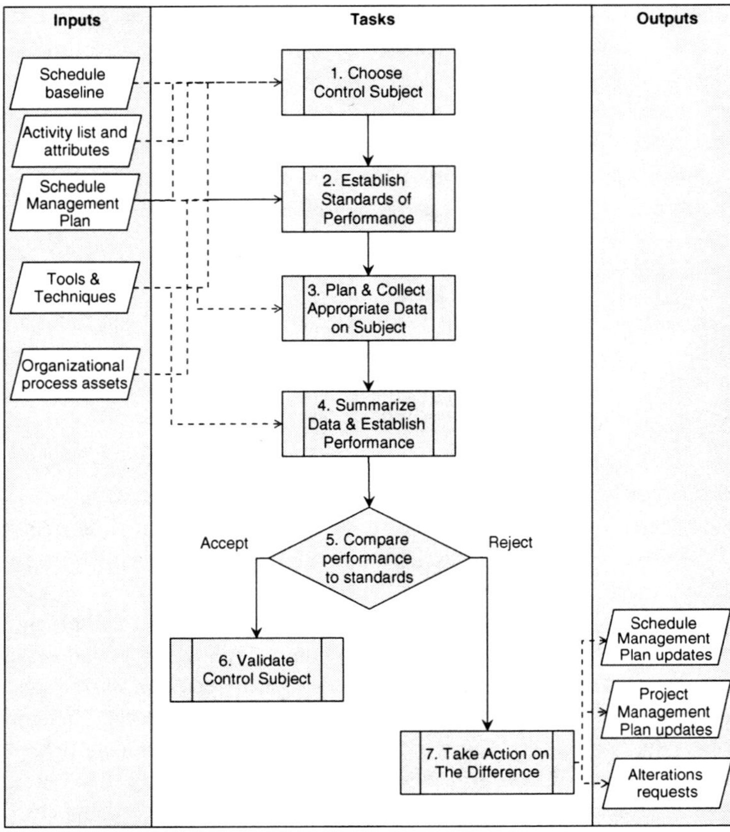

Fig. 11.4 The project schedule control process

11.3 Develop Schedule Control Plan

This is the project management process for planning a set of systematic observation techniques and activities focused on scheduling of project activities to monitor and record project scheduling in order to:

1. Assess schedule performance of the "process improvement" project; and
2. Recommend necessary alterations to the project objectives and/or "process to be improved" goals.

A generic form of the "Project Schedule Control" process is shown in Fig. 11.4.

11.3.1 Choose Control Subject

The first step of the "Project Schedule Control Process" is "Choose the Control Subject"—Each critical activity that has been identified from the project activities is a control subject; a center around which the schedule control process is built.

11.3.2 Establish Standard Performance

The second step of the "Project Schedule Control Process" is "Establish Standard of Performance"—It relates to collecting the standards of schedule performance baseline accepted and approved by the project team as the schedule baseline with baseline start dates and baseline finish dates for the accomplishment of the project activities. For each control subject it is necessary to know its standard of schedule performance.

11.3.3 Plan and Collect Appropriate Data

The third step of the "Project Schedule Control Process" is "Plan and Collect Appropriate Data" on the chosen "Control subject"—It relates to progress reporting or to establishing the means of tracking established schedule on the project activities in order to determine the actual schedule performance of the project.

The progress reporting and current schedule status includes information such as actual start and finish dates, and the remaining durations for unfinished schedule activities. If progress reporting data collected such as earned value (to be discussed in a next section on "Cost Control") is also used, then the percent complete of in-progress schedule activities can also be included in the data collection outcomes. To facilitate the periodic reporting of project progress, a template created for consistent use across various project organizational components can be used throughout the project life cycle. The template can be paper-based or electronic.

Schedule tracking begin with the collection of information needed to accomplish the prescribed schedule analyses. Data collection can be specified to occur at some recurring point in time when data is needed for schedule analysis purposes, or it may be accomplished as an ongoing activity over a period of time where data is collected regardless of when schedule analyses are performed. An ongoing data collection approach is recommended, particularly if schedule performance analyses are conducted infrequently, for example, only monthly or quarterly. This removes the burden of trying to capture or recreate past data that may have been replaced by current data. Also, ongoing data collection (even without formal schedule analysis) can sometimes provide indicators of potential project schedule performance issues or problems that would not otherwise surface in a timely manner.

The schedule tracking effort considers the project baseline schedule that were created during the "Develop Project Schedule" process (and that are currently aligned with work activities in the project work plan) and examines them relative to actual schedule that has been incurred by the project.

11.3.4 Summarize Data and Establish Actual Performance

The fourth step of the "Project Schedule Control Process" is "Summarize Data and Establish Actual Performance" of the chosen "Control subject"—Schedule tracking information is typically summarized weekly for shorter projects and at least monthly for larger projects. To ensure proper schedule control, the project manager (or a qualified designee) should review and approve all schedules incurred by the project. Such approval should not be "rubber stamped." Rather, the schedule approval process should prompt a detailed examination of planned project schedule versus acquired schedule, in conjunction with verifying completion date and resource demands.

A "Schedule Performance Index" is a performance indicator that can be used by the project team to summarize schedule performance data on the project. We will discuss it in a coming section on "Cost Control."

11.3.5 Compare Actual Performance to Standard

The fifth step of the "Project Schedule Control Process" is "Compare Actual Performance to Standards"—The act of comparing the actual schedule performance of the chosen "Control subject" to standards is performed by carrying out any or all of the following activities:

1. Compare the actual schedule performance to the project completion date goal.
2. Interpret the observed difference; determine if there is conformance to the goal.
3. Decide on the action to be taken.
4. Stimulate corrective action.

During project implementation, one of the key responsibilities of project manager is to measure schedule performance. This responsibility entails monitoring schedule performance to detect and analyze deviation from the established schedule baseline.

11.3.6 Validate Control Subject

The sixth step of the "Project Schedule Control Process" is "Validate Control Subject"—It relates to acceptance decisions from the schedule control results, which will indicate how well the chosen "Control subject" has been absorbed by the project and how much work has been completed.

11.3.7 Take Action on Difference

The last step of the "Project Schedule Control Process" is "Take Action on the Difference." It relates to actuate alterations, through corrective actions, which restores conformance with the project schedule goals.

The decision to issue corrective or preventive actions is to ensure that the observed non-conformance to schedule requirements are repaired and brought into compliance with baseline schedule requirements. A corrective action here is anything done to bring expected future project schedule performance in line with the approved project schedule baseline. Corrective action in the area of time management often involves expediting, which includes special actions taken to ensure completion of a schedule activity on time or with the least possible delay. Corrective action frequently requires root cause analysis to identify the cause of the deviation. The analysis may address schedule activities other than the schedule activity actually causing the deviation; therefore, schedule recovery from the deviation can be planned and executed using schedule activities delineated later in the project schedule.

Develop Resources Management Plan 12

Resources are absolutely essential for the project manager and/or the project team leader to have at their disposal if they hope to be able to successfully complete a project and attain a level of results that are considered satisfactory. However, in order to assure that the project team is able to successfully complete a project and attain a level of results required, they must assess properly what exactly resources are. First of all, resources can consist of any and all groups or individual labor staffing resources. In this instance, resources are specifically the people who can be allocated to the respective project or to the particular work element within the project, and whose time can be allocated accordingly. Resources can, however, also refer to any number of inanimate objects that may be utilized by the project team in the management of the project, such as supplies, services, commodities, and budgets.

12.1 Defining Resource Management

Resource management is the efficient and effective deployment of an enterprise business' resources when they are needed. Such resources may include financial resources, inventory, labor skills, production resources, or information technology.

In the realm of project management, the "Resource Management Plan" is a global process for managing the allocation, application and utilization of resources (people, materials and equipment) throughout the project lifecycle with the goal to increase efficiency and productivity. It identifies, describes and documents the type and quantities of physical resource required to complete the project successfully. This includes a list of the types of resource required, such as labor, material, equipment, physical facilities, inventories, and supplies which have limited availabilities, as well as a schedule identifying when each resource will be utilized. As a whole, resource management involves the specification of requirements, planning roles and responsibilities, allocating resources according to the project schedule, managing resource work activities and responding to resource related issues.

The resources management plan is created after the project management plan has been defined. Although summarized resource information may be described in

A. van Aartsengel and S. Kurtoglu, *Handbook on Continuous Improvement Transformation*, 179
DOI 10.1007/978-3-642-35901-9_12, © Springer-Verlag Berlin Heidelberg 2013

the business case, feasibility study, terms of reference and project plan documents, a detailed resource management plan cannot be created until every activity and task in the project management plan has been identified.

Following the completion of the resource management plan, it will be possible to finalize the financial management plan, as the fixed cost portion of the project will have been identified. The following are steps to be undertaken to create a resource management plan:

1. List the general types of resources to be utilized on the project.
   - Identify the number and purpose of each type of resource required.
   - Identify when each resource will be utilized, by completing a resource schedule.
2. Assign the resources to project activities, by completing a resource usage table.
3. Acquire, develop and manage project team

For small projects, it is sufficient to take each activity listed in the project management plan and assign resources to it. This is relatively easy using a planning tool such as Microsoft Project. For larger more complex projects, a full resource management plan, the process steps of which are as described below, should be completed to ensure that the amount and type of allocated resources are both accurate and timely.

12.2 List the Resources to Be Consumed by the Project

To create a comprehensive resource management plan, the project manager will first need to list the types and number of resources required to complete the project successfully. A "resource" is defined as the labor, material, equipment, physical facilities, inventories, and supplies which have limited availabilities used to complete each activity in the project.

12.2.1 Labor

This is the most difficult type of resource to schedule and manage because most project plans specify only the skills required to perform the project work, when that skill is needed, and in what amounts. Labor resource planning is used to determine and identify human resources with the necessary skills required for project success. It is essential to have the right labor resources with the required skills. Also, these resources and requirements must be matched to develop individual talent and organizational capabilities. The enterprise business' desire is to build a successful, profitable organization with talented people who have the opportunity to fulfill their professional dreams while taking part in challenging, exciting project work for which they are appreciated and rewarded.

The labor resource plan is the process that involves the careful and deliberate identification, categorization, and ultimately, documentation of the entirety of all project roles assigned to all individual members of the project work team.

Table 12.1 Labor listing

Role	No.	Summarized responsibilities	Summarized skills	Start date	End date
Project Manager	1	Delivering the approved solution to meet the full requirements of the customer and the business	Time management Cost management Quality management People management	dd/mm/yy	dd/mm/yy
Procurement Manager	1	Ensuring that the entire procurement process is undertaken effectively.	Time management Cost management Quality management People management	dd/mm/yy	dd/mm/yy
Staff Member	10	Undertaking each delegated task to the best of their ability	...	dd/mm/yy	dd/mm/yy
...	...	...	...	...	...

Included among this documentation process is a careful delineation of all of the individual project team members' personal responsibilities in regards to management of the project, as well as all the specific reporting relationships among all members of the project team.

One additional essential component of the project management process of human resource planning is a careful and thoroughly executed encapsulation of all members of the project team through a complete project staffing management plan. This can be done through the use of either a formally written document, which can include a detailed graphically formatted chart, or it can be in the form of a less formalized document of sorts.

In developing the labor resource plan, the project manager could summarize the roles, responsibilities and skill-sets required to complete the project. This includes the roles of current staff appointed, further roles to be appointed, the roles of external business staff involved with the project and the roles of external suppliers. In short, every role in the project should be defined using Table 12.1.

In Table 12.1 the 'No.' column represents the number of full-time equivalent people required to undertake the role. For instance a project might require one project manager, one project administrator and 10 staff members. The "Start date" and "End date" columns identify how long the role will exist for. In the instance of the project manager, the start date will be during the project initiation phase, and the end date will be soon after the completion of the project closure report in the project closure phase.

Table 12.2 Project manager responsibilities in the HRM practice areas

HRM Practice Area	Roles of the Project Manager
Flows	Manage in- and out- flows of project staff. Match project staff and project task assignments. Planning of labor resource needs on project tasks. Build relationships with responsible functions to handle shifts and changes in project tasks.
Performance	Facilitate knowledge sharing within the project. Detect and work to minimize problems in project work conditions. Give recognition and feedback to individuals on their performance. Take part in performance appraisals with line managers
Performance	Clarify assignments. Clarify roles and responsibilities. Improve project staff opportunities to positively influence their work performance and work conditions. Clarify deliveries to trigger motivation.
Development	Identify needs for competence development. Support project staff in their work, improving their skill sets. Spread the word on positive experiences with project workers.

Table 12.3 Staff member responsibilities in the HRM practice areas

HRM Practice Area	Roles of the Project Manager
Flows	Know one's competence Adjust assignments to competence Build reputation
Performance	Share knowledge Clarify expectations Role carving and redefinition Seek feedback
Performance	Get involved actively or decide not to get involved. Actively influence work conditions and the content of work
Development	Search for new, challenging tasks. Learn from experience. Network to build social capital, learn from others, and share knowledge with others.

The following categorization of the four Human Resource Management practice areas, summarized in Tables 12.2 and 12.3, can also be used as a framework that can be applied for the management of labor resource.

Table 12.4 Facilities listing

Item	No.	Purpose	Specification	Start date	End date
Meeting Room	1	Facilitate the conduct of project member meetings and project review activities.	Equipped with a computer with DVD drive, LCD data projector, projection screen, wall-mounted white board, overhead projector, and VCR.	dd/mm/yy	dd/mm/yy
...	...	...	...	...	...

The project management team, also called the core, executive, or leadership team, is responsible for project planning, controlling, and closing and takes directives from the project team. Smaller project responsibilities can be shared by the team or designated by the project manager. The project team and the project sponsor work together to secure funding, simplify scope questions, and influencing team members.

12.2.2 Facilities

Project work takes place in specific locations. Planning rooms, conference rooms, presentation rooms, and auditoriums are but a few examples of facilities that projects require.

The exact specifications as well as the precise time at which they are needed are some of the variables that must be taken into account. The project resource management plan can provide the detail required. The facility specification will also drive the project schedule based on availability.

Each facility item should be listed, as illustrated in Table 12.4, including a description of the purpose of each item, the specification of the item and the period that the item is needed for the project.

In Table 12.4 the 'No.' column represents the number of facility items required. The 'Start date' and 'End date' columns identify how long the facility is required for.

Table 12.5 Equipment listing

Item	No.	Purpose	Specification	Start date	End date
Laptop	1	To enable the project manager to plan, monitor and control the project both on and off site.	High processing speed and wide screen.	dd/mm/yy	dd/mm/yy
...	...	...	...	...	...

12.2.3 Equipment

Now that the labor and facilities required to undertake the project have been identified, it is necessary to list in detail all of the items of equipment needed. This includes computers, furniture, building facilities, machinery, vehicles and any other items of equipment needed to complete the project. Equipment is treated exactly the same as facilities. What is needed and when drive the project schedule based on availability. Each item of equipment should be listed, as illustrated in Table 12.5, including a description of the purpose of each item, the specification of the item and the period that the item is needed for the project.

In Table 12.5 the 'No.' column represents the number of equipment items required. The 'Start date' and 'End date' columns identify how long the equipment is required for.

12.2.4 Materials

Parts to be used in the fabrication of products and other physical deliverables are often part of the project work, too. For example, the materials needed to build an automobile might include steel, plastic, aluminum, rubber, and glass. The project manager should identify all of the generic materials required to undertake the project, including stationery, computer consumables, building materials, power, water and gas. Each item of material should be defined by listing its components and the period of required usage, as illustrated in Table 12.6.

In Table 12.6, the "Amount" column describes the approximate quantity of each item of material. The "Start date" and "End date" columns identify how long the materials are required for.

Table 12.6 Materials listing

Item	Component	Amount	Start date	End date
Computer consumables	Printer cartridges Printer paper DVDs for file backup.	10	dd/mm/yy	dd/mm/yy
...	...	...	...	...

12.3 Assign the Resources to Project Activities

Keeping track of resource schedules is one of the most fundamentally important tasks that are the responsibility of the project team. One of the best ways to accomplish this feat is through the careful and well orchestrated use of calendars to keep track of the multitude of project related tasks events, occurrences, and dates that will take place during the project's life cycle. One of these types of calendars that is often utilized and implemented by the project team is that of the resource schedule calendar.

The resource schedule calendar refers to the specific calendar that lists all of the working days as well as all of the nonworking days that the project team needs to utilize in order to determine the specific dates on which a specific resource of element is being utilized or engaged, versus the dates on which they may in fact be inactive.

The resource schedule calendar enables a project manager to identify the quantity required of each type of resource on a daily, weekly or monthly basis. For simplicity, a sample monthly resource schedule calendar is shown in Table 12.7.

When the project team is attempting to create and or establish a schedule that the project needs to follow over the course of its project life cycle, often times, in fact nearly always, the timing of the start of varying elements of the project is not entirely based on what the project team feels is the ideal. Instead, typically there are constraints and assumptions involved based on the resources that may or may not be available at a given time. This is referred to as resource leveling. The project team should list any assumptions made during this resource schedule calendar planning exercise.

For instance: It is assumed that the resource requirements and the delivery dates will not change throughout the project. It is also assumed that resources listed will be available as required to undertake the associated project activities.

The project team should also list any risks identified during this resource planning exercise. For example:

1. Key staff resigns during the project;
2. Further training is required to complete the tasks allocated;
3. Budgetary constraints lead to inferior resources being allocated;
4. Equipment is not delivered on time, as per the resource schedule.

Table 12.7 Resource schedule calendar

	Month												Total
Resource	Jan	Feb	Mar	Apr	May	Jun	Jul	Aug	Sep	Oct	Nov	Dec	
Labor Labor #1 …													
Facility Facility #1 …													
Equipment Equip.#1 …													
Materials Material#1 …													
…													
…													
…													
Total													

12.4 Plan Project Team Development and Management

The planning of project team development refers to the early stage planning process in which the fundamental core team of the project is been developed. These individuals make up the project team, and the maximization of their performance is essential to the proper functioning of the project process and the ultimate successful completion of the project as well as the successful completion of all of the individual components within. The process of developing the project team refers to the specific activity of enhancing the performance of each individual member of the project team as well as the performance of the team as a whole by improving the individual competencies of the team as well as enhancing the communication and interactions among the team members. Essential to the entire life of the project, and imperative to the ultimate success of it, is effective and reliable communication.

As project manager or team leader, you rarely inherit a fully fledged and effective team at onset of a "process improvement" project. More often than not you will likely inherit one that is already misfiring or you will have to start by building your team from scratch. The practical constraints that you will encounter when assembling your team will make this a challenging task. Some of the following might sound familiar to you:

1. Budget constraints preventing much-needed recruiting. Or conversely a generous budget fuelling unrealistic expectations of a fast ramp up.
2. Projects being used as a dumping ground. Other colleague managers using your new team as a convenient home for staff that they are not really sure what to do with.
3. Selfish colleague managers who monopolize the best staff. They hold onto the enterprise business star performers even when their skills and experience are desperately needed elsewhere.

All this is invariably against a setting of an acute sense of urgency to get a team up and running for an improvement intervention. Building and developing the right team—as far as it is realistic—is one of the factors critical to the success of any project. Planning the development and management of the project team involves organizing and managing the project team. The team is usually made up of people with specific skills and responsibilities. The project team, also known as project staff, should be involved in plans and decision making from the beginning of the project. Team members should feel invested in the outcome of the project. This will increase loyalty and commitment to project goals and objectives. The number of team members and their responsibilities can change as the project develops.

Several models of team development have been proposed in the literature. The most widely and solidly established such model is Tuckman's "Forming, Storming, Norming, and Performing" model. First published in 1965 and revised in 1970 with the addition of a fifth stage—Adjourning. Tuckman's model explains the necessary and inevitable stages through which a group of individuals must grow before they can function as a cohesive and efficient tasks focused unit. The model has become the basis for subsequent models of group development and team dynamics and management theories frequently used to describe the behavior of existing teams. It has also taken a firm hold in the field of experiential education.

The value of Tuckman's model, as illustrated in our first book entitled "A Guide to Continuous Improvement Transformation: Concepts, Processes, Implementation," is that it helps enterprise executives and managers understand that groups and teams evolve. It also helps to consider the different problems that may be encountered at different stages of their development. The model also illustrates four main leadership and management styles, which a good leader is able to switch between, depending on the situation (i.e., the group maturity relating to a particular task, project or challenge).

Project Team management refers to the comprehensive set of activities followed to establish, implement and improve unity and coordination between the members of a group or a team working towards a common goal—achieve the activities resulting from the enterprise alignment. As project manager or a team leader, your aim is to help your team reach and sustain high performance as soon as possible. If you have opted for Tuckman's model for developing your team, then you will need to change your approach at each stage of the group/team development.

Project labor resource management planning may be required if more experienced members are added to the team. The project team should also prepare for risk management and changes to project duration.

Collating into one document all of the materials listed in the sections above creates the resource management plan document.

13 Develop Quality Management Plan

This chapter is concerned with the project management process required to ensure that the "process improvement" project includes all the quality policy and procedures required as well as the customer specifications to complete the project successfully. It describes how the project will ensure the level of quality required by the customer in its deliverables and "process to be improved." Quality management activities ensure that:

1. The "process to be improved" outcomes are built to meet agreed-upon standards and requirements;
2. Work processes are performed efficiently and as documented;
3. Non-conformances found are identified and appropriate corrective action is taken.

In conformity with the Project Management Body of Knowledge guidelines, the constituent processes of the "Develop Quality Management Plan" the project management process, illustrated in Fig. 13.1, which reflects a structure that mirrors the perspective of the Project Management Institute's PMBOK Guide, include:

1. *Develop Quality Plan*—This sub-process is concerned with identifying quality requirements and/or standards for the project and the "process to be improved," understanding how well the "process to be improved" meets its associated customer specifications, and documenting how the project will demonstrate achievement of quality requirements. Its purpose is to build, as precisely as possible, a factual understanding of existing "process to be improved" conditions and problems.
2. *Develop Quality Assurance Plan*—This sub-process is concerned with documenting a set of preventive and systematic activities, focused on processes used in the project, which can be demonstrated to show commitment to delivering and provide confidence that project execution and its deliverables will fulfill specified quality standards and objectives.
3. *Develop Quality Control Plan*—This sub-process is concerned with monitoring and recording results of executing the quality plan activities to assess performance of the "process to be improved" and recommend necessary changes.

A. van Aartsengel and S. Kurtoglu, *Handbook on Continuous Improvement Transformation*,
DOI 10.1007/978-3-642-35901-9_13,

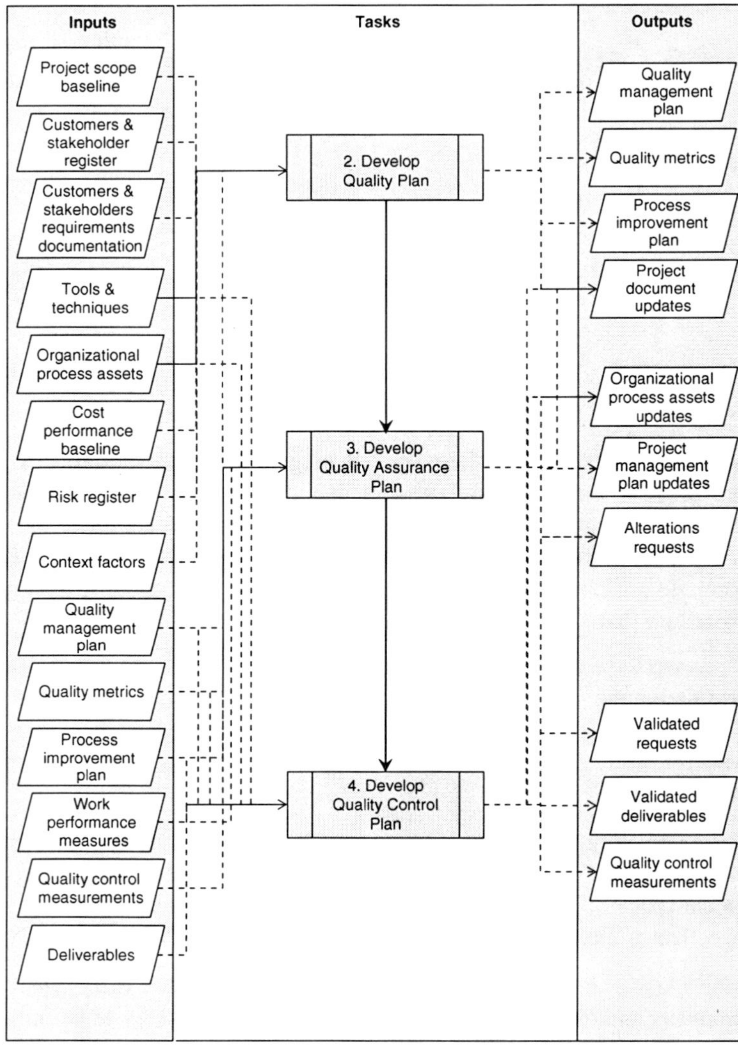

Fig. 13.1 "Develop Quality Management Plan" process

As the PMBOK Guide indicates, these processes interact with each other and with the processes in the other PDSA "Process Groups." Each aspect of executing any of these can involve effort from one or more persons or groups of persons based on the project requirements. Each constituent process occurs at least once in every project and occurs in one or more "process improvement" project phases.

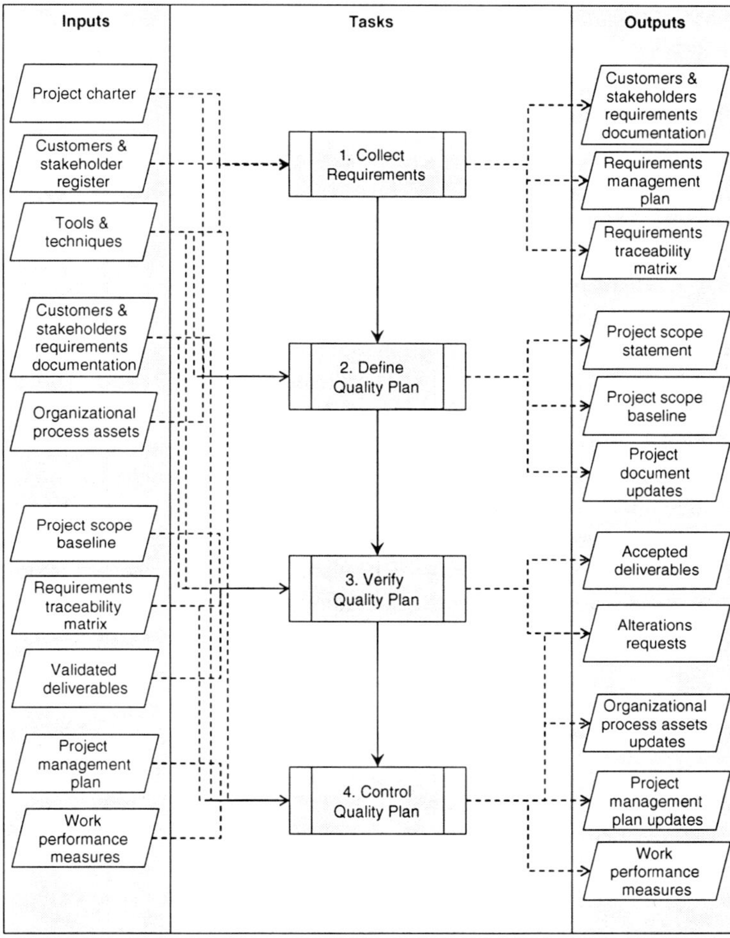

Fig. 13.2 "Perform Planning of 'Quality Plan'" process

13.1 Develop Quality Plan

This is the project management process required to ensure that the project includes all the quality related work to complete the "process improvement" project successfully. It is a key process of the PDSA Plan "Process Group" that elaborates on the characteristics of the "process to be improved" that are described in the project charter.

Managing quality of a "process improvement" project is primarily concerned with defining and controlling quality requirements and/or standards, "Process to be improved" requirements and characteristics, "Process to be improved" acceptance criteria, and what are not included in the project.

The constituent project management processes used during the development of the project quality planning, illustrated in Fig. 13.2, include the following:

1. Collect Requirements
2. Define Quality Plan
3. Verify Quality Plan
4. Control Quality Plan

These four constituent processes interact with each other and with the project management processes in the PDSA "Process Groups." Each aspect of executing any of these can involve effort from one or more persons, based on the needs of the project. Each aspect occurs at least once in every "process improvement" project and occurs in one or more project phases.

The project management constituent processes utilized to manage project quality plan as well as the supporting tools and techniques, vary by application area and are defined as part of the "process improvement" project life cycle. The approved detailed project quality plan statement must be included in the scope baseline for the "process improvement" project.

As indicated already in a previous section, this baseline scope should be monitored, verified and controlled throughout the lifecycle of the project. Furthermore, performance completion of the "process improvement" project quality plan is measured against the project management plan, while performance completion of the process scope is measured against the requirements of the actual "process to be improved."

13.1.1 Collect Requirements: V.O.P.

The first step in developing the project management quality plan is to "Collect Process Requirements." It relates to defining and documenting the "process improvement" project and "process to be improved" features and functions needed to fulfill the "process to be improved" needs and expectations (Voice of the Process—V.O.P.) only. The project's success is directly influenced by the care taken in capturing and managing these requirements.

The "process to be improved" must meet the requirements of the customers and stakeholders, and the ability of this process to meet these requirements is called Voice of the Process. It is a construct for examining what the "process to be improved" is telling about its inputs and outputs and the resources required to transform the inputs into outputs. Collecting the Voice of the Process is a practice used in process improvement undertakings to capture the process requirements, expectations, and entitlements. This is the subject of the next chapter.

13.1.2 Define Quality Plan

The second step in developing the project management process "Perform Quality Planning" is "Define Quality Plan." It relates to developing a detailed description of the extent of work and effort of the "process improvement" project and the "process to be improved" from the quality perspective. The preparation of a detailed project quality plan is critical to project success and builds upon the basic plan of action resulting from the V.O.P. data collection process, the major deliverables, assumptions, and constraints that are documented during project initiation and development of the preliminary project scope.

The project quality plan is used to guide the project team to perform critical oversight activities necessary to avoid rework, concentrate on improvements, and reduce costs by enhancing or avoiding schedule delays. It describes how the project team will implement the quality policy and practices, and the identified V.O.C. and V.O.P. quality requirements for direct and indirect project deliverables. It identifies technical and project management audits that may be scheduled and conducted as a part of the project quality oversight effort.

The project quality plan can reference applicable quality standards and specification documents, and adjunct technical plans having greater quality process and procedural details. Quality standards provide a basis for determining achievement and acceptability of project work and project deliverables. The quality standards used can originate from within an industry, a governing body, an organization, or an individual. When possible, the project quality plan should identify individuals and functions responsible for quality management within the enterprise business. Finally, the project quality plan should specify the procedures and criteria for customer acceptance of project deliverables as indicated by the V.O.C. key needs.

During this "Define Quality Plan" step, the preliminary project scope statement is refined and described with greater specificity as more information on the Voice of the Customer (V.O.P.) and the Voice of the Process (V.O.P.) from the quality perspective about the "process to be improved" and the project is known from the collected data. Existing risks, assumptions, and constraints are analyzed for completeness; and additional risks, assumptions, and constraints are added as necessary. Key tools and techniques used in defining the quality plan include but are not limited to:

1. Expert judgment
2. Process analysis
3. Alternative identification
4. Facilitated workshop

Expert Judgment—Expert judgment is often used to analyze the quality related information on the V.O.C. and the V.O.P. needed to develop or refine the project scope statement. Such judgment and expertise is applied to any technical details. Such expertise is provided by any group or individual with specialized knowledge or training in quality process improvement, and is available from many sources, including:

1. Other functions or business units within the enterprise;
2. Consultants;
3. Stakeholders, customers, and sponsors;
4. Professional and technical associations;
5. Industry groups; and
6. Subject matter experts.

Process Analysis—Each application area has one or more generally accepted methods for translating high-level process descriptions into tangible deliverables. Process analysis includes techniques such as process breakdown, systems analysis, systems engineering, value engineering, value analysis, and functional analysis.

Alternatives Identification—Identifying alternatives is a technique used to generate different approaches to execute and perform the quality work associated with the project.

The key outcomes of the preparation of a detailed project quality plan should be included, but are not limited to:

1. Process Improvement Plan
2. Project quality management plan
3. Project quality performance measures
4. Project quality check lists
5. Improvement plan for the "process to be improved"
6. Project quality objectives baseline

These key outcomes should serve as basis for updating the project management plan through the inclusion of a subsidiary quality management plan and process improvement plan.

The process improvement plan builds on the plan of action obtained from a V.O.P. data collection process and must include the following activities, which will be performed to actually improve the "process to be improved":

1. Identify and Quantify Assignable Causes of Variations
2. Explore Cause-and-Effect Relationship
3. Verify identified assignable causes
4. Generate Improvement Solutions
5. Assess Risk and Pilot Solution(s)
6. Devise Controls Measures

13.1.3 Verify Quality Plan

The third step in developing the project management process "Perform Quality Planning" is "Verify Quality Plan." It relates to formalizing acceptance of the relevant and identified quality standards and requirements. Verifying the project quality plan includes reviewing the "process to be improved" deliverables to ensure that each is completed satisfactorily and taking into account the identified V.O.C. key needs. If the project is terminated early, the project quality plan, through the project scope verification process, should establish and document the level and extent of completion.

Quality plan verification is performed through inspection. Inspection comprises activities such as measuring the performance level of the "process to be improved" outcomes, examining, and verifying to determine whether work and deliverables meet quality standards and "process to be improved" acceptance criteria. Inspections are sometimes called gate reviews, process reviews, audits, and walkthroughs. In some application areas, these different terms have narrow and specific meanings.

Quality plan verification differs from quality control in that quality plan verification is primarily concerned with acceptance of the "process to be improved" deliverables, while quality control is primarily concerned with correctness of the "process to be improved" deliverables and meeting the quality requirements specified for the deliverables. Quality control is generally performed before scope verification, but these two processes can be performed in parallel.

The "Verify Quality Plan" project management process also documents those completed "process to be improved" deliverables that have been formally accepted. Through the "Verify Scope" project management process, those completed deliverables that have not been formally accepted are documented, along with the reasons for non-acceptance.

13.1.4 Control Quality Plan

The last step in developing the project management process "Perform Quality Planning" is "Control Quality Plan." It relates to monitoring the status of the performance of the "process to be improved" outcomes and controlling their alterations. Controlling the project quality plan ensures that all requested alterations and recommended corrective actions on the "process to be improved" outcomes are taken into account and processed. The "control quality plan" is also used to manage the actual alterations of the "process to be improved" quality related outcomes when they occur and is integrated with the other control processes. Uncontrolled alterations of the project are often referred to as process creep, hope creep, effort creep, or feature creep.

13.2 Develop Quality Assurance Plan

This is the project management process for documenting a set of preventive and systematic activities, focused on processes used in the project, which can be demonstrated to show commitment to delivering and provide confidence that project execution and its deliverables will fulfill specified quality standards and objectives. Quality standards include project processes and product goals.

It represents the proactive side of the "Develop Quality Management Plan" process. It effectively selects, defines, prepares, integrates, coordinates and documents all subsidiary assurance plans into one document in order to:

1. Prevent quality problems from occurring.
2. Ensure appropriate quality standards and operational definitions will be used effectively to produce quality project deliverables.

The project quality assurance plan is the primary source of information that documents how assurance of quality on the project execution and its deliverables will be demonstrated, monitored and controlled. The project quality assurance plan can be either summary level, broadly framed or highly detailed based on the requirements of the project.

In any case, the quality assurance plan is a composite document containing the information related to the quality control activities. It schedules the reviews and audits[1] that will be used for assessing the processes used in the project to achieve the project goals and to produce project quality deliverables. Planning the project "Quality Assurance Plan" is highlighted by the following activities:

1. Define the quality goals for the processes
2. Identify all relevant organizational process assets
3. Define the roles and responsibilities of "quality assurance" activities
4. Identify the tasks and activities for "Quality Control"

13.2.1 Define the Quality Goals for the Processes

The first step in planning the quality assurance plan is to define the quality goals for the processes to be used by the "process improvement" project. These goals must be described with greater specificity because more information about the enterprise business intended strategic goals, the voice of the customer and the voice of the process are known. The project team might also set a standard to define the goals. If possible, the quality assurance plan can also describe the quality goals in terms of performance measures. This will ultimately help to measure the performance of the processes. Processes have two sets of quality goals:

1. To produce outcomes which do meet the identified CTXs. Ideally, each and every unit of process outcome should meet the identified CTXs.
2. To operate in a stable and predictable manner. Each process should be in a state of "statistical control."

These goals may be directly related to the costs of producing the process outcomes.

13.2.2 Identify All Relevant Organizational Process Assets

The second step in planning the quality assurance plan is to identify all organizational process assets, which have references to any of the processes to be used in the

[1] A quality audit is a structured, independent review to determine whether project activities comply with organizational and project policies, processes, and procedures. The objective of a quality audit is to identify missing, inefficient or ineffective policies, processes, and procedures in use on the project. The subsequent effort to correct these deficiencies should result in a reduced cost of quality and an increase in sponsor or customer acceptance of the project's product. Quality audits may be scheduled or random and may be conducted by internal or external auditors.

project. The organizational process assets include formal and informal policies, procedures, plans, and guidelines whose effects can influence the processes to be used in the "process improvement" project. These subsidiary process assets are related to the quality standards of several business components and how they are related to each other in achieving the collective qualitative objective.

The quality level that a "process improvement" project can achieve depends upon the efficiency and efficacy of the organizational process assets available. The "garbage in"—"garbage out" philosophy works very well here. Hence, in developing the quality assurance plan, it is very important that the project team understand the various processes to be used in the project. This can be done through process analysis. This analysis examines problems experienced, constraints experienced, and non-value-added activities identified during operation of the selected processes to be used in the project. The inputs and outputs of each selected process should be well defined. The controls that are in place to ensure the quality of these inputs and outputs should be analyzed. This helps in understanding where and why a process can go wrong and also assists in addressing those areas in the respective procedures. This information also helps to determine the different types of reviews and audits to be performed and how often they will be performed during the project lifecycle.

13.2.3 Define Roles and Responsibilities of "Quality Assurance" Activities

The third step in planning the quality assurance plan is to define the organization and the roles and responsibilities of the "quality assurance" activities that will be undertaken during the project lifecycle. It should include a clear definition of the reporting system for the outcome of the quality reviews and audits.

13.2.4 Identify Tasks and Activities for "Quality Control"

The third step in planning the quality assurance plan is to identify the task and activities of the quality control team. The quality assurance plan should clearly explain the inspections and testing procedures for quality control, their frequencies and how they will be conducted at the various stages of the project lifecycle. Generally, identify the task and activities of the quality control team will include, but are not limited to:

1. Reviewing project plans to ensure that the project abide by the defined processes.
2. Reviewing project to ensure that the performance of its outcomes according to the specified plans.
3. Endorsement of deviations from the identified standard processes and procedures.
4. Assessing the improvement of the identified processes.

The project manager and a quality manager within the enterprise business should fix a detailed timetable for all scheduled reviews and audits. This schedule should also be documented in the quality assurance plan.

Thus, the entire process of quality control is documented within the quality assurance plan. For any future reference, this could be used as a practical evidence of total quality control.

The key outputs of planning the project quality assurance include, but are not limited to:

1. *Organizational process assets updates:* Elements of the organizational process assets that may be updated include but are not limited to the quality standards.
2. *Alteration requests:* Quality improvement includes taking action to increase the effectiveness and/or efficiency of the policies, processes, and procedures of the performing enterprise business. Alteration requests are created and used to allow full consideration of the recommended enhancements. Alteration requests can be used to take corrective action or preventative action or to perform defect repair.
3. *Project management plan updates:* Elements of the project management plan that may be updated include but are not limited to:
   - Quality management plan,
   - Schedule management plan, and
   - Cost management plan.

13.3 Develop Quality Control Plan

This is the project management process for planning a set of systematic observation techniques and activities, focused on outcomes of the project (i.e., project deliverables and project management processes used to produce the outcomes), to monitor and record results of executing the quality assurance plan in order to:

1. Assess performance of the "process improvement" project and "process to be improved" outcomes; and
2. Recommend necessary alterations to the project objectives and/or "process to be improved" goals.

"Develop Quality Control Plan" process represents the reactive side of the "Develop Quality Management Plan" process. The set of preventive and systematic activities documented in the quality assurance plan must be performed throughout the project lifecycle using the "Quality Control Process." A generic form of the Quality Control Process is shown in Fig. 13.3.

13.3.1 Choose Control Subject

The first step of the "Quality Control Process" is "Choose the Control Subject"—Each feature of the "process to be improved" outcomes documented in the quality

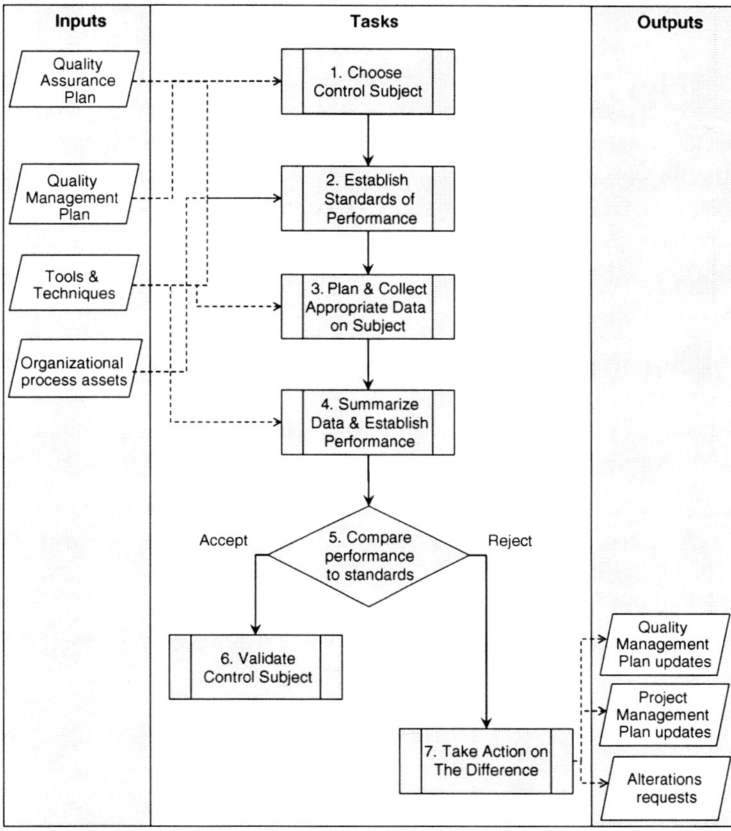

Fig. 13.3 The quality control process

assurance plan is a control subject; a center around which the quality control process is built. Control subjects are derived from the collected and identified CTXs.

13.3.2 Establish Standard of Performance

The second step of the "Quality Control Process" is "Establish Standard of Performance"—It relates to collecting the standards of performance ("process to be improved" goals as well as its outcomes goals) documented in the quality assurance plan. For each control subject it is necessary to know its standard of performance.

13.3.3 Plan and Collect Appropriate Data

The third step of the "Quality Control Process" is "Plan and Collect Appropriate Data" on the chosen "Control subject"—It relates to establishing the means of collecting the V.O.P and the project data, and collecting these data through inspection and testing, as illustrated in the previous sections of this chapter, in order to determine the actual performance of the "process to be improved" or the quality level of characteristics of its outcomes.

13.3.4 Summarize Data and Establish Actual Performance

The fourth step of the "Quality Control Process" is "Summarize Data and Establish Actual Performance" of the chosen "Control subject"—It relates to:

1. Providing answer to the questions: "what can be learned from the collected data?" and "Does the process conform to its quality goals?" The answer to this question is an understanding and a summary of the collected data in some meaningful graphical formats as indicated in a previous section.
2. Determining the actual "process to be improved" capabilities and performance indices.

13.3.5 Compare Actual Performance to Standards

The fifth step of the "Quality Control Process" is "Compare Actual Performance to Standards"—The act of comparing the actual performance of the chosen "Control subject" to standards is often seen as the role of the quality control function with the enterprise business called on to carry out any or all of the following activities:

1. Compare the actual quality performance to the quality goal.
2. Interpret the observed difference; determine if there is conformance to the goal.
3. Decide on the action to be taken.
4. Stimulate corrective action.

13.3.6 Validate Control Subject

The sixth step of the "Quality Control Process" is "Validate Control Subject"—It relates to acceptance decisions from the quality control results, which will indicate how well the chosen "Control subject" has achieved quality assurance specifications and project quality objectives.

13.3.7 Take Action on the Difference

The last step of the "Quality Control Process" is "Take Action on the Difference." It relates to actuate alterations which restores conformance with quality goals. This step is popularly known as "troubleshooting" or "firefighting." It involves rework decisions, process adjustments decisions, and quality improvement decisions.

1. *Rework* Decisions—In contrast to the previously mentioned acceptance decisions, quality control results may also indicate the need for deliverable rework, that is, the additional work needed to enable a defective or nonconforming observed characteristic of the process outcomes or project deliverable to become compliant with quality specifications.
2. *Process Adjustments*—Quality control results may indicate a process that is hindering the achievement of expected project quality objectives. The questionable process must be examined and activity steps adjusted to make the process more aligned with the project or quality of deliverable needs.
3. *Quality Improvements*—Quality control results will provide an indication of need for quality improvement. Improvement areas (e.g., process, materials, skill, etc.) can be identified for the current "process improvement" project and for, and quality improvement solutions can be implemented.
   The decision to issue corrective or preventive actions is to ensure that the observed defect (i.e., non-conformance to quality requirements as specified in the quality assurance plan) are repaired and brought into compliance with quality assurance requirements or specifications. Two situations should be distinguished:
4. The quality control process is well designed to eliminate sporadic nonconformance due to "assignable causes" of variations in the process under consideration. The focus here is on finding "What has changed." Sometimes the causes are not obvious, so the main obstacle to the corrective action is diagnosis. The diagnosis makes use of methods and tools such as:
   - Autopsies to determine with precision the symptoms exhibited by the process under consideration and its outcomes.
   - Comparison of outcomes of the process under consideration made before and after the occurrence of "assignable causes" of variation to see what has changed; also
   - Comparison of good and bad process outcomes made since the occurrence of "assignable causes" of variation began.
   - Comparison of process data before and after the occurrence of "assignable causes" of variation began to see what process conditions have changed.
   - Reconstruction of the chronology, which consists of logging on a time scale (of hours, days, etc.): (1) the events which took place in the process before and after the occurrence of "assignable causes" of variation, that is, rotation of shifts, new employees on the job, maintenance actions, etc., and (2) the time-related process outcomes information; that is, date codes, cycle time for processing, waiting time, move dates, etc.

- Analysis of the resulting collected data usually sheds a good deal of light on the validity of the various theories of causes. Certain theories are denied. Other theories survive to be tested further.
- Operating personnel who lack the training needed to conduct such diagnoses may be forced to shut down the process and request assistance from specialists, the maintenance department, etc. They may also run the process under consideration "as is" in order to meet schedules and thereby risk failure to meet the quality goals.

5. The quality control process is not well designed to deal with the area of unacceptable level of variations, which occur beyond the specification limits. The process improvement methodology must be fully applied.

13.4 Conclusion

As shown above, quality control management involves measuring project quality expectations against project actual quality results. Every project team member has quality control responsibility. In turn, the project manager must ensure that team members have sufficient knowledge and skill to properly apply quality control methods to evaluate outcomes of the project (i.e., projects include deliverables and project management results, such as cost and schedule performance). In some enterprise businesses, this activity is assigned to a dedicated quality control team that contributes quality control expertise across all projects. Nevertheless, each project manager still retains the responsibility for making adjustments based on quality control results.

The key outputs of performing a quality control at any stage of the project lifecycle include, but are not limited to:

1. *Quality Control Collected Data*—Quality control collected data are the documented results of quality control activities in the format specified during quality planning.
2. *Validated Alterations*—Any altered or repaired control subjects are inspected and will be either accepted or rejected before notification of the decision is provided. Rejected control subjects may require rework.
3. *Validated Deliverables*—A goal of quality control is to determine the correctness of deliverables. The results of the execution quality control processes are validated deliverables.
4. *Organization Process Assets Updates*—Elements of the organizational process assets that may be updated as a result of execution of quality control processes.
5. *Alterations Requests*—If the recommended corrective or preventive actions or a defect repair requires an alteration to the project management plan, an alteration request should be initiated accordingly.

14 Collecting V.O.P. Requirements

Collecting the process requirements is as much about defining and managing the "process to be improved" expectations as any other key project deliverables and will be the very foundation of completing to "process improvement" project. It is also about focusing the improvement effort by gathering information on the current situation. Its purpose is to build, as precisely as possible, a factual understanding of existing "process to be improved" conditions and problems or causes of underperformance. Cost, schedule, and quality planning are all built upon these requirements. In other words, the purpose of collecting the process requirements is to get sufficient and accurate information to complete improvement of the "process to be improved."

The constituent project management processes used during the capturing of the voice of the process, illustrated in Fig. 14.1, include the following:

1. Plan V.O.P. Data Capturing
2. Collect Data
3. Display Data and Patterns
4. Establish Process Performance
5. Set/Revise Process Quality Targets

14.1 Plan V.O.P. Data Capturing

The first step in collecting the "process to be improved" requirements is "Plan V.O.P. Data Capturing." In much the same as with the voice of the customers, this is the project management process for documenting the actions necessary to define, prepare, integrate, and coordinate all subsidiary V.O.P. data capturing actions into one document.

Planning for V.O.P. data collection includes, but is not limited to the following steps:

1. Identify V.O.P. data and clarify goals
2. Develop operational definitions and procedures
3. Develop sampling strategy
4. Validate data collection system

A. van Aartsengel and S. Kurtoglu, *Handbook on Continuous Improvement Transformation*,
DOI 10.1007/978-3-642-35901-9_14,

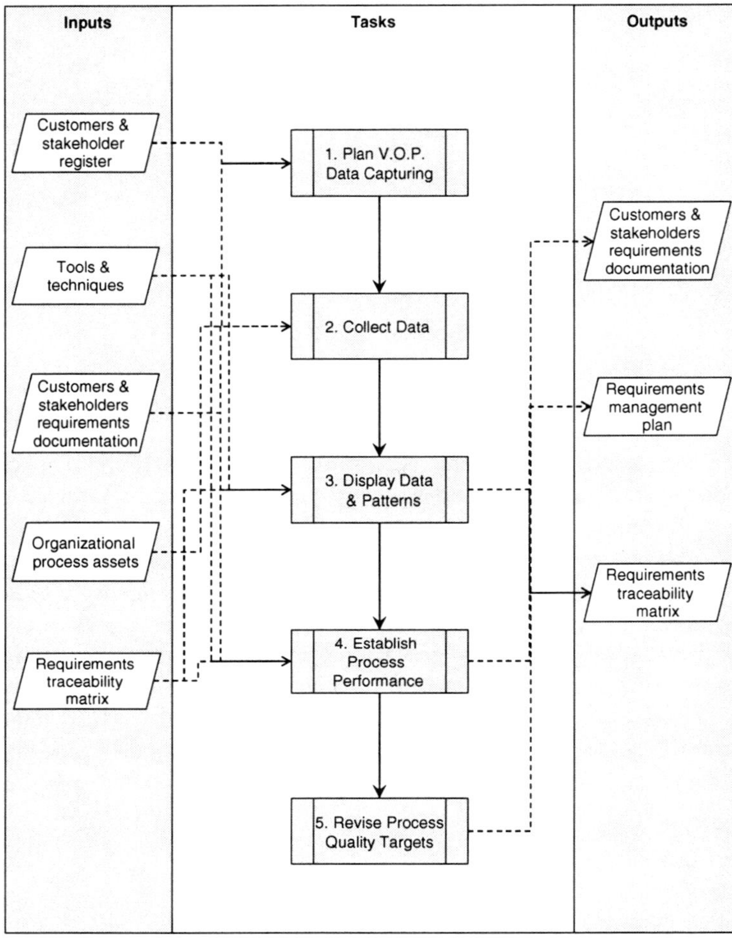

Fig. 14.1 The V.O.P. management process

14.1.1 Identify V.O.P. Data and Clarify Goals

The first step in planning for V.O.P. data collection, as with any data collection, is to identify the V.O.P. data and clarify goals. The purpose here is to ensure that the V.O.P. data, which the project team collects, will provide the answers needed to carry on the "process improvement" project successfully. Knowing what type of data the project team will be dealing with also tells which tool should be used to capture it. The right V.O.P. data should:

1. Describe the issue or problem that the "process to be improved" is facing;
2. Describe related conditions that might provide clues about causes of underperformance of the "process to be improved";
3. Lead to analysis in ways that answer the project team questions.

Desired V.O.P. data characteristics are: sufficient, relevant, representative, and contextual. As with customers' requirements, there are two types of data: qualitative and quantitative data.

Qualitative V.O.P. data are obtained from description of observations or measures of the process outcomes in terms of words and narratives statements. They can be grouped by highlighting key words, extracting themes, and elaborating concepts.

Quantitative V.O.P. data are obtained from description of observations or measures of the process outcomes in terms of measurable quantity and in which a range numerical values are used without implying that a particular numerical value refers to a particular distinct category. Nevertheless, data originally obtained as qualitative information about description of observations of the process outcomes may give rise to quantitative data if they are summarized by means of counts; and conversely, data that are originally quantitative are sometimes grouped into categories to become qualitative data.

As recommended during the process of collecting customers requirements, one of the most important things that the project team should also do in planning for V.O.P. data collection is to draw and label the graph that will communicate the findings before the actual V.O.P. data collection begins. This directs the project team to exactly what V.O.P. data is needed. Moreover, it raises questions that the project team might not have though of, which it can add to the planning. This will prevent having to return for V.O.P. data that the project team had not though of.

14.1.2 Develop Operational Definitions and Procedures

An operational definition for a V.O.P. data is a description of term as applied to a specific situation of the "process improvement" project to facilitate the collection of meaningful (standardized) V.O.P. data.

When collecting V.O.P. data it is important to define terms very clearly in order to assure that all the appraisers or people collecting and analyzing the data have the same understanding. Any V.O.P. data for which an "operational definition" has not been defined often will lead to inconsistencies and erroneous results.

With processes, as with customers, it is easy to assume that those collecting the data understand what and how to complete the task. However, appraisers or people collecting data have different opinions, views, and working habits and these will affect the V.O.P. data collection. As a result, operational definitions should also be very precise for the V.O.P. and be written to avoid possible variation in interpretations and to ensure consistent and quality data collection.

The procedures associated with an operational definition for a V.O.P. defines exactly how the project team will proceed to collect and record the V.O.P. data. The template shown in Fig. 8.2 is equally valid for use for V.O.P. data collection. During this planning step, the following must also be considered by the project team:

1. Importance of the Voice of the Process (V.O.P.) data;
2. Accuracy of V.O.P. data;
3. Completeness of V.O.P. data capturing.

Importance of the Voice of the Process (V.O.P.) data—Whereas the V.O.C. communicates customers' needs and expectations, the V.O.P. communicates information about the performance of the process under consideration for improvement. In detailed product (resp. Service) improvement and development, the V.O.P. is key the source of information on the performance of the product (resp. service) characteristics; also, enough information must be gathered so that the mappings with the customer requirements in the product (resp. Service) improvement and development process can be carried out flawlessly. Therefore, operational definitions and procedures for the V.O.P. needs should be developed as these are the source of information for both the strategically important business value proposition, and all the building blocks in the product (resp. Service) design.

Accuracy of V.O.P. data—The V.O.P. data must be captured accurately because of the importance of the performance of the process been considered. Only with accurate V.O.P. data can an accurate business value proposition and "process to be improved" outcome performance specification be developed.

Completeness of V.O.P. data capturing—Not only accurate capturing of the voice of the process must be performed, but a sufficient amount of V.O.P. information needed to define process characteristics. The project team must collect enough V.O.P. information to build, as precisely as possible, a factual understanding of existing "process to be improved" conditions and problems or causes of underperformance.

14.1.2.1 V.O.P Data Collection Sources

Unless the "process to be improved" is highly unstable, there is ideally only one voice of the process. Information on the "process to be improved" performance outcomes can only come from observational studies and/or experimental studies.

Experimental Studies

An experimental study is a methodical procedure carried out with the goal of observing, verifying, explaining, or establishing the validity of a hypothesis. Experimental studies vary greatly in their goal and scale, but always rely on repeatable procedure and logical analysis of the collected results. This is described in the following chapter.

Observational Studies

A "process improvement" project is concerned with optimizing a system or getting the "process to be improved" to a higher performance. Therefore, the project team cannot depend solely upon experimental results which are always obtained in a limited context. The project team has to deal with the response variable in the presence of all of the factors that have an impact upon it. The project team cannot simply study some factors and ignore others. But, of necessity, every experiment will choose some factors and exclude other factors. So while the project team may begin with a set of experiments, it needs to remember that limited results and conditional relationships do not tell the whole story. Eventually the project team will need a holistic approach, and this is what observational studies provide.

With an observational study, the data arise as a side effect of some continuing operation or on-going execution of the "process to be improved." It may take longer to discover things with an observational study, but all the possible interactions and all the various factors are present and are allowed to make their contribution to the results of the study. When a factor makes its presence known in an observational study, the project team can be certain that it is a dominant factor.

With an observational study the clues to the source of any particular behavior will come from the context for each observed event. Here the key to discovery is the connection between context and the observed behavior. The V.O.P data will have to be interpreted in terms of their context. Moreover, since any of the input variables is not been ignored in an observational study, there is no need for any insurance device, like randomization. In fact, any attempt to impose randomization on an observational study will merely result in confusion. In an observational study some of the most important information may consist of the time order sequence for the data. Therefore, with an observational study, any careful data to be collected must preserve the time-order sequence of the data.

14.1.2.2 Prioritize V.O.P Data

Since data collection can consume a tremendous amount of time, it is critical that the project team focus on the inputs that matter the most. We have defined a process as "a set of logically related discrete elements (tasks, actions, or steps) taken in order to achieve a particular end." Furthermore, most process outcomes (products and services) result from a complex system of interaction among its inputs (i.e. people, equipment, procedures, methods, equipment, materials, and environment). Once all discrete elements of the "process to be improved" have been assessed and broken down into their critical elements, two funneling tools can be used to prioritize the V.O.P. data: a prioritization matrix and a FMEA matrix.

Prioritization Matrix

There are two applications for a prioritization matrix: linking response variables to identified process key requirements and linking input and process variables to response variables. A prioritizations matrix, as shown in Table 14.1, can be used when the project team has determined that too many input variables might have an impact of the response variable and collecting data on all possible variables would cost too much time and resources (including money).

The following are steps to be followed to construct a prioritization matrix:

1. List all response variables, as shown in Table 14.1.
2. Rank and assign weights to the response variables.
3. List all input variables.
4. Evaluate the strength ρ of the relationships between responses and inputs variables.
5. Cross multiply weight and strength of relationships. The combinations with the highest total are the inputs on which the project team needs to focus the improvement efforts on.
6. Highlight the critical few variables that matter the most from the computed totals.

Table 14.1 Prioritization matrix template

Response variables		Y_1	Y_2	...	Y_j	...	Y_M	Total
Weights		w_1	w_2	...	w_j	...	w_M	
Inputs variables	X_1	p_{11}			p_{1j}			
	X_2							
	...							
	X_i	p_{i1}			p_{ij}			$\sum_{j=1}^{M} w_j * p_{ij}$
	...							
	X_N							

Failure Mode and Effect Analysis (FMEA)

The Failure Mode and Effect Analysis (FMEA) is an effective step-by-step approach for focusing the data collection effort on those inputs variables that matter the most (i.e. those related to the identified process critical elements) for the current "process to be improved." It is a structured approach to identify, estimate, prioritize and evaluate risk associated with execution of the identified "process to be improved" critical elements.

Failures are unwanted features of a characteristic of a "process to be improved" outcomes; it is any errors or defects, especially ones that affect the customer, and can be potential or actual. "Effects analysis" refers to studying the consequences of those failures.

In the FMEA approach, failures are prioritized according to how serious their consequences are, how frequently they occur and how easily they can be detected. The purpose of the FMEA in performing quality planning is to focus the data collection effort by documenting current knowledge and actions about the risks associated with failures.

The project team should use this approach when there is not a clear understanding about what the important variables are and how they affect the response variable. The following are steps to be followed to construct a FMEA matrix; specific details may vary with standards of the enterprise business or industry.

1. *Identify potential failure modes.* These are all the ways in which outcomes of the "process to be improved" is failing to meet performance requirements. It is not unusual for an FMEA to list 50 to 200 different potential failures. If an FMEA has over 200 potential failures it is a good sign that the product or process under investigation should be broken into subunits, each with its own FMEA. For example, automotive companies do not conduct FMEAs on the entire car, but rather on individual components and sub-components of the car.

2. *Identify potential effects or consequences of each failure*. For each failure mode, the project team should identify all the consequences on the system, related systems, process, related processes, product, service, customer or regulations within the enterprise business. To achieve this, the project team should find the answers to the questions "What does the customer experience because of this failure?" and "What happens when this failure occurs?"
3. *Rate the severity of effects or consequences of each failure, or S*. Severity is usually rated on a scale from 1 to 10, where 1 is insignificant and 10 is catastrophic. If a failure mode has more than one effect, the project team should write on the FMEA table only the highest severity rating for that failure mode.
4. *Identify potential root causes of these effects*. The project team could use root causes analysis tools, as well as the best knowledge and experience of the team to achieve this. List all possible causes for each failure mode on the FMEA form shown in Table 14.2.
5. *Rate the likelihood of occurrence of the potential root causes of these effects*. The likelihood of occurrence, which is denoted by or O, estimates the probability of failure occurring for that reason during the lifetime of the "process to be improved" scope. Occurrence is usually rated on a scale from 1 to 10, where 1 is extremely unlikely and 10 is inevitable. On the FMEA table, list the occurrence rating for each cause.
6. *For each cause, identify current process controls*. These are tests, procedures or mechanisms that the enterprise business currently has in place to keep failures from reaching the customer. These controls might prevent the cause from happening, reduce the likelihood that it will happen or detect failure after the cause has already happened but before the customer is affected.
7. *Rate the project team's ability to detect failure modes*. For each control, determine the detection rating, or D. This rating estimates how well the controls can detect either the cause or its failure mode after they have happened but before the customer is affected. Detection is usually rated on a scale from 1 to 10, where 1 means the control is absolutely certain to detect the problem and 10 means the control is certain not to detect the problem (or no control exists). On the FMEA table, list the detection rating for each cause.
8. *Calculate the risk priority number, or RPN, which equals* $S \times O \times D$. Also calculate Criticality by multiplying severity by occurrence, $S \times O$. These numbers provide guidance for ranking potential failures in the order they should be addressed. The combinations with the highest RPNs are the potential failures that the project team needs to focus the improvement efforts on.
9. *Identify recommended actions*. These actions may be design or process changes to lower severity or occurrence. They may be additional controls to improve detection.

This procedure formally documents standard practice, generates a historical record, and serves as a basis for future improvements.

The result of prioritizing is a set of selected inputs that matter the most.

Table 14.2 FMEA matrix template

Project:________								Date:________(original)							
Team:________								________(revised)							
Process step	Potential Failure Mode	Potential Effects of Failure	Severity	Potential Causes	Occurrence	Current Controls	Detection	RPN	Recommended Actions	Responsibility & Target Data	(After) Action Taken	Severity	Occurrence	Detection	RPN
Total Risk Priority Number:									After Risk Priority Number:						

14.1.3 Develop Sampling Strategy

From all the possible interactions between selected inputs and extraneous factors the totality of all associated responses and about which the data should be collected can still be relatively large and it might not be possible, nor it is necessary, to collect information from the total population considered. It is incumbent on the project team to clearly define the target population. There are no strict rules to follow, and the project team must rely on logic and judgment. The population is defined in keeping with the questions to be answered and the objectives of capturing the V.O.P.

Sometimes, the entire population will be sufficiently small, and the project team can include the entire population in the study. Collecting the V.O.P. data in this case is called a "census V.O.P. data collection" because data is gathered on every input and associated response of the target population.

Usually, the target population is too large for the project team to attempt to observe and record all data. A small, but carefully chosen sample can be used to represent the population. The sample should reflects the characteristics of the population from which it is drawn and the goal in choosing a sample is to have a picture of the population, disturbed as little as possible by the act of gathering information. The sampling methods described in a previous section can be used to achieve this purpose.

14.1.4 Validate V.O.P. Data Collection System

The "data collection system" consists of data sample, appraisers or people executing the data collection tasks, operational definitions and procedures followed to collect the data. The events associated with any one of these constituents are not conveyed to the other constituents; that is, the constituents of a "data collection system" are statistically independent. Validation of the V.O.P data collection system follows the procedure described in the validation of the V.O.C. data collection system in a previous section.

14.2 Collect Data

Once the plan for collecting the V.O.P. data is established, the next step is to begin the actual data collection from the determined sample.

14.3 Summarize Data & Display Patterns

The major purposes of summarizing the collected V.O.P. data and displaying their patterns within the "Perform Quality Planning" project management process are:

1. To help get the "process to be improved" into a "satisfactory state," which one might then be content to monitor if not persuaded by arguments for the need of improvement.
2. To provide preliminary route to investigate what can be accomplished by operating the current "process to be improved" up to its full potential. To get a "process to be improved" to operate up to its full potential it is necessary to operate a process predictably. To operate a process predictably is to operate that process with minimum variance. Unpredictable operation will inevitably increase the variation, which will lower the capability indexes and increase the effective cost of production and use of the process outcome. As Donald

J. Wheeler indicates in his column (Wheeler, The Effective Cost of Production and Use: How to turn capability indexes in dollars, 2010a; Wheeler, What Is the Zone of Economic Production? And how can you get there?, 2010b), the effective cost of production and use of the process outcome is defined to be the ratio of the actual cost of production and use of the process outcome by its nominal cost of production. The actual cost of producing and using a process outcome will consist of the nominal cost of production plus the average excess costs per unit associated with producing and using such process outcomes. These excess costs can be broken down into three components: the costs of scrap; the costs of rework; and the excess costs associated with the use of conforming process outcome. Since experience has shown that the predictable operation of a process will generally cost little or nothing, this particular piece of low-hanging fruit is the cheapest type of process improvement possible. Capital expenditures are seldom required when operating a process predictably. Moreover, operating a process predictably and on target is observed in practice when the process capability index is in the neighborhood of 1.5 to 2.0 or larger, making further improvements unnecessary. Operating a "process to be improved" on-target is a necessity simply because, regardless of how large the process capability ratio might be, operating off-target can increase the effective cost of production and use of the process outcomes. Operating a process predictably requires a learning enterprise business; i.e. one where knowledge is both gained and shared. This happens through continuous practice of a way of thinking rather than through implementation of the "right" technique. Without the practice in the way of thinking, simply developing a control chart on a new process and displaying it over the wall to production will not result in predictable operation.

Process behavior charts will help to measure what the "process to be improved" is doing and to determine when the "process to be improved" is not operating up to its full potential. Process behavior charts also help to identify opportunities for process improvements and provide a way to continue to operate a "process to be improved" up to its full potential in the future (i.e. to control it).

In the process of developing process behavior charts, the first question to be asked once the V.O.P data have been collected is "What can be learned from these data?" The second question to be asked is "Does the process conforms to its quality goals?" The answer to this question is an understanding and a summary of the collected data in some meaningful graphical formats. A well-chosen graphical format conveys an enormous amount of quantitative information that a trained eye can detect quickly and extract salient features. Even for small sets of data, there are many patterns and relationships that are considerably easier to discern in graphical display. The commonly used graphical formats are, but are not limited to:

1. Control Charts
2. Run Charts
3. Scatter Diagrams
4. Frequency Plots
5. Pareto Charts

Outcomes of business activities can be products, transactions, services delivered, sub-parts or particular features of these entities. In the remaining of this chapter, we will also use the term "element" as a generic term to designate a measurable feature or a measurable characteristic of these entities. We will also considering each element as a balanced sum of a large enough number of unobserved random events acting additively and independently, each of which with finite mean and variance. As a consequence, the central limit theorem tells us that the occurrence pattern of these elements will tend to follow a normal distribution in nature.

14.3.1 Control Charts

A "process to be improved" will either display "controlled variation" or it will display "uncontrolled variation." If it displays controlled variation then, according to Deming (Shewhart, Economic Control of Quality of Manufactured Product, 1931), "it will not be profitable to try to determine the cause of individual variations." When a process displays controlled variation its behavior is indiscernible from what might be generated by a "random" or "chance" process (i.e. tossing coins or throwing dice). When a process displays controlled variation the individual variations may be thought of as being created by a constant system of a large number of "chance causes" in which no cause produces a predominating effect. On the other hand, when a process displays uncontrolled variation, then "it will be profitable to try to determine and remove the cause of the uncontrolled variation."

Given this distinction, the control chart is a technique used to indicate the quality status of the "process to be improved" by detecting which type of variation is displayed. The objective is provide the project team a guide for taking appropriate action—to look for "assignable causes" of variations when the data display uncontrolled variations, and to avoid looking for "assignable causes" when the data display controlled variations.

A control chart is a graph used to study how a process changes over time. Control charts graphically answer the question: "Is the variance of the 'process to be improved' within acceptable limits?" The pattern of data points on a control chart may reveal random fluctuating values, sudden process jumps, or a gradual trend in increased variation.

By monitoring the output of a "process to be improved" over time, a control chart can help assess whether the application of process changes resulted in the desired improvements. When the observed variations of a characteristic of the outcome of the "process to be improved" are within acceptable limits, the "process to be improved" is said to be in a "state of control" and does not need to be adjusted. Conversely, when the observed variations of a characteristic of the outcome of the "process to be improved" are outside acceptable limits, the "process to be improved" should be adjusted.

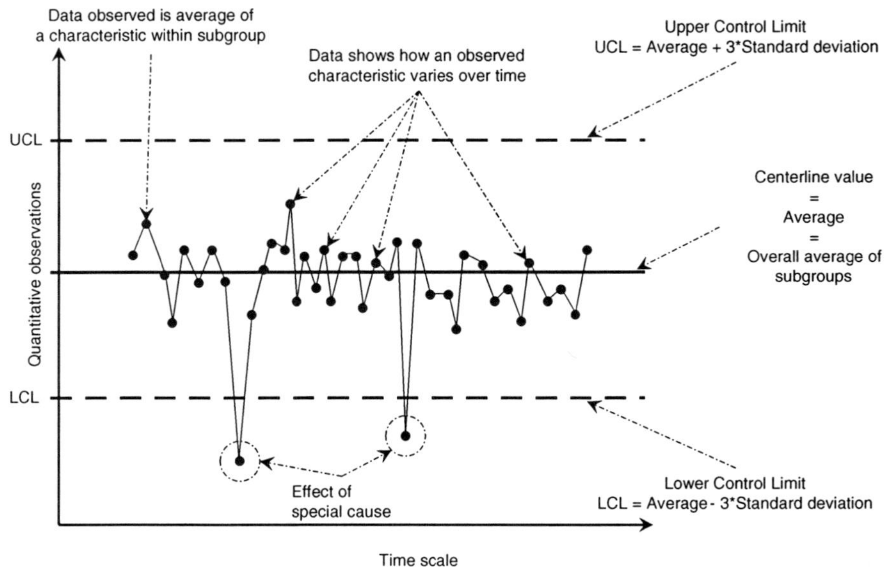

Fig. 14.2 Example of control chart

A control chart, illustrated in Fig. 14.2, consists of:

1. Time-ordered points representing estimates of characteristics or parameters (e.g., a mean, a range, a proportion, a standard deviation) of subgroups of outcomes of the "process to be improved" in samples taken from the V.O.P. data.
2. A center line drawn at the overall average value of estimates of characteristics or parameters (e.g., a mean, a range, a proportion, a standard deviation) of subgroups of outcomes of the "process to be improved" in samples taken from the V.O.P. data.
3. Upper and lower control limits that indicate the threshold at which the process output is considered statistically "unlikely," drawn typically at ± 3 standard deviations from the center line.
   The chart may have other optional features, including:
4. Upper and lower warning limits, drawn as separate lines, typically two standard deviations above and below the center line
5. Division into zones, with the addition of rules governing frequencies of observations in each zone
6. Annotation with events of interest, as determined by the project team.

There are several types of control charts and a correct control chart selection is a critical part of creating a control chart. If the wrong control chart is selected, the control limits will not be correct for the collected data. The type of control chart to be used is determined by the type of data to be plotted and the format in which these data has been collected. The following are description of the commonly used control charts.

14.3.1.1 P-Charts

A p-chart is a control chart on attributes data, used to monitor the proportion of nonconforming elements in a sample; with the sample proportion nonconforming been defined as the ratio of the number of nonconforming elements to the sample size. The binomial distribution is the basis for the p-chart and requires the following assumptions:

1. The probability of nonconformity p is the same for each considered element;
2. Each element is independent of its predecessors or successors;
3. The treatment is same for each sample and is carried out consistently from sample to sample.

The binomial distribution is the discrete probability distribution of the number of successes in a sequence of n independent yes/no experiments, each of which yields success with probability p. In general, if a random variable K follows the binomial distribution with parameters n and p, we write $K \sim B(n,p)$. The probability of getting exactly k successes in n trials is given by the following probability mass function:

$$f(k;n,p) = \frac{n!}{k!(n-k)!}p^k(1-p)^{n-k}, \quad \text{for k} = 0, 1, 2, \ldots, \text{n}$$

The mathematical expectation and the variance associated with the binomial distribution are given by:

$$Average = E(K) = np, \qquad Variance = Var(K) = np(1-p)$$

Using an estimate $\hat{p}$ of the expectation of collected data from a sample of size n, the control limits for the p-chart type are given by:

$$\hat{p} \pm 3\sqrt{\frac{\hat{p}(1-\hat{p})}{n}}$$

Naturally, if the lower control limit is less than or equal to zero, process observations only need be plotted against the upper control limit. Note that observations of proportion nonconforming below a positive lower control limit are cause for concern as they are more frequently evidence of improperly calibrated test and inspection equipment or inadequately trained appraisers than of sustained quality improvement.

14.3.1.2 NP-Charts

An np-chart is a control chart on attributes data, used to monitor the number of nonconforming elements in a sample. It is an adaptation of the p-chart and used in situations it easier to interpret the "process to be improved" performance in terms of concrete numbers of elements rather than the somewhat more abstract proportion of elements. The np-chart differs from the p-chart in only the three following aspects:

1. The control limits given by:

$$n\hat{p} \pm 3\sqrt{n\hat{p}(1 - \hat{p})}$$

2. The number nonconforming, rather than the fraction nonconforming p, is plotted against the control limits.
3. The sample size, n, is constant.

14.3.1.3 C-Charts

The c-chart is a control chart on attributes data, used to monitor "count"-type data, typically total number of nonconformities per element. It is also occasionally used to monitor the total number of events occurring in a given unit of time. The c-chart differs from the p-chart in that it accounts for the possibility of occurrence of more than one nonconformity per inspection element. The p-chart models "pass"/"fail"-type or "yes"/"no"-type inspection only. Nonconformities may also be tracked by type or location, which can prove helpful in tracking down assignable causes. Examples in the automobile industry of processes suitable for monitoring with a c-chart include:

1. Monitoring the number of voids per inspection unit in injection molding or casting processes;
2. Monitoring the number of discrete components that must be re-soldered per printed circuit board;
3. Monitoring the number of product returns per day.

Poisson's distribution is the basis for the c-chart. It requires the following assumptions:

1. The number of opportunities or potential locations for nonconformities is very large;
2. The probability of nonconformity at any location is small and constant;
3. The inspection procedure is same for each sample and is carried out consistently from sample to sample.

Poisson's distribution (or Poisson law of small numbers) is a discrete probability distribution that expresses the probability of a given number of events occurring in a fixed interval of time and/or space if these events occur with a known average rate and independently of the time since the last event. In general, if a random variable K follows Poisson's distribution where the expected number of occurrences in a given interval is λ, we write $K \sim P(\lambda)$. The probability that there are exactly k occurrences (k being a non-negative integer, k = 0, 1, 2, ...) is equal to:

$$f(k;\ \lambda) = \frac{\lambda^k e^{-\lambda}}{k!}$$

Where:

1. e is the base of the natural logarithm (e = 2.71828...);
2. k is the number of occurrences of an event—the probability of which is given by the function;
3. λ is a positive real number, equal to the expected number of occurrences during the given interval. For instance, if the events occur on average four times per minute, and one is interested in the probability of an event occurring k times in a 10 min interval, one would use a Poisson distribution as the model with $\lambda = 10 \times 4 = 40$.

Poisson's distribution can be derived as a limiting case of the binomial distribution. The mathematical expectation and the variance associated with the binomial distribution are equal to the parameter λ. The control limits for this c-chart type are $\widehat{\lambda} \pm 3\sqrt{\widehat{\lambda}}$ where $\widehat{\lambda}$ is the estimate of λ.

14.3.1.4 U-Charts

The u-chart is a control chart on attributes data, used to monitor "count"-type data where the sample size is greater than one, typically the average number of nonconformities per elements. The u-chart differs from the c-chart in that it accounts for the possibility that the number or size of inspection elements for which nonconformities are to be counted may vary. Larger samples may be an economic necessity or may be necessary to increase the area of opportunity in order to track very low nonconformity levels. An example in the automobile industry of a process suitable for monitoring with a u-chart is the monitoring of the number of nonconformities per lot of raw material received where the lot size varies.

As with the c-chart, Poisson's distribution is the basis for the u-chart and requires the same assumptions. Using an estimate $\hat{u}$ of the expectation of collected data from a sample of size n, the control limits for the u-chart type are given by:

$$\hat{u} \pm 3\sqrt{\hat{u}/n}$$

Using the u-chart, observations plotted against these control limits are the ratios of nonconformities in subgroup to the number of inspection elements in the subgroup.

14.3.1.5 $\bar{X}$ and R Charts

The $\bar{X}$ and R Charts are of control charts on variables data, used to monitor averages and ranges when samples are collected in rational subgroups. In the rational subgroup, all the common causes of variation are assumed to have been represented and none of the assignable causes of variation. These charts allow to clearly separate changes in process average from changes in process variability. They are advantageous in the following situations:

1. The sample size is relatively small (say, n $\leq$ 10—and $\bar{X}$ and s charts are typically used for larger sample sizes);
2. The sample size is constant;
3. Humans must perform the calculations for the charts.

Within the $\bar{X}$ and R charts, the R chart is used to monitor the range (as approximated by the sample moving range) of a characteristic of the "process to be improved" outcomes and the $\bar{X}$ chart is used to monitor the average of a characteristic of the "process to be improved" outcomes.

The normal distribution is the basis for $\bar{X}$ and R charts and requires the following assumptions:

1. The quality characteristic to be monitored is adequately modeled by a normally-distributed random variable;
2. The parameters μ and σ for the random variable are the same for each element and each element is independent of its predecessors or successors;
3. The treatment procedure is same for each sample and is carried out consistently from sample to sample;

Using the overall average range $\bar{R}$ of the subgroups considered, the lower and upper control limits for monitoring the range of a characteristic of the "process to be improved" outcomes are given by:

$$LCL = D_3\bar{R}, \qquad UCL = D_4\bar{R}$$

The control limits for monitoring the average of a characteristic of the "process to be improved" outcomes are given by:

$$\bar{X} \pm A_2\bar{R}$$

Where A_2, D_3, and D_4 are sample size-specific anti-biasing constants given in Tables 8.2–8.5.

Decisions made based on the results of the $\bar{X}$ and R charts can be used only when the variability within the samples considered is constant. Thus, the practice is to examine the R chart before examining the $\bar{X}$ chart; if the R chart indicates that the sample variability is in statistical control, then the $\bar{X}$ chart is examined to determine if the sample mean is also in statistical control. If on the other hand, the sample variability is not in statistical control, then the entire process is judged not to be in statistical control regardless of what the $\bar{X}$ chart indicates.

14.3.1.6 $\bar{X}$ and s Charts

The $\bar{X}$ and s Charts are of control charts on variables data, used to monitor averages and standard deviations when samples are collected in rational subgroups. In the rational subgroup, all the common causes of variation are assumed to have been represented and none of the assignable causes of variation. These charts allow to clearly separate changes in process average from changes in process variability. They are advantageous in the following situations:

1. The sample size is relatively large (say, n > 10—and $\bar{X}$ and R charts are typically used for small sample sizes);
2. The sample size is variable;
3. Computers can be used to ease the burden of calculation.

Within the $\bar{X}$ and s charts, the s chart is used to monitor the standard deviation (as approximated by the sample moving standard deviation) of a characteristic of the "process to be improved" outcomes and the $\bar{X}$ chart is used to monitor the average of a characteristic of the "process to be improved" outcomes.

The normal distribution is the basis for $\bar{X}$ and s charts and requires the following assumptions:

1. The quality characteristic to be monitored is adequately modeled by a normally-distributed random variable;
2. The parameters μ and σ for the random variable are the same for each element and each element is independent of its predecessors or successors;
3. The treatment procedure is same for each sample and is carried out consistently from sample to sample;

Using the overall average standard deviation $\bar{s}$ of the subgroups considered, the lower and upper control limits for monitoring the standard deviation of a characteristic of the "process to be improved" outcomes are given by:

$$LCL = B_3\bar{R}, \qquad UCL = B_4\bar{R}$$

The control limits for monitoring the average of a characteristic of the "process to be improved" outcomes are given by:

$$\bar{X} \pm A_3\bar{s}$$

Where A_3, B_3, and B_4 are sample size-specific anti-biasing constants given in Tables 8.2–8.5.

Decisions made based on the results of the $\bar{X}$ and s charts can be used only when the variability within the samples considered is constant. Thus, the practice is to examine the s chart before examining the $\bar{X}$ chart; if the s chart indicates that the sample variability is in statistical control, then the $\bar{X}$ chart is examined to determine if the sample mean is also in statistical control. If on the other hand, the sample variability is not in statistical control, then the entire process is judged not to be in statistical control regardless of what the $\bar{X}$ chart indicates.

14.3.1.7 Steps to Construct a Control Chart

The following are steps to be used to construct a control chart:

1. *Choose the process outcome quality characteristic to be charted.* In making this choice, there are several things to consider:

- Choose a process outcome quality characteristic that is currently experiencing a high number of nonconformities or items that do not conform. A Pareto analysis is useful to assist the process of making this choice.
- Identify the process variables contributing to the end-product characteristics to identify potential charting possibilities.
- Choose process outcome quality characteristics that will provide appropriate data to identify and diagnose problems. In choosing characteristics, it is important to remember that attributes provide summary data and may be used for any number of characteristics. On the other hand, variables data are used for only one characteristic on each chart but are necessary to diagnose problems and propose action on the characteristic.
- Determine a convenient point in the process considered to locate the chart. This point should be early enough to prevent nonconformities and to guard against additional work on nonconforming items.

2. *Choose the type of control chart.*
   - The first decision is whether to use a variables chart or an attributes chart. A variables chart is used to control individual measurable characteristics, whereas an attributes chart may be used with go no-go type of inspection. An attributes chart is used to control percentage or number of nonconforming items or number of nonconformities per item. A variables chart provides the maximum amount of information per item inspected. It is used to control both the level of the process and the variability of the process. An attributes chart often provides summary data that can be used to improve the process by then controlling individual characteristics.
   - Choose the specific type of chart to be used. If a variables chart is to be used, decide whether the average and range or the average and standard deviation are to be charted. If small shifts in the mean are important, a cumulative sum or exponentially weighted moving average chart may be used. The disadvantage of these two latter charts is that they are more difficult for the practitioner to use and understand. If subgroups are not possible, individual readings may be used, but these are to be avoided if possible. For attributes charts, the percentage nonconforming or number of nonconforming items may be charted. In some cases, the number of nonconformities per inspection item may be preferable.
3. Choose the center line of the chart and the basis for calculating the control limits. The center line may be the average of past data, the average of data yet to be collected, or a desired (standard) value. The limits are usually set at ± 3 standard deviations, but other multiples of the standard deviation may be used for other risk factors. As indicated in the previous chapter, during his work on "Economic Control of Quality of Manufactured Product" (Shewhart, Economic Control of Quality of Manufactured Product, 1931), Shewhart created the control chart with 3 standard deviations around the central tendency as a performance permissible limit of variations. Shewhart's use of 3 standard deviations limits, as opposed to any other multiple of standard deviations, did not stem from any specific mathematical computation. Rather, the choice of 3 standard deviations limits

was seen to be an acceptable economic value, and it was also justified by "empirical evidence that it works." Furthermore, the use of 3 standard deviations results in a negligible risk of looking for problems that do not exist, i.e., false alarms. However, this multiple may result in an appreciable risk of failing to detect a small shift in the parameter being studied. Smaller multiples increase the risk of looking for a false alarm but reduce the risk of failing to detect a small shift. The fact that it is usually much more expensive to look for problems that do not exist than to miss some small problems is the reason that the ± 3 standard deviations limits are usually chosen.

4. Choose the rational subgroup or sample. It should be pointed out that the term sample is usually used, but sample could mean an individual value, and samples of more than one are desirable for control charts if feasible. For variables charts, samples of size 4 or 5 are usually used, whereas for attributes charts, samples of 50–100 are often used. Attributes charts in fact may be used with 100 % inspection as a reflection of the underlying process involved. In addition to the size of the sample, the samples should be selected in such a way that the chance of a shift in the process is minimized during the taking of the sample (thus a small sample should be used); whereas the chance of a shift, if it is going to occur, is at a maximum between samples. This is the concept of rational sub-grouping. Thus it is better to take small samples periodically than to take a single large sample. Experience is usually the best method for deciding on the frequency of taking samples. That is, the known rate of a chemical change or the known rate of tool wear should be considered when making these decisions. If such experience is not available, samples should be taken frequently until such experience is gained.
5. Provide a system for collecting the data. If control charts are to become a shop tool during the "process improvement" project, the collection of data must be an easy task. Data collection must be made simple and relatively free of error. Data collection systems must give quick and reliable readings. If possible, the data collection system actually should record the data, since this will eliminate a common source of errors. Data sheets should be designed carefully to make the data readily available. The data sheets must be kept in a safe and secure place, free from dirt or oil.
6. Calculate the control limits and provide adequate instruction to all concerned on the meaning and interpretation of the results.

14.3.2 Run Charts

A run chart, also known as a run-sequence plot refers to a graph used to display observed data in a time series, and it typically represents an aspect of the performance of characteristics of the "process to be improved" outcomes. The run sequence plot displays data samples taken over a specific period of time.

Time is generally represented on the horizontal (x) axis and the property under observation on the vertical (y) axis. Often, some measure of central tendency (mean or median[1]) of the data is indicated by a horizontal reference or center line. A "run" is a sequence of consecutive observations on the same side of the median and that are each increasing larger or smaller than the previous observation. Because one often count runs on a time plot, these plots are called "run charts."

Run sequence plot charts are analyzed to focus attention on truly vital changes in the "process to be improved." This is done by detecting anomalies in data, or unusual data, which occur during a time series and suggest shifts in a process over time or special factors that may be influencing the variability of a process. Factors involved in analyzing anomalies include abnormally long series of consecutive decreases or increases in data above or below the centerline, and the total amount of such series in a data set.

Signals of assignable causes, which indicate that something in the "process to be improved" has changed, often include:

1. Too few or too many runs. Runs can be caused by faulty data collection instruments and equipment, calibration issues, and cumulative effects, among other things.
2. Six (6) or more points in a row continuously increasing or decreasing (i.e., indication of a trend in the process). A trend is a steady, gradual increase or decrease in the central tendency of the observed characteristic of the "process to be improved" over time. If all the conditions in the system where the "process to be improved" is operated stay constant, then the level of performance of the observed characteristic of the "process to be improved" will also stay constant. The presence of a trend in a graphical behavior plot is evidence that something out of the ordinary has happened to move the location of the behavior of the observed characteristic of the "process to be improved." In production environment, trends in performance are almost always caused by system factors that gradually change over time, like temperature, tool wear, machine maintenance, rising costs, etc.. . .
3. Eight (8) or more points in a row on the same side of the centerline (i.e., indication of a shift in the process). Shifts are sudden jumps, up or down, in the center of variation of the observed characteristic of the "process to be improved." They are evidence that something in the system considered changes permanently—a piece of equipment, a new operator, a change in material, a new procedure, etc.. . .
4. Fourteen (14) or more points in a row alternating up and down.

The number of runs expected to appear in a stable process depends on the number of data points used.

[1] The median of a set of collected data is the point along the scale of measure associated with the collected data where half of the data are below and half are above. It is the preferred measure of variation location when the collected data contains outliners or extreme data points occurring well outside of the range of the rest of the data.

The plots of periodic fluctuations exhibited by time series, or a statistical sequence of data points measured at uniform time intervals may be used after a run sequence plot is constructed to detect periodic fluctuation differences between group patterns and within group patterns. Plots of periodic fluctuations exhibited by time series, or a statistical sequence of data points measured at uniform time intervals use a horizontal axis to display time ordered by month. The vertical axis represents a time variable, or values directly dependent on time.

Run charts are similar in some regards to control charts, but do not show the control limits of the process. They are therefore simpler to produce, but do not allow for the full range of analytic techniques supported by control charts. In production environment, run charts can be used as a quick test of system performance. Start ups and short runs in manufacturing settings often produce too little data for conventional control chart analysis, but are easily analyzed in a run chart. Inspection data generated in these situations should be plotted immediately on a run chart to enable quick diagnosis of system changes over time or to identify signs that the process has begun to stabilize.

Run charts are also good tools to illustrate and share information with other departments. They are often used to post sales figures for all to see. Because run charts can be easily constructed, they are especially useful for one-time analysis of historical data.

14.3.3 Scatter Diagrams

A scatter plot or scatter graph is a type of mathematical diagram using Cartesian coordinates to display values for two variables for a set of data. The data is displayed as a collection of points, each having the value of one variable determining the position on the horizontal axis and the value of the other variable determining the position on the vertical axis.

The purpose of using scatter plots is to look at the relationship between the variables and determine if there are any problems/issues with the data or if the scatter plot indicates anything unique or interesting about the data, such as:

1. How is the data dispersed?
2. What does this imply about the questions and/or data in your study?

A scatter plot is used when a variable exists that is under the control of the appraiser. The inputs variable that is systematically incremented and/or decremented is also called the control parameter or independent variable and is customarily plotted along the horizontal axis. The output variable is also called dependent variable and is customarily plotted along the vertical axis. If no dependent variable exists, either type of variable can be plotted on both axis and a scatter plot will illustrate only the degree of correlation (not causation) between the two variables.

A scatter plot can suggest various kinds of correlations between variables with a certain confidence interval. Correlations may be positive (rising), negative (falling), or null (uncorrelated). If the pattern of dots slopes from lower left to upper right, it suggests a positive correlation between the variables being studied. If the pattern of

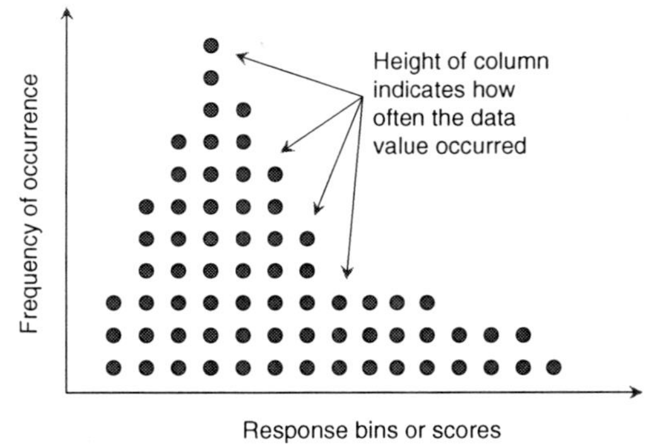

Fig. 14.3 Example of frequency (dots) plot

dots slopes from upper left to lower right, it suggests a negative correlation. A line of best fit (alternatively called "trend line") can be drawn in order to study the correlation between the variables. An equation for the correlation between the variables can be determined by established best-fit procedures. For a linear correlation, the best-fit procedure is known as linear regression and is guaranteed to generate a correct solution in a finite time. No universal best-fit procedure is guaranteed to generate a correct solution for arbitrary relationships.

A scatter plot is also very useful when the project team wishes to see how two comparable data sets agree with each other. In this case, an identity line, i.e., a $y = x$ line, or a 1:1 line, is often drawn as a reference. The more the two data sets agree, the more the scatters tend to concentrate in the vicinity of the identity line; if the two data sets are numerically identical, the scatters fall on the identity line exactly.

One of the most powerful aspects of a scatter plot, however, is its ability to show nonlinear relationships between variables. Furthermore, if the data is represented by a mixture model of simple relationships, these relationships will be visually evident as superimposed patterns.

14.3.4 Frequency Plots

A frequency plot is a graph or data set organized to show the frequency of occurrence of each possible outcome of a repeatable event observed many times. It summarizes how often different scores occur within a sample of scores. A frequency plot is constructed by dividing the response variable into equal sized intervals (or bins) and then, counting the number of occurrences of the response variable for each bin. The frequency plot, as shown in Fig. 14.3, then consists of:

1. A vertical axis, which consist of frequencies or relative frequencies;
2. A horizontal axis, which consist of response variable.

A histogram is the most commonly used frequency plot. It is a representation of a frequency distribution by means of rectangles whose widths represent class intervals and whose areas are proportional to the corresponding frequencies.

The relative heights of the bars represent the relative density of observations in the intervals. The total area of the histogram is equal to the number of data. A histogram may also be normalized to display relative frequencies. It then shows the proportion of data that fall into each of several intervals.

Dots plots and histograms are used for summarizing extremely large sets of data by reducing them to a single graph that can show primary, secondary and tertiary peaks in data as well as give a visual representation of the statistical significance of those peaks. They are widely used and thus are familiar even to most nontechnical people and without extensive explanation. This makes it a convenient way to communicate distributional information to general audiences. Common shapes of frequency plots are shown in Fig. 14.4.

If a frequency plot shows a bell-shaped, symmetric distribution, then the project team could conclude that no assignable causes of variations are indicated by the distribution. The collected data may come from a stable process or assignable causes of variation may be detected by a control chart or a run chart.

If a frequency plot shows a two-humped, bimodal distribution, then the project team could conclude that the "process to be improved" operates like two processes; two sets of operating conditions with two sets of outputs. The project team can use stratification to seek out the causes of these two humps.

If a frequency plot shows a long-tailed distribution, then the project team could conclude that the normality assumption on the outcomes of the "process to be improved" cannot be easily explained by the collected data. The project team must exercise caution when using data analysis techniques on the collected data as they might lead to erroneous conclusions.

If a frequency plot shows a basically flat distribution, then the project team could conclude that the outcomes of the "process to be improved" may be a mix of many operating conditions or these outcomes may be drifting over time. The project team should use run charts time series to track the "process to be improved" outcomes over time and look for possible stratifying factors.

If a frequency plot shows one or more outliners, then the project team could conclude that something unusual is happening to the "process to be improved" as outliners are often the result of clerical errors. The project team should confirm that these outliners are not clerical errors and treat them as assignable causes of variation.

If a frequency plot shows fewer distinct values, then the project team could conclude that the data collection system is not sensitive enough or the interval scales used for the frequency plot is not fine enough. The project team should refine the interval scales accordingly.

If a frequency plot shows a large pile-up of data points, then the project team could conclude that a sharp cut-off point occurs if the data collection system is incapable of collecting data across the complete range of data, or when appraisers ignore data that goes beyond a certain limit. The project team should improve the data collection system and also eliminate fear of reprisals for collecting "unacceptable" data.

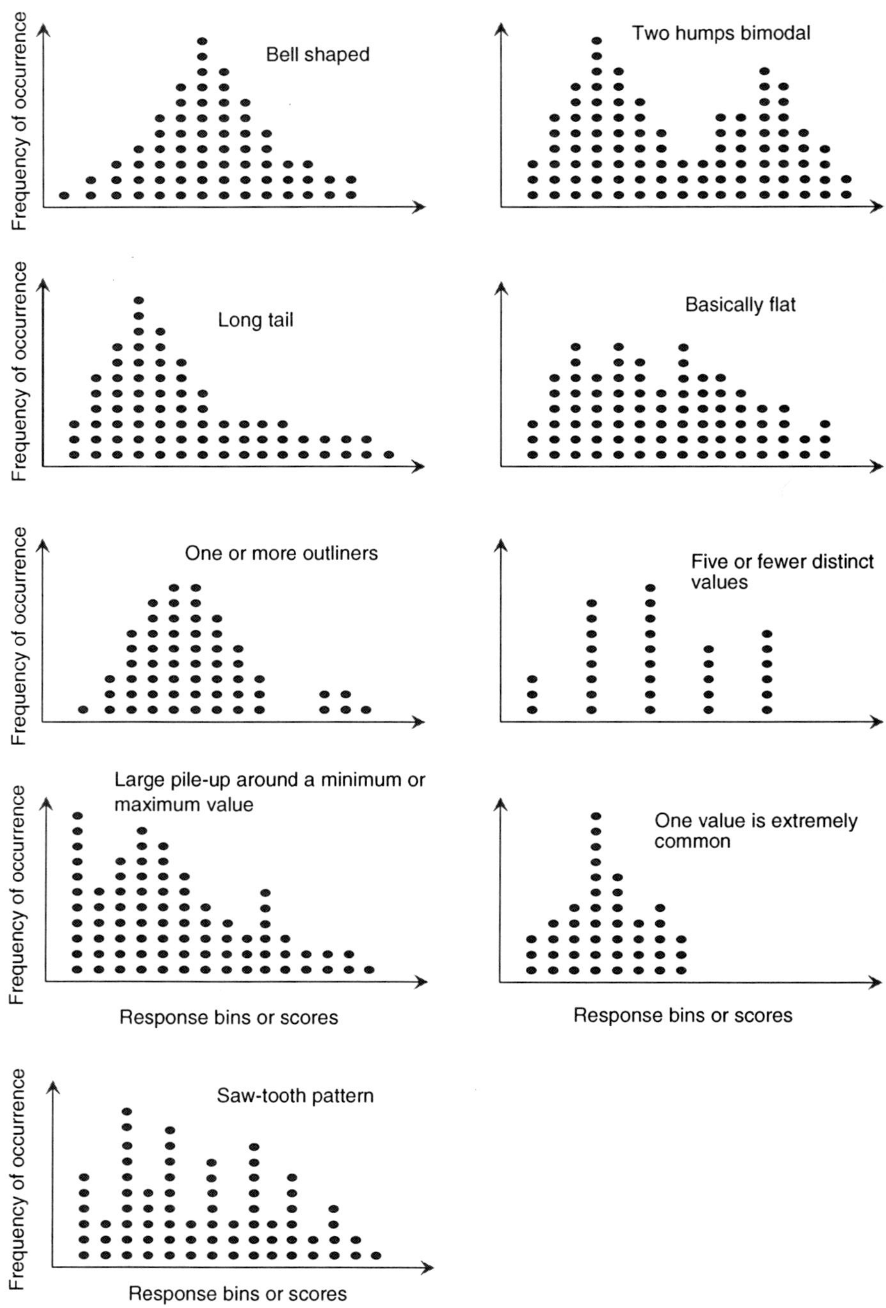

Fig. 14.4 Common shapes of frequency plots

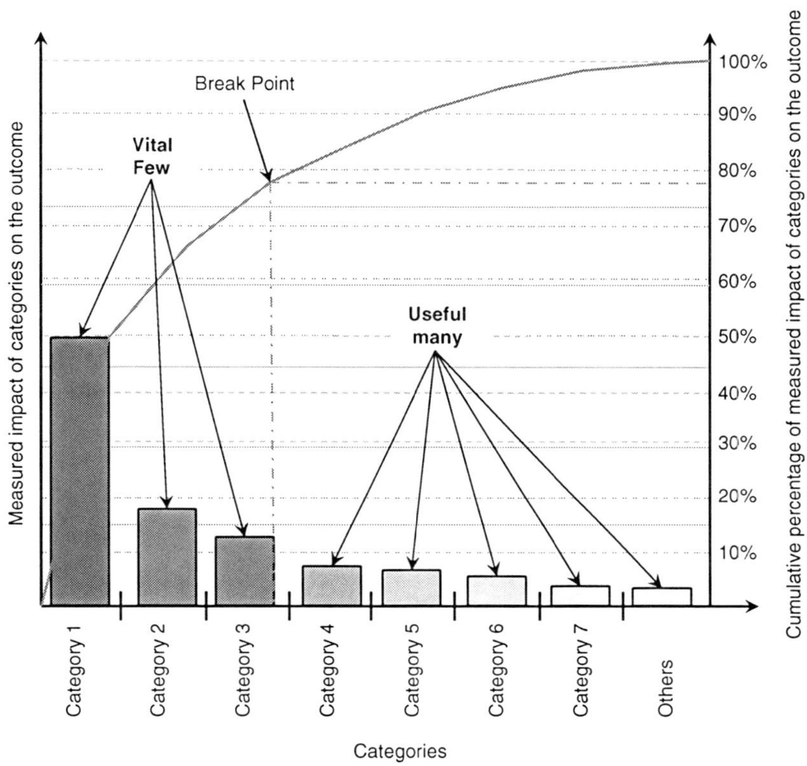

Fig. 14.5 Example of Pareto chart

If a frequency plot shows one value that is extremely common, then the project team could conclude that the appraiser may have a subconscious bias or that the data collection instrument used may be damaged. The project team should check the data collection procedures and the data collection instrument.

If a frequency plot shows a saw-tooth pattern, then the project team could conclude that the appraiser may have a subconscious bias for even (or odd) numbers or that the data collection instrument used may be easier to read at even (or odd) numbers. The project team should check the data collection procedures and the data collection instrument.

14.3.5 Pareto Charts

According to the "Pareto Principle," in any group of things which contribute to a common effect, relatively few contributors account for the majority of the effect.

A Pareto diagram, shows in Fig. 14.5, is a type of bar chart in which the various factors which contribute to an overall effect are arranged in order according to the magnitude of their effect. This ordering helps to identify the "vital few"—the factors or inputs that warrant the most attention—from the "useful many" factors or inputs that, while useful to know about, have a relatively smaller impact or effect on the "process to be improved" outcomes. Using a Pareto diagram helps a team concentrate its efforts on the factors or inputs that have the greatest impact on the "process to be improved" outcomes. It also helps a team communicate the rationale for focusing on certain areas.

The purpose of a Pareto diagram is to separate the significant aspects of a problem from the trivial ones. By graphically separating the aspects of the outcome of the "process to be improved," the project team will know where to direct its improvement efforts. Reducing the largest bars identified in the diagram will do more for overall improvement than reducing the smaller ones.

The Pareto chart is used to help you focus your improvement efforts on those issues that:

1. Cost the most;
2. Pose the highest risk/liability;
3. Occur the most often.

The following are steps to construct a Pareto chart:

1. Collect data about the contributing factors to the particular impact/effect on a characteristic of the response of the "process to be improved" and group them into categories.
2. Order the categories according to the magnitude of effect. If there are many insignificant categories, they may be grouped together into one category labeled "others." Make sure the "others" category (if the project team has chosen to have one) does not become unreasonably large. If the "other" category accounts for more than 25 % of the measured impact on the "process to be improved" outcome, then the project team should try to break it down.
3. Write the measured impact or magnitude of contribution next to each category and determine the grand total. Calculate the percentage of the total that each category represents.
4. Working from the largest category to the smallest, calculate the cumulative percentage for each category with all of the previous categories.
5. Draw and label the left vertical axis with the unit of comparison.
6. Draw and label the horizontal axis with the categories, largest to smallest from left to right.
7. Draw and label the right vertical axis "Cumulative Percentage," from 0 to 100 %, with the 100 % value at the same height as the grand total mark on the left vertical axis.
8. Draw a line graph of the cumulative percentage, beginning with the lower left corner of the largest category (the "0" point).
9. Analyze the diagram to indicate the cumulative percentage associated with the "vital few."

The Pareto principle implies that the project team can frequently solve a problem by identifying and attacking its "vital few" sources. There are two ways to analyze Pareto data depending on what the project team wants to know:

1. *Counts Pareto*: The project team should use this type of Pareto analysis to learn which category occurs most often.
2. *Cost Pareto*: The project team should use this type of Pareto analysis if it wants to know which category of problem is the most expensive in terms of some cost. A cost Pareto provides more details about the impact of a specific category, than a count Pareto can. Based on a count Pareto, the project team would be likely to tackle the problems that occurred very often first. However, suppose the problem that occurs very often financially costs less (total of all its occurrences) during a period of time than the problem that occurs few times. Based on the cost Pareto, the project team may want to tackle the more expensive problem first. To create a cost Pareto, the project team will need to know the categories, how often each occurred, and associate cost for each category.

Despite its simplicity, a Pareto chart is one of the most powerful of the tools for summarizing collected data. Getting the most from a Pareto chart includes making subdivisions, multi-perspective analyses, and repeat analyses. In most cases, two or three categories will tower above the others. These few categories which account for the bulk of the measured impact on the "process to be improved" outcome will be the high-impact points on which to focus. If in doubt, here are things that the project team should look for on a Pareto chart:

1. The project team should look for a break point in the cumulative percentage line. This point occurs where the slop of the line begins to flatten out. The categories under the steepest part of the curve are the most important.
2. If there is not a fairly clear change in the slope of the line, the project team should look for the categories that make up at least 60 % of the measured impact. These few can always been improved by redoing the Pareto analysis.
3. If the bars are all similar sizes or more than half of the categories are needed to make up the needed 60 %, the project team should try a different breakdown of categories that might be more appropriate.

14.4 Establish Process Performance

A process performance refers to how well the process achieves its goals. As indicated in the previous chapter, a process "performance measure" is a criterion of success stated in relation to the enterprise business intended strategy and the goal of a "performance measure" is to enable improvement. Three performance measures are often used during process improvement activities. These are:

1. Improve Process Yield;
2. Reduce Process Defect Rate;
3. Improve Process Capability.

14.4.1 Process Yield: Rolled Throughput Yield

Process Yield is a criterion used to control process performance. We can think of it as a percentage of "process to be improved" outcomes passing the compliance check (their key parameters fall within certain range of tolerance), in other words these outcomes will not be rejected as defective ones, so additional costs for repairing or scrapping the defective "process to be improved" outcomes will not be incurred to the enterprise business. A process yield uses the concepts of upper and lower specification limits (and a target limit between them) which are boundaries defining acceptable performance level.

All the outcomes of the "process to be improved" suiting the range between the upper and lower specification limits or precisely meeting a target limit will make up a process yield rate (the fluctuation of characteristics of these outcomes can be depicted on control charts). Yield loss (quality gaps) is caused by certain faults in the "process to be improved," entailing different deficiencies or shortcomings in the "process to be improved" key parameters. Yield loss rate can be classified by deficiency or defect types and this helps to pinpoint the problematic areas of the "process to be improved."

We have defined a process as "a set of logically related discrete elements (tasks, actions, or steps) taken in order to achieve a particular end." In this definition, a discrete element, the performance of which is measurable, is meant to be the smallest identifiable and essential piece of activity that serves both as a unit of work and as a means of differentiating between the various aspects of a project or an operation work. Each discrete element is designed to create unique outcomes by ensuring proper control, acting on and adding value to the resources that support the work being completed.

From the perspective of this definition, we can define the first pass yield (FPY) of a discrete process element as the number of deficiency free or defect free outcomes with no rework resulting from execution of the discrete element divided by the number of raw inputs going into execution of the discrete element over a specified period of time. Here, only deficiency free or defect free outcomes with no rework are counted as outcomes of the discrete element.

Also related, we can define the first time yield (FTY) of a discrete process element as the number of deficiency free or defect free outcomes including rework resulting from execution of the discrete element divided by the number of raw inputs going into execution of the discrete element over a specified period of time. Unlike the first pass yield, the first time yield captures the harsh reality (including rework) of the effectiveness of work associated with the discrete element considered.

The process first time yield is defined as the overall first time yield of the string of its logically related discrete elements. It is computed by multiplying the first time yields for each discrete element, creating what is also called the process rolled throughput yield (RTY):

$$RTY = \prod_{i} FTY_i,$$

Where FTY_i is the first time yield of the i-th discrete element of the process.

Like a chain that is only as strong as its weakest link, a process rolled throughput yield can never be greater than the lowest first time yield within the set of its logically related elements. Furthermore, a process rolled throughput yield erodes as the number of discrete element making up the process increases. To immediately improve the process performance, the project team should focus first on the individual discrete element with the lowest first time yield, then move onto the next discrete element with the lowest first time yield, and so on.

In the automobile industry, a very high individual first time yield must be achieved in order to have any hope of achieving an acceptable rolled throughput yield.

The purpose of calculating the "process to be improved" rolled throughput yield is to establish a process performance baseline. Once calculated, the project team should revisit and update the scope of the "process improvement" project. Significant differences in yields for the process discrete elements suggest creating a new map for the elements with the lowest yield.

14.4.2 Process Defect Rate

As we indicated in Chap. 2, in business applications which operate at a performance permissible limit of variations of z standard deviations, every process outcome within those business applications is intended to add value to the enterprise (businesses & customers) as a whole. It has a set of requirements or descriptions of what an element needs to add value to the enterprise. When a particular element meets those requirements, it is said that it has achieved quality, provided that the requirements accurately describe what the businesses and the customers actually need. Those process outcomes whose characteristics are falling beyond z standard deviations of the expected central tendency are often regarded as flaw, defective, unacceptable, or in non conformance quality. They will undergo more or less corrective actions: rework, scrapping (of whatever can not be reworked) and conformance use.

Establishing the rate at which defects occur on a characteristic of the "process to be improved" outcomes with respect to the number of "process to be improved" outcomes inspected is complementary to establishing the process yield. This defect rate or defect per ubiquitous outcome inspected (DPU) is often expressed as:

$$DPU = \frac{Total\ number\ of\ defects\ observed}{Total\ number\ of\ process\ outcomes\ inspected}$$

If the observed characteristic of the "process to be improved" outcomes is approximately normally distributed, then defect per ubiquitous outcome inspected is approximately equal to the area under the probability density function defined by the relation:

Table 14.3 12 Digits microsoft excel calculations of process yield p(z) and process fall out (1−p(z))

z	% falling within zσ Process yield	% falling beyond zσ Process fall out	Amount falling beyond zσ out of:		
			One thousand occurrences	One million occurrences	One billion occurrences
1.0	0.682689492137	0.317310507863	317.3	317310.5	317310507.9
1.5	0.866385597462	0.133614402538	133.6	133614.4	133614402.5
2.0	0.954499736104	0.045500263896	45.5	45500.3	45500263.9
2.5	0.987580669348	0.012419330652	12.4	12419.3	12419330.7
3.0	0.997300203937	0.002699796063	2.7	2699.8	2699796.1
3.5	0.999534741842	0.000465258158	0.5	465.3	465258.2
4.0	0.999936657516	0.000063342484	0.1	63.3	63342.5
4.5	0.999993204654	0.000006795346	0.0	6.8	6795.3
5.0	0.999999426697	0.000000573303	0.0	0.6	573.3
5.5	0.999999962021	0.000000037979	0.0	0.0	38.0
6.0	0.999999998027	0.000000001973	0.0	0.0	1.97
6.5	0.999999999920	0.000000000080	0.0	0.0	0.08
7.0	0.999999999997	0.000000000003	0.0	0.0	0.0

$$F(\sigma) = \frac{1}{\sqrt{2\pi}} \int_{-\sigma}^{+\sigma} exp\left(-\frac{t^2}{2}\right) dt$$

Mathematically, the defect per ubiquitous outcome inspected is linked to the process yield through the relation:

$$RTY = exp(-DPU), \qquad or\ DPU = -ln(RTY)$$

The defect per ubiquitous outcome inspected, shown in Table 14.3, is viewed differently from business application to business application. Indeed, a defect per ubiquitous outcome inspected of 0.375 for an automobile application is viewed differently than the same per ubiquitous outcome defect rate for a post delivery application. That is because the automobile, with all its thousand of components, parts, dimensions, and integrated systems has many opportunities for occurrence of defects than the post delivery application has. A defect per ubiquitous outcome inspected of 0.375 for an automobile application is evidence of a much lower defect rate than the same defect rate on the post delivery application.

To contrast the defect rates per ubiquitous outcome inspected of business applications that have very different levels of complexity, the defect rate must be transformed into terms that are common to any observed characteristic of a process outcome, whatever it is or however complex it may be create. The common ground is number of opportunities of occurrence of a defect; that is, the set of circumstances that makes it possible for a defect to occur on a characteristic of a process outcome. The number of opportunities inherent to a characteristic of a process outcome, regardless of the observed characteristic, the process outcome and the business application, is a subjective measure of the complexity of the characteristic considered.

Using the number of opportunities inherent to a characteristic of a process outcome, a process defect per opportunity (DPO) is defined as:

$$DPO = \frac{Total\ number\ of\ defects\ observed\ on\ an\ outcome}{Total\ number\ of\ opportunities\ on\ an\ outcome}$$

A process defect per opportunity (DPO) depends upon how the continuum associated to a characteristic of a process outcome is subdivided into potential opportunities.

14.4.3 Process Capability & Process Performance Indices

Process capability is the measured inherent reproducibility of the outcome of a process. Before a process begins operation, it must be demonstrated to be capable of meeting its quality goals. Any project quality planning must measure not only the capability of its processes, but primarily the capability the process that the project wished to improve with respect to the key quality goals. Failure to achieve process capability should be followed by systematic diagnosis of the root causes of the failure and improvement of the process to eliminate those root causes before the process becomes operational.

The most widely adopted formula for process capabilities is the one associated with 3 standard deviations as a performance permissible limit of variations around the central tendency. Hence a width of variations equal to two times 3 standards deviations.

$$Process\ capability = 6\sigma$$

Where σ is the standard deviation of the observed characteristic of the process outcomes under a state of statistical control, i.e., under no drift and no sudden changes.

As indicated in Chap. 1, Shewhart's use of 3 standard deviation limits, as opposed to any other multiple of sigma, did not stem from any specific mathematical computation. Rather, the choice of 3-sigma limits was seen to be an acceptable economic value, and it was also justified by "empirical evidence that it works." No calculations from the normal distribution, or any other distribution, were involved in the choice of the multiplier of 3. Certainly, Shewhart did then check that this multiplier turned out to be reasonable under the artificial conditions of a normal distribution—and plenty of other circumstances as well.

If the process is centered at the nominal specification and follows a normal probability distribution, 99.73 % of production will fall within $\pm 3\sigma$ of the nominal specification.

Some industrial processes do operate under a state of statistical control. For such processes, the computed process capability of 6σ can be compared directly with specification tolerances, and judgments of adequacy can be made. However, the majority of industrial processes exhibit drift and/or sudden changes. These departures from the ideal are a fact of life, and the project team must deal with them.

Nevertheless, there is great value in standardizing on a formula for process capability based on a state of statistical control. Under this state, the product variations are the result of numerous small variables (rather than being the effect of a single large variable) and hence have the character of random variation. It is most helpful for the project team to have such limits in quantified form.

A major reason for quantifying the process capability (i.e., process variation) is to be able to compute the ability of the process to hold its outcomes specifications (including V.O.C. and V.O.B.). For processes that are in a state of statistical control, a comparison of 6σ to the specification limits permits a ready calculation of the percentage of defective characteristic of process outcomes by conventional statistical theory. These two factors are expressed in a capability index C_p, defined to be:

$$C_p = \frac{\textit{specification range}}{\textit{process capability}} = \frac{USL - LSL}{6\sigma}$$

Where *USL* is the upper specification limit and *LSL* is the lower specification limit.

The capability index C_p measures whether the process variability can fit within the specification range. It does not indicate if the process is actually running within the specification, because the index does not include a measure of the average of the process characteristic under observation (this is addressed below through the process performance index).

Figure 14.6 shows four of many possible relations between process capability and specification limits and the likely courses of action for each. Note that in all these cases the average of the process characteristic observed is nearly at the midpoint between the specification limits. In the figure above, the process capability index will be greater than one for (a) and (b) observations, equal to one for (c) and will be less than one for (d). The higher the value of the process capability index, the lower will be the amount of process outcomes that is outside the specification limits.

For a process that is in a state of statistical control, the process capability is a measurable property of the process and it summarizes how much variation there is in the process relative to a set of customer and business specifications. It also allows different processes to be compared with respect to how well an enterprise business controls them. Therefore, the process capability represents the capability of the process to meet its purpose as defined by the enterprise business intended strategy and process definition structures.

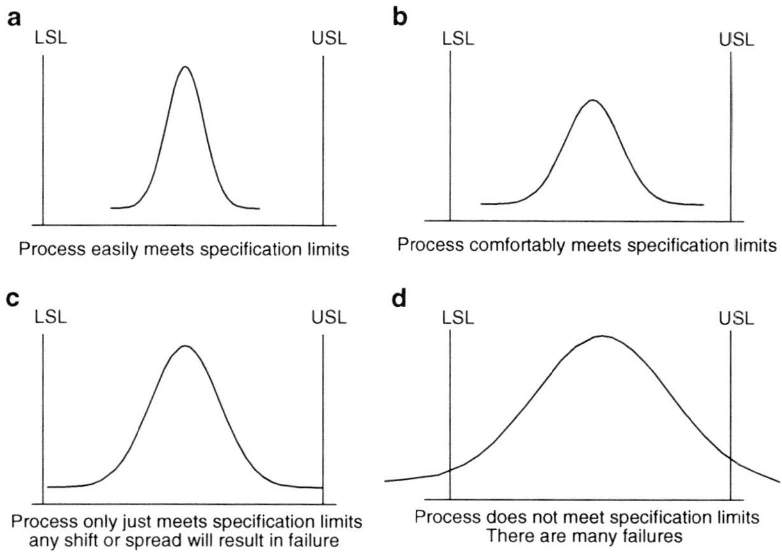

Fig. 14.6 Four examples of process capability

If a process is out of control and the causes cannot be eliminated economically, the standard deviation and process capability limits nevertheless can be computed (with the out-of-control points included). These limits will be inflated because the process will not be operating at its best. In addition, the instability of the process means that the prediction is approximate.

The comparison of process capability with specification limits leads to broad plans of action, summarized in Table 14.4.

It is important to distinguish between a process in a state of statistical control and a process that is meeting specifications. A state of statistical control does not necessarily mean that the outcomes from the process conform to specifications. Statistical control limits on sample averages cannot be compared directly with specification limits because the specification limits refer to individual units.

For some processes that are not in control, the specifications are being met, and no action is required; other processes are in control, but the specifications are not being met and action is needed. In summary, the project team will need to have the "process to be improved" in both stable (in statistical control) and capable (meet product specifications) states.

In most processes, not only are there departures from a state of statistical control but the process is not necessarily being operated to secure optimal process yields; e.g., the average of the process is not centered between the upper and lower tolerance limits. To allow for these realities, it is convenient to try to select processes with the 6σ process capability well within the specification range.

Under the normality assumption on the observed characteristic of the "process to be improved" outcomes, the collected data are arranged into subgroups of specific

Table 14.4 Plan of action for process capability

	Process outcomes meet specifications		**Process outcomes do not meet specifications**	
	Process variations small relative to specifications	Process variations large relative to specifications	Process variations small relative to specification	Process variations large relative to specification
Process in a state of statistical control	Consider cost reduction through less precise process. Consider value to improvement with tighter specifications.	Closely monitor process settings.	Process is misdirected to wrong average. Generally easy to correct	Process may be misdirected and also too scattered. Correct misdirection. Consider economics of more precise process versus wider specifications versus sorting the process outcomes.
Process not in a state of statistical control	Process is erratic and unpredictable. Investigate the causes of lack of control. Take the decision to correct the causes based on economics of corrective action.		Process is misdirected or erratic or both. Correct misdirection. Discover cause for lack of control. Consider economics of more precise process versus wider specifications versus sorting the process outcomes.	

period of time. If the upper and lower specification limits of the process are USL and LSL, the target process mean is T, the estimated expectation of the observed characteristic of the "process to be improved" is $\hat{\mu}$, the estimated variability of the process (expressed as a standard deviation) within a subgroup is $\hat{s}$, and the estimated overall variability of the process (expressed as a overall standard deviation) is $\hat{\sigma}$, then commonly-accepted estimates of process capability indices within subgroups and overall process performance indices are given in Tables 14.5 and 14.6. The process capability indices serve multiple purposes:

1. Predict the extent of variability that a process will exhibit.
2. Help choose from among competing processes or equipment those that are best suited for use to meet the specifications.
3. Provide guidance in planning the inter-relationship of the sequential discrete elements of a process. For example, one discrete element may distort the precision achieved by a predecessor element, as in hardening of gear teeth. Quantifying the respective process capabilities often points the way to a solution.

Table 14.5 Common process capability indices

Estimate of capability index	Description
$\hat{C}_p = \frac{USL - LSL}{6\hat{s}}$	Estimates what the process is capable of producing if the expectation of the observed characteristic of the "process to be improved" outcomes is centered between the specification limits.
$\hat{C}_{p,lower} = \frac{\hat{\mu} - LSL}{3\hat{s}}$	Estimates process capability for specifications that consist of a lower specification limit only.
$\hat{C}_{p.upper} = \frac{USL - \hat{\mu}}{3\hat{s}}$	Estimates process capability for specifications that consist of an upper specification limit only.
$\hat{C}_{pk} = min\left(\frac{USL - \hat{\mu}}{3\hat{s}}, \frac{\hat{\mu} - LSL}{3\hat{s}}\right)$	Estimates what the process is capable of producing, considering that expectation of the observed characteristic of the "process to be improved" outcomes may not be centered between the specification limits. If the expectation of the observed characteristic is not centered, $\hat{C}_p$ overestimates the process capability. $\hat{C}_{pk} < 0$ if the expectation of the observed characteristic falls outside of the specification limits.
$\hat{C}_{pm} = \frac{\hat{C}_p}{\sqrt{1 + \left(\frac{\hat{\mu} - T}{\hat{s}}\right)^2}}$	Estimates process capability around a target, T. $\hat{C}_{pm}$ is always greater than zero. This estimate is also known as the Taguchi capability index.
$\hat{C}_{pkm} = \frac{\hat{C}_{pk}}{\sqrt{1 + \left(\frac{\hat{\mu} - T}{\hat{s}}\right)^2}}$	Estimates process capability around a target, T, and accounts for an off - center process mean.

Table 14.6 Common process performance indices

Estimate of capability index	Description
$\hat{P}_p = \frac{USL - LSL}{6\hat{\sigma}}$	Estimates process performance if the expectation of the observed characteristic of the "process to be improved" outcomes is centered between the specification limits.
$\hat{P}_{p,lower} = \frac{\hat{\mu} - LSL}{3\hat{\sigma}}$	Estimates process performance for specifications that consist of a lower specification limit only.
$\hat{P}_{p.upper} = \frac{USL - \hat{\mu}}{3\hat{\sigma}}$	Estimates process performance for specifications that consist of an upper specification limit only.
$\hat{P}_{pk} = min\left(\frac{USL - \hat{\mu}}{3\hat{\sigma}}, \frac{\hat{\mu} - LSL}{3\hat{\sigma}}\right)$	Estimates process performance, considering that expectation of the observed characteristic of the "process to be improved" outcomes may not be centered between the specification limits. If the expectation of the observed characteristic is not centered, $\hat{P}_p$ overestimates the process performance. $\hat{P}_{pk} < 0$ if the expectation of the observed characteristic falls outside of the specification limits.
$\hat{P}_{pm} = \frac{\hat{P}_p}{\sqrt{1 + \left(\frac{\hat{\mu} - T}{\hat{\sigma}}\right)^2}}$	Estimates process performance around a target, T. $\hat{P}_{pm}$ is always greater than zero. This estimate is also known as the Taguchi performance index.
$\hat{P}_{pkm} = \frac{\hat{P}_{pk}}{\sqrt{1 + \left(\frac{\hat{\mu} - T}{\hat{\sigma}}\right)^2}}$	Estimates process performance around a target, T, and accounts for an off-center process mean.

4. Provide a quantified basis for establishing a schedule of periodic process control checks and readjustments.
5. Provide a mean for testing theories of causes of defects at other stages of the "process improvement" project.
6. Serve as a basis of quality performance requirements for "process to be improved" outcomes.

The process capability index estimate, $\hat{C}_p$, varies from subgroup to subgroup. The measure of dispersion represents the inherent Voice of the Process (V.O.P.) while the width of the specifications $USL - LSL$ represents the Voice of the Customer (V.O.C.). The process centered capability index estimate, $\hat{C}_{pk}$, can also be expressed as:

$$\hat{C}_{pk} = \hat{C}_p \cdot (1 - d),$$

Where d is a scaled distance between the mid-value of the specification range, m, and the estimated expectation of the observed characteristic of the "process to be improved," $\hat{\mu}$. The mid-value of the specification range is defined to be:

$$m = \frac{USL + LSL}{2}$$

The distance between the estimated expectation of the observed characteristic of the "process to be improved," $\hat{\mu}$, and the optimum, which is m, is defined to be equal to $\hat{\mu} - m$. The scaled distance is defined to be:

$$d = \frac{2|\hat{\mu} - m|}{USL - LSL}, \qquad 0 \leq d \leq 1$$

Thus, the process centered capability index estimate, $\hat{C}_{pk}$, is always less or equal to the process capability index estimate, $\hat{C}_p$.

As illustrated in Fig. 14.7, the process capability index estimate, $\hat{C}_p$, compares the gap available within the specifications with the gap required by the process. The process performance index estimate, $\hat{P}_p$, compares the gap available within the specifications with the gap used by the process in the past.

The only difference between these two indices is the manner in which their denominators are computed. The process capability index estimate uses an estimate of a measure of dispersion (expressed as a standard deviation) within a subgroup, $\hat{s}$, while the process performance index estimate uses an estimate of global standard deviation, $\hat{\sigma}$.

When a process under consideration is operated predictably these two measures of dispersion tend to converge and the two process indices will be quite similar. However, when a process under consideration is operated unpredictably the global measure of dispersion will be inflated relative to the within-subgroup dispersion, which will deflate the process performance index.

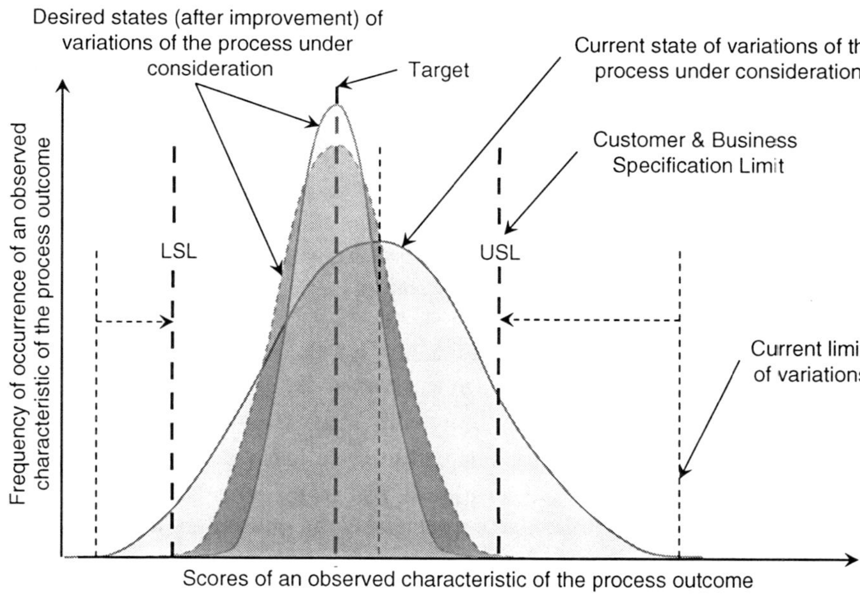

Fig. 14.7 Aligning V.O.C. (specifications) with V.O.P.

Similarly, the process centered capability index estimate, $\hat{C}_{pk}$, compares the width of the specifications to the nearest specification with the gap required by the process, while the process centered performance index estimate, $\hat{P}_{pk}$, compares the width of the specifications to the nearest specification with the gap used in the past.

As the process under consideration is operated closer to the conditions where the observed characteristic of the process outcomes is closer to its expectation, the process centered capability index, $\hat{C}_{pk}$, approaches $\hat{C}_p$, and the process centered performance index, $\hat{P}_{pk}$, approaches $\hat{P}_p$.

As the process under consideration is operated more predictably, the centered process performance index estimate, $\hat{P}_{pk}$, approaches the process centered capability index estimate, $\hat{C}_{pk}$, and the process performance index estimate, $\hat{P}_p$, approaches the process capability index estimate, $\hat{C}_p$.

The process capability index estimate, $\hat{C}_p$, and the process centered capability index estimate, $\hat{C}_{pk}$, represent estimates of the actual capability of a predictable process, or the hypothetical capability of an unpredictable process. The centered process performance index estimate, $\hat{P}_{pk}$, and the process performance index estimate, $\hat{P}_p$, represent estimates of the actual past performance of a process.

The process capability index estimate, $\hat{C}_p$, and the process performance index estimate, $\hat{P}_p$, describes the potential or the performance of a process that is centered at the mid-value of the specifications, while the process centered capability index estimate, $\hat{C}_{pk}$, and the process centered performance index estimate, $\hat{P}_{pk}$ describe how the potential or performance suffers when the process under consideration is not centered within the specifications.

A process capability index estimate equal to 1, $\hat{C}_p = 1$, means that the voice of the customer (V.O.C.) is equal to the voice of the process (V.O.P.). A process capability index estimate less than 1, $\hat{C}_p < 1$, means that the process variations goes beyond the specifications, with defects spilling out over the edges. A process capability index estimate greater than 1, $\hat{C}_p > 1$, means that the effective width of the process variations is less than the width of the specifications, with fewer defects occurring.

When a process is operated predictably and on target these four indexes will be four estimates of the same quantity.

When a process is operated predictably but is not centered within the specifications there will be a discrepancy between the process capability index estimate, $\hat{C}_p$, and the process centered capability index estimate, $\hat{C}_{pk}$, on one hand, and a discrepancy between the process performance index estimate, $\hat{P}_p$, and the process centered performance index estimate, $\hat{P}_{pk}$, on the other hand.

When a process is being operated unpredictably the process performance index estimate, $\hat{P}_p$, and the process centered performance index estimate, $\hat{P}_{pk}$, will be substantially smaller than the corresponding process capability index estimate, $\hat{C}_p$, and the process centered capability index estimate, $\hat{C}_{pk}$.

Finally, when a process is operated unpredictably and off target the four indices estimates will be estimates of four different things. While the process capability index estimate will be the best-case value, the process centered performance index estimate will be the worst-case value, and the gap between these two values will define the opportunities for improvement connected with operating the process under consideration up to its full potential. In the case of this chapter, the gap between these two values will define the opportunities for improvement connected with operating the "process to be improved."

In order to make decisions based these results the project team must understand and verify some important assumptions that are essential for statistical validity of these indices. Four key assumptions are:

1. *Process stability*: This means a state of statistical control with no drift or oscillation.
2. *Normality of the characteristic of the process outcome being observed*: This is needed to draw statistical inferences about the population.
3. *Representativeness of samples*: This includes random sampling.
4. *Independence of the collected data*: This means that consecutive collected data cannot be correlated.

In practice, these assumptions are often not verified. Examination likely would reveal that one or more of the assumptions are not realistic. These assumptions are not theoretical refinements; they are important conditions for properly applying capability indices.

Capability indices allow us to characterize the relationship between the process potential and the specifications. Performance indices characterize the past

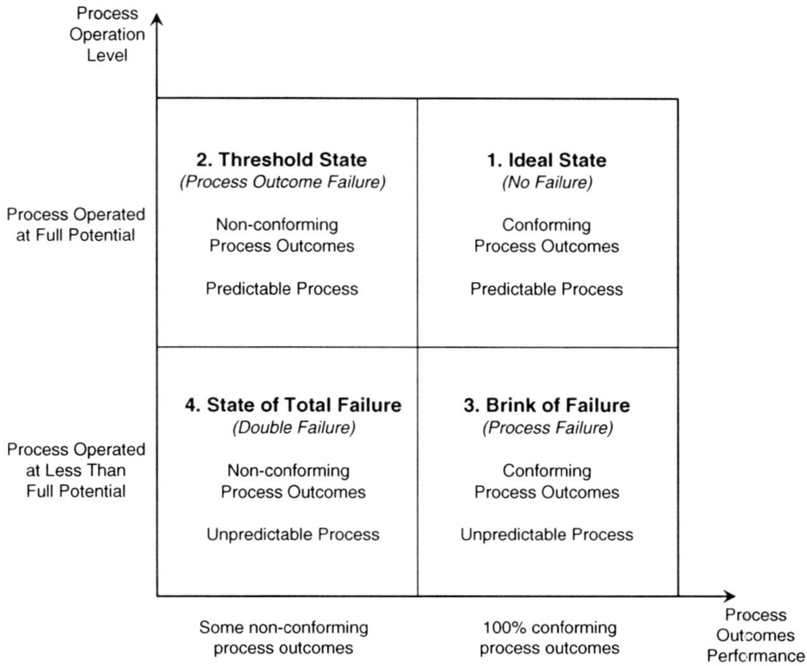

Fig. 14.8 Process outcomes versus process operation grid

performance relative to the specifications. Capability indices serve a role in quantifying the ability of a process to meet customer quality goals. The emphasis, however, should be on improving processes and not just determining a capability index for a product characteristic. Achieving customer quality goals (particularly for quality levels of 1–10 ppm) means meeting requirements on all variables and attributes characteristics.

14.5 Characterize Process & Revise Process Quality Targets

The purpose of calculating the "process to be improved" rolled throughput yield, defect rate per ubiquitous outcome inspected (DPU), process capability indices and process performance indices, is to characterize the "process to be improved" and establish a performance baseline for the "improved" process. Once calculated, the project team should revisit and update the scope of the "process improvement" project. Significant differences in yields for the process discrete elements suggest creating a new map for the elements with the lowest yield.

Any process can be characterized by one of four categories (Wheeler, Two Definitions of Trouble, 2009b), each occupying one space in a four-space grid—the process outcomes conformance versus the process operation—as illustrated in Fig. 14.8:

1. Conforming Process Outcomes and a Predictable Process (No Failure)
2. Non-conforming Process Outcomes and a Predictable Process (Process Element Outcomes Failure)
3. Conforming Process Outcomes and an Unpredictable Process (Process Element Failure) and
4. Non-conforming Process Outcomes and an Unpredictable Process (Double Failure).

As shown in Fig. 14.8, the process operation level from less than full potential to full potential runs along a line from the bottom to the top of the grid, and the process outcomes from some non-conforming to 100 % conforming runs along a line from left to right.

All processes belong to one of these four states. But processes do not always remain in one state. In time, it is possible for a process to move from one state to another if the correct preventive actions for continuous improvement are not taken.

14.5.1 The Ideal State (No Failure)

The "Ideal State," toward which every process aspires to, appears in the upper right quadrant. A process in this state is predictable and all its outcomes are in full conformance. The predictability of the process will be the result of purposeful continuous efforts on the part of the enterprise business and the personnel who operate the process. A predictable process is an achievement, requiring constancy of purpose and the effective use of process behavior charts. The conformity of the process outcomes will be the result of having natural process limits that fall inside the specification limits. When the process is operating in the "Ideal State," its centered capability index estimate $\hat{C}_{pk}$ will be close to, or greater than, 1.00. A process in the "Ideal State" satisfies four conditions:

1. The process must be inherently predictable over time.
2. The enterprise business personnel must operate the process in a predictable and consistent manner. The operating conditions cannot be selected or changed arbitrarily.
3. The process central tendency must be set at the proper level.
4. The natural process limits must fall inside the specification limits for its outcomes.

Whenever one of the four conditions above is not satisfied, the possibility of producing non-conforming outcomes exists. When a process satisfies these four conditions, the enterprise business can be confident that nothing but conforming products or services are being produced. Furthermore, the conformity of the process outcomes should continue as long as the process behavior remains predictable.

Therefore, a process that is in the "Ideal State" does not need further improvement. Since the process outcomes stream for a predictable process can be thought of as being homogeneous, the measurements taken to maintain the process behavior chart will also serve to characterize the process outcomes produced by the predictable process.

14.5.2 The Threshold State (Process Outcome Failure)

A process in this state will be predictable, but it will be producing some nonconforming products or services. When a process is operating in the "Threshold State" its centered capability index estimate $\hat{C}_{pk}$ will be less than 1.00.

As with the "Ideal State," the predictability of the process will be the result of purposeful and persistent efforts on the part of the enterprise business and the personnel who operate the process—a predictable process does not occur by accident. Moreover, because the process is predictable, it must be thought of as operating as consistently as it currently can operate. As Nevertheless, the existence of some non-conforming process outcomes will be the result of one or both of the Natural Process Limits falling outside the specification limits.

As Donald J. Wheeler demonstrates, the fact that the process is predictable puts a new twist on process outcomes failure. First of all, as long as the process remains predictable, the non-conforming process outcomes will persist. Therefore the process owner cannot wait for things to spontaneously improve. Second, the ultimate solution to the problem of non-conforming process outcomes will require moving this process up to the "Ideal State." This will only happen when the process either changed or its specifications changed.

If the process capability index estimate $\hat{C}_p$ is greater than 1.00, then the process has enough elbow room to operate in the "Ideal State" and the non-conforming process outcomes are likely to be due to a faulty process aim. In this case the personnel who operate the process will need to tweak the process inputs to adjust the process aim. Here the process behavior chart can be used as a feedback loop to help to determine how various adjustments affect the process central tendency.

If the capability index estimate $\hat{C}_p$ is less than 1.00, then the process is not likely to have enough elbow room to meet the specifications even if it is operated on target. Here there will be a need to reduce the process variation. Since a predictable process is already operating as consistently as it currently can operate, reduction of the process variation will require a major change in the process itself. Therefore, a process in the "Threshold State" is one that needs to be reengineered.

14.5.3 The Brink of Failure (Process Failure)

Processes in the "Brink of Failure" state are unpredictable even though they are currently producing 100 % conforming outcomes the process is changing unpredictably, and the 100 % conformity can disappear at any time. Such processes will usually have a centered performance index estimate $\hat{P}_{pk}$ value that is close to, or greater than, 1.00. A process in the "Brink of Failure" is often incorrectly perceived to be operating well, even though it is in need of being operated predictably.

Any unpredictable process is subject to the effects of assignable causes. So while the conformity to specifications may lull the personnel who operate the process into thinking all is well, the assignable causes will continue to change the process until it will eventually produce some non-conforming outcomes. The personnel who operate the process will suddenly discover that the process is in outcomes failure, yet no indication will be perceived on how this occurs and how to correct the process outcomes failure. The change from 100 % conforming process outcomes to some non-conforming process outcomes can occur at any time, without the slightest warning. When this change occurs the process will be in the "State of Total Failure."

There is no way to predict what an unpredictable process will yield in time. Since the unpredictability of such processes is due to assignable causes, and since assignable causes are dominant causes that are not being controlled by the personnel who operate the process, the only way to move out of the "Brink of Failure" is to first eliminate the assignable causes, which can be readily identified through the process behavior charts.

14.5.4 The State of Total of Failure (Double Failure)

The "State of Total Failure" exists when an unpredictable process is producing some non-conforming outcomes. The unpredictable process means that the personnel who operate the process is confronted with a changing level of non-conformity in the process outcomes stream. So even though the personnel who operate the process may know that non-conforming process outcomes are been produced, the percentage non-conforming process outcomes cannot be reliably predicted in time. Here the centered performance index estimate $\hat{P}_{pk}$ value will usually be less than 1.00.

A process owner whose process is in the "State of Total Failure" knows that he has a problem, but he usually does not know what to do to correct it. If he/she tries to address the problem of non-conforming process outcomes directly he/she is likely to be frustrated by the random changes in the process which result from the presence of the assignable causes. When a needed modification to the process is made, the effect will be short-lived because the assignable causes continue to change the process. When an unnecessary modification is made, a fortuitous shift in the process rolled throughput yield by the assignable causes may mislead the personnel who operate the process. A process in the "State of Total Failure" is simply a mystery where the clues keep changing.

To make any progress in moving a process out of the "State of Total Failure," the assignable causes must be identified and eliminated. This will require the use of process behavior charts. Whenever a process is in either the "Brink of Failure" or the "State of Total Failure" the first step should always be to learn how to operate that process predictably. A process is operated up to its full potential only when it is operated predictably. Predictable operation is not something that is beyond the capability of a process. It is merely the realization of operating a process as it could and should be operated.

Table 14.7 Basic plan of action for process improvement

	Threshold State Process Outcomes Failure Adjust Process Aim or Redesign Process	**Ideal State** No Failure Ignore or Gently Tweak Process
Predictable Process		
Unpredictable Process	**State of Total Failure** Double Trouble Find Assignable Causes and Remove Their Effects	**Brink of Failure** Process Trouble Find Assignable Causes and Remove Their Effects
	Some non-conforming process outcomes	**100% conforming process outcomes**

14.5.5 Summary of Process Characterization

The characterization of the "process to be improved" leads to broad plans of action, shown in Table 14.7:

1. Find assignable causes for an unpredictable "process to be improved" and remove their effects;
2. Upgrade or adjust a predictable "process to be improved" in the "Threshold State"; and
3. Ignore or tweak the predictable "process to be improved" in the "Ideal State."

Process improvement approaches focused on moving from the left side to the right side of Table 14.7. Operating a process up to its full potential focused on moving from the bottom to the top of Table 14.7. Economic operation requires that both actions be performed to ascertain that the realized or accomplished outcome of the "process improvement" project be recognized as is an improvement.

The first line of the course of action Table 14.7 defines how to operate the process with minimum variance. When a process is operated on target with minimum variance it is operating up to its full potential. The second line deals with what must be done when that is not enough. Thus, this course of action is one that will guide the "process improvement" project team through the complexities of improving the existing "process to be improved." It uses the four possibilities as a matrix to triage the process improvement efforts.

Experimental Study: Design of Experiments 15

An experimental study is a study relating to, based on, or having the nature of an experiment. Experiments are studies involving intervention by the researcher beyond that required for measurement. The usual intervention is to manipulate some variable in a setting and observe how it affects the subject being studied under conditions constructed and controlled by the researcher.

This chapter provides an overview of how the components of an experimental design fit and work together. If you are not familiar with experimental design, this chapter will provide the necessary "road map" for placing the subsequent chapters, concerned with experimental design into proper context.

15.1 Designing and Conducting and Experimental Study

In every experimental study, a response, or a performance characteristic of an output, or outcome variable of a process will be defined—let denote this varying outcome as Y. Out of the large collection of input variables, to the "process to be improved," that may have some impact upon this response, a subset, consisting of a few variables, will be selected to be observed or studied in the experiment—let denote this subset as X. The project team must gain knowledge on this selected subset in order to improve the "process to be improved." The remaining input variables—also called extraneous input variables—are then excluded from the study.

We can express the impact of the selected subset X upon the response Y by the relation:

$$Y = f(X) + \varepsilon$$

1. Y is the response, a performance characteristic of an output or outcome of an execution of the "process to be improved";
2. X represents the selected subset, selected inputs, or selected factors that are needed to produce the response Y. In enterprise business applications, Inputs are further classified according to a number of criteria as follows:

A. van Aartsengel and S. Kurtoglu, *Handbook on Continuous Improvement Transformation*, 247
DOI 10.1007/978-3-642-35901-9_15, © Springer-Verlag Berlin Heidelberg 2013

(a) *Controllable inputs*—are those that management can control in some manner such as employees, data collection systems, and types of materials;
(b) *Uncontrollable inputs*—are inputs such as weather conditions that must be considered when improving process capabilities and reducing defects;
(c) *Critical inputs*—are critical to a process must be marked as such so that they are given a priority before the process begins;
(d) *Incidental inputs*—are considered "noise" that does not affect the process in a significant way. Inputs are further divided into a number of categories that can be easily remembered as they all begin with the letter M:
   - *Man*—This refers to human resources and intervention that is necessary for a process to be completed successfully. From employees to management, this input clarifies the roles and responsibilities of every person involved in the process.
   - *Machines*—The performance of individual machines is important for the assessment of a process and whether any improvement will be necessary. To reduce process variation amongst different machines, it is important to provide regular maintenance and replacement as part of the process.
   - *Methods*—Methods and procedures used in every step of the process are an important component of inputs. To assess process variation from one production unit to another, the project team will need to assess whether production methods are being adhered to or not.
   - *Mother Nature*—While the environment cannot be controlled in many instances, enterprise businesses must assess its impact on processes. The environment, for instance, impacts the availability and transportation of raw materials and products.
   - *Management*—Management systems and methodologies are important inputs in processes. Whether formal or informal, a management system ensures that an enterprise business functions as a single unit with a shared vision.
   - *Materials*—Materials refer to both raw and manufactured elements of process inputs. When making furniture, for example, materials include wood products, metal screws, paint, paper and labeling products, and many more production materials. The quality, availability, ease of transportation of materials has a strong impact on a process and its success in producing services or products.
   - *Measurement systems*—Every process dictates the type of data collection system that needs to be put in place. Using the right type of data collection system ensures that the appropriate data and information are collected.

3. f represents the "process to be improved" by which the selected inputs X are transformed into output Y;
4. ε is the uncertainty in depending upon the selected inputs X and the "process to be improved" to actually produce the desired output Y.

15.2 Basic Concepts

15.2.1 Replication

"Replication" is the repetition, the rerunning, of an experiment or measurement in order to increase precision or to provide the means for measuring precision. A single replicate consists of a single observation or experimental run. Replication provides an opportunity for the effects of uncontrolled factors or factors unknown to the experimenter to balance out and thus, through randomization, acts as a bias-decreasing tool. Replication also helps to detect gross errors in the measurements. In replications of groups of experiments, different randomizations should apply to each group. Rerun experiments are commonly called "replicates." However, a sequence of observations made under a single set of experimental conditions, under a single replicate, are simply called "repeated observations."

15.2.2 Extraneous Input Variables

If the "process improvement" project team thinks that some of these extraneous input variables may have an impact upon the response, then the project team might hold these input variables constant during the course of the experiment, however, most of these extraneous input variables will be ignored during the course of the experiment. Consequently, there are three things that the project team can do with an input variable during the course of an experiment:

1. Study the factor by changing its levels and observing the different responses;
2. Hold the factor constant during the experiment; or
3. Ignore the factor altogether.

15.2.3 Blocking (Planned Grouping)

When we hold a factor constant (or block on that factor) we are assuming that it does not interact with the factors that we are studying. When we ignore a factor we are assuming that it has minimal impact upon the response and minimal interactions with the factors we have held constant or studied. When we randomize over one or more of the ignored factors we are simply buying some insurance that if the prior assumptions are not correct, then perhaps the contamination will be averaged out within each treatment.

When an experiment is performed while holding one or more input variables constant at some levels, the experimental results will only characterize what happens when these constant factors are at those particular levels. The experiment will not tell what happens at other values of these constant factors. Since the fixed factor levels do not allow to detect the interactions with other input variables, the experimental results might be of limited value.

When an experiment is performed while ignoring one or more input variables, it is implicitly assumed that such factors do not have a pronounced effect upon the response variable. However, the project team cannot ascertain that the factors that have been ignored or held constant do not have important effects upon the response variable.

15.2.4 Randomization

For industrial experiments, randomization is a process of performing experimental trials in a random order in which they are logically listed. It is generally recommended because an experimenter cannot always be certain that all important input variables affecting a response has been included and considered in the experiment. The purpose of randomization is to safeguard the experiment from the influence of extraneous input variables. It is essential to quantify the effect of the overall extraneous input variables and then to reduce it to its acceptable limits prior to carrying out the actual experimentation. Randomization protects against various forms of bias that can creep into an experimental study, even indirectly, and therefore it has become a very useful research tool.

However, it is important to note that in any one experiment there is no guarantee that randomization will prevent the effect of an ignored factor from showing up in later analysis. The mechanism used by randomization is that of averaging. When an experiment uses each treatment combination several times, the randomization of which experimental units receive each treatment combination, or the randomization of the order in which the different treatment combinations are studied, will tend to average out the effects of the various factors that are not included in (and are therefore ignored by) the study.

Since the basic mechanism of randomization is averaging, it should be apparent that randomization becomes less effective as the number of observations per treatment combination gets smaller. This will happen simply because, as the subgroup size gets smaller, there will be less chance for the effects of extraneous inputs to average out within each subgroup.

When performing an experiment in which each treatment will be applied to a large number of experimental units, and when there is no opportunity to conduct subsequent, confirmatory experiments, then randomization is both an insurance policy and a necessary part of good scientific experimentation. Randomization works when the project team cannot resort to the primary confirmatory tool of the scientific method—the nontrivial replication of results. It is useful when:

1. The analysis is confirmatory (rather than exploratory) in nature.
2. There are multiple observations per treatment combination; and
3. Experiments are conducted in circumstances that do not demonstrate statistical control;

If these three conditions are not present, then randomization loses much of its usefulness.

15.2.5 Randomized Block Design

In industrial practice, the experimental units are often not completely homogeneous. Usually, a grouping of these units according to a stratification factor can be observed. If we have such prior information then a gain in efficiency compared to the completely randomized experiment is possible by grouping experimental units into blocks. The experimental units are grouped together in homogeneous groups (blocks) and the treatments are assigned randomly to the experimental units within each block. Hence the block effect (differences between the blocks) can now be separated from the experimental error. This leads to a higher precision. The strategy of building blocks should yield variability within each block that is as small as possible and variability between blocks that is as high as possible.

Block design is a way of holding a factor locally constant while ignoring the variation between the blocks. The purpose of block designs is to reduce the variability of response by removing part of the variability as block numbers. If in fact this removal is illusory, the block effects being all equal, then the estimates are less accurate than those obtained by ignoring the block effects and using the estimates of treatment effects. On the other hand, if the block effect is very marked, the reduction in basic variability may be sufficient to ensure a reduction of the actual variances for the block analysis.

The most widely used block design is the randomized block design. Here s treatments with r repetitions each (i.e., balanced) are assigned to a total of $n = r \times s$ experimental units. First, the experimental units are divided into r blocks with s units each in such a way that the units within each block are as homogeneous as possible. The s treatments are then assigned to the s units at random, so that each treatment occurs only once per block.

15.2.6 Incomplete Block Designs

In many situations the number of treatments to be compared is large. Then, large number of blocks is needed to accommodate all the treatments and in turn more experimental material. This may increase the cost of experimentation in terms of money, labor, time etc. The completely randomized design and randomized block design may not be suitable in such situations because they will require large number of experimental units to accommodate all the treatments. In such cases when sufficient number of homogeneous experimental units are not available to accommodate all the treatments in a block, then incomplete block designs are used in which each block receives only some and not all the treatments to be compared.

The designs in which every block receives all the treatments are called complete block designs whereas the designs in which every block does not receive all the treatments but only some of the treatments are called incomplete block designs. In incomplete block designs, the block size is smaller than the total number of treatments to be compared.

With incomplete block designs two types of analysis can be conducted: intra-block analysis and inter-block analysis. In intra-block analysis, the treatment effects are estimated after eliminating the block effects and then the analysis and test of significance of treatment effects are conducted further. If the blocking factor is not marked, then intra-block analysis is sufficient enough and the derived statistical inferences are correct and valid. There is a possibility that the blocking factor is important and the block totals may carry some important information about the treatment effects. In such situations, one would like to utilize the information on block effects (instead of removing it as in the intra-block analysis) in estimating the treatment effects to conduct the analysis of design. This is achieved through inter-block analysis of an incomplete block design by considering the block effects to be random. When intra-block and inter-block analysis have been conducted, then two estimates of treatment effects are available from each of the analysis.

15.2.7 Balanced Incomplete Block Designs

A balanced incomplete block design is an arrangement of m treatments in b blocks, each containing k experimental units $(k < m)$ such that:

1. Every treatment occurs at most once in each block,
2. Every treatment is observed r times in the design and
3. Every pair of treatment occurs together in exactly $p_,$ of the b blocks.

The quantities m, b, r, k and $p,$ are called the parameters of the balanced incomplete block design. The balanced incomplete block design is a proper, binary and equi-observable design. The parameters m, b, r, k and $p,$ are integers which are not chosen arbitrarily and are not at all independent. They satisfy the following relations:

$$b \times k = m \times r$$

$$p \times (m - 1) = r \times (k - 1)$$

$$b \geq m \,(and\, hence\, r \geq k)$$

The last relationship above is also called as Fisher's inequality.

15.2.8 Factorial Designs

In practice, for most designed experiments it can be assumed that the response Y is not only dependent on a single variable but on a whole group of prognostic factors. If these variables are continuous, their influence on the response is taken into account by so called factor levels. These are ranges (e.g., low, medium, high) that classify the continuous variables considered as ordinal variables.

Experimental studies that analyze the response for all possible combinations of two or more factors are called factorial experiments or cross-classification. Suppose that we have k factors $X_1, X_2, \ldots, X_k$ with $r_1, r_2, \ldots r_k$ factor levels. The complete factorial design then requires $r = r_1 \times r_2 \times \ldots \times r_k$ observations for one trial. This shows that it is important to restrict the number of factors as well as the number of their levels.

15.2.9 2^k-Factorial Designs

In industrial practice, factorial designs at the first stage of a data collection and analysis are usually conducted with only two factor levels for each of the included factors. The idea of this procedure is to make the important effects identifiable so that the analysis in the following stages can test factor combinations more specifically and more cost-effectively. A complete analysis with k factors, each of two levels, requires 2^k observations for one trial. This fact leads to the nomenclature of the design: the 2^k experiment. The restriction to two levels for all factors makes a minimum of observations possible for a complete factorial experiment with all two-way and higher-order interactions.

15.2.10 Confounding

If the number of factors or levels increase in a factorial experiment, then the number of treatment combinations increases rapidly. When the number of treatment combinations is large, then it may be difficult to get the blocks of sufficiently large size to accommodate all the treatment combinations. Under such situations, one may use either connected incomplete block designs, e.g., a balanced incomplete block design where all the main effects and interaction contrasts can be estimated or use unconnected designs where not all these contrasts can be estimated. Non-estimable contrasts are said to be confounded.

Develop Cost Management Plan 16

"Cost" is the most important term in the entire field of business. All concepts and meanings related to continuous improvement are ultimately tied to the term cost. "Cost" can be defined as "a resource sacrificed or forgone to achieve a specific objective." It is usually measured by the monetary amount that must be paid to acquire goods and services. As an aid to decision-making, a project manager must know how much the use of a given project resource costs. Consequently, the objective of this chapter is to address the "cost" topics that will assist the project manager to speedily, efficiently, and effectively achieve the "process improvement" project's goals. It is a large topic, which leads to it being a long chapter.

"Develop Cost Management Plan" is the project management process required to ensure that the project resources are used efficiently and the project is financially viable and worthwhile undertaking. It describes a set of activities for collecting cost data in an organized manner, allocating the appropriate type of accumulated costs to project resources, and controlling spending for the purpose of ensuring that the project is performed under approved budget.

"Develop Cost Management Plan" is concerned with two basic aspects related to cost: cost accumulation and cost assignment. Cost accumulation refers to the process and methods used to collect cost data in an organized manner. This is typically accomplished through managerial accounting methods. Because the information included in managerial accounting systems is relied on so heavily for management decision-making, it is important to ensure that the costs associated with the use of any given project resource are as accurate and valid as possible. To accomplish this, all costs must be properly assigned. Cost assignment consists of tracing or allocating the appropriate type of accumulated costs to a project resource.

The constituent project management processes used during the development of the project cost management plan, illustrated in Fig. 16.1, include the following:

1. Plan Cost Data Collection
2. Collect Cost Data
3. Allocate Cost To Activities
4. Control Spending

A. van Aartsengel and S. Kurtoglu, *Handbook on Continuous Improvement Transformation*, DOI 10.1007/978-3-642-35901-9_16, © Springer-Verlag Berlin Heidelberg 2013

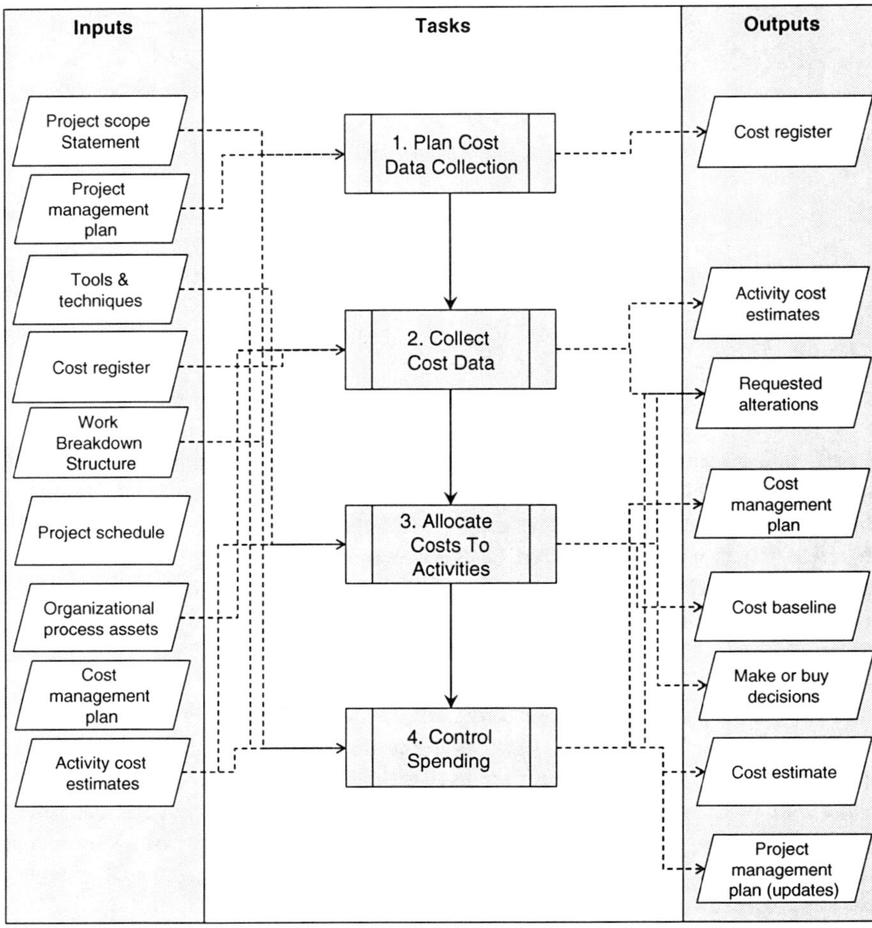

Fig. 16.1 The cost management plan process

These four constituent processes interact with each other and with the project management processes in the PDSA "Process Groups." Each aspect of executing any of these can involve effort from one or more persons, based on the needs of the project. Each aspect occurs at least once in every "process improvement" project and occurs in one or more project phases.

16.1 Plan Cost Data Collection

Planning the cost data collection builds on a number of key outputs of other project management processes in the PDSA Plan "Process Group." These include, but are not limited to:

1. The project management plan
2. The project scope statement
3. The work breakdown structure
4. The Schedule management plan
5. The organizational process assets

The purpose of planning for cost data collection, as with any data collection, is to identify the cost types to be collected and their behavior.

Having a solid understanding and strong working knowledge of costs behavior and the many types of costs will prove invaluable to the project manager in developing the project cost management plan for performing cost accumulation and cost assignment. Whether a cost is traced or allocated relates directly to whether it is a direct cost or an indirect cost.

16.1.1 Cost Classifications for Assigning Costs

The distinction between direct and indirect costs has widespread application in the manufacturing and production industries. It also relates directly to the methods that the project manager should use to correctly estimate project costs. In turn, proper project cost estimation has an even broader implication to the enterprise business, because project costs directly influence the estimated value of the assets created through project execution and used by the enterprise business to conduct business.

16.1.1.1 Direct Cost

A direct cost is a cost that can be easily and conveniently traced to a specified project resource. It reflects work, equipment and materials employed directly in a work effort. Direct costs can be reasonably measured and directly attributed to a work activity or a specific output. The concept of direct cost extends beyond just direct materials and direct labor. For example, if an enterprise business is assigning costs to its various regional and national offices through the project, then the salary of the manager in a regional office would be a direct cost of that office.

Examples of direct project costs are team wages and expense on materials used during the project. Another example of direct costs in the context of manufacturing operations would be the cost of direct materials that are directly used in making the product and the cost of direct labor or the hourly pay rate, which is labor directly involved in manufacturing the product, such as a mechanic. This includes people working with their hands or operating machines used to manufacture the product. These costs are easily traced to the cost of the final product; hence, the term cost tracing is used to describe the assignment of direct costs to a specific project resource.

16.1.1.2 Indirect Cost

An indirect cost is a cost that cannot be easily and conveniently traced to a specified cost object. It is incurred to support the directly productive work effort. Indirect costs cannot be reasonably attributed to a specific output, but they are still considered part of doing business. Overhead (e.g., the price of electricity, office rent, the cost of maintaining a secretarial pool) and fringe benefits (e.g., life insurance, pension plans, and profit sharing schemes) are examples of indirect costs. In the context of manufacturing operations, an example of an indirect cost is the cost of quality control personnel who conduct tests on all aspects of an assembled car released for distribution. In this case, it is difficult to trace the exact amount of quality-control support to a specific project resource. Instead, a calculated, proportional amount of additional cost is allocated to the final cost of that car product.

The use of indirect cost practice is very consistent with the product costing approach that many enterprise businesses use called Activity-Based Costing. Activity-Based Costing is an accounting procedure for allocating the cost of indirect and overhead expenses (the cost of an enterprise business' resources) to specific activities in proportion to the use of a given resource by that activity. This is in contrast to conventional accounting practice, which allocates indirect and overhead expenses in proportion to direct costs incurred by an activity. The aim of Activity-Based Costing is to improve overall cost effectiveness through focus on key cost elements. This approach is particularly compatible with quality objectives which target these costs for reduction. It is also used to determine charge rates for project personnel, thus allowing for the more accurate and appropriate valuation of the assets created by projects and owned by the enterprise business. Using activity-based costing methods allows accountants to be able to properly or balance financial accounting records with managerial accounting records.

In aggregate, all indirect costs tend to break down into two major categories: overhead costs and fringe benefit costs. Overhead costs include labor-related expenditures required to provide the environment in which project work are carried out, such as the staff needed to perform facilities maintenance activities. Overhead costs may also include a certain non-labor support, such as office supplies and utility costs. Fringe benefit costs are additional non-salary, employee-based expenditures incurred by the company as an ordinary part of maintaining a workforce. These may include the enterprise business payments toward employee health insurance, stock options, pension plans, or tuition-aid programs.

The most common application of indirect costing occurs when an enterprise business adds all additional indirect costs to worker salaries to establish a fully-loaded charge-out rate. This is how most enterprise businesses allocate their overhead costs to project work. Therefore, the fully-loaded charge-out rate, typically a per hour figure, is used whenever the project manager is estimating the cost associated with the use of any internal resources on projects. If this is not done, the financial accounting and the managerial accounting system will not properly reconcile.

The calculation of a predetermined overhead rate is determined as follows:

$$Overhead\,Rate = \frac{Estimated\,Total\,company\,Overhead\,costs}{Estimated\,Activity\,Base}$$

Although the actual calculation of the load rate is more complex than this, the principle is essentially the same. The fully loaded charge-out rate is a figure that is ordinarily given to the project manager by the enterprise business financial department.

An objective of cost-conscious enterprise business is to reduce indirect costs to the extent possible.

16.1.2 Cost Classifications for Predicting Cost Behavior

The proper accumulation, collection, and control of costs are also related to the concept of cost behavior patterns. Quite frequently, it is necessary to predict how a certain cost will behave in response to a change in activity. Cost behavior refers to how a cost reacts to changes in the level of activity. As the activity level rises and falls, a particular cost may rise and fall as well or it may remain constant. For planning purposes, a project manager must be able to anticipate which of these will happen; and if a cost can be expected to change, the project manager must be able to estimate how much it will change. To help make such distinctions, costs are often categorized as variable or fixed costs. Both variable and fixed costs are characterized with respect to a specific project resource, and for a prescribed period.

16.1.2.1 Variable Cost

A variable cost is a cost that varies, in total, in direct proportion to changes in the level of activity. If the activity level doubles, the total variable cost also doubles. If the activity level increases by only 10 %, then the total variable cost increases by 10 % as well. The activity can be expressed in many ways, such as units produced, units sold, hours worked, and so forth. A variable cost is a cost that is directly tied to carrying out a work effort. It changes with the quantity of output, volume, or other measure of activity. An increase or reduction of project scope causes a respective change in a variable cost. A good example of a variable cost is direct materials. The cost of direct materials used during a period will vary, in total, in direct proportion to the number of units that are produced.

While total variable costs change as the activity level changes, it is important to note that a variable cost is constant if expressed on a per unit basis. The idea that a variable cost is constant per unit but varies in total with the activity level is crucial to understanding cost behavior patterns.

In a manufacturing company, variable costs include items such as direct materials, shipping costs, and sales commissions and some elements of manufacturing overhead such as lubricants. We will also usually assume that direct labor is a variable cost, although direct labor may behave differently in some situations. In a merchandising company, the variable costs of carrying and selling products include items such as cost of goods sold, sales commissions, and billing costs. In a hospital, the variable costs of providing health care services to patients would include the costs of the supplies, drugs, meals, and perhaps nursing services.

When we say that a cost is variable, we ordinarily mean that it is variable with respect to the amount of goods or services the enterprise business produces.

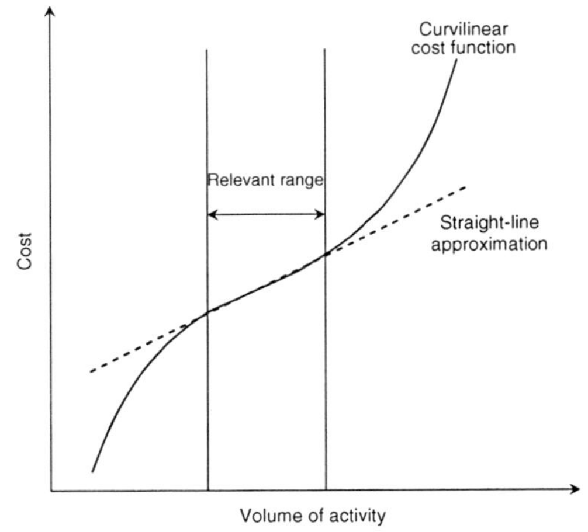

Fig. 16.2 The linearity assumption and the relevant range of variable cost

However, costs can be variable with respect to other things. For example, the wages paid to employees at an automobile factory plan will depend on the number of hours the plant is open and not strictly on the number of car produced. In this case, we would say that wage costs are variable with respect to the hours of operation. Nevertheless, when we say that a cost is variable, we ordinarily mean it is variable with respect to the amount of goods and services or project outcomes produced.

Not all variable costs have exactly the same behavior pattern. Some variable costs behave in a continuous variable or proportionately variable pattern. Other variable costs behave in a step-variable pattern.

Continuous Variable Costs—In manufacturing industry, direct materials is a true or proportionately variable cost because the amount used during a period will vary in direct proportion to the level of production activity. Moreover, any amounts purchased but not used can be stored and carried forward to the next period as inventory.

Step-Variable Costs—The cost of a resource that is obtained in large chunks and that increases or decreases only in response to fairly wide changes in activity is known as a step-variable cost. For example, the wages of skilled repair technicians are often considered to be a step-variable cost. Such a technician's time can only be obtained in large chunks if it is difficult to hire a skilled technician on anything other than a full-time basis. Moreover, any technician's time not currently used during the course of a project cannot be stored as inventory and carried forward to the next period. If the time is not used effectively, it is gone forever. Furthermore, a repair technician can work at a leisurely pace if pressures are light but intensify his or her efforts if pressures build up. For this reason, small changes in the level of production may have no effect on the number of technicians used by the project.

Except in the case of step-variable costs, a strictly linear relationship between cost and volume is often assumed. Economists correctly point out that many costs that the accountant classifies as variable actually behave in a curvilinear fashion; that is, the relation between cost and activity is a curve. A curvilinear cost is illustrated in Fig. 16.2.

Although many costs are not strictly linear, a curvilinear cost can be satisfactorily approximated with a straight line within a narrow band of activity known as the relevant range. The relevant range is that range of activity within which the assumptions made about cost behavior are reasonably valid. For example, note that the dashed line in Fig. 16.2 approximates the curvilinear cost with very little loss of accuracy within the shaded relevant range. However, outside of the relevant range this particular straight line is a poor approximation to the curvilinear cost relationship. Project managers should always keep in mind that assumptions made about cost behavior may be invalid if activity falls outside of the relevant range.

16.1.2.2 Fixed Cost

A fixed cost is a cost that remains constant, in total, regardless of changes in the level of activity. It will be incurred whether or not an asset is actually used by the project. A fixed cost is unaffected by reasonably large changes in activity or volume over some feasible range of operation and period. Consequently, as the activity level rises and falls, total fixed costs remain constant unless influenced by some outside force, such as a price change. However, because fixed costs remain constant in total, the average fixed cost per unit becomes progressively smaller as the level of activity increases. Typical fixed costs could include:

1. Interest on borrowed capital
2. Insurance and taxes
3. General management and administrative salaries
4. Facility leasing arrangements

Very few costs are completely fixed. Most will change if activity changes enough. When we say a cost is fixed, we mean it is fixed within some relevant range. The relevant range is the range of activity within which the assumptions about variable and fixed costs are valid.

Fixed costs can create confusion if they are expressed on a per unit basis. This is because the average fixed cost per unit increases and decreases inversely with changes in activity.

Fixed costs are sometimes referred to as capacity costs because they result from outlays made for buildings, equipment, skilled professional employees, and other items needed to provide the basic capacity for sustained operations. For planning purposes, fixed costs can be viewed as either committed or discretionary.

1. *Committed Fixed Costs*—Investments in facilities, equipment, and the basic organization often cannot be significantly reduced even for short periods of time without making fundamental changes. Such costs are referred to as committed fixed costs. Examples include depreciation of buildings and equipment, real estate taxes, insurance expenses, and salaries of top management and operating personnel. Even if operations are interrupted or cut back, committed fixed costs remain largely unchanged in the short term.

   During a recession, for example, an enterprise business will not usually eliminate key executive positions or sell off key facilities; the basic organizational structure and facilities ordinarily are kept intact. The costs of restoring them later are likely to be far greater than any short-run savings that might be realized.

Once a decision is made to acquire committed fixed resources, the enterprise business may be locked into that decision for many years to come. Consequently, such commitments should be made only after careful analysis of the available alternatives.

2. *Discretionary Fixed Costs*—Discretionary fixed costs, often referred to as managed fixed costs, usually arise from annual decisions by an enterprise business management to spend on certain fixed cost items. Examples of discretionary fixed costs include advertising, research, public relations, management development programs, and internships for students.

Two key differences exist between discretionary fixed costs and committed fixed costs. First, the planning horizon for a discretionary fixed cost is short term, usually a single year. By contrast, committed fixed costs have a planning horizon that encompasses many years. Second, discretionary fixed costs can be cut for short periods of time with minimal damage to the long-run goals of the enterprise business. For example, spending on training necessary to acquire skills to execute the project can be reduced because of poor economic conditions. Although some unfavorable consequences may result from the cutback, it is doubtful that these consequences would be as great as those that would result if the enterprise business decided to economize by laying off key personnel.

Whether a particular fixed cost is regarded as committed or discretionary may depend on the enterprise business intended strategy. For example, during recessions when the level of home building is down, many construction companies lay off most of their workers and virtually disband operations. Other construction companies retain large numbers of employees on the payroll, even though the workers have little or no work to do. While these latter companies may be faced with short-term cash flow problems, it will be easier for them to respond quickly when economic conditions improve. And the higher morale and loyalty of their employees may give these companies a significant competitive advantage.

The most important characteristic of discretionary fixed costs is that the enterprise business management is not locked into its decisions regarding such costs. Discretionary costs can be adjusted from year to year or even perhaps during the course of a year if necessary.

The concept of the relevant range, which was introduced with variable costs, is also important in understanding fixed costs, particularly discretionary fixed costs. The levels of discretionary fixed costs are typically decided at the beginning of the year and depend on the needs of planned programs such as advertising and training. The scope of these programs will depend, in turn, on the overall anticipated level of activity for the year. At very high levels of activity, programs are often broadened or expanded. For example, if the an enterprise business hopes to increase sales by 25 % through a set of "process improvement" projects, it would probably plan for much larger advertising costs than if no sales increase were planned. So the planned level of activity might affect total discretionary fixed costs. However, once the total discretionary fixed costs have been budgeted, they are unaffected by the actual level of activity. For example, once the advertising budget has been established and spent, it will not be affected by how many units are actually sold. Therefore, the cost is fixed with respect to the actual number of units sold.

Discretionary fixed costs are easier to adjust than committed fixed costs. They also tend to be less "lumpy." Committed fixed costs consist of costs such as buildings, equipment, and the salaries of key personnel. It is difficult to buy half a piece of equipment or to hire a quarter of a product-line manager, so the step pattern behavior is typical for such costs. As an enterprise business expands its level of activity, it may outgrow its present facilities, or the key management team may need to be expanded. The result, of course, will be increased committed fixed costs as larger facilities are built and as new management positions are created.

There are two major differences between the step-variable costs and the fixed costs.

The first difference is that the step-variable costs can often be adjusted quickly as conditions change, whereas once fixed costs have been set, they usually cannot be changed easily. A step-variable cost such as the wages of repair technicians, for example, can be adjusted upward or downward by hiring and laying off technicians. By contrast, once a company has signed a lease for a building, it is locked into that level of lease cost for the life of the contract.

The second difference is that the width of the steps for step-variable costs is much narrower than the width of the steps for the fixed costs. The width of the steps relates to volume or level of activity. For step-variable costs, the width of a step might be 40 h of activity per week in the case of repair technicians. For fixed costs, however, the width of a step might be thousands or even tens of thousands of hours of activity. In essence, the width of the steps for step-variable costs is generally so narrow that these costs can be treated essentially as variable costs for most purposes. The width of the steps for fixed costs, on the other hand, is so wide that these costs should be treated as entirely fixed within the relevant range.

16.1.2.3 Mixed Cost

Finally, mixed costs are a combination of fixed and variable costs, often expressed in conjunction with a single entity. A mixed cost contains both variable and fixed cost elements. Mixed costs are also known as semi-variable costs.

Understanding the types of behavior exhibited by costs is necessary to make valid estimates of total costs at various activity levels. Cost accountants generally separate mixed costs into their variable and fixed components so that the behavior of these costs is more readily apparent. Using the linearity assumption and the relevant range of variable cost, the following equation for a straight line can be used to express the relationship between a mixed cost and the level of activity:

$$Y = a + b \cdot X,$$

Where:

1. Y is the total mixed cost
2. a is the total fixed cost (the vertical intercept of the line)
3. b is the variable cost per unit of activity (the slope of the line)
4. X is the level of activity

Because the variable cost per unit equals the slope of the straight line, the steeper the slope, the higher the variable cost per unit. This equation also makes it easy to calculate the total mixed cost for any level of activity within the relevant range.

When step variable or step fixed costs exist, the project manager must choose a specific relevant range of activity that will allow step variable costs to be treated as variable and step fixed costs to be treated as fixed. Whether variable costs are traded for fixed or vice versa, a shift in costs from one type of cost behavior to another change the basic cost structure of an enterprise business and can have a significant impact on profits.

By separating mixed costs into their variable and fixed components and by specifying a relevant range for step costs, the project manager force all costs into either variable or fixed behavior as an approximation of true cost behavior. Assuming a variable cost to be constant per unit and a fixed cost to be constant in total within the relevant range can be justified for two reasons. First, the assumed conditions approximate reality and, if the enterprise business operates only within the relevant range of activity, the cost behaviors selected are appropriate. Second, selection of a constant per-unit variable cost and a constant total fixed cost provides a convenient, stable measurement for use in planning cost accumulation and cost assignment.

16.1.3 Cost Classifications for Management and Operations

Several cost terms are tied to general business management and the management of operations:

1. Allowable and Unallowable Costs
2. Controllable and Non-controllable Costs
3. Recurring Costs and Nonrecurring Costs
4. Standard Costs

Allowable and Unallowable Costs—These costs relate most often to the world of contracts and contracting, but it can also be applicable to a wide range of internal costs, primarily expenses. An allowable cost is a cost that the parties of a contract agree to include in the costs that will be reimbursed. Some contracts also specify how allowable costs are to be determined. Conversely, some contracts identify costs that are unallowable, that is, not reimbursable.

Controllable and Non-controllable Costs—Although nearly all types of costs can be controlled by someone with an enterprise business, all are not controllable at the same level of management. For any given level of management, controllable costs are costs that can be influenced by the project manager at that level. For example, project managers often have significant control over the expenses associated with their project resources. However, they have virtually no control over the insurance or taxes associated with the facility in which their human resources work. These are referred to as non-controllable costs. This distinction is particularly relevant to the concept of responsibility centers.

Recurring Costs and Nonrecurring Costs—Recurring costs are repetitive in nature, and occur when an enterprise business produces or provides similar goods and services on an ongoing, but discrete, basis. Nonrecurring costs do not repeat. They are sometimes referred to as "one-time costs." Nonrecurring costs often involve the development or establishment of a capability or capacity to operate.

Standard Costs—Standard costs are those costs of a unit of physical output that is estimated and developed in advance of any actual production or delivery of services. The practice of developing standard costs is commonly applied in operational settings. Standard costs are developed by combining direct labor costs, material costs, and overhead costs. Standard costs serve a useful role in cost control and other management functions, particularly related to operations. They can be used to evaluate operating performance levels, prepare bids, and establish inventory values.

16.1.4 Cost Classifications for Quality

As indicated in Chap. 1, these types of costs focus on a wide variety of efforts aimed at trying to ensure that the outcomes of a "process improvement" project conform to predetermined quality standards.

For the purpose of identifying exactly where and how money is being spent to achieve a given level of conformance, costs related to quality are typically subdivided into three categories.

1. Prevention Costs
2. Appraisal Costs
3. Failure Costs

16.1.4.1 Prevention Costs

Prevention costs are the costs of all activities specifically designed to prevent poor quality in element. These costs can be divided into two categories: costs related to non-conforming elements and costs incurred because the business activities to produce them are themselves less than adequate.

There are those costs that may be regarded as an essential part of business activities, for example field testing, design proving, failure modes and effect analysis. These are really costs associated with performing good business practice; they would be incurred regardless of the failure and appraisal costs and are not to be considered in this definition of prevention costs.

Costs that are considered in the definition of prevention costs are those that must be incurred if the current cost of failure and appraisal is to be reduced. These represent an investment in the "Continuous Improvement" initiative and, if effective, should result in a significant reduction of the overall costs. Obviously, these costs are likely to be small otherwise the failures would not occur and relevant appraisal cost would not be necessary.

16.1.4.2 Appraisal Costs

These are costs associated with measuring, evaluating or auditing elements to assure conformance to quality standards and performance requirements. These costs can be divided into two categories: costs related to non-conforming elements and costs incurred because the business activities to produce them are themselves less than adequate.

There are those costs that must be incurred regardless of the likelihood of occurrence of the associated adverse risk event, because the consequences of such an event are severe and potentially life threatening. Such is the case for many of the controls and procedures at power stations. This form costs are not to be considered in this definition of appraisal costs. Because they will always be incurred regardless of the likelihood of occurrence of a threatening risk event.

Costs that are considered in the definition of appraisal costs are those that are related directly to the likelihood of occurrence of error or failure. In this case, the amount of appraisal costs increases as the likelihood of occurrence of error increases more or less in direct proportion and vice-versa. Business activities which are included embrace all the costs of: incoming and source inspection/test of purchased material; in-process and final inspection/test; product, process or service audits; calibration of measuring and test equipment; and associated supplies and materials; which are carried out for no other reason than that the related failure or non achievement of an element quality occurred.

Failure Costs

These are costs resulting from elements not conforming to requirements or customer/ user needs. Failure costs are divided into internal and external failure categories.

Internal Failure Costs

These are failure costs occurring prior to delivery or shipment of an element to the customer. Internal failure costs can be many and varied. They include all costs and losses due to performing again what has already been done, or repairing or modifying the result of an activity, the cost of post mortems and all other consequential costs together with the waste of resources performing the business activities that need to be redone.

The consequential costs will include the effect on the balance sheet of excessive inventory and work-in-process (WIP) resulting from quality related deficiencies. In service industries, the equivalent problems do not show in inventory, but are hidden in direct costs. Most inventory and work-in-process, other than work actually being processed, can be regarded as Quality-Related costs. These include:

1. Reworking, redoing or repeating activities already performed because of inadequate performance at the first attempt. Costs of modification resulting from previous undetected design or planning weaknesses. These costs include the associated design or planning business activities, changes to tools and cost of retraining if procedures and methods are changed.
2. Retro design of a business activity element with a known design fault and all of the associated new features, fixtures and tools. Extra space in stores to accommodate

replacement parts with different issue numbers. Revisions to parts lists, instruction manuals and the increased complexity of related service activities.
3. Increases to inventory and work-in-process due to disruptions to the smooth flow of work.
4. Modifications due to poor quality design.
5. Storage space

External Failure Costs

These are failure costs occurring after delivery or shipment of the product—and during or after furnishing of a service—to the customer. These costs can be further subdivided into residual and random categories.

The residual non conformances of produced elements to requirements or customer/user needs include the underlying costs of warranty calls, servicing, complaints, etc.. . .

Some of the more spectacular costs may be found in the random category which, if they occur, can produce catastrophic results. These will include product recall or product withdrawal. Enterprise businesses often spend fortunes on advertising how good their products or services are; then suddenly they are plunged without warning into huge expenditure telling the public that they have put their lives at risk. In many cases, this negative publicity is overwhelmed by media attention, which places the very survival of the enterprise business at stake.

Other external costs which can also be included in the records include:

1. Failed product (resp. service) launches which are due to deficiencies in the product (resp. service) and identified and exposed by its first customers. These costs are invariably incurred when an enterprise business is overzealous in its attempt to obtain prior franchise with an innovative new product (or service) and is a common problem. In these cases of failed product (resp. service) launches, the enterprise business tries to take shortcuts and fails to test and prove the product (or service) performance characteristics prior to launch. This results in the customer unwittingly being the first inspector of the product (or service).
2. Failure to meet either the emotional or specified needs of the customer: this is usually caused by poor voice of the customer capturing, poor market research and poor competitor-related information, inadequate and misdirected promotion, wrong launch time, short shelf-life in the case of chemical, food and pharmaceutical products, contamination, poor packaging and consequent adverse publicity.
3. Customer complaints and the recording and analysis of customer complaints, and the cost of running a customer service department (i.e. a euphemism for customer complaints department).
4. Excessive after-delivery, service or maintenance support. Excessive costs including storage, delivery and all related administration, particularly those that infer, conceal from or mislead the public.

The failure costs go far beyond the internal and external costs indicated above. They include the devastatingly demotivating impact on employees within an enterprise. Employees want to feel good about the quality of their work. But regrettably, some enterprise businesses make decisions, and design systems, that deprive employees of their right to pride in workmanship, a prerogative that Edwards Deming considered one of the keys to motivation in the workplace (Deming, The New Economics: For Industry, Government, Education, 1994; Deming, 1982).

16.1.4.3 The Cost of Quality

The Cost of Quality is the total of the costs of the above costs. As indicated already, it is a measure of the costs specifically associated with achievement or non achievement of an element quality—including all elements requirements established by the business and its contracts with its customers. It is not the cost of creating a quality element; it is the cost of NOT creating a quality element. It represents the difference between the actual cost of an element and what the reduced cost would be if it did not deviate from the central tendency within the group as a whole. It is the total of the costs incurred by:

1. Investing in the prevention of nonconformance to requirements.
2. Appraising an element for conformance to requirements.
3. Failing to meet requirements.

Generating the quality-cost figures is not always easy, because most quality-cost categories are not a direct component in the accounting records of the enterprise business. Consequently, it may be difficult to obtain extremely accurate information on the costs incurred with respect to the various categories. The enterprise business' accounting system can provide information on those quality-cost categories that coincide with the usual business accounts, such as, for example, product testing and evaluation. In addition, many enterprise businesses will have detailed information on various categories of failure cost. The information for cost categories for which exact accounting information is not available should be generated by using estimates, or, in some cases, by creating special monitoring and surveillance procedures to accumulate those costs over the study period.

The reporting of quality costs is usually done on a basis that permits straightforward evaluation by management. Managers want quality costs expressed in an index that compares quality cost with the opportunity for quality cost.

16.1.5 Cost Classifications for Buying and Selling

Several types of costs are associated with the practice of selling goods and managing inventory. Purchasing costs are the actual costs of goods acquired from suppliers, including freight and transportation costs. Ordering costs include expenditures related to preparing, issuing, and paying the purchase orders needed to acquire the goods. Ordering costs also include expenses related to receiving and inspecting the items included in orders. Carrying costs occur whenever an enterprise business maintains an inventory of goods for sale; carrying costs includes two basic components. First, an obvious cost is associated with the storage of the goods, such as space rental, insurance, and spoilage. However, an opportunity cost also is related to inventory, because the money invested in inventory could be used for other investments.

The last type of cost associated with goods for sale is stock out cost. A stock out occurs when an enterprise business runs out of a specific item for which demand exists. This may result in the need to expedite a special order from a supplier, which frequently results in added expenditures. A stock out situation could also result in lost sales, and even lost future sales, as a result of customer dissatisfaction caused by the stock out situation.

16.1.6 Cost Classifications for Project Economics

Perhaps one of the most important, yet not well understood, cost issues on projects is the practice of determining whether a given project expenditure is a capital cost or an expense cost. It is an extremely important, yet tricky, area related to asset management.

16.1.6.1 Capital Costs

Capital costs are the one-time costs associated with a project, including the price of purchased assets such as land, equipment, or other supplies, and the cost of going into debt or issuing stock in order to fund the project. Capital costs are incurred on the purchase of equipment or other supplies to be used in the project for the production of goods or the rendering of services; in other words, the total cost needed to bring a project to a commercially operable status. For example, the purchase of a new machine that will increase production and last for years is a capital cost. Capital costs do not include labor costs except for the labor used for construction. Unlike operating costs, capital costs are one-time expenses, although payment may be spread out over many years in financial reports and tax returns. Capital costs are fixed and are therefore independent of the level of output.

Generally speaking, capital costs comprise the bulk of the costs that the project manager will be required to estimate and include in any project financial analysis performed.

16.1.6.2 Expense Costs

Project-related expense costs are expenditures associated with the supporting environment of the project, but not directly attributable to the creation of a specific asset. Cost items that fall in the expense category may include items such as travel, secretarial staff assigned to support the project, computer usage time (data gathering, labor charging, etc.).

A number of issues add significant concern to the question of correctly categorizing items as capital or expense costs. First, considerable confusion and a lack of knowledge often exist on the part of project team members regarding the rules on how to classify some kinds of project expenditures. Second, different people within an enterprise business can frequently have different motivations for classifying a given project cost as capital or expense. Third, it's very important to classify project costs correctly, because the practice directly impacts the proper valuation of the enterprise business asset base and enables compliance with corporate income tax regulations.

16.1.7 Cost Classifications for Decision Making

Costs are an important feature of many business decisions. In making decisions, it is essential to have a firm grasp of the concepts of differential cost, opportunity cost, and sunk cost.

16.1.7.1 Differential Cost and Revenue

Decisions involve choosing between alternatives. In business decisions, each alternative will have costs and benefits that must be compared to the costs and benefits of the other available alternatives. A difference in costs between any two alternatives is known as a differential cost. A difference in revenues between any two alternatives is known as differential revenue. Differential costs can be either fixed or variable.

A differential cost is also known as an incremental cost, although technically an incremental cost should refer only to an increase in cost from one alternative to another; Differential cost is a broader term, encompassing both cost increases and cost decreases between alternatives.

The differential cost concept can be compared to the marginal cost concept. In speaking of changes in cost and revenue, the terms marginal cost and marginal revenue are often used. The revenue that can be obtained from selling one more unit of product is called marginal revenue, and the cost involved in producing one more unit of product is called marginal cost. The marginal concept is basically the same as the differential concept applied to a single unit of output.

16.1.7.2 Opportunity Cost

An opportunity cost is the cost of any activity measured in terms of the value of the best alternative that is not chosen (that is foregone). Opportunity cost is a key concept in economics, and has been described as expressing "the basic relationship between scarcity and choice." The notion of opportunity cost plays a crucial part in ensuring that scarce resources are used efficiently.

The "opportunity cost" of a resource refers to the value of the next-highest-valued alternative use of that resource. If, for example, the project uses time a given resource, let say "Resource A," it cannot use the alternative resource, let say "Resource B," simultaneously. If the project next-best alternative to using "Resource A" is using "Resource B," then the opportunity cost of using "Resource A" is the cost spent plus the consequence forgone by not using "Resource B."

The true cost of using a resource is what the project manager gives up to get it. This includes not only the cost spent in acquiring the resource, but also the economic benefits (effectiveness) that the project manager did without because the project acquired that particular resource and thus can no longer acquire the alternative.

16.1.7.3 Sunk Cost

A sunk cost is a cost that has already been incurred and that cannot be changed by any decision made now or in the future. It is a cash outlay that has already occurred or has been committed. Because sunk costs cannot be changed by any decision, they are not differential costs. And because only differential costs are relevant in a decision, sunk costs can and should be ignored.

To illustrate a sunk cost, assume that an enterprise business paid a certain amount of cash several years ago for a special-purpose machine. The machine was used to make a product that is now obsolete and is no longer being sold. Even though in hindsight purchasing the machine may have been unwise, the amount of paid cash has already been incurred and cannot be undone. And it would be folly to continue making the obsolete product in a misguided attempt to "recover" the original cost of the machine. In short, the amount of cash originally paid for the machine is a sunk cost that should be ignored in current decisions.

The concept of sunk costs is particularly relevant when making a decision regarding a future investment, such as the decision on whether to approve the usage of a resource for the project. Sunk costs must not be included in any financial analysis.

16.1.7.4 Total Cost and Incremental Cost

As the name suggests, total cost is merely the overall, bottom-line expression of an item of cost. Total costs primarily are used in reference to other cost forms.

An incremental cost represents a change in costs, typically a change in enterprise business cash flows that occurs as a direct result of accepting a certain decision. Typically, this decision relates to either selecting a particular alternative solution to a problem dealt by the project. Every item of costs included in a project financial analysis is expressed in terms of incremental costs. These costs represent the difference in every existing cash flow within the enterprise business that comes as a result of approving that project.

16.2 Collect Costs Data

Once the costs types and their behavior for the project have been identified, the actual cost data collection can be performed. It involves developing approximations or estimates of the identified costs needed to complete the project successfully.

16.2.1 Personnel Costs

These costs often include:

1. *Salaries*—The first step is to develop an estimate of salaries paid to employees directly involved in the project. Salaries for direct supervisory personnel should be prorated to the extent that their time is spent supervising individuals involved in the project.

In addition, any cost of regular, estimated overtime should be included. If seasonal or part-time employees are required, their salaries should be included in the collected cost data.

2. *Benefits*—Standard benefits can be calculated as a percentage mark-up of the direct salary costs. The appropriate financial function within the enterprise business should provide an estimate of the total employer cost for benefits expressed as a percentage of salary.
3. *Other Allowances*—Estimates should also be made for those allowances that can be expected to be payable to employees performing the project activity work, based on past experience. If employees are entitled to performance pay or a bonus in addition to their regular salaries, an estimated amount (normally based on past experience) should be included as well in the collected cost data.
4. *Training*—Any costs for training (e.g. vocational, skill training, retraining and technology transition) that would be expected to be incurred if the project activities work were provided in-house should be included. This should include any travelling expenses associated with attending a training course or program.

16.2.2 Operating and Maintenance (O&M) Costs

These costs are often incurred and accumulated at higher organizational levels than the "process improvement" project in question. Only the portion of costs applicable to the project should be computed or estimated when calculating operating and maintenance costs. Operating and Maintenance (O&M) costs include, but are not limited to:

Materiel and Supply Costs—Any materials and supplies needed to provide the project outcomes or perform the project activities work should be included. Examples are raw materials, repair parts, subassembly components, office automation costs and operating supplies.

The required quantities of materials and supplies can be taken from records of previous fiscal years and adjusted to reflect any expected increases or decreases in quantity, as identified in the output specifications. Unit prices can be taken from the same sources and adjusted to reflect the estimated costs of materials and supplies for the period of contract performance.

Maintenance and Repair—These are costs for maintaining equipment in normal operating condition during the fiscal year, which can reasonably be expected to be realized given the changes in requirements.

Travel—Any costs of travel that are related to the actual performing of the project activities work and that would be incurred if the activities work were performed in-house should be included in the collected cost data.

Other Costs—Any other O&M expenditures that have not been specifically included above should be included here. As a minimum, there should be an estimated cost for replacing minor items. Minor items are durable items that are paid by the O&M budget. Rather than attempting to assess the replacement cost of a large number of minor items, an aggregate amount could be included in the collected cost data.

16.2.3 Capital Costs

Capital costs are the net cost of assets of significant value that have a life cycle of more than one year. To determine the cost of assets used by the project, one needs to consider the out-of-pocket costs when an asset is acquired at the beginning of or during the project time period, less the present value of the money that could be realized when the asset is either disposed of or salvaged.

It should be noted that not all assets are newly acquired. In fact, most assets that would be used in performing project activity work in-house would already exist within the enterprise business. Even though these assets have already been paid for, their costs should not be considered as "sunk costs" and thus be excluded. Because these resources may be used elsewhere within the enterprise business, the value of the usage that is foregone should be considered (i.e. opportunity cost).

For existing assets that could be reused elsewhere, a current market price (i.e. net of disposal costs) for the assets in that condition needs to be established. For assets that cannot be reused elsewhere, the current value would be the scrap value less disposal costs.

Thus, the general approach to determining the cost of the capital assets is as follows: determine the current value (i.e. the net amount realizable if sent for disposal) of any existing assets that will not be used if performing project activities work in-house ceases; during the project life time period; and deduct any disposal value (discounted) of assets at the end of the project life time period.

16.2.4 Overhead Costs

These costs include the portion of corporate and administrative (C&A) overhead within each department of the enterprise business and the project support overhead within the unit responsible for "process improvement" project that can be expected to be saved.

Overhead costs are allocated to the project in a two-step process. First, determine the costs of services provided by specific C&A overhead components. Second, allocate each project's share of these costs, plus its own project support costs to the specific activities that make up the project.

Overhead costs often include:

1. *Corporate and Administrative Overheads*—These are incurred outside operational branches in support of operating programs and activities. They may include the costs of such functions as: Executive Management, Communications, Administration, Personnel, Finance and Informatics.

   A simple departmental factor can be calculated and applied to all items in personnel costs and to the project support overhead. This factor is equal to the

C&A costs divided by all other departmental costs, excluding transfer payments or flow through charges.

2. *Project Support Overhead*—These are salary, benefits and O&M costs incurred in performing functions that are not directly involved with the project activities, but which support these activities. They include all supervisory and management personnel within a project branch that relate to the activity, but that have not been included in the scope.

16.2.5 Additional Costs

These include the costs of unusual or special circumstances that arise under execution of the project activities or costs that do not fit appropriately in the previous categories. Any item included as an additional cost should be supported by defining the type of cost, documenting the way the cost is computed and listing the component elements. An example may be the value of any additional inventory that would have to be maintained.

Cost estimates are dynamic and at different stages of a project execution, they take on different forms and purposes. For example, at the initiate stage, there is insufficient information to develop a detailed and accurate cost estimates. As execution of the PDSA "PDSA Plan" Process Group starts and is ongoing, more information is uncovered and made available, so the cost estimates become more detailed. Once the project is fully established and basic processes of the "PDSA Plan" Process Group have been developed, the estimates are even further refined. It is important to recognize that the "Estimating Costs" process continues throughout the lifecycle of the project.

Table 16.1 shows a cost estimate summary as it might be set out for a typical project.

16.2.6 Why Estimate Costs

There are several reasons why cost estimates, i.e. quantitative assessment of the likely amount of funds, are established:

1. *To serve as a basis for control:* The estimates are prepared as baseline measures against which the project expenditures will be controlled. For this purpose the estimates may need to be quite detailed.
2. *To assess the project viability:* The project viability must be assessed throughout the expected duration of the project. The rationale for the project is set out in the business case which is expressed in terms of a set of benefits which contribute towards the enterprise business intended strategic goal(s). The project framework and planning is written to ensure that achievement of those benefits is maximized. However, a project can change at any time during its life, moving gradually or swiftly towards non-viability. The project can lose value for a number of reasons, some of which include:

Table 16.1 Generic summary layout of a project costs

Above-the-line items	Direct (variable) costs	Direct labor	The wages and salaries of people employed on the project, for time that can be wholly and specifically attributed to the project. These times should be estimated using the standard cost rates applicable to each grade of staff.
		Direct materials	Equipment, materials and bough-out services used specifically on the project.
		Direct expenses	Travel, accommodation and other costs chargeable specifically to the project. These can include the hiring of external consultants.
	Indirect (fixed) costs	Overhead costs	A portion of the costs of running the business, such as general management and accommodation. Usually calculated as a proportion of the total direct costs. Overheads costs are not applicable if the project is itself charged as an overhead.
Below-the-line items		Contingency funds	An addition, usually calculated as a small percentage of the above-the-line costs, in an attempt to compensate for estimating errors and omissions, unfunded project changes and unexpected costs.
		Escalation	An addition to allow for costs that increase with time as a result of annual cost inflation. Particularly important for long-duration projects in time when national cost inflation rates are high.
		Mark-up for profit & Selling price	These two items apply on only to projects sold to external clients. There are various ways in which they can be calculated and their levels are often judged according to the strength of the competition and what the market will stand. These are management decisions, not part of the cost estimating process. Such decisions are always more easy to make when there is confidence in the cost estimating accuracy.
		Provisional sums	The estimated costs of items that are not included in the quoted price which have to be charged extra if the need for them is revealed as project work proceeds.

- The costs escalate and outweigh the benefits.
- The delivery is delayed beyond the point where it has sufficient benefits.
- There is a change in the business environment with the market moving to new products or services.
- The intended business strategy changes direction, making the project less relevant.
- The project cannot deliver the benefits originally expected, making the pay-off marginal.
- The enterprise business is unable to adequately resource to all the projects in its current portfolio.

- The project is reliant on a technology or capability which may not materialize.
- There is internal competition for limited resources that are locked up in the project.
- As the project unfolds, the risk/return ratio begins to look worse than expected.
- There is doubt about the true cost and measures of progress.
- The project team cannot cope with the constant changes to scope or requirements.
- There is lack of confidence about the "fitness for purpose" of the final deliverables.
- It may become clear during the project life cycle that the original quality expectations cannot be met. This can have an impact on the acceptability and hence the usability of the project's outcomes by the customer. Changes to quality must be assessed against the benefits.

Consequently, the project manager or the project sponsor should assess the project's viability at the earliest sign of significant changes to markets, change to the enterprise business intended strategy or change to the project costs.

3. *To obtain funding:* After approval has been obtained, the project must be financed. Funding will be awarded on the basis of the appraisal of cost estimates prepared.
4. *To manage cash flow:* Once funding has been obtained, and project work started, the project must be managed so that work takes place and consumes cash no faster than the rate agreed upon.
5. *To allocate resources:* Human resources are a special form of project funding. The business plans their allocation in advance against the cash-flow estimate. They will be assigned to the project week by week against the control estimate.
6. *To estimate durations:* The duration of a work element is calculated by comparing the estimate of work content to resource availability, and so the cost estimates form an input to time estimating. Time estimating, which was described in a previous section, is performed for similar reasons to cost estimating.
7. *To prepare tenders:* Contracting firms tendering for bespoke contracts need to prepare estimates for the tender.

16.2.6.1 Categories of Cost Estimates

Cost estimates can be productively categorized into three levels: conceptual estimates, preliminary estimates, and definitive estimates.

Conceptual Cost Estimates

For most projects, at the "Initiate" stage of the project life cycle, little thought has been given to the details of what it will take to execute a project. In fact, a decision has not yet been made as to whether the enterprise business should support the project. The function of a cost estimate at this point is to provide the enterprise business executives and managers with the information they need to take the appropriate decision.

Obviously, a problem with making estimates at the "Initiate" stage is the lack of adequate information on just about all aspects of the project, including its duration, specifications, and tasks. Yet it is important that a solid cost estimate be developed so that the enterprise business executives and managers will have at least a rough sense of the resources that need to be employed to execute the project.

When an estimate is made at the "Initiate" stage, it is called a conceptual estimate. It is also referred to as an order of magnitude estimate, which offers a good sense of the rough magnitude of project costs. Very early in the project life history, there will be outline proposals for the nature and scope of the project, but certainly no detailed task list or comprehensive work breakdown. Thus cost estimates can be made only on a global comparative basis. That means trying to assess the cost of the whole project by comparing it with similar projects that have been completed in the recent past and for which their actual cost records can be accessed.

If the project can be divided into a few major parts at this early stage, it should be possible to distribute the total estimate over those parts, remembering to leave something in reserve in the form of a separate contingency item. The specific approaches to making conceptual cost estimates are the analogous cost estimating and the top-down cost estimating.

With the analogous approach, the project sponsor or the project manager takes, through expert judgment, the actual costs of similar projects and uses these costs as estimates for the current project. Analogous cost estimating is frequently used to estimate costs when there is a limited amount of detailed information about the project (e.g., in the early phases). Analogous cost estimating is generally less costly, but it is also generally less accurate. It is most reliable when previous projects are similar in fact, and not just in appearance, and the project sponsor or the project manager preparing the cost estimates have the needed expertise.

With the top-down approach, the project sponsor or the project manager takes a look at the cost of similar projects, make a number of logical assumptions, and develop a quick estimate of project costs. This information becomes part of the body of data needed to decide whether or not to go ahead with the project.

It is apparent that top-down estimates must usually be ballpark. They have the disadvantage of not being based on a detailed project specification. They cannot take into account many factors that will not become known until much later in the project life history and their inherent accuracy will not be high. However, because top-down estimates are often based on comparisons with completed projects, there is less risk (when compared to bottom-up estimating) of forgetting to include items and thus arriving at dangerously low estimates.

Top-down cost estimates tend to be "quick and dirty." They are carried out simply to develop a rough sense of what level of resources a project will consume. They can be quite accurate if the enterprise business has carried out many projects of a similar nature and has a rich "organizational process assets" in which previous cost experiences are reflected.

The best-known top-down cost estimating technique is called parametric cost estimating. Parametric cost estimating is a technique that uses a statistical relationship between historical data and other variables (e.g., square footage in construction, lines of code in software development, required labor hours) to calculate a cost estimate for a scheduled activity resource. This technique can produce higher levels of accuracy depending upon the sophistication, as well as the underlying resource quantity and cost data built into the model. A cost-related example involves multiplying the planned quantity of work to be performed by the historical cost per unit to obtain the estimated cost.

Preliminary Cost Estimates

Preliminary estimates are developed once a decision has been made to pursue a project. For example, in pricing a project to be included in a bid, the project sponsor or the project manager needs more detailed and more accurate figures than can be derived from a conceptual cost estimate. While the preliminary estimate is more accurate than the conceptual estimate, it still remains rather crude.

Preliminary estimates may use either a refined top-down estimating procedure or a crude bottom-up procedure. Bottom-up cost estimates are derived from work breakdown structures (WBSs) described in a previous section. This technique involves estimating the cost of individual work packages or individual schedule activities with the lowest level of detail. This detailed cost is then summarized or "rolled up" to higher levels for reporting and tracking purposes. The cost and accuracy of bottom-up cost estimating is typically motivated by the size and complexity of the individual schedule activity or work package. Generally, activities with smaller associated effort increase the accuracy of the schedule activity cost estimates.

Bottom-up cost estimates are generally more accurate than top-down cost estimates, because they have been developed with substantial care and have explicitly taken into account all the cost elements of a project. However, it takes an enormous amount of time and resources to create a good bottom-up estimate.

Definitive Cost Estimates

Definitive cost estimates are the most accurate cost estimates that the project sponsor or the project manager can make. They usually are not fully developed until the project is underway. They are created as part of the funded project planning effort. At this time, fairly detailed work breakdown structure (WBS) and project schedules have emerged. The work is now fairly well defined. The estimate invariably arises as the consequence of a bottom-up estimating effort.

An important function of definitive estimates is to provide the basis of developing a detailed project budget.

16.3 Allocate Costs to Activities

This is the project management process for aggregating and approving the estimated costs of individual scheduled project activities or work packages in order to establish a total cost baseline for measuring the project cost performance.

The most fundamental concept used to allocate costs to the project activities or work breakdown structure components is that work is worth the planned or negotiated value (scope in financial terms) of the work. Depending on the industry, this value might be set by a formal contract, an annual budget, or a price to be paid for the project outcomes. This concept flows down into the project activities as team members, subcontractors and suppliers agree to provide materials or services to achieve the project objectives.

Once at the work packages level of the project work break down structure, scheduled activity cost estimates are aggregated, make or buy analysis performed, and the planned value of project activities must be established. The work package cost estimates are then aggregated for the higher component levels of the work breakdown structure and ultimately for the entire project. An important output of this "Allocate Costs to Activities" process is the "Make or Buy" decisions document, which also contains a listing of buys, representing those materials, products, services, or results needed from outside the project boundaries which must be procured for the project.

16.3.1 Make or Buy Analysis

Make or Buy Analysis is an effort to identify the most efficient and cost effective manner for performing schedule project activities. It relates to comparing the cost of in-house work with the cost of procuring a schedule project activity work.

A basic and critical concept underlying the "Make or Buy" analysis is that the assessment of the costs of performing a schedule project activity under both in-house work and procurement options should be based on the same well-defined level and quality. The comparative costing should be based on three basic principles:

1. *Relevance of the cost*—Only those costs that differ between the two options (in-house work or procurement) are taken into consideration in the analysis. Costs that remain the same regardless of the mode of performance need not be calculated.
2. *Fairness of comparison*—Costs are included or excluded to ensure that the comparison between the in-house work and procurement options is as fair as possible. For example, in the case of overhead costs, those portions of the costs that are sensitive to the tendering decision and savings that are realistically achievable should be accounted for in the analysis. Overhead will normally be restricted to costs within a department of the enterprise business. However, if other significant overhead costs can be attributed to performing an activity work (e.g. legal work), they should be specifically identified.

   When it is difficult to decide whether to include a particular cost or where a particular cost warrants further scrutiny, a sensitivity analysis of that cost may help the project manager to make the ultimate make or buy decision.

   It should also be recognized that there is a target return on equity in the private sector, which will not be included in the cost analysis of performing an activity work in-house.

3. *Same level of service compared*—For cost comparison, procurement and in-house activity work options should provide the same level and quality of the outcomes under consideration. There should be no significant difference in the level and quality of the outcomes when comparing the costs of performing an activity work under the two options.

The cost comparison should cover a long term period of time as specified by the enterprise business policies, even though the initial contract may cover only medium or short term periods. The long term time frame allows conversion costs to be spread over a reasonable period of time. It also allows more in-house costs to be considered relevant, since some expenditures (particularly capital) may not be planned to occur in a shorter term. If significant costs are expected to occur beyond the long term period, the analysis could be extended to include them (e.g. Work Force Adjustment Directive).

The costs of activities with a known life span of less than the specified long term period of time should, of course, appear in the calculations only for those periods of time the activities exist. Project managers are in the best position to determine what time period is appropriate for the analysis. Although long term period of time of five years may be a rule of thumb, circumstances may warrant a shorter (but normally not less than three years) or longer period. The choice should be justifiable.

Cost comparisons also involve considerations of cash flows over time. Because money has a time value, equal amounts received or paid out over different periods are not equivalent. Therefore, all cash flows should be discounted to present value to permit costs to be compared without any bias for different timing.

The make or buy costing methodology involves four steps:

1. Specifying procurement items costs on which cost are assigned;
2. Determining the avoidance costs of in-house work;
3. Determining the cost of procurement; and
4. Comparing the difference between net costs and making the final decision.

16.3.1.1 Specifying Procurement Items on which Costs are Assigned

The procurement items specifications are typically descriptions of the service and the level of service that a supplier is required to deliver. These specifications describe the quantity and quality of service to be delivered and could include levels of service, performance standards, measurement systems, and methods and frequency of reporting. Specialized equipment required to perform the service (e.g. technology platform architecture) should be identified in the specifications.

Procurement items specifications should not describe how their delivery should occur. Significant savings may be realized by allowing prospective suppliers to develop more efficient ways of delivering the service. Procurement items specifications should be sufficiently detailed to allow potential supplier(s) to develop a reasonable estimate of the cost of providing the service.

Defining procurement items specifications is an important step as procurement items represent the target for all costs and generally influences the costing process. The specifications should be as clear as possible. For example, output specifications for information technology might include common networks, common systems and so on. Poorly written procurement items specifications will often lead to deficient services and to considerable disputes between the project manager and suppliers. This step should include input from project manager who has the most knowledge about the nature of the project outcomes as well as the quality and level of service required.

Procurement items specifications should be developed as early as possible in the process to obtain consensus on outcomes and the level of service to which costs should be assigned. When establishing the output costs, financial specialists within the enterprise business can provide help and valuable advice.

16.3.1.2 Determining the Avoidable Cost of In-House Work

The basis to be used for costing the in-house activity work should be the "most efficient" scenario. This may be, in the project manager's view, the current way of operations. Alternatively, it may be some other modus operandi (as a result of a special study or in the project manager's estimation) that could well be the "most efficient."

If the alternative scenario is used for in-house costing, regardless of the decision on contracting out, the project manager is expected to implement necessary changes on a timely basis that will lead the "process improvement" project to that most efficient scenario.

16.3.1.3 Determining the Cost of Procurement

These costs comprise:

1. *Contract Administration Costs*—These costs are incurred by the responsible contract administration office function to ensure that the contract is properly executed by both parties: the project manager and the supplier. Included are costs for initiating the contract, reviewing the Request for Proposal, evaluating and selecting the winning proposal, executing the Quality Assurance Plan, following up on any problems, dealing with disputes, processing payments (including holding back payments if required) and negotiating change orders. Estimating the cost of contract administration should be done on the basis of the rates established by the enterprise business.
2. *Costs of Converting to Contracted Service Delivery*—A conversion to contract may require some materials-related or labor-related one-time conversion costs (OTCC).
   - *Materials-Related One-Time Conversion Costs*—These costs arise from disposing of or transferring expendable procurement items that are on hand and being used in the activity, including the costs of packing, crating and shipping these items. The amount of money that can be recovered from their disposal or transferred may or may not exceed disposal costs (i.e. the materials-related OTCC may be a gain or loss).
   - *Labor-Related One-Time Conversion Costs*—These costs do not include severance pay, but do include removal and relocation expenses and retraining costs required to place employees in alternative positions.

Based on various factors (such as marketability of the skills of affected employees, past experience in placing surplus employees and likely vacancies in the department), the project manager should estimate and include work force adjustment costs based on the number of employees who can be expected to receive these benefits. Staff in the human resources function should be able to provide advice and assistance to the project manager.

- *Additional One-Time Conversion Costs*—These costs are neither materials-related nor labor-related. Examples are penalty fees for terminating leases or supply contracts, or loss of a volume purchasing discount.

3. *Net Salvage Value on Disposal or Transfer of Assets*—Contracting out through procurement may result in ending the need for certain property, materials or equipment. When these assets are transferred or otherwise disposed of, there will be a net salvage value which may be positive or negative.
   The net salvage value is the estimated disposal value of the assets minus the costs of their disposal or transfer, with the latter including the costs of packing, crating and shipping. The amount should be included under contract performance and allocated entirely to the time period of cost comparison, or an adequate explanation should be provided where this assignment of cost is not reasonable.
4. *Contract Performance Costs*—The price of the final contract cannot be included until all proposals have been received and evaluated. At that time, the best price proposed by a qualified contractor offering the best value is added to the total cost that was estimated for Contract Administration Costs, Costs of Converting to Contracted Service Delivery, and Net Salvage Value on Disposal or Transfer of Assets Costs.
5. *Total Contract Costs*—The total cost for contract performance should be estimated and the present value of the cash flows calculated.

16.3.1.4 Difference Between Net Costs: Making the Final Decision

The project manager should ensure that all relevant costs are identified and included to the extent appropriate in calculating the in-house costs. The cost of procurement is compared to those costs of performing the project activities work in-house that would be saved by ceasing to perform the activities. The costs of risk for making or buying should be added to each side of the calculation, as appropriate. Taxes have not been included in this cost comparison methodology, because it is not possible to make an adequate estimate of tax flows from every enterprise business.

The final decision, a make or buy decision, is the act of making a strategic choice between performing project activities work internally (in-house) or outsourcing it externally (from an outside supplier). Make-or-buy decisions usually arise when an enterprise business has developed a product or part—or significantly modified a product or part—is having trouble with current suppliers, or has diminishing capacity or changing demand.

As a rule of thumb for procurement or for out-sourcing, an enterprise business often outsources all activities that do not fit one of the following three categories:

1. The activity is critical to the success of the project outcomes, including customer perception of important project outcomes attributes;
2. The activity requires specialized design and production skills or equipment, and the number of capable and reliable suppliers is extremely limited; and
3. The activity fits well within the enterprise business core competencies, or within those the enterprise business must develop to fulfill its intended strategic plans.

   Project activities that fit one of these three categories are considered strategic in nature and should be performed internally if at all possible. Factors that may influence to take a buy decision include:

1. Lack of expertise.
2. Suppliers' research and specialized know-how exceeds that available in-house to perform a project activity.
3. Cost considerations (less expensive to procure the activity outcome).
4. Small-volume requirements
5. Limited production facilities or insufficient capacity
6. Desire to maintain a multiple-source policy
7. Indirect managerial control considerations
8. Procurement and inventory considerations
9. Activity not essential to the enterprise business intended strategy

   In much the same vein, factors that may influence to take a make decision:

1. Cost considerations (less expensive to perform the activity in-house)
2. Desire to integrate enterprise business operations
3. Productive use of excess enterprise business capacity to help absorb fixed overhead (using existing idle capacity)
4. Need to exert direct control over production or service and/or quality
5. Better quality control
6. Design secrecy is required to protect proprietary technology
7. Unreliable suppliers
8. No competent suppliers
9. Desire to maintain a stable workforce (in periods of declining sales)
10. Quantity too small to interest a supplier
11. Control of lead time, transportation, and warehousing costs
12. Greater assurance of continual supply
13. Provision of a second source
14. Political, social or environmental reasons (union pressure)
15. Emotion (e.g., pride)

The two most important factors to consider in a make-or-buy decision are cost and the availability of production capacity within the enterprise business. As indicated in the previous sections, cost considerations should include all relevant costs and be long-term in nature. While cost is seldom the only criterion used in a make-or-buy decision, simple break-even analysis can be an effective way to quickly deduce the cost implications within a decision. Indeed, suppose that an automobile manufacturing plant can purchase equipment for in-house use for 250,000 monetary units and produce the needed automobile parts for 10 monetary units each. Alternatively, suppose that a supplier can produce and ship the same automobile part for 15 monetary units each, including the operational costs. Disregarding the cost of negotiating a contract with the supplier, the project manager can set up an equation that shows the number of parts required for the in-house use cost to equal the procured cost. In this way, the project manager can determine when it makes sense financially to purchase equipment and produce the parts in-house rather than buying them from the supplier. In the equation that follows, x is the number of automobile parts:

$$250,000 + 10 \cdot x = 15 \cdot x$$

Subtracting $10 \cdot x$ on both sides leads to:

$$250,000 = 15 \cdot x - 10 \cdot x = 5 \cdot x$$

Dividing both sides by 5 gives:

$$50,000 = x$$

This means that the cost of purchasing equipment and producing parts in-house equals the cost of buying the same parts from the supplier for 50,000 parts. Therefore, it would be more cost effective for the automobile manufacturing plant to buy the part if the demand from the "process improvement" project is less than 50,000 units. It would be more cost effective for the automobile manufacturing plant to purchase the equipment and make the parts in-house if the demand from the "process improvement" project exceeds 50,000 units. However, if the automobile manufacturing plant had enough idle capacity to produce the parts, the fixed cost of 250,000 monetary units would not be incurred (meaning it is not an incremental cost), making the prospect of making the part too cost efficient to ignore.

16.3.2 Planned Value of Project Activities

The planned value of project activities is the value of all project activities; i.e. the cost of resources to be applied to project activities over their time frame. It is the sum of each activity's time-phased aggregated and agreed cost estimates, created by associating the applicable cost to each detailed work activity and the work schedule. Here, each piece of work is considered to possess two attributes: its value or authorized funds (i.e., its aggregated and approved cost estimates), and its planned completion date or allocated time. The sum of the planned value for all project activities should equal the total project agreed funds, less any funds set aside for risk management or not yet allocated to a specific activity.

For example, if an activity is planned for four months and is expected to use one person during the first month and two people during the remaining three months,

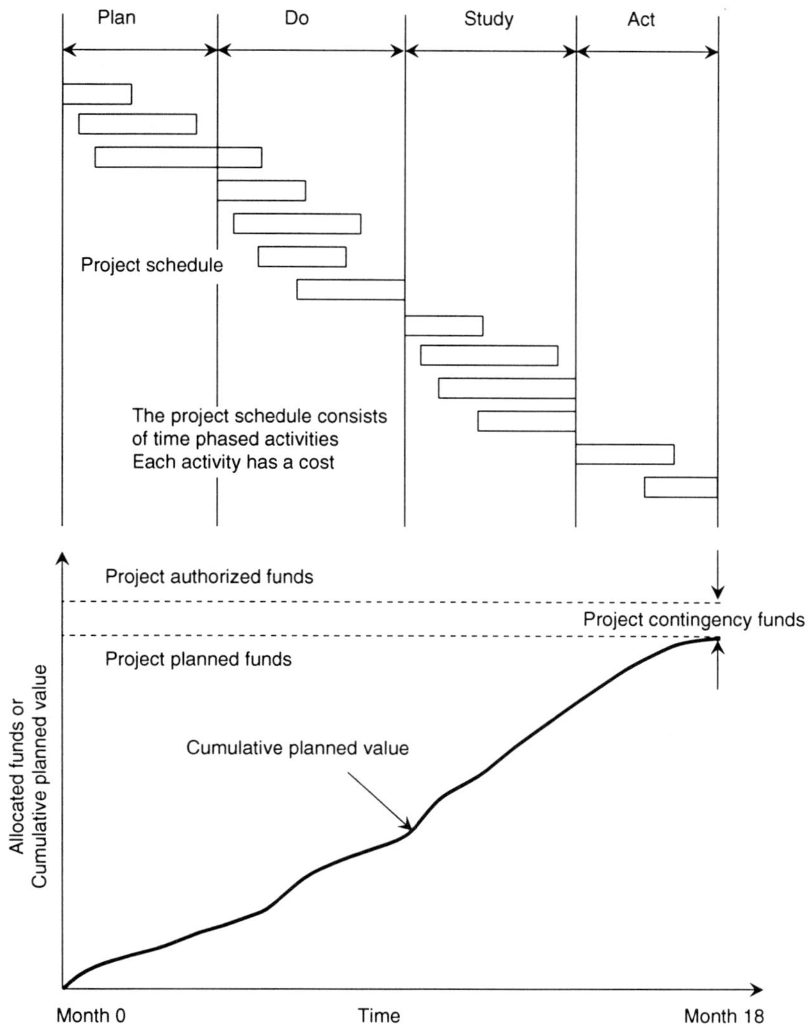

Fig. 16.3 The cost performance baseline chart

then the planned value for the activity is the cost of seven staff-months. Further, the expectation is to spend one staff-month during the first month and two staff-months in each of the final three months.

Plotting the cumulative value of all project activities and their allocated funds on the vertical axis with time on the horizontal axis reveals a cumulative planned value increasing from zero to the total value of the project, as shown in Fig. 16.3. The total planned value of project activities is the value of all work stated in the project plan, thus it also represents the project scope in financial terms.

The plot shown in Fig. 16.3 reflects the value of activities to be completed each period of time. This curve, called the cost performance baseline, is the cumulative planned value of all activities that should be accomplished each period of time. It captures both value and time frame of all the project work. It also represents the reference against which periodic cost performance will be measured.

The cost performance baseline is relatively easy to create when a project is properly planned. If resources with expected aggregated costs have been properly allocated to the identified project activities in the work breakdown structure, the expected cost of resources for each activity sets the value of the activity. Summing the value of all activities to be completed each period of time and then adding that sum to all the previous periods of time planned value gives the cumulative planned value.

Allocating funds to project activities is very important since it forms the foundation for the measurement of the project cost performance. Any error occurring in the amount of funds allocated to a particular activity in the project or the timing of that expenditure will result in over or understating the performance of that activity in the project.

If an activity in the project is allocated a certain cost amount, let say "*cost amount A*," when it should have been allocated a less cost amount, let say "*cost amount B amount A*," the performance on this over allocated cost will be unjustifiably high. Worse, the actual cost of the activity could actually be "*cost amount A*" because work tends to fill the time allowed and spend the amount for which it was allocated. Most project managers will not initiate a corrective action when activities are being done within their predicted funds.

Allocating funds to project activities also has an effect on the enterprise business itself. All enterprise businesses have financial departments who must concern themselves with the timing of the expenditures of the enterprise business and making sure that there are funds available to pay the bills. Allocating too much funds for the project activities means that excessive funds that are not required will be at hand. Not allocating enough funds for the project activities means that funds will have to be found for the project on short notice.

Indeed, as shown in Fig. 16.3, at start of the project, all authorized funds are available and no work has been completed, thus the project planned funds is zero. At the project completion, all authorized funds have been spent and all work expected to be completed. The final cumulative planned value at the end of the project is the total value for all the activities from the start of the project to the end; this is called the project's budget at completion. It might not be the project value; instead, it is the expected cost of all the resources needed to complete the planned work. Depending on the nature of the "process improvement" project, this value might be internal authorized funds for the project or the cost to a customer for a cost-reimbursable contract.

The final cumulative planned value can be used to determine a bid amount in a competitive environment. Profit and contingency might also be added to the bid amount. Once the project is awarded and negotiated, the final cumulative planned value may be less than estimated. This final cumulative planned value is the value and cost that should not be exceeded if profit and cost to the customer are to be met.

When a financial reserve or profit has been set aside for the project, two distinct authorized funds can be distinguished:

1. The contract authorized funds, which is an amount agreed to by buyer and seller or sponsor and project manager.
2. The project manager's plan, which sets aside some funds for contingencies, profits, or other categories.

The project plan expects to consume all the final cumulative planned value, but hopes not to use contingency or profit funds to complete the work.

16.4 Control Spending

A generic form of the "Control Spending" process is shown in Fig. 16.4. This is the project management process for planning a set of systematic observation techniques and activities focused on allocated costs to monitor and record cost allocation in order to:

1. Assess cost performance of the "process improvement" project; and
2. Recommend necessary alterations to the project objectives and/or "process to be improved" goals.

16.4.1 Choose Control Subject

The first step of the "Control Spending Process" is "Choose the Control Subject"—Each aggregated estimated cost that has been allocated to project activities is a control subject; a center around which the control spending process is built.

16.4.2 Establish Standard Performance

The second step of the "Control Spending Process" is "Establish Standard of Performance"—It relates to collecting the standards of cost performance baseline required by the financial function within the enterprise business and documented in the organizational process assets. For each control subject it is necessary to know its standard of cost performance.

16.4.3 Plan and Collect Appropriate Data

The third step of the "Control Spending Process" is "Plan and Collect Appropriate Data" on the chosen "Control subject"—It relates to establishing the means of tracking allocated costs and current spending on the project activities in order to determine the actual cost performance of the project.

Cost allocation and spending tracking begin with the collection of information needed to accomplish the prescribed cost analyses. Data collection can be specified

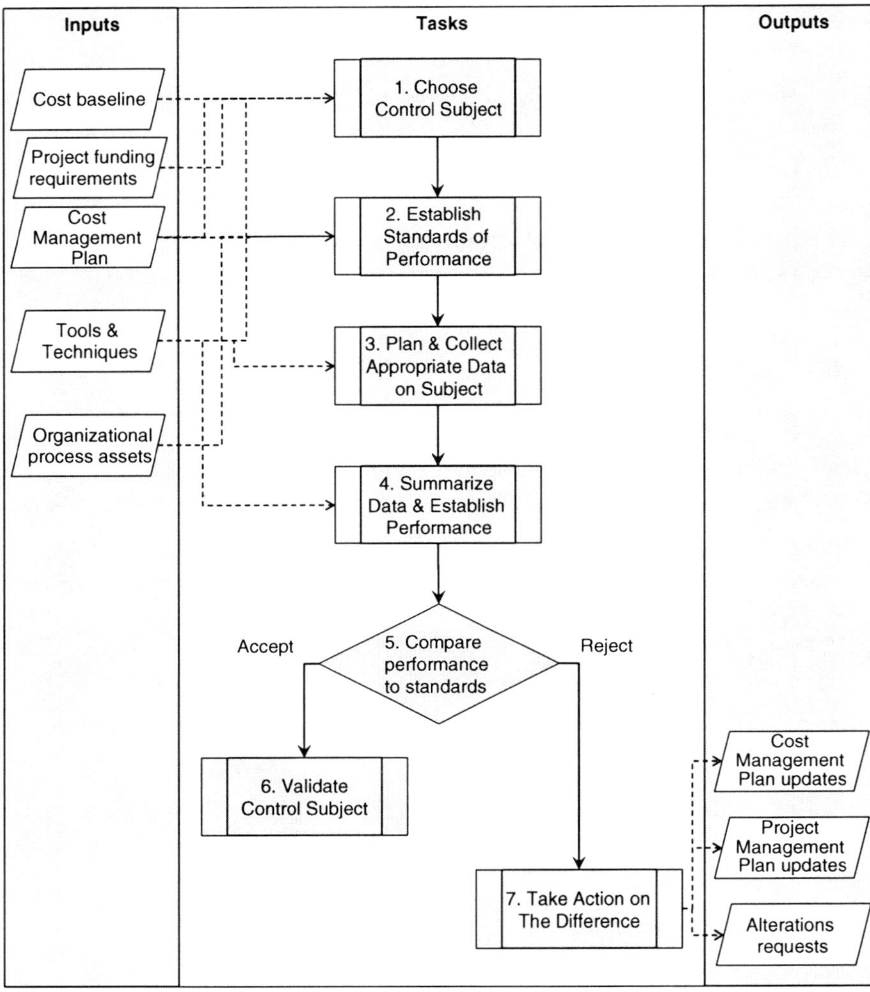

Fig. 16.4 The control spending process

to occur at some recurring point in time when data is needed for cost analysis purposes, or it may be accomplished as an ongoing activity over a period of time where data is collected regardless of when cost analyses are performed.

An ongoing data collection approach is recommended, particularly if cost performance analyses are conducted infrequently, for example, only monthly or quarterly. This removes the burden of trying to capture or recreate past data that may have been replaced by current data. Also, ongoing data collection (even without formal cost analysis) can sometimes provide indicators of potential project cost performance issues or problems that would not otherwise surface in a timely manner.

The cost tracking effort considers the baseline cost estimates that were created during the "Allocate Cost to Activities" process (and that are currently aligned with work elements in the project work plan) and examines them relative to actual costs and expenditures that have been incurred by the project.

16.4.4 Summarize Data and Establish Actual Performance

The fourth step of the "Control Spending Process" is "Summarize Data and Establish Actual Performance" of the chosen "Control subject"—Cost tracking information is typically summarized weekly for shorter projects and at least monthly for larger projects. To ensure proper cost control, the project manager (or a qualified designee) should review and approve all spending incurred by the project. Such approval should not be "rubber stamped." Rather, the cost spending approval process should prompt a detailed examination of planned project expense versus acquired value, in conjunction with verifying authorization for the expenditures.

Several performance indicators are used to summarize cost performance data on projects. These include, but are not limited to:

1. Percent Spent
2. Earned Value
3. Actual Cost
4. Estimate to Complete Cost
5. Estimate at Completion Cost
6. Cost Variance and Schedule Variance
7. Cost Performance Index and Schedule Performance Index

16.4.4.1 Percent Spent

The percent spent represents an estimate, expressed as a percent, of the amount of funds that has been spent on an activity or a work breakdown structure component. It makes a statement about the project's financial expenditure; it is not an indicator of project progress toward completion.

16.4.4.2 Earned Value

Earned value represents the value of work performed expressed in terms of the authorized funds allocated to that work for a schedule activity or work breakdown structure component. Earned value accrues as a result of completing project activities. Completion of each activity or work breakdown structure component contributes to the project earned value. Here, a project activity falls into one of the following three categories:

1. Not started
2. Started and underway
3. Completed

If a project activity has not yet started, its earned value is zero since nothing has been accomplished. If an activity is completed, then its earned value is equal to its planned value. If an activity is underway, then its earned value is approximated based on the amount of work done. This approximation is done without regard to the funds spent or time elapsed.

Thus, at start of the project, the total earned value is zero. As the project progresses, completion of each individual activity adds to the project's total earned value. When lengthy activities are carried out partial earned value can be determined until the total activity is completed. At completion of the project, the total

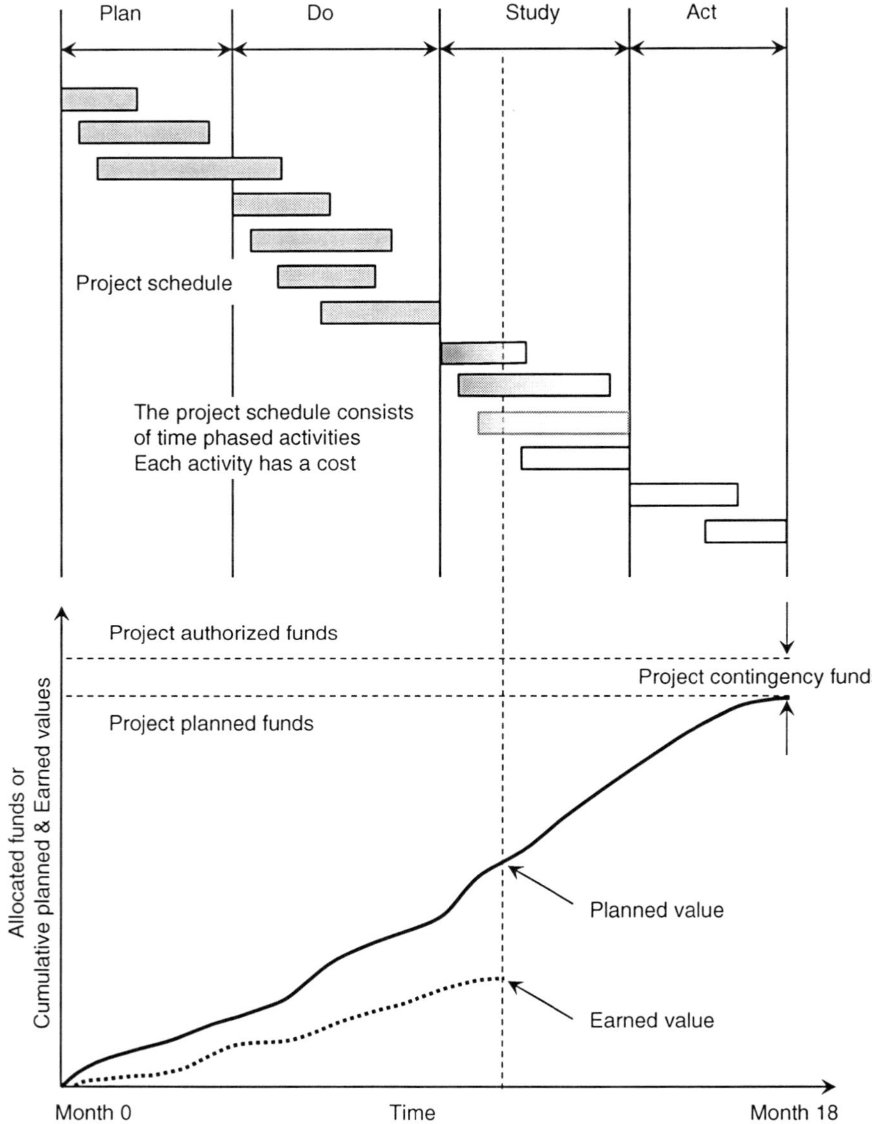

Fig. 16.5 Earned value on the cost performance baseline chart

earned value equals the cumulative planned value introduced in a section above. Thus a key element of earned value is that work has value equal to its authorized funds, not what was spent to complete the work.

Once the value of completed work is known, it can be plotted on the cost performance baseline chart, as indicated in Fig. 16.5. Comparing the planned value to the earned value reveals whether the planned work is being completed at a sufficient rate.

16.4.4.3 Actual Cost

The actual cost is the cost of completing a scheduled activity, a work breakdown structure component, or the project. Comparing the actual cost to the earned value at a given point in time indicates whether the project expenditures were correct for the amount of work completed. The ability to perform this comparison is one of the key benefits of the earned value management approach.

16.4.4.4 Estimate to Complete Cost

Estimate to complete cost represents the expected cost needed to complete all the remaining work for a schedule activity, work breakdown structure component, or the project.

16.4.4.5 Estimate at Completion

Estimate at completion cost represents the expected total cost of a schedule activity, a work breakdown structure component, or the project when the defined scope of work will be completed. It is equal to the actual cost plus the estimate to complete cost for all of the remaining work. The estimate at completion cost may be calculated based on performance to date or estimated by the project team based on other factors, in which case it is often referred to as the latest revised estimate.

16.4.4.6 Cost Variance and Schedule Variance

A completed work has an economic value, often expressed in monetary unit or in staff-hours of earned value, and that value can be compared with the value of actual cost and planned value. Cost Variance (respectively Schedule Variance), illustrated graphically in Fig. 16.6, is the algebraic difference between earned value and actual cost (respectively planned value).

The variances reflect deviation from the approved cost baseline plan. A positive value indicates an under-run condition and a negative value indicates an over-run condition. A favorable (positive value) variance tells the project manager that if everything else stays constant the project's actual profit will likely exceed the planned profit. An unfavorable (negative value) variance tells the project manager that if everything else stays constant the project's actual profit will be less than planned. The sooner the project manager detects a cost variance, the sooner the project manager can direct attention to the difference from the planned amounts.

To understand the concept of schedule variance in unit of time, the point on the cost performance baseline whose planned value is equal to the current earned value should be determined and projected on the time line, as illustrated in Fig. 16.7. The projected point on the time line is called the "Earned schedule." It represents the amount of "schedule time" that the project has earned by completing work. From this projected point on the time line, a statement can be made about being ahead or behind schedule of the planned work needed to complete the project. On this figure, the actual time represents the time that has expired since project initiation. The Schedule Variance on the time line is the algebraic difference between earned schedule and actual time.

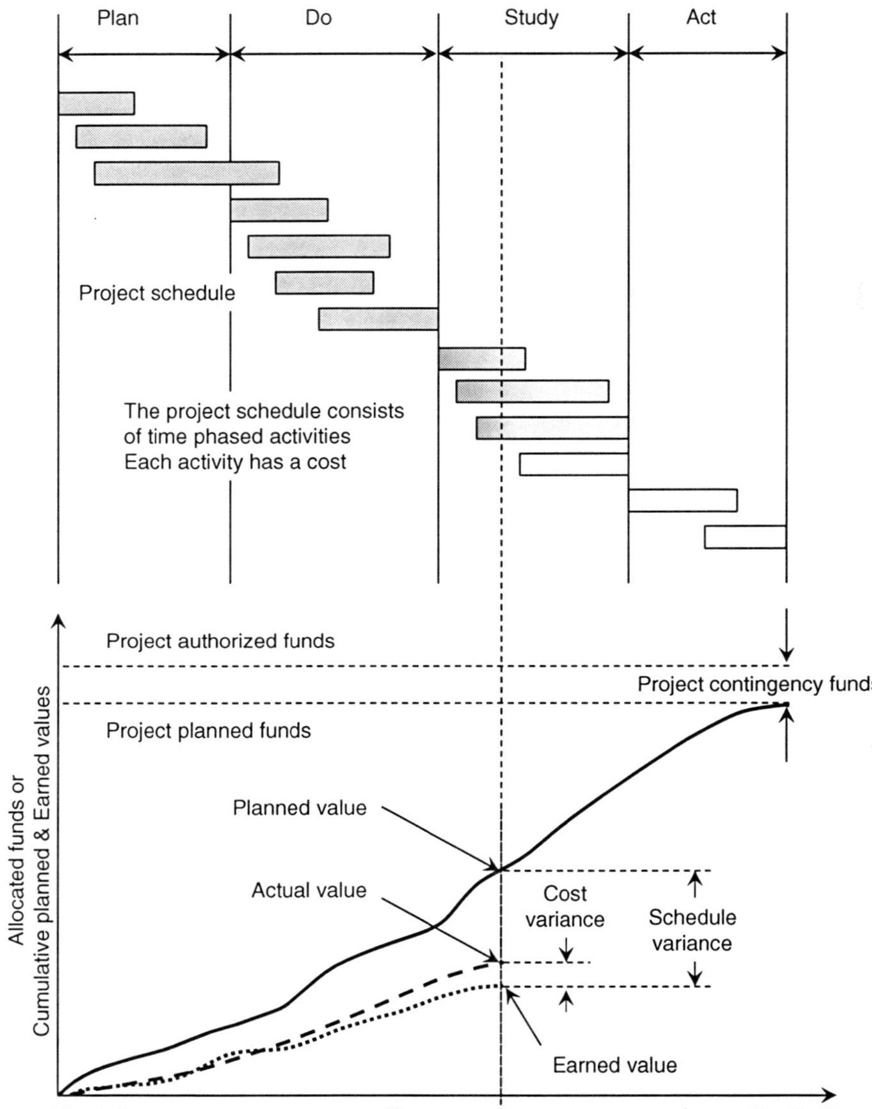

Fig. 16.6 Illustration of cost variance and schedule variance

These variances will give an "early warning" signal of impending problems and are used to determine whether or not corrective action needs be taken in order to stay within the commitments to management.

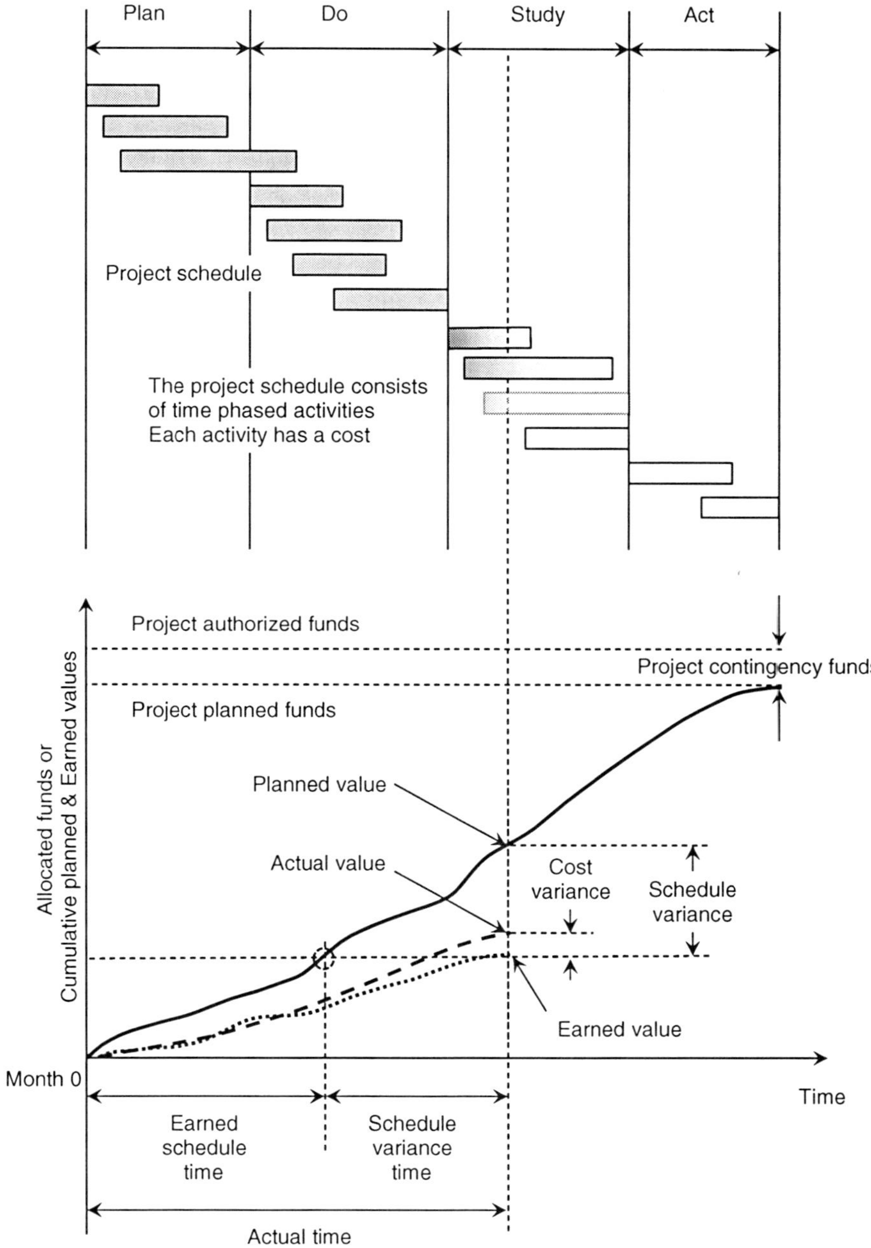

Fig. 16.7 Cost performance baseline and earned schedule concept

There are five reasons why the project manager must measure schedule and cost variances:

1. *Catch deviations from the curve early.* As shown in Fig. 16.7, the cumulative actual cost or actual duration can be plotted against the planned cumulative cost or cumulative duration. As these two curves begin to display a variance from one another, the project manager will want to put corrective measures in place to bring the two curves together. This reestablishes the agreement between the planned and actual performance.
2. *Dampen oscillation.* Planned versus actual cost and schedule performances should display a similar pattern over time. Wild fluctuations between the planned and actual performances are symptomatic of a project that is not under control. Such a project will get behind schedule or overspent in one period, be corrected in the next, and go out of control in the next report period. Variance measures can give an early warning that such conditions are likely and give the project manager an opportunity to correct the anomaly before it gets serious. Smaller oscillations are easier to correct than larger oscillations.
3. *Allow early corrective action.* As just suggested, the project manager would prefer to be alerted to a schedule or cost problem early in the development of the problem rather than later. Early problem detection may offer more opportunities for corrective action than later detection.
4. *Determine weekly schedule variance.* In our experience, we found that progress on activities open for work should be reported on a weekly basis. This is a good compromise on report frequency and gives the project manager the best opportunity for corrective action plans before the situation escalates to a point where it will be difficult to recover any schedule slippages.
5. *Determine weekly effort (person hours/day) variance.* The difference between the planned effort and actual effort has a direct impact on both planned cumulative cost and schedule. If the effort is less than planned, it may suggest a potential schedule slippage if the person is not able to increase his or her effort on the activity in the following week. Alternatively, if the weekly effort exceeded the plan and the progress was not proportionately the same, a cost overrun situation may be developing.

Early detection of out-of-control situations is important. The longer we have to wait to discover a problem, the longer it will take for our solution to bring the project back to a stable condition.

16.4.4.7 Cost Performance Index and Schedule Performance Index

While any reasonable means of tracking project cost performance will produce some form of cost and schedule variance, the use of schedule performance index and cost performance index as project cost performance indicators is unique to the earned value management approach.

The cost performance index is the ratio of earned value to actual costs of a schedule activity, a work breakdown structure component, or the project. It reveals the efficiency with which the project is using funds or staff-hours.

Thus, allowing comparison of cost performance of different projects. A value equal to or greater than one indicates a favorable condition; a value less than one indicates an unfavorable condition.

Similarly, the schedule performance index is the ratio of earned value to planned value of a schedule activity, a work breakdown structure component, or the project. It is a representation of how much of the original scheduled work has been accomplished at a particular point in time. The cumulative schedule performance index is a reflection of the work being done according to plan, or falling behind the plan. A value equal to or greater than one indicates that the project is ahead of schedule and a value less than one indicates that the project is behind. This ratio, however does not allow a proper comparison of different projects since it takes the value one toward completion of the project, regardless of how early or late the project completes. While the schedule performance index helps convey how well the time given to complete the project is being used, it fails to communicate how far ahead or behind schedule the project is in terms of time.

The schedule performance index on the time line is the ratio of earned schedule to actual time of a schedule activity, a work breakdown structure component, or the project. It reveals the efficiency with which the project is using time. Thus, allowing comparison of time performance of different projects. A value equal to or greater than one indicates that the project is ahead of schedule and a value less than one indicates that the project is behind.

Using cost performance index and schedule performance index on the time line, and by using extrapolation on historical and similar projects data, it is possible to predict when the project will be completed and how much will be spent getting there.

The cost performance index, the schedule performance index, and the schedule performance index on the time line are important indicators to watch, but the cost performance index is clearly the more sensitive one. Indeed, a cost performance index less than one is likely to be non-recoverable by the project. Whereas the schedule performance index will eventually drift back up to a full value 1.0 when all of the project tasks have been completed, any cost performance index of less than one will rarely be improved by the project. Thus, it is imperative that the project manager monitor closely the trend and rate of the cost performance index, as well as the completion of all those tasks that are on the critical path. The schedule performance index is important to monitor during the planned phases of a project, but becomes less significant as the project nears completion.

By contrast, over-runs meaning performance at less than a cost performance index of one will typically constitute a permanent loss of funds to the project. The only issue is at what rate the project will perform the remaining work: at the full budgeted value of a cost performance index equal to one or at a lesser rate. There is strong empirical evidence suggesting that projects will most likely continue to perform at their cost performance index rate for all remaining tasks, and sometimes they will even deteriorate further. Only with the recognition that there is a cost problem, and through the aggressive management of all remaining tasks, can there ever be an improvement in the cost performance index.

Empirical evidence also suggests that while over-runs can be improved by the project, they cannot be recovered in total. While the project team may dramatically improve the performance of its schedule performance indices, it will be doing non-recoverable damage to its cost performance index. The schedule performance index will eventually correct itself back to a full value of one when all the tasks have been completed. But any funds spent that cause an over-run inflict permanent damage to the cost performance index and cannot be subsequently recovered. Rather than the indiscriminate use of overtime, the project might better focus on the aggressive management of all late tasks along its critical path, and let the schedule performance index eventually recover over time. Any project that is operating with limited funds must maintain a careful balance between its schedule performance indices and critical path schedule, along with achieving its cost objectives.

The additional benefit monitoring the cost performance index, the schedule performance index, and the schedule performance index on the time line, is that these indices can be used to statistically forecast the estimated final costs for the project.

16.4.5 Compare Actual Performance to Standard

The fifth step of the "Control Spending Process" is "Compare Actual Performance to Standards"—The act of comparing the actual cost performance of the chosen "Control subject" to standards is often seen as the role of the financial control function with the enterprise business called on to carry out any or all of the following activities:

1. Compare the actual cost performance to the financial goal.
2. Interpret the observed difference; determine if there is conformance to the goal.
3. Decide on the action to be taken.
4. Stimulate corrective action.

During project implementation, one of the key responsibilities of project manager is to measure cost performance. This responsibility entails monitoring cost performance to detect and analyze variance from the allocated funds using the tools described in the previous section.

16.4.6 Validate Control Subject

The sixth step of the "Control Spending Process" is "Validate Control Subject"—It relates to acceptance decisions from the financial control results, which will indicate how well the chosen "Control subject" has been used by the project and how much work has been completed.

16.4.7 Take Action on Difference

The last step of the "Control Spending Process" is "Take Action on the Difference." It relates to actuate alterations which restores conformance with financial goals. This step is also known as "troubleshooting" or "firefighting."

The decision to issue corrective or preventive actions is to ensure that the observed non-conformance to financial requirements are repaired and brought into compliance with financial requirements or specifications. Corrective action here often involves adjusting schedule activity authorized funds and/or planned funds to balance cost variances.

The following actions can be considered for application to reduce project cost variation or otherwise to correct project cost performance:

1. High Labor Costs
   - Ensure accuracy on staff time sheets, review earlier staff time sheets or reports, and correct them as necessary.
   - Review resource utilization levels for underutilized resources that can either leave the project or be applied to broader work efforts.
   - Examine the schedule and use resource leveling to optimize resource utilization.
   - Examine labor cost estimates for accuracy; re-estimate work elements as needed.
   - Review the schedule for too much or too little level of work effort; examine use of overtime or multiple work shifts; ensure that only authorized work is being performed.
   - Examine time sheets and reports for indications of "too much, too frequent" overtime.
   - Conduct routine meetings with staff leaders and individual project team members to review their perspectives on labor costs, and examine ways to reduce such costs.
2. High Material and Supply Costs
   - Examine material and supply quantity use for their being more or less than originally estimated; either condition could reflect a higher-than-expected cost, so re-examine quantity estimates.
   - Review the unit cost for materials and supplies; if higher than expected, consider alternate sources and suppliers, and, if as expected, find ways to reduce cost.
   - Examine material consumption to determine if any materials are being used prematurely in the schedule; place controls on materials use and make schedule adjustments as needed.
   - Examine material and supply cost estimates for accuracy; re-estimate materials and supplies as needed.
3. High Supplier Quality Costs
   - Examine supplier invoices for comparison to the established supplier milestones and deliverables schedule and confirm that payments are due; if not, return or hold vendor invoices until payment is due.

   - Review the invoice schedule to determine whether the subcontractor is requesting payments prematurely; adjust the supplier's invoice and payment schedule as needed.
   - Review the supplier's scope of work in contrast to work performed to see if there are any indications of "scope creep"; if so, use project change control procedures to correct the incident-change the supplier's scope, or disallow the added and unauthorized work.
   - Identify any supplier billing patterns (e.g., frequency, timing, amounts, etc.) that might affect the perception or confirm the reality of higher supplier costs.
   - Conduct routine meetings with supplier to review their fees and expenses, and examine ways to reduce supplier costs.
4. High General Costs
   - Examine individual and group expense reports to identify and rectify any unauthorized expenditures.
   - Review the scope of work in contrast to work performed to see if there are any indications of "scope creep"; if so, use project change control procedures to rectify the cause of higher costs.
   - Review travel expense reports in an active and timely manner to approve all travel-related expenses.
   - Apply rigorous examination and approval processes for project-related requisitions.
   - Identify any risk events that have occurred to influence higher-than-expected project expenditures; revisit the project risk management plan to determine if there are any additional, lingering threats to the project cost.
5. Project Out-of-Control Costs
   - Revise the project scope—review cost, schedule, and resource utilization estimates, and revise the project authorized funds as necessary in collaboration with the customer.
   - For cost over-runs, identify any costs that can be passed to or otherwise shared with the customer.
   - Consider performing "project recovery" actions, including an examination of current project conditions, identification of causes for those conditions, and review of potential reassignment of the project manager and project team members, and conduct a major project re-planning effort.

Document any cost control actions taken. Examine areas where proactive controls can be used to prevent cost over-runs. Revisit these control actions and their results when project lessons learned are being identified and update the organizational process assets accordingly.

Develop Procurement Management Plan 17

Project procurement is the process of obtaining or procuring materials, products, services, or results needed from outside the project boundaries to perform the project work. It commonly involves purchase planning, standards determination, specifications development, supplier research and selection, value analysis, financing, price negotiation, making the purchase, supply contract administration, inventory control and stores, and disposals and other related functions.

An important distinction between the work of a project which is assigned inside of one's own enterprise business for performance (the in-house or make work of a "Make or Buy" decision) and sending work outside of one's own enterprise business (the buy work of a "Make or Buy" decision) is the "legal" relationship that the purchased work creates. With in-house work one can describe what is required in broad general terms. The tolerance for error is quite generous and self adjusting with internal funds allocations. But with the purchased effort one must better know what is required, and be able to describe those requirements precisely to the supplier, possibly to the supplier's attorneys, and ultimately to the courts so that they can understand. In procurement relationships it sometimes turns out to be the enterprise business attorneys versus the supplier's attorneys. Project managers need not become attorneys, or even trained in the law. But they should have a broad general understanding of certain fundamental legal concepts, if for no other reason than to be able to discuss their procurement requirements intelligently with legal counsel.

The term procurement is generally used by governments' agencies; many private companies use the terms purchasing and outsourcing. Organizations or individuals who provide procurement services are referred to as suppliers, vendors, contractors, subcontractors, or sellers. To keep consistency throughout this handbook, we will use the term "suppliers."

Procurement is a vital function of most "process improvement" projects. The decision to obtain or procure materials, products, services, or results needed from outside the project boundaries to perform the project work depends on a mix of criteria including price, quality, delivery performance and reliability in all these categories. Moreover, this decision is not simply a one-off, but part of a continuing

A. van Aartsengel and S. Kurtoglu, *Handbook on Continuous Improvement Transformation*, DOI 10.1007/978-3-642-35901-9_17, © Springer-Verlag Berlin Heidelberg 2013

series of such decisions which may be taken on a purely open market basis, or a longer-term agreement may be reached whereby both supplier and buyer accept a degree of interdependence in which the supplier is given a contract which excludes other suppliers in return for reliable deliveries which enables the purchaser to minimize their holdings of stocks of , products, services, or results needed from outside the project boundaries to perform the project work.

Procurement is a greedy function, because it consumes time and money in prodigious amounts. Procured materials, products, services, or results needed from outside the project boundaries account for over half the total costs of most "process improvement" projects. Therefore, the aim of procurement management is to minimize the long-term costs which arise not only from the prices paid and stock levels needing to be carried but also the reliability of continuous supply and the quality of procured items. Efficient Procurement management is essential to avoid serious over-expenditure or delays through shortages and the acquisition of the materials, products, services, or results needed from outside the project boundaries that are unfit for their intended purpose.

17.1 When to Develop a Procurement Plan?

If the "process improvement" project requires any materials, products, services, or results needed from outside the project boundaries to perform the work, then the project manager should develop a project procurement plan.

As a project moves from the initial phase through subsequent planning phases, knowledge of the project increases. Decisions are made about the project outcomes requirements, when they are needed, and what funds might be available to produce them. This is the ideal time to consider the procurement strategy or strategies that might be best suited to deliver the required project outcomes. By the time the project reaches the project risk management planning phase, project knowledge will be at a level where a preferred procurement strategy can be readily identified.

17.2 Developing the Procurement Management Plan

"Develop Procurement Management Plan" is the project management process required to ensure that the materials, products, services, or results needed from outside the project boundaries to perform the work are efficiently procured. Activities for all significant project procurements start other project management planning, well before a procurement order is placed and do not end until the materials, products, services, or results needed from outside the project have been delivered and put to use. Where international freight movements are needed, the procurement department typically makes the arrangements, either directly or (more usually) by engaging a shipping or freight forwarding agent. Routine procurement functions often include the establishment of preferred suppliers and vendors lists, and the rating of suppliers and vendors performance. In the remaining of this handbook, we will refer to the materials, products, services, or results needed from outside the project boundaries as the "procurement items."

The procedures for any procurement event will depend to a very large extent on the value and importance of the procurement items. If the project desperately needs a steel sheet material for stamping, the procedure can be as simple as sending someone to the nearest storage facility to acquire it, without even involving the procurement department. But most projects procurements need far more care and attention, to ensure that procured items are obtained at the right price, of the right quality, to be available at the right place and the right time.

In most enterprise businesses, the procurement of special or expensive items can be regarded as a mini-project in itself. Project managers rarely have delegated procurement authority, and this is deliberate. Rather projects are assigned someone who has such authority typically a buyer, a procurement agent, or a procurement manager by analogy to the project manager. Therefore, the decision making structure in the procurement process involves three main participants:

1. *The project manager*—He/she is responsible for the bidding process, for commissioning the work, running the client's side of the contract and maintaining the day-today relationship with the contractor. The project manager specific tasks are to:
   - Develop business case for use of suppliers
   - Obtain authority to procure supplier services
   - Prepare a detailed Request for Proposal specification
   - Agree selection and evaluation criteria
   - Evaluate tenders and recommends contract award
   - Negotiate contract
   - Administer contract, variation control and fee accounts
   - Prepare periodic cost reports and cash flow forecasts
   - Process invoices for payment
   - Coordinate, monitor and assess work of suppliers
   - Administer close-out of the contract(s)
2. *A procurement manager*—He/she assists the project manager in dealing with contractual and commercial issues. The procurement manager specific tasks are to:
   - Develop supplier identification and market knowledge
   - Forecast and plan contracting requirements
   - Develop optimum sourcing and contracting strategies
   - Assess abilities and resources of suppliers
   - Maintain lists of approved/preferred suppliers
   - Process requests for supplier services
   - Undertake appointment process for supplier services
   - Monitor liabilities and processing of fee accounts
   - Collate performance data, analyses and records results
   - Provide feedback to suppliers and supply chain partners

3. *Technical specialists*—They provide advice on detailed technical aspects of the work, including health and safety and quality assurance. The specific tasks of technical specialists are to:
   - Advise on bid specification
   - Advise on technical aspect of the project
   - Advise on selection and evaluation criteria
   - Participate in evaluation of bids

Activities in the procurement cycle vary somewhat according to the type of procured items involved (especially their cost or uniqueness) and the industry. Figure 17.1, which reflects a structure that mirrors the perspective of the Project Management Institute's PMBOK Guide, illustrates the principal steps in a typical procurement management process.

The constituent project management processes, used during the development of the project procurement management plan, include the following:

1. Plan Procurement
2. Plan Contracting
3. Call for Suppliers Bids
4. Select Optimal Suppliers
5. Administrate Contracts
6. Close Contracts

These six constituent processes interact with each other and with the project management processes in the PDSA "Process Groups." Each aspect of executing any of these can involve effort from one or more persons, based on the needs of the project. Each aspect occurs at least once in every "process improvement" project which involves procurement of some of its activities and occurs in one or more project phases.

17.2.1 Plan Procurement

This is the project management process for identifying and specifying the materials, products, services, or results needed from outside the project boundaries to perform the work.

The "Plan Procurement" process is perhaps the most critical of all the work done in procurement management. If not performed properly the project will likely suffer the consequences for the duration of the project. The procurement manager (or the project manager) must specify procurement items, plan, organize, and administer the procurement activities. This principle is crucial in large operations, especially chain enterprise businesses, in which more than one person may be involved in procurement. Small operations cannot afford to be casual about procurement, however; even a procurement manager (or project manager) needs a definite plan of action.

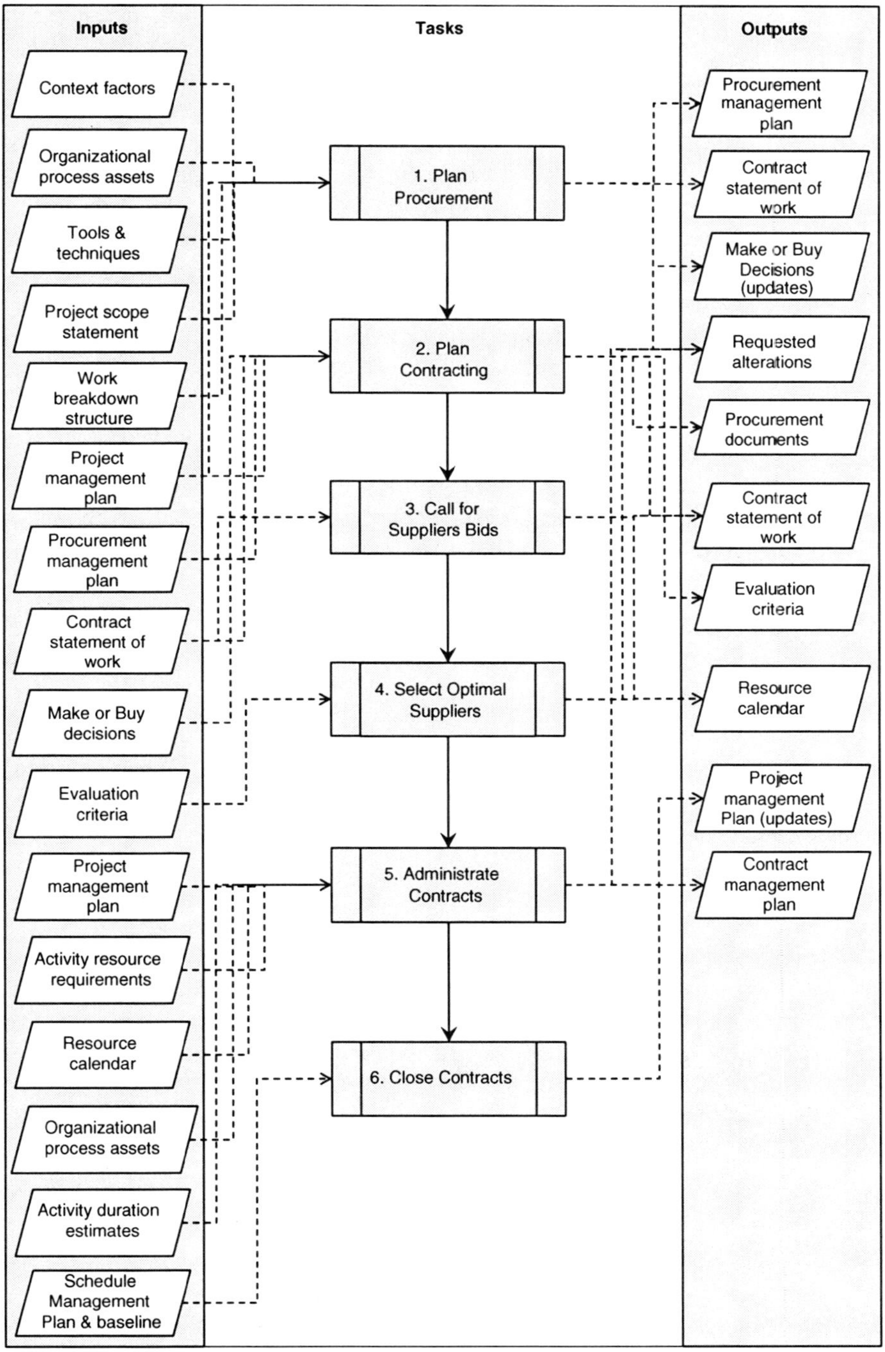

Fig. 17.1 The project procurement management process

The "Plan Procurement" builds on the project scope baseline, requirements documentation, teaming agreement, the risk register, risk-related contract decisions, activity resource requirements, the project schedule, activity cost estimates, the cost performance baseline, the "make or buy" decisions obtained from the "Cost Management Process" as to who will perform the project activities work, the enterprise business environment factors, and the organizational process assets.

The procurement planning process should culminate with the release of a formal document called a Procurement Management Plan. This plan should be coordinated and endorsed by all key functions supporting the project. Ideally, each major function within the enterprise business impacted by the procurement must contribute to the creation of this document.

17.2.1.1 Factors That Influence Procurement Decisions

Before selecting the procurement strategy for a "process improvement" project, whether at a strategic or detailed level, it is necessary to first identify the factors which will determine the most suitable procurement strategy for the project. These factors are:

1. The key objectives and constraints of the project;
2. The risks that may arise during the delivery of the project and how those risks might best be dealt with; and
3. The level of complexity of the required process improvement.

In order to meet the "process improvement" project objective of achieving value for money, the project manager must consider the above factors that influence procurement strategy selection together with the factors that drive value for money.

Achieving Value for Money

Achieving value for money invested in the project typically involves comparing alternatives for the procurement of any materials, products, services, or results needed from outside the project boundaries to get the best mix of quality and effectiveness for the lowest cost over the required term. Importantly, it involves an appropriate allocation of risk, making the selection of a suitable procurement strategy and contract a critical factor in determining whether value for money is achieved.

Assessing value for money involves more than a consideration of price alone, as done during the planning of the cost management process. Other factors to be considered include:

1. Compliance with relevant policy requirements;
2. Contribution to the advancement of the enterprise business intended strategy;
3. Cost-related factors such as transaction costs;
4. Non-cost related factors such as fitness for purpose, and the quality, service and support offered by the suppliers;

In terms of project procurement, there are a number of strategies that typically contribute to value-for-money outcomes, including:

1. Optimizing risk allocation between the project and the suppliers;
2. Using performance specifications, where appropriate, to encourage maximum innovation;
3. Ensuring the flexibility to secure scope changes at a reasonable cost;
4. Using incentives to reward "better than business as usual" outcomes;
5. Setting an appropriate contract period;
6. Ensuring suppliers have the required skills and capabilities to deliver the planned project procurement outcomes;
7. Adopting a procurement strategy appropriate to the complexity of the project.

The impact of these strategies on the achievement of value for money will depend upon the nature and specific circumstances of each "process improvement" project. The project manager should adopt the strategy/strategies that can best achieve value for money and ensure probity and accountability.

Other Factors Influencing Procurement Decision

Several factors that further influence procurement strategy selection include, but are not limited to:

1. Key objectives and constraints of the project;
2. The project risks;
3. The project level of complexity.

Key Objectives and Constraints

The key objectives of the "process improvement" project identified during the development of the project scope management plan are precursor to procurement strategy selection. The objectives are related to:

1. Scope (i.e. what is to be delivered) together with any required provision for flexibility in this regard;
2. Cost, including transaction costs;
3. Time, including an appropriate allowance for the contract period;
4. Quality, including fitness for purpose considerations of the project outcomes;
5. Innovation, encouraged through the use of performance, rather than prescriptive, specifications;
6. Customers and stakeholder needs and expectations;
7. Contribution to the achievement of the enterprise business intended strategy;
8. "Better than business as usual" outcomes, encouraged through performance incentives.

Constraints are aspects of the project that limit, restrict or otherwise impact upon the project objectives in some manner. Constraints are typically unique to each project and may include:

1. Time constraints;
2. Budget constraints;
3. Physical constraints;
4. Availability of resources, including labor resources;
5. Skills, capability and capacity of the project participants to deliver the planned project outcomes;
6. Market or industrial conditions;
7. Policy requirements.

The objectives and constraints of each "process improvement" project are frequently interdependent, and will therefore need to be considered concurrently. This approach will highlight the objectives and constraints that critically impact upon the planned delivery of the project and facilitate the selection of the most suitable procurement decision. In some cases, however, it will be clear that one objective or constraint takes precedence over all others due to its critical impact upon project. This critical objective or constraint should then be used to determine the most suitable procurement decision for the project.

Project Risks

The second factor that may influence selection of a procurement strategy is the risk associated with the "process improvement" project. The development of a project risk management plan is outlined in a later section. The nature of the risks to the project, and their impact on project outcomes if they occur, are often determined by the key objectives and constraints of the project. For example, if a project has a particularly tight timeframe for completion, delays to the supply of materials, products, services, or results needed from outside the project boundaries to perform the work will be a risk to securing the timely completion of the project. Once the key objectives and constraints of the project have been defined, the risks can also be identified.

Responsibility for managing or mitigating particular risks is broadly determined by the procurement strategy adopted for the project. Therefore, the project manager should consider and determine the most suitable method to deal with the identified risks prior to selecting a procurement strategy for the particular "process improvement" project.

As a guiding principle, responsibility for managing a particular risk in the context of procurement should be allocated to the party best able to deal with that risk. Inappropriate risk allocation is likely to result in project budget overruns (as suppliers can reasonably be expected to make allowances in their tenders for the risks for which they are responsible) and increase the likelihood of contractual disputes and litigation.

Project Level of Complexity

The level of complexity of a project must be considered when selecting an appropriate procurement strategy. The complexity of a project is determined by a combination of factors, including:

1. The size of the project;
2. The duration of the project;
3. The scope of the project;
4. The number of stakeholders involved;
5. The level of technology to be incorporated in the project;
6. The degree of innovation required by the customer;
7. The market conditions.

While contractually complex procurement strategies may sometimes be required for complex projects, the additional resources needed to administer a complex procurement strategy are likely to be wasted if a simple strategy can achieve the same outcomes. The inappropriate selection of a complex procurement strategy can also lead to unsatisfactory project outcomes in terms of cost, as suppliers may make allowances in their tenders for additional administration costs and the possibility of contractual disputes which might otherwise not have arisen.

17.2.1.2 The Procurement Decision Has Been Made. What Next?

Once the procurement decision has been made, the project manager (or the procurement manager) must develop the procurement model document and determine what type of procurement vehicle is best for the procurement that needs to be made. A simple procurement order may suffice, or a contract may be required. For contract work, a contract statement of work (CSOW) must be developed to define exactly what work the supplier is being asked to deliver. The contract statement of work is incorporated in a solicitation document that is distributed to the suppliers who will be bidding on the work.

Developing the Procurement Model Document

The procurement model document defines, in legally enforceable terms, what procurement items are expected from suppliers. For a given procurement item, the procurement model document has two components: the business definition, which builds on the organizational process assets, and the specification, which present the greatest risk to the project.

The Business Definition of Procurement Items

The definition of the business requirements is typically a routine effort that captures the requirements of the project. It often includes the following elements:

1. A supplier statement of work;
2. The Terms & Conditions;
3. The Requirements for Management Oversight.

Supplier Statement of Work—A supplier statement of work is a description of procurement items under a contract; a statement of requirements. It is essentially an interface (i.e. input/output) document that refers to a narrative description of the products, results, and services that are expected to be delivered by the prospective suppliers at the conclusion of a contract. It defines, for the selected procurement items, just the portion of the project scope that is included within the related contract.

The supplier statement of work must be written to describe in clear, concise, and as complete as possible terms what procurement items are to be delivered by the prospective supplier. Preparation of an effective supplier statement of work requires an understanding of the procurement items that are needed to satisfy the project requirements. A supplier statement of work prepared in explicit terms will facilitate effective supplier evaluation after contract award. As such, the supplier statement of work becomes the standard for measuring the supplier performance.

Each individual procurement item requires a supplier statement of work. However, multiple products or services can be grouped as one procurement item within a single supplier statement of work.The supplier statement of work will typically be developed at the start of the project and serves to encompass the entirety of the deliverables. It can be revised and refined as required as it moves through the procurement process until incorporated into a signed contract. For example, a prospective supplier can suggest a more efficient approach or a less costly product than that originally specified.

Most enterprise businesses have templates in their organizational process assets for creating a supplier statement of work. These templates ensure that all required procurement items are covered, and provide consistent information to suppliers.

A typical supplier statement of work might thus include the following items:

1. The objectives of the procurement
2. A listing of the procurement deliverables, hardware, software, and reporting
3. Performance standards
4. Commitment of specific personnel
5. A schedule or period of performance
6. Location(s) description of where work will be performed
7. Documents incorporated into the supplier statement of work by reference, including terms and conditions
8. The order of precedence of all specified documents
9. Other items of importance to the project

In the request for proposals (RFP), the supplier statement of work is the only official description of the work requirement. Accordingly, it must provide the supplier with enough information to develop and price the proposal without the need for further explanation.

The Terms & Conditions—Terms and conditions can cover a wide range of issues that include: supplier statement of work and changes to; specifications; order of precedence; testing and inspection; delivery, warranty; governing laws; terminations; force majeure; alternate dispute resolution; back charges; bonding and Insurance; payments; etc.... Terms and conditions can cover such critical issues as how one makes changes to an existing procurement, if there are conflicts in contractual requirements what is the order of precedence, how does one terminate a relationship, when does title pass from the supplier to the project manager, payments or no payments to the supplier, etc. It is critical that the appropriate terms and conditions be incorporated into the legal relationship.

Requirements for Management Oversight—The project manager and the project team should assess each major procurement document and determine what will be needed to properly manage the selected critical procurement item. The routine procured items will normally be tracked adequately by automated electronic systems (MRP, MRP-II, ERP, etc...) available within the enterprise business. However all complex and major critical procurement items require special management oversight. A number of factors will need to be considered by the project team including: an assessment of the known risks, a determination of how technically challenging the work may be, any past experiences with the proposed supplier, etc.. . .

The Specifications of Procurement Items

The purpose of the specifications is to define the features and functions that the procurement item product will perform, the physical attributes, the limitations, design requirements, constraints, and the environment in which the procurement item will be used. Specifications have several basic purposes and advantages, the primary ones being that:

1. They serve as quality assurance document for quality control standards and as cost control standards;
2. They help to avoid misunderstandings between suppliers, procurement manager, users, and other enterprise business officials;
3. In the absence of the procurement manager, they allow a qualified designee to fill in temporarily;
4. They serve as useful training devices for project manager trainees; and
5. They are essential when an enterprise business wants to set down all relevant aspects of items it wants to procure, to submit a list of these aspects to two or more suppliers, and to ask these suppliers to indicate (bid) the price they will charge for the specific product or service.

Specifications in procurements constitute the greatest risks for cost growth to any project which is procuring a complex new item, something which does not exist. Since procurements create legal relationships between a project manager (representing the project or the enterprise business) and a supplier, it is critical that the project manager defines well what the supplier is expected to do to satisfy the contract, and then not change the requirements once stated. Each change constitutes an opportunity for claims.

Specifications formats vary from enterprise business to enterprise business and can range from a variety of written descriptions to detailed drawings or even actual samples. The key in developing specifications is to convey the procurement items requirements so that they cannot be misunderstood by the supplier. The old carpenter's adage, "Measure twice, cut once," also applies to the value of well-developed specifications. It is far less costly to develop a clear description of the procurement items requirements in the first place than to have to go through the return and repair process because they were not specific enough or presented clearly.

Table 17.1 Example of technical specifications

Article	Specification
Engine location	Front
Engine alignment	Transverse
Drive Wheels	Front wheel drive
Fuel Supply System	SMPFI
Max Power	177HP (131 kW) @ 6000 rpm
Max Torque	233Nm @ 4500 rpm
CO2 Emissions	0.01ppm
Displacement	2488 cc
Bore	89mm
Stroke	100mm
Cam Design	Double Overhead Camshaft (per bank)
Cylinders	4 INLINE
Valves per cylinder	4
Total Valves	16
Compression Ratio	9.7:1

There are two elements to consider when developing a specification. First, there is the actual description of the procurement item in terms of its physical characteristics, what it looks like, or how it functions. Second, there is an element of quantification that evaluates the level of performance. Certain measures of quality, such as the frequency or mean time between failure for equipment and the allowable number of rejected parts per million for procured parts, are typically systematized into an inspection and audit process for determining acceptance at delivery and subsequent payment.

Specifications are typically created using one of three approaches, depending on the "process improvement" project's objectives:

1. *Technical Specifications.* Technical specifications describe the physical characteristics of the material or product being purchased, such as dimensions, grade of materials, physical properties, color, finish, and any other data that defines an acceptable product. In the production industry sector, written technical specifications are often supplemented by drawings or samples. Table 17.1 illustrates an example of a technical specification.
2. *Functional Specifications.* The function of a procured item can be defined in terms of its actual role and what it is intended to do. Functional specifications define the work to be done rather than the method by which it is to be accomplished. Typically, functional specifications do not limit the supplier to providing a specific solution, as in the case of a technical specification, thus

enabling the supplier to create the best possible solution. Functional specifications are typically used to solicit suppliers' proposals for further evaluation when a specific solution is not known. They are often combined with performance specifications to create a more detailed requirement.

3. *Performance Specifications*. While technical specifications define the procurement item's physical characteristics, and functional specifications describe what role the procurement item plays, neither describes just how well the procurement must perform. This is the purpose of a performance specification, which describes the parameters of actual performance the procurement item must meet. With a performance specification, we are primarily interested in results rather than in method. Performance specifications can be described by a virtually unlimited choice of criteria. However, they must be capable of some clearly stated measurement. Some of the more common parameters include:
   - Speed. Product must travel at 20 miles per hour.
   - Output. Product must produce 400 acceptable parts per hour.
   - Quality. Product must be capable of 2,000 operational hours before failure.
   - Efficacy. Product must reduce rejected parts by 20 %.

   In developing procurement items specifications, project manager must address issues that could increase cost unnecessarily. Some of which are:
4. *Customization*—Customization typically adds cost, so the project manager is well advised to investigate if this is truly required. An internal change in process with little or no resulting cost can often eliminate the need for customization. The term standardization also refers to the methods used to reduce or eliminate custom, one-time, and seldom-used components and processes that introduce variability and can potentially create added cost and quality problems.
5. *Disregarding Performance Requirements*—Specifications unnecessarily stricter than actual performance would require simply add cost without adding benefits. They may also eliminate potential suppliers who are unable to perform to the higher requirements and thus eliminate price-reducing competition. Conversely, specifications that are too open or loose, or with important details missing, tend to invite unacceptable quality and can create costly mistakes. The supplier can provide a procurement item that meets specification precisely but will not perform in its intended function.
6. *Brand Name*—Specifying a brand name limits competition and thus increases the likelihood of higher prices. Brand names may or may not improve the chances of receiving better quality; nevertheless, they typically cost more as a result of higher advertising costs to create the brand name in the first place and because of the perception that users have that substitutes will not perform as well. One way to avoid unnecessary cost increase with brand name is to specify the brand name and include the verbiage "or equivalent" to allow for greater competition. This means that any product meeting the same specifications as the brand name will be acceptable to the project manager.

17.2.2 Plan Contracting

This is the project management process for identifying and preparing contract documents needed to support tenders and to select optimal suppliers.

Having defined the key objectives and constraints of the "process improvement" project, identified the risks, broadly determined the preferred risk allocation and identified the level of complexity of the project, developed procurement items specifications and contract statement of work, the project manager (or procurement manager) should start identifying and selecting contract models best suited to the project.

The contract models best suited to the project will be the one that best aligns with the key objectives and constraints of the project, that deals most appropriately with the identified risks, and that suits the level of complexity of the project. There is a wide selection of contract types available to help project managers or procurement managers to make informed choices and better manage their project procurements. Quentin Fleming has made an excellent summary of these contract types, which we condense in the following sub-section (Fleming, 2003). These contract types can be grouped into two broad families, fixed price contracts and cost reimbursable contracts, each having its unique characteristics. It is important to understand the general distinction between these two generic families because a dozen or so unique contract variations have sprung from use of these two major groups.

17.2.2.1 Fixed Price Contracts

Fixed-price contracts involve a firm-fixed price or, in appropriate cases, an adjustable price. Fixed-price contracts involving an adjustable price may include a ceiling price, a target price (including target cost), or both. Unless otherwise specified in the contract, the ceiling price or target price is often subject to adjustment only by operation of contract clauses providing for equitable adjustment or other revision of the contract under stated circumstances.

Typically under fixed-price contracting the price will be set at the outset of the relationship between the procurement manager and the suppliers. Most contracts in this category will be classified as firm-fixed-price, wherein an absolute value is placed within the contract. However, sometimes the fixed price will be adjustable, to provide incentives to the supplier to complete the work by spending less money, portions of which the supplier may get to keep according to a stated formula in the contract. Other fixed price arrangements set a fixed price, but the price may be subject to adjustments caused by changes in economic conditions beyond the control of either the project manager or the supplier. These are typically contracts for services or commodities scheduled for performance over long periods of time.

The key feature of the fixed price contractual arrangement is the obligation it places on the supplier. It is absolute. Under the fixed price contract the supplier "must produce," in other words, is "obligated" to finish the work under contract, regardless of the circumstances that may happen later. If the work as stipulated in the contract involves more costs or risks or effort than was originally envisioned, so be it. No additional costs for the contracted work will be made available to a supplier merely because the work turns out to be more difficult than was originally anticipated, by either party. If the supplier does not finish the fixed price work, walks away from the obligation, the project manager (or procurement manager) can sue the supplier for any damages incurred.

The use of fixed price contracts typically requires less administrative oversight than cost type arrangements. However, if progress payments are included in the fixed-price relationship, typically monthly payments, oversight of supplier performance by the project manager is essential to make sure that progress is in fact being made prior to making such payments.

Because procurements made under fixed price arrangements place a higher risk on the supplier, logic would suggest that they should receive higher profits. However, this is not always the case, particularly in instances where aggressive competitions are held, as with publicly bid construction work.

There are a multitude of specific contractual arrangements which have evolved from the fixed-price generic contract family. These are:

1. Firm-Fixed-Price (FFP) Contracts;
2. Fixed Price Incentive (FPI) Contracts;
3. Fixed-Price Contracts with Award Fees; and
4. Fixed-Price: Indefinite-Delivery, or, Indefinite-Quantity Contracts.

Firm-Fixed-Price (FFP) Contracts

A firm fixed-price contract provides for a price that is not subject to any adjustments on the basis of the contractors cost experience in performing the contract. This contract type places upon the contractor maximum risk and full responsibility for all costs and resulting profit or loss. It provides maximum incentive for the contractor to control costs and perform effectively and imposes a minimum administrative burden upon the contracting parties.

Without question, the Firm Fixed Price (FFP) contract is the most favored contract type by most enterprise businesses in private industry. The FFP is appropriate whenever definitive design and product performance specifications are available. This contract type places absolute cost risks and incentives on the supplier to deliver the procured items in an efficient manner. The FFP contract type is not subject to subsequent price adjustments because of what a supplier may experience during performance, and the supplier is under an absolute obligation to finish the work under contract. Damages can be sought if a supplier fails to perform on a FFP contract. But the FFP is not without certain limitations, and there are times when this contract type may be totally inappropriate. It is important that the project manager and supplier know just when and when not to use the FFP contract.

Quentin Fleming (2003) shows that in order for the FFP contract to be effectively employed, three conditions must exist at the time of the procurement:

1. The project manager must know exactly what the procurement items are;
2. The project manager must be able to specify the desired procurement items in very precise terms, so as to agree on a price with the supplier; and

3. The project manager must have reasonable confidence (probable assurances) that the items being procured will not subsequently change in specifications, or performance requirements, or terms, so as to require a redirection to the supplier.

If these three requirements do not exist, it may well be advisable to consider some other form of contract type. Ambiguities in procurement items specifications, and subsequent changes, can sometimes make the FFP the wrong contract type for procurements, in deference to most enterprise businesses preferred type.

To properly use the FFP contract, the project manager must understand what is to be procured, and be able to define the desired procurement items in clear and legally enforceable terms. A project manager requires a definitive procurement specification from engineering/manufacturing/technical in order to use the FFP contract. Both parties to the contract must have the same understanding as to what is being procured. Conversely, if the desired item to be acquired cannot be specified except in very broad general terms, the use of the FFP contract type may well be unsuitable.

The other limitation with this most favored contract type is that the FFP leaves the project manager with, little (perhaps no) flexibility to later change direction without paying a high cost for each change. Since the items being procured must be specified in very precise terms, and all of the risks of cost growth are placed on the supplier, suppliers cannot and typically will not accept redirection from a project manager without requesting additional costs to accept any changes. Any supplier which has struck lean (initial profits) deals will often try to "get well" through contract changes or redirection. Also, the FFP price values can change (as with other contract types) for defective pricing, for liquidated damages provisions, for defective workmanship or materials, for latent defects, and so forth.

One distinct advantage of the FFP contract is that they typically require the least administration and management involvement from the project team of any of the contractual options available. However, if progress payment provisions are included in the contract, even the FFP contract will require performance oversight from the project manager.

Fixed Price Incentive (FPI) Contracts

A Fixed Price Incentive (FPI) contract is a fixed-price contract that involves adjusting profit and establishing the final contract price by application of a formula based on the relationship of total final negotiated cost to total target cost. The final price is subject to a price ceiling, negotiated at the outset.

This type of contract is typically used when a project's procurement description is available, but there are some open issues to be settled. Often there may not be sufficient procurement items specification data available to go directly with a Firm Fixed Price contract (FFP). The desired procurement items can be defined, but not to the point where responsible supplier would be willing to commit to a FFP obligation. And at the other extreme, there is sufficient procurement items data available so that the use of a cost reimbursement type contract would be inappropriate. The Fixed Price Incentive (FPI) contract gives both the project manager and supplier some flexibility, while providing strong incentives to the supplier to perform.

The Fixed Price Incentive (FPI) contract places positive incentives on the supplier to completely satisfy the procurement, while incurring the lowest possible costs. Under the FPI contract the performing supplier also participates in any cost savings, or even losses, according to a negotiated formula. The FPI contract is established with the understanding that the final contract profit and final price will be determined after performance, according to their agreed to formula. There will be a ceiling price specified in their contract, beyond which all costs are the responsibility of the supplier. The Fixed Price Incentive (FPI) is a fixed-price relationship.

Typically the incentives in a FPI will be costs. But in addition to cost incentives, FPI contracts may also incorporate other performance incentives such as: timely deliveries according to a schedule, or even bettering the specified schedule dates, reliability, warranty, maintenance, weight objectives, etc. Anything that can be quantified and objectively measured can be incorporated into the FPI contract.

At the time of FPI contract award, the project manager and the supplier must agree on certain provisions in their contract:

1. A target cost;
2. A target profit (without specifying either a profit ceiling or a floor);
3. A profit adjustment formula defining the cost sharing provisions; and
4. A price ceiling (which is the maximum amount which can be paid to the supplier, excluding any statement of work changes).

The cost sharing formula under incentive contracts may be set at any rate formula which adds to 100%, but typically they will fall in the ranges of 90/10; 80/20; 70/30; 60/40 and so forth. The higher values on the left side apply against the target costs, and the lower values on the right apply to adjustments in the supplier's target profit.

After the Fixed Price Incentive (FPI) contract has been completed, the project manager and supplier will then negotiate the final agreed to costs, and resulting profits based on the performance adjustment formula. Final costs, plus final profits result in a final established price.

Fixed-Price Contracts with Award Fees

Award-fee provisions may be used in fixed-price contracts when the enterprise business wishes to motivate a supplier and other incentives cannot be used because supplier performance cannot be measured objectively. Such contracts should:

1. Establish a fixed price (including normal profit) for the effort. This price will be paid for satisfactory contract performance. Award fee earned will be paid in addition to that fixed price; and
2. Provide for periodic evaluation of the supplier's performance against an award-fee plan.

Award fees can be used most effectively in any type of contract. An award fee provision in a contract is probably the strongest inducement provision that can be included between a project manager and a supplier. An award fee is given based on the supplier satisfying the project manager's broadly stated needs. They are based solely on the subjective determination by the project manager. All such project manager determinations are final, that is they cannot be appealed because the supplier typically waives that right in their contract.

An example of the use of an award fee with fixed price contracts might be where a supplier would not make a commitment (firm fixed-price) to an early delivery date of a commodity, but would agree to an award fee provision should they be able to deliver early. If they missed the earlier delivery date, they merely loose extra fees. Any broad project manager objectives may be incorporated into an award fee provision.

Fixed-Price: Indefinite-Delivery, or, Indefinite-Quantity Contracts

This is another category of fixed-price arrangements which are used frequently to procure items for projects in which the titles describe nicely their intended purpose. The first is a firm fixed-price contract with an indefinite delivery schedule. The project manager knows what he/she needs, but at the time of contract execution cannot prescribe the precise dates the procurement items will be needed. That definition will come later. The second is also a firm fixed-price contract covering procurement items, of an unspecified quantity, with the precise quantity to be later specified.

The set prices for either arrangement may vary based on the precise delivery dates or quantity later determined. Nevertheless, both contract types are considered firm fixed-price varieties.

17.2.2.2 Cost-Reimbursement Contracts

The second broad family of contract types is the cost reimbursable model. Cost-reimbursement types of contracts involve payment of allowable incurred costs, to the extent prescribed in the contract. These contracts establish an estimate of the total cost for the purpose of obligating funds and establishing a ceiling that the contractor may not exceed (except at its own risk) without the approval of the contracting officer.

By contrast, the key feature of the cost reimbursable type contract is the obligation of the supplier. Under the cost reimbursable type contract the supplier obligation is merely to provide a "best efforts" commitment to complete all of the work as stipulated in the contract. If the supplier incurs all of the costs as authorized in the contract, but does not finish the entire scope of work, the supplier cannot be sued for the difference. Their legal commitment is to provide their best effort only to finish the work as stipulated in the contract.

However, the project manager (or procurement manager) must fund the entire work to its completion, to the limit of their contractual arrangement. Suppliers are normally obligated to notify the project manager in advance if they anticipate exceeding the authorized funding levels, typically set at about the 70% to 80% point of contract value. Once notified, the project manager must decide whether or not to continue to fund the work, or to terminate the effort.

As Quentin Fleming pointed out (Fleming, 2003), the major concern, with the use of the cost reimbursable type contracts, is the "opportunity" they can provide to any unscrupulous supplier.

> *"Any supplier which has an under-utilized work force, or idle plant facilities, or surplus assets, or ambitious plans for growth, could, and sometimes have in the past, abused this type of arrangement by shifting their assets to the cost type contract in an attempt to keep their capital fully employed. Such practices border on the illegal, particularly with Government contracting, but such abuses with cost contracts have been known to happen in the past."*

It is a major concern of many enterprise businesses, and thus the use of cost reimbursable contracting is typically severely restricted, even when a cost type arrangement may be in the best interests of a particular project.

The advantages to a project with the use of cost type contracts are the flexibility they can provide. It is always easier to accommodate changes in the direction of a supplier when operating under a cost type arrangement, than with a fixed-price contract.

Since the risks of supplier performance on cost type contracts are lower, one would expect that suppliers should receive less fees for their work. Such is not always the case. Often, cost type contracts provide substantial profit opportunities to suppliers. One major disadvantage to the use of cost type contracts is the need for continuous oversight of supplier performance by the project team.

There are four popular variations of cost reimbursable type contracts in use. These are:

1. Cost-Plus-Fixed-Fee (CPFF) Contracts;
2. Cost-Plus Incentive Fee (CPIF) Contracts;
3. Cost-Plus Award Fee (CPAF) Contracts.

Cost-Plus-Fixed-Fee (CPFF) Contracts

A Cost-Plus-Fixed-Fee (CPFF) contract is a cost-reimbursement contract that involves payment to the supplier of a negotiated fee that is fixed at inception of the contract. The fixed fee does not vary with actual cost, but may be adjusted as a result of changes in the work to be performed under the contract. This contract type allows contracting for efforts that might otherwise present too great a risk to suppliers, but it provides the supplier only a minimum incentive to control costs.

The Cost-Plus-Fixed-Fee (CPFF) contract is the origin of all cost reimbursement type contracts. It allows for the reimbursement of all reasonable, allowable, and allocable costs incurred by a supplier up to the limits of the contract value. If a supplier needs to incur costs above the contract value in order to finish a work, they must advise the project manager and get the project manager approval to proceed. The supplier is under a "best efforts" only legal obligation to perform on a project, provided that the project manager reimburses all legitimate costs incurred.

The fixed fee is set as a percentage of the agreed to costs in the contract, and does not change based on the supplier's performance. The fee value remains constant and is only subject to change with an increase or decrease in the scope of contracted work. Should the supplier under-run the costs, the supplier's earned fee percentage will increase. Conversely, should their costs exceed the contract value, their fee percentage will decrease. Cost overruns or under-runs do not change the original fixed fee.

Cost-Plus Incentive Fee (CPIF) Contracts

The Cost-Plus Incentive Fee (CPIF) contract is a cost-reimbursement contract that involves the initially negotiated fee to be adjusted later by a formula based on the relationship of total allowable costs to total target costs. This contract type specifies a target cost, a target fee, minimum and maximum fees, and a fee adjustment formula. After contract performance, the fee payable to the supplier is determined in accordance with the formula. The formula provides, within limits, for increases in fee above target fee when total allowable costs are less than target costs, and decreases in fee below target fee when total allowable costs exceed target costs. This increase or decrease is intended to provide an incentive for the supplier to manage the contract effectively. When total allowable cost is greater than or less than the range of costs within which the fee-adjustment formula operates, the supplier is paid total allowable costs, plus the minimum or maximum fee.

The Cost-Plus Incentive Fee (CPIF) contract type is similar in concept to the Fixed Price Incentive (FPI), but with one important difference: CPIF contracts contain no price ceiling beyond which the supplier cannot recover costs. Fixed Price Incentive (FPI) contracts, by comparison, do have a ceiling. CPIF contracts are appropriate whenever there exists considerable technical risks, and the project manager would like to encourage the supplier to minimize costs.

Like the Fixed Price Incentive (FPI) contract, the Cost-Plus Incentive Fee (CPIF) contract type can be used by the project manager to incorporate additional performance incentives, in addition to cost incentives. Some examples of these performance incentives could be: deliveries of procurement items ahead of schedule, product reliability, maintenance costs, weight of hardware, and so forth. Anything that can be measured can be used as performance incentives. Should the supplier not meet the technical performance thresholds, fee would be lost. To be an effective performance incentive, fees must have the potential of being increased or decreased to the supplier.

CPIF contracts contain the following five elements:

1. A target cost;
2. A target fee;
3. A maximum allowable fee;
4. A minimum allowable fee; and
5. A fee adjustment formula based on supplier performance between the maximum or minimum fee ranges.

After the contract performance, the project manager and the supplier will negotiate the final contract fee between the allowable fee ranges, based on the actual performance of the supplier. A supplier who under-runs target costs will receive a greater fee percentage, and conversely, an overrun from target costs will reduce the fees paid to the supplier.

Just as it was with the Fixed Price Incentive (FPI) contract, under the Cost-Plus Incentive Fee (CPIF) contract the cost sharing formula may also be set with any rate formula which adds to 100 %, but typically will fall in the ranges of 90/10; 80/20; 70/30; 60/40. The higher values on the left side apply against the target costs, and the lower values on the right apply to adjustments in the supplier's target profit.

Cost-Plus Award Fee (CPAF) Contracts

A Cost-Plus-Award-Fee (CPAF) contract is a cost-reimbursement contract that involves a fee consisting of:

1. A base amount fixed at inception of the contract; and
2. An award amount that the supplier may earn in whole or in part during performance and that is sufficient to provide motivation for excellence in such areas as quality, timeliness, technical ingenuity, and cost effective management.

The amount of the award fee to be paid is determined by the enterprise business judgmental evaluation of the suppliers performance in terms of the criteria stated in the contract. This determination and the methodology for determining the award fee are unilateral decisions made solely at the discretion of the enterprise business.

The number of evaluation criteria and the requirements they represent will differ widely among contracts. The criteria and rating plan should motivate the supplier to improve performance in the areas rated, but not at the expense of at least minimum acceptable performance in all other areas.

A Cost-Plus Award Fee (CPAF) contract also involves evaluation at stated intervals during performance, so that the supplier will periodically be informed of the quality of its performance and the areas in which improvement is expected. Partial payment of fee generally corresponds to the evaluation periods.

This contract type provides maximum incentives on a supplier to perform to the fullest "satisfaction" of the project manager. The project manager's determination of award fee amounts is unilateral, and the supplier contractually waives their right to appeal such decisions. Thus the CPAF contract has a tremendous impact on the performance of the supplier. Award fee contracts have such an impact on a supplier's profit that some enterprise businesses have actually refused to accept award fee contracts. However, award fee contracts appear to be gaining in popularity, at least with the suppliers.

CPAF contracts typically contain six elements:

1. *The total estimated costs*;
2. *A "base fee," stated as a percentage of total estimated costs*—The "base fee" in a Cost-Plus Award Fee (CPAF) contract is not subject to change based on supplier's performance, and such fees are earned by performing to the statement of work. In concept base fees resemble a CPFF fee. Typically most base fees are limited to about three percentage points of estimated contract costs.

3. *An "award fee," also stated as a percentage of total estimated costs*—The "award fee" is conferred on top of the base fee. It is given out based on the periodic (yearly, semi-annually, quarterly), subjective and unilateral determination of a supplier's performance by the project manager's evaluation board. It is paid to a supplier according to a predetermined award fee schedule as defined in their contract. Because of the very high administrative effort required on both parties, award fee periods are best evaluated on an annual basis, certainly not more frequently than twice a year. Some enterprise businesses have tried award fees on a quarterly or even monthly basis, but typically abandon this practice because of the high administrative effort. Award fee determinations take considerable time from both the project manager and the supplier personnel to determine a fair award fee amount.
4. *The award fee's broadly defined "performance criteria";*
5. *An award fee "evaluation board"; and*
6. *A "fee determining official,"* typically a senior executive, often the program manager or even the next most senior executive.

Award fee contracts in the private sector may contain whatever the parties agree to, but typically follow the same general format.

The intent of the Cost-Plus Award Fee (CPAF) contract is to stimulate top performance from the supplier by exerting maximum influence over a supplier—into future periods. Any fee amounts not earned in a given evaluation period may, or may not, be carried over into a later period, at the sole discretion of the project manager's fee determining official, or as specified in the contract. The project manager's award fee evaluation board is normally comprised of multifunctional management personnel. To work properly, this board must be represented by all functions involved in the project management process.

Because the award fee evaluation board's findings are subjective—but always final—and are not subject to a disputes procedure, persons who serve on an evaluation board must be perceived by the supplier as being reasonable, impartial, and above reproach. Any hint of arbitrary or capricious findings from an evaluation board can destroy any benefits to be gained from the Cost-Plus Award Fee (CPAF) contract. If a supplier actually performs well, but award fee is arbitrarily withheld from them, such actions can have a negative impact on their performance in the later periods.

While the award fee type contract is normally used on cost reimbursable type contracts, it can also work well on any type of contract with the agreement of both parties.

17.2.2.3 Time and Materials (T & M) Contracts

A contractual variation, which is worthy of mention in order to round out the discussion of contract types, is the Time and Materials (T & M) Contract type.

Time and Materials (T&M) contract are hybrid type of contractual arrangement that contains aspects of both cost-reimbursable and fixed-price-type models. These types of contracts are considered appropriate in those circumstances when it is not possible at the time of award to estimate accurately the extent, or duration, or costs of a work with any degree of confidence. Labor costs typically carry all indirect expenses including profit, but material costs may carry material handing costs only, and are sometimes billed without profits. The T&M contracts are used primarily to procure emergency services, repairs, maintenance, overhauls, but are also used extensively to procure engineering and technical services as purchased labor, where the direct supervision of the purchased people is done by project staff.

Since a supplier in effect receives a cost plus percentage of costs fee type arrangement on at least the labor portion, the project manager must monitor closely the performance of the supplier. The T&M type of contract can be easily abused since there is almost a contractual incentive to increase labor costs to the maximum, and thereby increase a supplier's profits. Abuses may also take place whenever a supplier can substitute a lower-caliber of labor than was priced and envisioned in the negotiated hourly rate.

Because of the risks of cost growth, the T&M type contracts are normally discouraged without justification. Good business practices would suggest some form of restriction on their use; that they be used only with approval of senior management, and include a limitation of costs or a ceiling on the amount authorized for the contract.

17.2.2.4 Selecting the Appropriate Contract Type

The choice of contract type is a critical issue for both the project manager and the supplier. It should build on the consideration of many factors. Some of the more important issues to consider would include the life cycle of the project, the known risks facing the project, technology challenges, and of course, the ability of the project to describe what it wants to procure, without later changing these requirements.

Selecting the contract type is generally a matter for negotiation and requires the exercise of sound judgment. Negotiating the contract type and negotiating prices are closely related and should be considered together. The objective is to negotiate a contract type and price (or estimated cost and fee) that will result in reasonable supplier risk and provide the supplier with the greatest incentive for efficient and economical performance.

For some "process improvement" projects, certain objectives and constraints may be difficult to identify, or may be subject to change over time. As a consequence, the identified risk management strategy and hence the contract model, while previously well aligned with the key objectives and constraints, may become unsuitable for the project. In such cases, an alternative contract model should be selected and implemented.

Where there is a degree of uncertainty surrounding the key objectives and constraints of the project, the project team must remain flexible in order to rapidly address any misalignment between these objectives and constraints and the selected contract model. To facilitate such flexibility, it is necessary to monitor the key objectives and constraints as the project progresses and be prepared to adjust accordingly.

The careful selection of contract types can be used to balance project risks. What any project will want to do is to achieve a proper balance or parity between the obligations the project has accepted, and what they subsequently will want to obligate from their supplier. By carefully choosing the right contract for each procurement item, the project manager can mitigate risks with their supplier base.

17.2.2.5 Identify and Survey Potential Suppliers

Once it is determined that an item will be procured externally to the enterprise business, the procurement manager assigned to the project manager should start compiling a list of all of the possible suppliers, or at least a reasonable number of potential suppliers.

Prequalified supplier lists can be developed from the organizational assets if such lists or information are readily available. Most mature enterprise businesses which have had procurement operations for any length of time have found it advisable to create a database of approved suppliers for quick reference and use by their procurement managers. Some of these databases are quite sophisticated and contain a rating system reflecting the past performance of their suppliers. Such criteria as on-time deliveries, quality of products or service, cooperation, flexibility, responsiveness, etc., have been quantified and incorporated into these files.

Pre-qualification helps project managers secure an effective degree of competition by identifying the suppliers whose skills and experience most closely match the requirements of the work. It is a way of narrowing down the field to arrive at a select group of suppliers, chosen on the basis of their ability to satisfy a defined set of criteria. It means that ostensibly every identified supplier starts with the same degree of opportunity and no one reaches the stage of submitting a tender without getting through the preliminary rounds.

It is very important that the procurement manager considers augmenting the prequalified list by identifying new suppliers. Several factors make new suppliers important. First, there may exist new suppliers that are superior in some way to the enterprise business existing suppliers. For example, a new supplier may have developed a novel production technology or streamlined process which allows it to significantly reduce its production costs relative to pre-dominate production technology or processes. Or, a new supplier may have a structural cost advantage over existing suppliers, for example, due to low labor costs or favorable import/export regulations in its home country. Second, existing suppliers may have gone out of business, or their costs may have increased. Third, the procurement manager may need additional suppliers simply to drive competition, reduce supply disruption risks, or meet other business objectives such as supplier diversity. In recognition of these reasons, procurement managers and their internal customers may be obliged by enterprise business policy to locate a minimum number of viable, potential suppliers for every product or service procured.

Compiling an initial list of prequalified potential supplier can present three major problems. First, it may be difficult to determine which suppliers to include on the initial list. Many potential suppliers carry several product lines; consequently, the list can become larger than anticipated. A second problem stems from the first. In the procurement manager's haste to shorten the potential supplier list, they may stop adding suppliers when they reach a certain number. The longer the initial list of suppliers, the more time is required for interviewing, checking references, touring plants, and completing the other analytical work involved in culling the list. Indiscriminate culling can eliminate a good potential supplier, however. Furthermore, it tends to limit the pool of potential suppliers in the future when buyers stick with the original list. It can be costly to an enterprise business to lock out a good supplier in this way. The third problem is less common. It occurs when procurement managers need to procure a unique item. In such situations, the search for a supplier can be extremely time-consuming.

The exact specifications of procurement items may or may not be fixed at this stage, but their general nature and purpose are usually known. What is available on the market? Who makes such a product, or who can make it? Who provide such service? Who can supply the item most satisfactorily and most economically? These are the questions driving identification process.

The initial survey of potential sources to identify new suppliers should not overlook any possibility, provided they are reasonably accessible and there is some assurance that they meet required standards of quality, service, and price. The availability of electronic data sources located on the Internet greatly enhanced the procurement manager's ability to locate sources of supply. Most major suppliers now have homepages on the Internet that allow procurement manager to quickly scan the product and service offerings and list prices of the suppliers' goods. Trade directories also provide comprehensive and well-organized listings of the whole range of products, services and their suppliers on nationwide basis, usually with at least a general indication of size and commercial rating. The sources of information from which the procurement manager can extract potential suppliers and compile an initial list include, but are not limited to:

1. World Wide Web
2. Company intranets
3. Supplier home pages
4. National and regional trade directories
5. Suppliers' salespersons
6. Yellow Pages
7. Professional associations and meetings
8. Supplier catalogues and mailings
9. Trade shows and trade fairs
10. Advertisements in general circulation publications
11. Professional purchasing publications
12. Technical trade journals
13. Chamber of Commerce
14. Internal users
15. Other procurement managers within the enterprise business
16. Historical procurement records

Several procurement managers also maintain procurement items information files in which they have collected suppliers' mailing pieces and data sheets, advertisements, and new-product announcements from business magazines. Some of this information is so new that it has not yet found its way into the standard catalogs, but the alert procurement manager can have it on hand when needed.

Salespersons are an important source of information, both on their companies' products and capabilities and on their application to customers' processes. Experience has shown that the most successful salespersons do not limit their services to procurement managers merely to provide their products and services. They strive to meet procurement managers' needs, not only with their products but with whatever information, services, and technical advice are available from their companies.

The procurement manager can build a workable list of likely suppliers using information from the publications and persons mentioned above. Those that appear to be reliable and stable, have the needed capability and experience, and are conveniently located should be put initially at the head of the list. Conversely, those companies and firms that have low capitalization or credit ratings or whose products are not in the required quality range of the project procurement items should be excluded.

Of course, the extent of the identification efforts and the time expended on these efforts depend on the procurement items price, the criticalness of the procurement, and whether it is a new or routine procurement. If the procurement item required is of a routine nature, the procurement manager may issue the order to a continuing source or send out a request for bids from a list of preferred suppliers. If the procurement item is more important or more complex or one for which there is likely to be a continuing need, there will be a much more extensive inquiry and research into suppliers and their capabilities.

17.2.2.6 Develop a Request for Information

During the identification and survey of potential suppliers, the procurement manager will read white papers, brochures, and datasheets; attend conferences and demonstrations; and invite the prospective suppliers to give presentations or demonstrations. Even with all of this research, he/she may still not fully understand how the supplier solution will fit and work within the project. When more information is needed than is publicly available, the procurement manager may use a Request for Information (RFI) process.

A Request for Information (RFI) process is a way for procurement managers to determine what is available from suppliers who respond to procurement items requirements. It is also a way for procurement managers to determine whether the requested requirements are reasonable and whether appropriate solution is available.

Suppliers are encouraged to respond to the requirements and also to spell out where there may be potential problems, areas in which solution may not exist, or unrealistic goals and schedules of the portion of the project that is included within the related contract. The information gleaned from proposals may help guide the subsequent Request for Proposal (RFP) or cause it to be canceled if suppliers do not respond.

A Request for Information (RFI) is not a mandatory prerequisite to writing an RFP; many enterprise businesses write Requests for Proposal (RFP) without going through the RFI stage. RFIs may be considered as an inquiry stage:

1. When the goals of the portion of the project that is included within the related contract are in question;
2. When the solution for the portion of the project that is included within the related contract is new to the industry or the enterprise business; or
3. When the project team would like to explore a variety of potential solutions.

Trim Down or Qualify Suppliers

The request for information or inquiry stage can be seen as a prequalification of potential suppliers to narrow the initial list of possible suppliers to an acceptable list. More importantly, depending on the nature, size, and importance of the purchase, it is during the inquiry stage that decisions are formed concerning future projections on the potential for extended relationships, given the increased importance of relationships in the context of managing an entire procurement process.

The aim at this point is to find, from the initial list of suppliers, those suppliers: who are capable of producing the procurement items in the required quality and quantity; who can be relied on as a continuous source of supply under all conditions; who will keep their delivery promises and other service obligations; and who are competitive on price. Several steps can be taken to ensure that the identified suppliers are pre-qualified:

1. Suppliers must be technically capable. Do they have the correct products, or will they have to subcontract to other suppliers? If they subcontract portions of work, then determine who are their primary subcontractors?
2. Do they have the resources to manage the project properly?
3. Are they considered a "local" company? If not, how will you work with them? Do they need to travel every time a meeting takes place? How do they handle regular maintenance activities?
4. How many people does the supplier employ at how many locations? Where is the nearest location to your project site? If the project is to take place in many locations, can the supplier support multiple locations? Is there a need for international support?
5. How many projects is the supplier currently managing, and will the supplier be stretched too thin? Is the supplier managing other projects similar in size to yours, or are they typically much smaller or bigger?
6. Suppliers must be financially viable. Is the supplier in good shape financially and certain to remain in business and continue to support the product?
7. Suppliers are qualified for obvious reasons, but there are also some not-so-obvious reasons to consider. After a thorough review of their capabilities and previous projects and resources, it might happen that not all suppliers have the correct product mix and not all will be able to manage a large project.

Responses to Requests for Information (RFI) may show that:

1. Suppliers do not understand the RFI requirements;
2. The solution is available but far more costly than originally anticipated; or
3. The solution mandated by the procurement items requirements is not available.

If the suppliers are not responsive to a Request for Information (RFI), it could mean either returning to the procurement items requirements analysis phase or stopping further work on the portion of the project that is included within the related contract.

If suppliers' proposals are responsive to a Request for Information (RFI), the project manager must first review the procurement items requirements according to the information gained by reading these proposals. These requirements then become part of the Request for Proposal, and finally the Request for Proposal is developed and released.

Typically, a Request for Information should encompass all of the procurement items requirements. It is important that the procurement manager lists not only the technical issues but also the requirements for project management, maintenance, training, and support in the Request for Information. Thus, potential suppliers are allowed to comment on all aspects of the procurement and to establish what is possible and not possible from their point of view. Of course, the procurement manager and the project team will have to separate the wheat from the chaff since suppliers may try to say that solution other than their own does not exist. However, it is important not to combine competing solutions within a single requirement when rewriting requirements based on multiple suppliers' proposals.

If a Request for Information is poorly put together, has little focus, and demonstrates a fundamentally poor grasp of the solution, many suppliers will respond with datasheets and boilerplate text, or else not at all. Many potential Requests for Proposal are not released after a Request for Information, because the procurement manager and the project team have severely misjudged the solution, the implementation, and the cost. Suppliers are quick to grasp which projects are likely to move forward and which appear to be misguided "fishing expeditions."

On the other hand, a Request for Information is the best place for a supplier to try to influence the procurement items requirements and therefore have the inside track if and when the Request for Proposal is released. In the spirit of project team's education, the procurement manager should let suppliers provide as much information, help, support, and interaction as appropriate to the needs of that portion of the project that is included within the related contract.

17.2.2.7 Develop a Request for Proposal

A Request for Proposal (RFP) is a written document, which represents a certain amount of time, resources, and money, issued to prospective suppliers in order to communicate an understanding of the business needs of a project and to elicit bids from potential suppliers for a product or service. It represents an interpretation of those needs and involves the expenditure of a commensurate amount of time and resources on the supplier's part. It provides the structure that allows the project manager to take the project requirements for procurement that have been developed and put them into a form that suppliers can use, understand, and respond to. It also spells out how the portion of the project that is included within the related contract is to be implemented (the next phase), what the first steps will be, and how success will be measured on the supplier's part.

Proposals, by their very nature, are a supplier's interpretation of the Request for Proposal requirements. Therefore, a Request for Proposal is intended to promote a diversity of thinking, by establishing a competitive environment among suppliers, and to encourage suppliers to provide unique solutions based on their products and services.

Development of a RFP process can take six months or more to complete, and the team will be required to participate at varying levels during that time. The RFP project manager's responsibility includes securing the needed resources to complete the RFP.

A Request for Proposal is used when the following conditions apply:

1. Multiple solutions are available that will fit the project need.
2. Multiple suppliers can provide the same solution.
3. The project manager seeks to determine the "best value" of suppliers' solutions.
4. Products for the project cannot be clearly specified.
5. The project requires different skills, expertise, and technical capabilities from suppliers.
6. The problem requires suppliers to combine and subcontract products and services.
7. Lowest price is not the determining criterion for awarding the contract.
8. Final pricing is negotiated with the supplier.

The Request for Proposal can be seen an intermediate, but important, step in a "process improvement" project that involves procurement. It facilitates the project's aspiration to procure items for its successful completion. On the supplier's part, the Request for Proposal lays the groundwork for the portion of the project that is included within the related contract by allowing the project manager to state the project management requirements and to get the supplier's buy-in (in writing); thus ensuring good project controls. It is the unifying document that will lay the groundwork for how the portion of the project that is included within the related contract will be controlled from the time the contract is awarded to, perhaps, when the contract is finished.

Once a contract has been awarded to a supplier, the agreed-upon plan and schedule of that portion of the project that is included within the related contract, together with the winning proposal that defines the solution and establishes performance goals, will become the primary method for organizing and controlling the implementation tasks of procurement items.

As with the saying, "If you don't know where you are going, any road will get you there," the Request for Proposal not only tells suppliers where that portion of the project that is included within the related contract is going but also selects the road on which they will travel.

The advantages of using a Request for Proposal far outweigh the potential problems of dealing directly with suppliers and of not having a formal set of requirements to work from. Some of these advantages include:

1. A Request for Proposal requires the project team to examine the problems and issues concerning the project in greater detail than would normally occur.
2. A Request for Proposal (RFP) forces suppliers to create competitive solutions that not only respond to the RFP requirements but go beyond them, thus providing additional value for a given price.
3. A Request for Proposal does not favor one supplier over another, but allows all to compete fairly based on the same set of rules and requirements.
4. Because suppliers are working from the same set of rules and requirements, it will be easier to understand the differences between proposed solutions.
5. Having similar, but different, proposed solutions facilitates the evaluation.

During the planning phase of the Request for Proposal, the project manager (or procurement manager) should consider elaborating on the following key areas:

1. RFP personnel and organization.
2. Project schedule.
3. Technology and supplier education.
4. Budget estimation and development.
5. Return on investment (ROI) analysis (if required).
6. RFP development.
7. Proposal evaluation.
8. Contracts and awards.
9. Post-RFP activities.
10. Project personnel and organization for the new product

Each individual procurement item requires a Request for Proposal. However, in much the same vein as with a supplier statement of work, multiple products or services can be grouped as one procurement item within a single Request for Proposal.

When properly developed and written, a Request for Proposal is a powerful tool for selecting the most appropriate solution and developing straightforward relationships with suppliers. It represents a vehicle that allows both the project manager and the supplier to establish a dialogue and to work from the same set of rules, requirements, schedules, and information.

The opportunity to have this dialogue is an important element in the development of a RFP, because RFP requirements are often not clear and the supplier, as the expert on the particular product or service, is allowed to question and interpret what is being requested. Conversely, the project manager has the opportunity to clarify issues in supplier proposals. Proposals, by their very nature, are a supplier's interpretation of an RFP's requirements.

To enable suppliers to offer their best solutions, a RFP must represent a clear understanding of all the technical issues, must provide a method for implementing and managing those issues, and must provide the supplier with an acceptable method for doing business (contracts and price).

Many Requests for Proposals are not successful because they fail to communicate one or more of these requirements properly. While an RFP can be assembled in many different ways, the following is a suggested outline for the major RFP sections:

1. Cover Letter
2. Project Overview and Administrative Information
3. Procurement Items Specifications
4. Management Requirements
5. Supplier Qualifications and References
6. Supplier Additional Information
7. Pricing
8. Contract and License Agreement
9. Appendices

Cover Letter

The cover letter is an important first page to the RFP. It introduces the RFP project and provides some of the most vital dates. The cover letter may also include anything special about the RFP that should be noted by the suppliers. Special dates often include:

1. Proposal due date and time
2. Bidder's Conference
3. "Bidder's Intent to Respond" form due date Special notices and comments include such items as:
4. Brief introduction and description of the portion of the project that is included within the related contract.
5. Bidder's conference information
6. Information in the RFP is highly confidential and bidders may not share the RFP.
7. Who to contact about the RFP
8. Warning that suppliers may not contact other people on the RFP team.

The cover letter reinforces information that is found in the administrative requirements or other sections in the RFP. Information in the cover letter can help suppliers determine quickly what the RFP is about, who should receive the RFP and be responsible for the proposal, and what are important dates.

The cover letter may also contain instructions to the suppliers that they must return the "Intent to Bid" form or a Non-disclosure Agreement (NDA) prior to receiving the RFP itself. In some cases, the RFP may contain highly confidential and proprietary data, and you do not want to release it to the general supplier community.

Project Overview and Administrative Information

The project overview and administrative information section contains an overview or summary statement of the problem, similar to a proposal's executive summary, as well as the administrative information concerning the management of the

Request for Proposal (RFP). It provides suppliers with an overview of the enterprise business and a statement of the problem that the project manager hopes to resolve through this RFP. The statement of the problem must be detailed enough for suppliers to grasp both the business issues that are driving the RFP and the technical issues that may have precipitated the problem.

The administrative section contains all of the administrative requirements and information with which a supplier must comply in order to submit an acceptable proposal. It allows establishing the rules and setting the requirements for responding to the RFP. It is an important section for the RFP because it allows the project manager to establish how suppliers will contact him/her, how they will format their proposals, what rules must be adhered to during the RFP competition, and other important items that the project manager wants to specify. Some typical items in the administrative section include:

1. Introduction (basic overview of the RFP)
2. Schedule of events
3. Contact names and addresses where and when to submit the proposal
4. How questions will be handled
5. Information about the bidder's conference
6. Information about oral presentation and demonstrations
7. Requirements for preparing proposals
8. How proposals will be evaluated
9. The required format for vendor proposals
10. General proposal submission information, such as number of copies, printed or electronic, etc.
11. How alternate proposals will be handled
12. How to include subcontractors in the proposal
13. Other information that is required for a supplier to be responsive

While each of the instructions is important, there are three key requirements that must be followed to ease the task of reading and evaluating suppliers' proposals. These are:

Single point of contact—The first key requirement relates to directing all vendors to a single point of contact for any questions they may have. It is furthermore, necessary to ensure that all questions are written and sent to the designated person. The designated person can review and distribute questions to the appropriate resources. With everything in writing, there is no opportunity for a misunderstanding.

Standardization of requirements—The second key requirement relates to using a standard form for RFP requirements. Quite often, RFPs are written by a team of people who have different writing styles and different ways of specifying requirements. Using a standard form for RFP requirements reduces confusions for the suppliers.

Specifying Format for Supplier Proposal—The third key requirement relates to specifying how suppliers must organize their proposals. One example of a proposal format may be:

1. Cover letter
2. Executive Summary
3. Technical Response
4. Management Response
5. Pricing
6. Appendices

This outline requires the supplier to provide proposals in a consistent format, which will facilitate the evaluation process.

Administrative requirements are very important to keep suppliers moving forward with their proposals in a timely manner. If the instructions are missing or not clear, suppliers may overlook important meetings or milestones. For insightful suppliers, lack of instructions may signal a weak RFP team and a confused or conflicted project. This potential weakness may influence whether they decide to continue with their proposal.

On the other hand, failure to comply with the administrative requirements might be cause for rejecting that supplier's proposal. The purpose of this section is to lay down clear rules for responding to the RFP and to ensure that suppliers are aware of the penalties for not following them. If a supplier fails to abide by these rules, it may be a sign of carelessness and lack of attention to detail.

Procurement Items Specifications Requirements

The procurement items specifications section provides suppliers with specifications of the procurement items and enough information to enable them to understand the issues, the amount and type of work that has to be accomplished, and write a firm proposal. This section is the heart of the RFP. It contains all of the information and specifications of procurement items needed to enable suppliers to respond to the RFP. In describing procurement items specifications, the project manager should first summarize the problem or issue that is the basis for the RFP. This overview should address both the current business application and the technical environment (hardware, software, communications, etc.. . .).

Following the problem statement, the project manager (or procurement manager) should list and elaborate on the specifications (described in a previous section) to which a supplier must respond in the proposal, for example:

1. Goals and objectives for the project
2. Critical success factors
3. Technical specifications
4. Functional specifications for the current system
5. Functional specifications for the projected system
6. Performance specifications
7. Hardware requirements (if mandatory)
8. Resources requirements
9. Communications requirements (if mandatory)

The procurement items specifications are not only the foundation for a supplier's technical proposal, but also drive other sections such as project management and pricing.

This procurement items specifications section can be somewhat difficult to write because it is a balance between describing what the current needs are versus describing what procurement item is expected from the supplier.

It is appropriate to provide information on how the procurement items are expected to be used (in a benefits-oriented fashion) but not specify features that are unique to one supplier that other suppliers cannot satisfy. In some cases, it is acceptable to provide minimal requirements in the RFP in order to receive the most comprehensive and wide-ranging proposals. The down side to casting such a wide net is that the project manager may receive a bewildering number of proposals with solutions that are only marginally acceptable, making it almost impossible to evaluate them. Thus, instead of spending quality time on the suppliers with good solutions, the project manager will find himself/herself spending hours upon hours weeding out the non-compliant suppliers.

Since a Request for Proposal is intended to promote competition among suppliers, it is also appropriate to provide procurement items specifications that are not so tightly focused that only one or two suppliers can respond, so as not to limit choices in terms of products, prices, and suppliers. Tightly focused and overly restrictive procurement items specifications may be a signal to other suppliers that a solution has already been chosen and that the RFP is merely an exercise to justify that solution.

If possible, the procurement items specifications section of the RFP should reflect a reasonable understanding of the products and services that are being requested in the RFP. Procurement items specifications, as indicated in a previous section, should have the following three characteristics:

1. Identify a capability, characteristic, or quality factor of a system in order for it to have value and utility to a user.
2. Be measurable in some manner.
3. A product or service must exist in order to satisfy the requirement.

Procurement items specifications are the key components of a Request for Proposal and they should not be difficult to spot or understand. If a requirement is in a Request for Proposal it must represent something that is needed within the project. Procurement items specifications can range from the project specifications to the product specifications. In order to write an effective procurement items specifications section, the RFP team (project manager, procurement manager, and technical specialists) must have the following knowledge:

1. The team must know in detail how the current "process to be improved" operates and must be able to communicate that knowledge to the suppliers.
2. The team must know what is expected of the new "improved process" and provide suppliers with direction.

3. The team must know what constitutes acceptable solutions, given that it has several valid solutions from which to choose.
4. The team must be able to intelligently evaluate the differences between the acceptable solutions.

This section must be well documented and complete; otherwise, suppliers will have to ask questions in order to clarify statements or requirements.

Management Requirements

The management requirements section states the conditions for managing and implementing the portion of the project that is included within the related contract. It provides suppliers with the information they need to develop a plan that will cover the implementation, installation, training, maintenance, and other aspects of the portion of the project that is included within the related contract. The proposed plan should provide the needed assurance that the supplier has the resources required to perform the contract successfully. The plan typically contains the following:

1. Functional requirements.
2. Staffing requirements.
3. Site preparation responsibilities.
4. Delivery and installation schedule and plan.
5. Procurement item acceptance test requirements.
6. Procurement item maintenance requirements.
7. Procurement item training requirements.
8. Documentation requirements.

Development of this section is essential for ensuring that suppliers can meet the overall requirements of the portion of the project that is included within the related contract. It is possible that suppliers can meet the technical requirements but cannot meet the management requirements as evidenced in their poor or inadequate responses to the requirements in this section. It is possible that a supplier has put all of its energy into the procurement item development and little or no effort into determining how the item should be installed and maintained, specifying what type of training is needed, and providing good readable documentation. The management section will help to differentiate the suppliers with good management capabilities from those with little management capability.

Supplier Qualifications

The supplier qualifications and references section asks the supplier to describe qualifications, list references, and to provide information about their company and financial status and the customers who will serve as references for their proposal effort. These qualifications and references are as important as the technical and management requirements.

It is important not to bury the supplier qualifications and references section and to ensure that suppliers do not take it lightly or simply say that the information requested is already provided in their annual report. The following are examples of what is typically required in this section:

1. A brief history of the supplier's company.
2. The supplier's installation and maintenance offerings and capabilities.
3. A description of the relationship between the supplier and each manufacturer, and how long this relationship has been in existence.
4. Evidence that the supplier has the necessary technical skills, technical staff, and financial resources to perform the contract.
5. A list of the currently installed systems or developed items.
6. Names of customers with similar configurations and/or applications who can provide references, including contact names and telephone numbers.

Supplier Additional Information

The A suppliers' additional information section allows suppliers to include information they feel is relevant although not required or requested in the RFP. It reserves a place in the RFP for suppliers to provide information that they feel is necessary but was not requested. Suppliers can also discuss potential issues that are relevant to the RFP and to their proposal in this section.

For example, a supplier may have additional product features to demonstrate that are outside the scope of the RFP. Suppliers may also comment on the procurement items requirements they feel are missing from the RFP, or present a unique solution that was not anticipated by the project manager.

This section is also an appropriate place for suppliers to discuss issues they believe are relevant to the project and that have not been covered in the RFP. The RFP's instructions to suppliers will direct them to use the suppliers' section for any additional information outside the scope of the RFP.

A supplier might provide a solution to a problem evident in the RFP that other suppliers did not consider. Even if this particular supplier does not win, the explanation of the problem and the potential solution will still be worth considering for use with the winning supplier

Pricing

The pricing section specifies how suppliers are to provide pricing information. It provides a detailed format for suppliers to follow in developing their price proposals. Instructions should be in a clear format to ensure that price proposals from different suppliers can be compared on an equal basis. To facilitate this comparison, the project manager (or procurement manager) may consider providing a sample spreadsheet that breaks the proposed system into components such as the following:

1. Hardware.
2. Procurement item software.
3. Application development software.
4. Installation.
5. Maintenance.

6. Training.
7. Documentation.
8. Project management.
9. Integration of unique hardware or software.
10. License fees (ongoing).

An area deserving of particular attention involves onetime costs versus recurring costs. The initial price of a software package for example is a onetime cost; annual maintenance and software licensing fees are recurring costs. Recurring costs need to be identified if the project manager is developing a life-cycle cost for the "process improvement" project that is expected to have a long term valid life time. Pricing is not usually the sole determinant for winning but should be used to break a tie between two suppliers with equally good technical and management proposals.

Contract and License Agreement

The contract and license agreement section contains the purchase contract, nondisclosure agreements, and other legal documents. It provides basic guidance to the supplier on how to respond to contracts and agreements. It can either become part of the pricing section or stand alone.

Contracts are provided to suppliers, who can begin to study them along with the RFP requirements. If contract provisions are such that suppliers cannot respond, suppliers may either choose not to bid on the RFP or take exception to the contract provision in their proposal. For example, a contract may state that custom procurement items must pass a 90-day acceptance test period prior to the first payment. A supplier may agree to only a 30-day test, or may not agree to any acceptance test that is tied to payments.

The project manager should identify showstopper issues during the proposal evaluation period because it is possible to select a supplier who will not accept the contract. The project manager should not spend time and resources on an unproductive supplier, as this takes time away from working with the potential winners.

Appendices

The appendices section contains bulky but relevant information such as network diagrams, technical requirements studies, and project plan outlines. If the RFP team generates detailed information that is too lengthy for the body of the RFP, this information should be placed it in an appendix. Examples include the following:

1. Workflow diagrams and studies.
2. Spreadsheets with statistical information.
3. Communications network drawings and plans.
4. List of current equipment.
5. Standards used within the enterprise business.
6. Tentative project plan with dates.

The information is then available to the supplier but does not distract from the narrative portion of the RFP.

17.2.2.8 Plan Evaluation Criteria of Supplier Proposals

As the requirements of that portion of the project that is included within the related contract are confirmed and agreed upon, the criteria for evaluating procurement items requirements, hence rating or scoring suppliers proposals, must also be established. These evaluation criteria can be objective or subjective.

Evaluation criteria are often included as part of the RFP documents. They can be limited to purchase price if the procurement item is readily available from a number of acceptable suppliers. Purchase price in this context includes both the cost of the item and ancillary expenses such as delivery. Other selection criteria can be identified and documented to support an assessment for a more complex product or service. While specific suppliers' offerings and quotations may be sought, questions about the following areas will make up a significant portion of the evaluation criteria:

1. *Technical requirements.* Does the supplier have, or can the supplier be reasonably expected to acquire, the technical skills and knowledge needed?
2. *Management requirements.* Does the supplier have, or can the supplier be reasonably expected to develop, management processes and procedures to ensure a successful completion of that portion of the project that is included within the related contract?
3. *Price.* Will the selected supplier produce the lowest total cost (i.e., purchase cost plus operating cost)?
4. *References.* Can the supplier provide references from prior customers verifying the supplier's work experience and compliance with contractual requirements?
5. *Qualifications/Technical approach.* How well does the supplier's proposal address the contract statement of work? Do the supplier's proposed technical methodologies, techniques, solutions, and services meet the procurement documentation requirements or are they likely to provide more than the expected results?
6. *Production capacity and interest.* Does the supplier have the capacity and interest to meet potential future requirements?
7. *Financial capacity.* Does the supplier have, or can the supplier reasonably be expected to obtain, the necessary financial resources?
8. *Business size and type.* Does the supplier's enterprise meet a specific type or size of business, such as small business, women-owned, or disadvantaged small business, as defined by the procurement manager or established by governmental agency and set as a condition of being award a contract?
9. *Intellectual property rights.* Does the supplier assert intellectual property rights in the work processes or services they will use or in the products they will produce for the project?
10. *Proprietary rights.* Does the supplier assert proprietary rights in the work processes or services they will use or in the products they will produce for the project?

On major proposals enterprise businesses have found it advisable to secure independent evaluation criteria of proposals to provide assurances to management that the suppliers proposed costs are reasonable.

17.2.2.9 Review and Formalize the Request for Proposal

Before an RFP is tendered to selected suppliers, it should be reviewed by people outside the primary RFP team and formalized by the procurement manager and the project team. This review group may look at different aspects of the RFP, for example:

1. Are the network structures current and accurate?
2. Are there any basic system architectural concerns?
3. Is the current working environment description accurate?
4. Are the functional requirements stated clearly and accurately?
5. Does the pricing section meet the enterprise business standards?
6. Is the plan of the portion of the project that is included within the related contract achievable?

The most critical aspects of this review are to ensure that the RFP is valid, reliable, and repeatable. Valid means that the RFP must accurately reflect what is needed to be developed or produced by the supplier. Reliable means the RFP should produce response proposals that are all priced within a reasonably narrow range, say, ± 10 %. Repeatable means the RFP should produce response proposals that are sufficiently similar in technical understanding and work approach from the different selected suppliers.

This "objective" review will help team members to see the strengths and weaknesses of the Request for Proposal. Once the review is complete and the review issues responded to, the RFP should be a stronger document and should be formalized.

17.2.3 Invite Tenders

This is the project management process used to officially publish the RFP by send a formal written invitation to pre-qualified suppliers to bid or tender to undertake work or provide services as described in the RFP. Its primary purpose is to obtain viable responses from prospective pre-qualified suppliers, sufficient to satisfy the project's listing of procurement work, the components, subsystems, services, support, purchased labor, whatever, which will be supplied from other suppliers. Another secondary objective of this process is to make sure that all suppliers are treated fairly so that no one gains an unfair advantage over others.

In most sectors of procurement, competitive tendering/bidding is the norm for all except small, low-value and low-risk procurements. Tendering invitation to a single procurement supplier is generally considered acceptable only if the work is a logical extension of a previous or existing contract and continuity is required, or if only one supplier is qualified to undertake the work, or if a contract has to be awarded quickly in an emergency. This type of invitation is used quite frequently in commercial businesses where the reputation of the supplier and the long-term relationship which may exist between the project manager and the supplier is paramount in the selection decision. The rationale for deliberately going to a single procurement supplier must be documented properly in the procurement file.

Competitive tendering is a process that enables the project manager to leverage several potential sources of supply through a single activity to obtain the most favorable business terms. In order for this process to be successful, a number of conditions, such as those outlined below, must be met.

1. *Provide clear content.* A solicitation for tendering should provide sufficient information about the procurement requirements so that a supplier will be able to offer exact pricing and provide whatever other detailed information is required to successfully obtain the procurement order. Typically, this will include facts such as the exact specification of the required procurement items or a contract statement of work for a service, the quantity required, payment terms, the expected time of performance, necessary quality levels, and shipping or performance location. The invitation to tender must also include the deadline for submission.
2. *Determine compressible spending.* Before engaging in the invitation to tender process, the project manager is responsible for determining if market conditions will support a reduction in price or an improvement in terms. Unless favorable market conditions are present, competitive tendering will not be worthwhile. While there is no precise way to ensure this under all conditions, benchmarking industry trends, whenever possible, might provide some guidance.
3. *Ensure responsive, responsible competition.* When selecting potential suppliers or candidates to which invitation to tender will be sent, it is important that the project manager qualifies them to ensure that the tenders returned will be responsive to the "process improvement" project needs.

   Qualified suppliers are those that have successfully completed a formal screening process but may not yet have been qualified for the approved supplier list or that may be supplying a product or service that does not require the stringent supplier site inspection criteria used to establish eligibility for the approved supplier list. These are usually suppliers who meet all of the business requirements of the organization and are approved by the Procurement Department for future business as may arise. This means that the supplier has the means to fully understand the project manager's needs and can, under normal business conditions, fulfill the requirements.

   The project manager should ensure that the suppliers are in a position to meet any procurement requirements; that, for example, they have the necessary financial means to produce the product being specified or that they have the equipment needed to meet the requirement in a timely manner. If tooling is required, the project manager must be careful to ensure that the supplier does absorb the cost of the tooling as a way of buying the business.

4. *Enable fair and ethical tendering/bidding processes.* The project manager's task is to properly ensure ethical conduct in the solicitation and acceptance of tenders/bids, making sure that all suppliers are provided with exactly the same information and have an equal amount of time to respond. Answers to questions submitted by one supplier need to be distributed to all suppliers bidding to further enhance the competitive process.

   Suppliers should also be made aware of the process for awarding procurement by the enterprise business, whether it is the lowest price or some combination of terms, as well as the criteria for making the final selection. Many organizations use a weighted-average scoring process developed by a cross functional internal team to select suppliers for complex procurement, since it can be extremely difficult to unilaterally evaluate and select the most appropriate supplier.

The typical objective of competitive tendering is to ensure that the "process improvement" project receives the most appropriate tender for a given procurement, with all other terms and conditions remaining equal. To do this, the project manager needs to ensure that a number of conditions are present:

1. *Competition.* The qualified suppliers are willing to compete. The more suppliers available (within manageable degrees), the greater the competition will be. Competition is the project manager's best friend.
2. *Value.* The procurement items have significant enough value to make the tendering process worthwhile.
3. *Savings.* The tendering has the potential to result in lower prices.
4. *Requirements.* A clear specification, statement of work, or industry standard is available to all suppliers.
5. *Contract.* The suppliers have the capability and are willing to commit to furnishing the procurement items as specified in the chosen contract type.
6. *Time.* There is sufficient time to conduct a fair and impartial process.
7. *Corrections and clarifications.* A process exists to provide suppliers with answers to questions or corrections to specifications. Answers to questions asked by one supplier must be shared with all others.

During the solicitation for tenders/bids process, the project manager, the procurement manager and the technical specialists must be prepared to hold a RFP conference and to provide answers to suppliers' questions as they come in. In some cases suppliers may not be able to move forward with their proposals until they receive answers to their questions.

Thus, great care must be taken in responding to questions from prospective suppliers requesting a clarification of language, or perhaps more technical detail. The reason: suppliers who ask for and get more information are placed at a competitive advantage over those who have not raised the question and received either an official or informal response.

Some enterprise businesses allow for questions from prospective suppliers to be addressed to anyone in the project manager's team. Typically it is not the project manager who will get such questions, but rather the technical persons who wrote procurement specifications. Suppliers' questions should be controlled and funneled through the project manager. A fair way to handle such questions is to require that they be sent directly to the project manager, who will obtain an "official" response, and then send both the question and response to all prospective suppliers who have expressed an interest in the procurement. Everyone must remain on equal footing in the solicitation for bid process.

The solicitation for bid process culminates with:

1. *Qualified Suppliers List*—The qualified suppliers list are those suppliers who have been asked to submit a proposal and have responded.
2. *Procurement Document Package*—The procurement document package is an enterprise business-prepared formal request sent to each prequalified supplier and is the basis upon which a supplier prepares a bid for the requested products, services, or results that are defined and described in the RFP included in the procurement documentation.
3. *Suppliers Proposals*—These are supplier-prepared documents that describe the supplier's ability and willingness to provide the requested products, services, or results described in the RFP included in the procurement documentation. Suppliers Proposals are prepared in accordance with the requirements of the relevant procurement documents and reflect the application of applicable contract principles. The supplier's proposal constitutes a formal and legal offer in response to an invitation to bid.

   After a supplier proposal is formally submitted, the project manager sometimes requests the supplier to supplement its proposals with an oral presentation. The oral presentation is meant to provide additional information with respect to the supplier's proposed staff, management proposal, and technical proposal, which can be used by the project manager in evaluating the supplier's proposal.

17.2.4 Select Optimal Suppliers

Once proposals are submitted, the actual selection of the optimal supplier is the next logical step. The selection is based primarily on both the evaluation criteria established and included as part of the RFP documents and on an assessment of the supplier's willingness to participate in additional aspects related to the bidding process. Proper supplier selection, despite requiring a strong measure of distinctly human intuition, must be performed systematically and to the most objective criteria you are capable of developing.

17.2.4.1 Evaluate Proposals

Before selecting an offer, every project manager should employ an evaluation process to ensure adequate consideration that all aspects of the "process improvement" project needs are being optimized. Evaluating a supplier's offer means not only evaluating its bid or proposal from a cost perspective, but it also means evaluating the supplier's ability to perform to the required level of speed and quality. The project manager should evaluate offers in terms of potential risk as well as potential benefits. In providing incentives to obtain the contract by reducing the price, for example, will the supplier continue to maintain the level of quality the "process improvement" project requires? Issues such as this will be merely one among the many you will have to consider during the supplier selection process.

In performing proper due diligence, the project manager reviewing a supplier offer should evaluate three key criteria before reaching a decision to award contracts to specific suppliers: responsiveness, capability, and competitive value. Because of the inherent subjectivity of much of evaluating suppliers, there is strong evidence to show that a thorough review process produces the most reliable results when it involves several individuals from a cross section of functional departments within the enterprise business, performing separate evaluations then developing a consensus opinion.

Responsiveness

Most obviously, the basic criteria for selection will be the supplier's ability to perform to the specification or scope of work contained in the Request for Proposal (RFP). The thoroughness of the supplier's response and the level of detail the supplier provides generally signify the supplier's level of understanding of the RFP requirements and its expertise in providing workable solutions, services, or conforming procurement items. Here, evaluation of proposals encompasses many different areas, including but not limited to:

1. *Basic adherence and compliance with the RFP administrative requirements.* Was the proposal submitted on time? Was the suggested supplier proposal format followed? Did the supplier acknowledge and incorporate changes to the RFP as requested? Was the proposal readable?
2. *Overall understanding of the RFP issues.* Was it evident that the supplier understood the basic issues driving the RFP? Did the supplier's proposal respond to those issues with a logical and understandable solution?
3. *Technical requirements.* Did the vendor respond to each of the technical requirements and were the responses adequate?
4. *Management requirements.* Was a reasonable and acceptable project and implementation plan submitted with the proposal? Did the plan demonstrate an understanding of the RFP needs?
5. *Pricing.* Was the pricing reasonable compared to the estimated budget and other proposals that were submitted? Was the pricing broken into the component parts as requested, or was pricing presented as a single total?

In high-value or high-profile contracts, it is wise to actually visit the supplier's facility and physically inspect the facility to qualify the supplier and determine its ability to meet the "process improvement" project requirements. If site visits were part of the RFP, these should take place prior to any final evaluation. The site visit is to a reference site designated by the supplier and is generally only for the final two suppliers in the competition. In a very close competition, site visits can make the final difference in the choice of supplier. In some cases the project manager may want to visit the suppliers' factories or headquarters and meet management teams. This allows making certain that suppliers are financially sound and that all support groups proposed actually exist.

In many situations, however, site visits may be physically or financially impractical, so other methods to confirm the supplier's ability to respond should be used. For instance, the project manager might consider contacting a supplier's references to ask about similar work performed in the past. This is frequently an effective way of determining overall supplier competency. The project manager may also want to review the response document to ensure that the supplier has answered all the questions in the RFP and successfully addressed any mandatory requirements set forth. While oversights sometimes occur, it is not a good sign to discover that some of the key elements of the RFP remain unaddressed. Offers that do not answer the RFP specific questions should be considered nonconforming and rejected.

The project manager should also determine the extent to which the supplier's proposal conforms to the enterprise business and environmental and ethical policies and procedures. Does the proposal appropriately address warranty and replacement issues? Does it conform to the enterprise business policy regarding commercial liability and damages? Is it signed by the proper authority? Are the correct documents, such as evidence of insurance certificates and copies of applicable licenses, attached? Perhaps even more significantly, how closely do the supplier's terms and conditions match those of the enterprise business?

Capability

While many capable suppliers may respond to the invitation to tender, the project manager's task will be to determine which supplier is the most qualified for this particular contract award. In the capability evaluation, the project manager should consider several critical factors amongst which:

1. *The Operational Capacity.* One of the key considerations in supplier selection will be the supplier's physical capacity to meet RFP needs as identified.

   It is not advisable to select a supplier that could have difficulty meeting the required volume due to capacity constraints or conflicts with the scheduling of other tasks. A simple ratio of current output to capacity can provide a valuable indication of this ability. The likelihood of on-time delivery failure increases when a supplier's loading for the procurement items may exceeds 90 percent of capacity, especially in industries where skilled labor or production capacity can be difficult to obtain.

   The project manager will also need to ensure that the supplier has the ability and systems to properly schedule orders and keep track of current production operations to meet its customers' commitments. With little or no technology to assist in the scheduling process, the supplier may have difficulty keeping track of its customer order obligations and may prove unreliable in meeting delivery promises. The project manager should be able to benchmark this through the customer references the supplier provides.

Past performance, while not necessarily a clear predictor of future performance, can provide further insight into the supplier's operational capability. The project manager may be able to develop data on this from the enterprise organizational process assets records such as supplier delivery efficiencies or production lead times within the enterprise business, if applicable. If not, the project manager may need to perform some benchmarking activities and, certainly, check with as many referenced accounts as possible.

2. *The Technical Capability*. Another key capability to be evaluated is the supplier's technology and technical ability. Does the supplier possess the necessary equipment, tools, and talent to meet the RFP requirements? This can be determined not only through site visits but also through historical performance records and active participation in industry events. How many patents does the supplier's company hold in comparison to its competition? How often does it lead the market with the introduction of new products? To what extent is it funding its research and development efforts?

   The project manager might also consider supplier certification as an adjunct to technical enablement. Does the supplier possess the necessary licenses, insurance, and certifications required to ensure regulatory compliance? This not only reduces the supplier's liability, but in many cases it may reduce the liability of the enterprise business, too, because lawsuits directed at the supplier while it is performing a contract issued by the enterprise business will potentially bring the enterprise business into potential litigation as well.

3. *The Financial Ability*. A key indication of the supplier's ability to service the RFP needs is its history of profitability and cash flow management. When a company's profit trend spirals downward at a faster rate than its competitors, it is usually an indication that it will soon begin to experience financial strain. This may also affect its ability to meet current schedules, to effectively invest in new equipment, to employ new technologies for future efficiency, or to hire the best talent available.

Competitive Value

Most of all, the project manager should expect to gain the greatest value possible through the award of procurement. Here, competitive value can be considered the optimal combination of a number of factors. Most importantly, these factors include price, service, technology, and quality.

1. *Price*. Price is driven by a number of factors, not the least of which is the current supply-and-demand ratio situation in the particular marketplace. While there is in general an identified tendency for supply and demand to seek equilibrium, the condition where one exactly matches the other, it is rarely the case that markets ever reach this condition for long. More often, the two factors are in continual flux.

   When supply of available procurement items exceeds the demand generated by project managers, prices usually drop. Conversely, when demand exceeds supply, that is, when many project managers are chasing fewer available products or services, prices traditionally rise. However, faced with declining prices, suppliers usually move away from marketing the product or service and on to other, more profitable, offerings. In the same vein, faced with rising prices, consumers tend to move to less costly alternatives. In today's economy, conditions are rarely uniform from industry to industry or cycle to cycle. In addition, in the more complex industries, such as electronics, numerous factors beyond supply and demand come into play. This makes price trend predictions virtually impossible.

Price, of course, is relative to other considerations as well. As the project manager analyzes price, he/she should consider two factors: competition and return on investment (ROI). First, how does the price offered by the supplier compare to prices commonly found in the open market for other products or services of a similar nature? A project manager, negotiating price, wants to achieve at the very least a fair and reasonable price. Second, does the price paid provide a reasonable ROI? That is, does the price paid reduce costs substantially enough to justify the initial expenditure? An Enterprise business typically looks for a return on investment in less than one year or adds profit at a rate above what is traditionally earned.

When considering price, the project manager may also want to consider overall lifecycle costs, which include all of the costs associated with procurement and using the procurement outcomes (products or services) for the duration of its useful life. This consideration is also known as total cost of ownership (TCO) and includes various other costs, such as maintenance or storage, which might affect the comparison of suppliers' offers. At the very minimum, an analysis of the cost of materials should include the expense of transportation. The cost of goods, which include transportation, is known as landed price.

2. *Service*. When the project manager evaluates the service aspect, he/she should look at a number of factors:
   - Full support for just-in-time delivery
   - The flexibility to accommodate rush orders
   - Strong engineering and design support
   - An accommodating credit policy or a guarantee of satisfaction

   The project manager should also evaluate how well the supplier responds to unexpected situations, such as accepting return of slow-moving or obsolete procurement items components. From the customers' perspective, the service aspect is the element that bonds the enterprise business to the supplier. In developing relationships with customers, the supplying sales team generally strives to develop a perception of responsiveness to problems and issues. But in evaluation for suppliers' selection, the project manager should evaluate the supplier's proactive efforts in avoiding problems in the first place.
3. *Technology*. In any consideration of value, two questions regarding the use of technology are important: first, how effectively does the proposed technology meet current RFP requirements? Second, how long into the future will the technology continue to be viable? In answering these questions, the project

manager evaluation should rely heavily on input from engineering and other user groups familiar with the technical qualities in the requirements.

Technological innovation can provide the enterprise business with a competitive advantage. Therefore, the project manager should also take into account the reputation of the supplier as a technological leader in the market.

4. *Quality*. The evaluation of quality involves both the supplier's ability to conform to specifications and the perceived satisfaction of the user. In the automobile industry, this concept parallels the argument that a Ford or Nissan is as functional as any of the luxury automobiles, yet buyers are willing to pay substantial premiums to own the latter. Clearly, variations in such intangibles as comfort and appearance can be hard to evaluate mathematically, yet consumers continue to value and pay for them.

   Key to any supplier evaluation will be an analysis of the systems the supplier has in place to control the quality of its output and the programs it utilizes to maintain continuous improvement. Certifications, such as those issued in accordance with the standards of the International Organization for Standardization (ISO), also provide assurances that the supplier has programs in place that will reasonably ensure continued levels of quality.

17.2.4.2 Eliminate the First Round of Suppliers

The basic steps in evaluating proposals to eliminate the first round of suppliers are as follows:

1. *Perform a review of all proposals submitted, and eliminate any proposals that do not satisfy the three key criteria described above*. The first proposals eliminated may be poorly written, priced significantly above or below other suppliers by a factor of 50 percent, or on closer review lack the right product.
2. *For the remaining proposals, read them in-depth and score them against the evaluation criteria*. Eliminating first round of suppliers' efforts at this point should result in a list of several acceptable suppliers, all capable of furnishing the requirements, with whom the project manager would be willing to place a procurement order.
3. *The approved-listed suppliers resulting from the step above, called the short list of suppliers, comprises suppliers with the potential to win procurement contracts*. The last step in the evaluation is to evaluate the remaining suppliers according to their references, demonstrations, and presentations. At this point, pricing can be the determining factor between two suppliers with equal evaluation scores, given that both have good references.

There are several schools of thought on how to score proposals. They range from assigning a simple "meets/does not meet" scoring to a complex scoring system using points and weights for each section. A middle-of-the-road scoring system assigns numerical values to the different RFP sections, and the point values are divided among the requirements within a section.

Overtime suppliers on the approved list may grow into much more valuable partners through superior performance. Depending on their performance, suppliers on the approved list can be classified into categories such as conditional, approved, certified, or partnered. These are discussed further in the experience stage.

Suppliers who perform well and have the ability to meet or exceed the FRP requirements become valuable members of the enterprise business supply base. Unfortunately, not all suppliers have the capabilities or performance to merit higher status, and thus a categorization of suppliers based on their proposal scores is developed in many enterprise businesses. An example of such categorization can consist of:

1. *Conditional:* An existing supplier whose performance does not meet minimum standards or a new supplier who has not yet established a performance history
2. *Approved:* Meets minimum standards and can supply components for existing products but not new ones.
3. *Preferred/key:* Has proven ability to meet procurement objectives and a mutual commitment to a continuing long-term relationship.
4. *Strategic alliance:* Characterized by integrated management planning and scheduling; shared technology and plans; access to each other's financial information;

At conclusion of this process step, eliminated suppliers should be notified and should have the opportunity to understand why they were eliminated. Notifications need not wait until the contract is awarded. It is important to document the rationale for eliminating a supplier, as this information may be used later when justifying the winning proposal.

17.2.4.3 Call References

For suppliers on the shortlist, it is now time to call references. This call should be scheduled, and all of the RFP team members should participate in it. Only references for the shortlisted suppliers should be called.

In many industries companies have found it a useful practice to hold what is called bidders conferences on major procurements to allow for questions to be asked from suppliers on the shortlist. These conferences are to reinforce the requirements specified in the Request for Proposal (RFP). The bidders' conferences need not be called if the RFP were always a perfect document, containing everything needed to respond to the request. Since most Requests for Proposal are compiled in a rather hasty and disorganized manner, these meetings are often a good idea.

One of the ground rules on a bidder's conference is that every bidder on the shortlist be kept on equal footing. It is a controlled meeting, typically with the project manager acting as chairperson, supported by the procurement manager and the technical specialists. Questions from bidders on the shortlist are solicited in advance of the meeting so intelligent answers can be presented to all present. Sometimes additional questions may be allowed from the bidders on the shortlist in attendance, but sometimes not.

The practice of holding a bidders' conference is a good one on major procurements. Sometimes the questions from bidder on the shortlist prove to be so insightful, that in some cases the project manager may in fact choose to modify the official RFP to incorporate additional or clarifying materials.

17.2.4.4 Host Demonstrations

The RFP may require that suppliers on the shortlist demonstrate their products or services, either at the supplier's factory or on site, so that the RFP team and other users in the community can get direct experience with the supplier and the products or services.

17.2.4.5 Best and Final Offer

As part of the give and take during the evaluation period, there is a reasonable chance that a supplier overestimated a requirement's impact or overscheduled part of the implementation. The best and final offer allows suppliers the opportunity to rethink and fine-tune their pricing by submitting their best and final offer.

17.2.4.6 Suppliers Selection

This is the final step of the supplier selection process. When selecting suppliers at this stage, project managers are advised not to think in terms of lowest cost, but appreciate that in working with suppliers they get what they pay for, and that suppliers who offer services at rates cut to the bone may be offering also low quality, poor performance and minimal standards of professionalism.

Thus, the first concern of the project manager must be to identify the proposals that offer the best overall value for money, display the most business like approach to meeting the "process improvement" project objectives, and respond best to the procurement items specifications. The project manager should also consider the proposal to see: prices that are realistic in relation to the scale of the contract; quality in the inputs and resources proposed for the work; outcomes and deliverables clearly defined; and evidence of distinctive added value and reliable performance.

In a business context particularly, there are other factors that come into play, and these can influence decisively the project manager's views about the proposals that are right for the "process improvement" project. These include, but are not limited to:

1. *Insight into the enterprise business operating environment:* Does the bidder appear informed about the sectors of activity in which the project manager's enterprise business is engaged and the factors that influence its market environment and profitability?
2. *Partnering and synergy:* Is there a sense that the bidder is the one best placed to work with the project manager in a productive team effort? Are the corporate values and policies of the enterprise business understood and supported?

3. *Risk and professional accountability:* Has the supplier's proposal addressed these concepts? Does it indicate an understanding of their significance for successful contract performance?
4. *Innovation:* New ideas, fresh thinking and solutions that competitors will find it hard to match are ingredients that can win the day, but innovation needs to be dependable. Has the bidder taken account of the risks associated with innovation?
5. *Flexibility and responsiveness:* Does the supplier's proposal communicate a willingness to adapt methods and procedures in response to unforeseen changes in the requirements of the contract?
6. *Cost and efficiency savings:* Suppliers are expected to be able to offer cost and efficiency savings as well as continuity of personnel, and to have the capacity to get up to speed rapidly on a new contract so as to start producing useful output without mobilization delays or steep learning curves.

The final step in the selection process is to recommend winning suppliers. The recommendation report reviews why the chosen supplier was selected and why the second place supplier was not selected. This report is given to management and procurement function, which will begin contract negotiations with winning suppliers.

Once selections are completed, several steps, described in the sub-sections below, must be completed prior to notifying the winning supplier. These steps will help the project manager to organize and close the RFP phase of the project.

17.2.4.7 Review Selection Process with Management

Once a winning supplier has been selected, the project manager must produce an evaluation report for the organizational process assets and review the selection process with the enterprise business management. The report reviews which suppliers were considered, how they were evaluated, and why the winning supplier was selected over other suppliers on the short list. This report is then delivered to senior management in the enterprise business unit, information technology function, and procurement function. If needed, or desired, the RFP team may be asked to provide a presentation of the evaluation results. Producing an evaluation report is beneficial for a number of reasons:

1. It allows the RFP team to review all events leading up to the selection and demonstrate that a fair and objective decision was made.
2. By having the enterprise business unit, IT function, and procurement function in agreement with the decision, the project manager ensures their participation in the project when it is started.
3. It provides the enterprise business unit with an "audit trail" of the project if there are changes within the enterprise business that necessitate a review of the project.
4. If it is needed, the project manager can defend his/her decision should a losing supplier file a protest.
5. The evaluation report finalizes the decision and closes the RFP phase of the project.

The evaluation report should contain, but not be limited to, the following:

1. Summary of why the RFP was initiated
2. What were the project goals and objectives
3. List of participants on the RFP Team and their roles
4. How suppliers were selected to participate
5. A list of the suppliers that submitted proposals and their scores
6. Review of evaluation criteria, including which were most important
7. Which suppliers made the short list and why they were not selected
8. Review of why the winning suppliers were selected

The next step, if the evaluation report is accepted, is to notify the winning suppliers and schedule a meeting to begin the contract negotiations, and get the contract signed so that the project can begin.

Contract negotiation clarifies the structure and requirements of the contract so that mutual agreement can be reached prior to signing the contract. The final contract language reflects all agreements reached. Subjects covered include responsibilities and authorities, applicable terms and law, technical and business management approaches, proprietary rights, contract financing, technical solution, overall schedule, payments, and price. Contract negotiations conclude with a document that can be signed by both the project manager and the supplier, that is, the contract. The final contract can be a revised offer by the supplier or a counter offer by the project manager.

For complex procurement items, contract negotiation can be an independent process with inputs (e.g., an issues or open items list) and outputs (e.g., documented decisions) of its own. For simple procurement items, the terms and conditions of the contract can be fixed and non-negotiable, and only need to be accepted by the supplier.

The project manager may not be the lead negotiator on the contract. The project manager and other members of the project team may be present during negotiations to provide, if needed, any clarification of the project's technical, quality, and management requirements.

As a cautionary note, it is possible that negotiations with a winning supplier break down and that winning supplier is deselected. Reasons for this can be varied, but they may revolve around contract issues, such as liquidated damages, ownership of custom developed products, and payment terms. To mitigate these types of contract issues, the project manager must ensure that he/she has included the procurement contract with the RFP and request that supplier review it; and, as part of their proposal, point out specific contractual issues that they may have trouble with should they be selected. This allows the project manager to prepare for these issues prior to making a final selection, but it also prevents a contractual surprise at the most inopportune time.

After contract negotiations, a final contract is awarded to each winning supplier. The final contract can be in the form of a complex document or a simple purchase order. Regardless of the document's complexity, a final contract is a mutually binding legal agreement that obligates the supplier to provide the specified products, services, or results, and obligates the project manager to pay the contracted supplier. A final contract is a legal relationship subject to remedy in the courts. The major components in a final contract document generally include, but are not limited to, section headings, statement of work, schedule, period of performance, roles and responsibilities, pricing and payment, inflation adjustments, acceptance criteria, warranty, product support, limitation of liability, fees, penalties, incentives, insurance, performance bonds, subcontractor approval, change request handling, and a termination and disputes resolution mechanism.

17.2.4.8 Notify Suppliers Who Were Not Selected

If not previously done, the project manager should notify the losing suppliers in writing that he/she has selected another supplier. The content of such a notification will draw heavily from the evaluation forms and the notes taken during the meetings to compare evaluations. Many suppliers are truly interested in why they did not win and how they could improve their proposal writing or products and services being proposed. They seek an above-board, timely evaluation process. They want to be advised of any negative comments being entered into official reports and given ample opportunity for a rebuttal. They fear inflated assessments as much as poor assessments because inflated assessments help poor suppliers and hurt good suppliers.

17.2.4.9 Lessons Learned and Documentation of Proposals

Developing even a straightforward procurement process can mean that the following are accumulated: a mass of disparate information about the suppliers, the contract and its context—faxes and e-mails, Request for Information, Request for Proposals, documents of all kinds, notes of meetings jotted down on scraps of paper, contact details entered into diaries and organizers, not to mention the lessons learned and knowledge that never gets written down but survives perilously in people's memory. All this information may be highly relevant, but the next time the need to develop or plan a procurement process arises, will it be readily available to hand to make the task easier? Not unless the enterprise business has a system for recording and managing the information and a means of storing it conveniently and securely.

This closing step of the bidding process is intended to record, in the organizational process assets, the massive amount of information generated by the procurement process. The closing activities that the project manager may consider at this closing step include:

1. *Filling away at least one copy of each of suppliers' proposals and evaluation criteria.* These proposals should be kept at least until the project begins or classified according to the enterprise business records retention criteria. It is advisable to keep at least one copy of all proposals in the enterprise business organizational process assets for at least six months. While this is not a legal obligation (for commercial companies), it is possible that a losing supplier will

question the final selection decision three or four months after the award of the contract. Also, there may be good information in these losing proposals that the project manager may want to review and profit from. Quite often, a supplier may raise a valid point about procurement items requirements, contingencies, or scheduling that the project manager will want to incorporate into the "process improvement" project.

2. *Offering to notify suppliers that were not chosen.* These suppliers, especially the ones on the short list, spent time and resources on this effort and should be given a reasonable explanation as to why they were not selected.
3. *Reviewing the proposals that lost for potential information that can be incorporated into the "process improvement" project.*

17.2.5 Administer Contracts

Contract Administration involves those activities performed by the project manager (or a qualified designee) after a contract has been awarded to determine how well the portion of the project that is included within the related contract is been implemented and how well the supplier(s) performs to meet the requirements of the contract. It encompasses all dealings between the project manager and the supplier(s) from the time the contract is awarded until the work has been completed and accepted or the contract terminated, payment has been made, and disputes have been resolved. As such, contract administration constitutes that primary part of the procurement process that assures the enterprise business financial the "process improvement" project gets what it paid for.

In contract administration, the focus is on obtaining procurement items, of requisite quality, on time, and within budget. While the legal requirements of the contract are determinative of the proper course of action of the project manager in administering a contract, the exercise of skill and judgment is often required in order to protect effectively the interests of both the enterprise business and the supplier(s).

How well the project manager administers in-process contracts and discusses with suppliers their current performance determines to a large extent how well the portion of the project that is included within the related contract will be implemented and provide value to the enterprise business. By increasing attention to supplier performance on in-process contracts, project managers are reaping a key benefit: better current performance because of the active dialog between the supplier and the project manager.

The specific nature and extent of contract administration varies from contract to contract. It can range from the minimum acceptance of a delivery and payment to the contractor to extensive involvement by program, audit and procurement officials throughout the contract term. Factors influencing the degree of contract administration include the nature of the work, the type of contract, and the experience and commitment of the personnel involved.

Contract administration starts with developing clear, concise performance based statements of work to the extent possible, and preparing a contract administration plan that cost effectively measures the supplier's performance and provides documentation to pay accordingly. It culminates with close monitoring and controlling closely monitoring of the contract performance to ensure compliance and fulfillment of the contract conditions.

17.2.5.1 Prepare Contract Administration Plan

The development of a contract administration plan is essential for good contract administration. The plan should be designed to facilitate effective and efficient contract administration considering:

1. The required level of contract monitoring;
2. Contract terms and conditions related to administration;
3. Supplier performance milestones;
4. Project manager performance milestones (e.g., responding to contractor plans and other required submissions);
5. Supplier reporting procedures;
6. Supplier contract quality requirements;
7. Name, position, and authority of contract administration team members; and
8. Milestones for any reports required from contract administration team members.

The Contract Administration Plan can be simple or complex but must specify what the performance outputs of the statement of work are, and describe the methodology to conduct the inspections. This saves time and resources because the project manager is not monitoring the mundane, routine portions of the contract; instead the project manager is focusing on the major outputs of the contract.

The contract administration plan should contain a performance assurance plan as a subpart. Development of a plan is important since it provides a systematic structured method for the project manager to evaluate services and products that suppliers are required to furnish. The performance assurance plan of the contract should focus on the performance of the procurement items to be delivered by the supplier and not on the steps taken or procedures used to provide those items. It includes appropriate use of pre-planned inspections and audits, validation of complaints and random unscheduled inspections and audits.

The project team must work closely with the supplier to reach the goal of satisfying the customer in terms of cost, quality, and timeliness of the delivered procurement items.

The project manager should communicate often with the supplier, starting with a good post award conference. This part of the process ensures that everyone has the same vision of successful performance. Members of the project team should read the contract and clearly establish the expectations of the portion of the project that is included within the related contract. Everyone should understand how contract performance information will be recorded. The project manager and the supplier should agree on how often they will discuss contract performance.

Status meetings should be planned at least monthly on large contracts. The focus should be on the supplier's performance against cost, schedule, and performance goals. The project manager and the supplier should discuss the supplier's performance deficiencies, corrective actions, areas where the supplier is meeting expectations, and any project manager deficiencies. This process applies to smaller contracts as well, adjusting the meeting frequency to match the relative complexity of the contract requirement. In successful enterprise businesses, project managers are also encouraged to have an open door policy that allows suppliers to voluntarily discuss performance problems as they arise. These meetings should be a complete discussion on the supplier's performance, both good and bad, and the project manager's compliance with contract requirements.

The goal of contract administration planning is to achieve excellent contract performance that provides products or services at the best value for the "process improvement" project! This goal cannot be achieved unless the project manager does some homework:

1. Track and document contract performance closely;
2. Read and understand the supplier's cost, schedule and performance reporting data;
3. Know how well the supplier is meeting its other contract requirements;
4. Know if the enterprise business policies contributed to performance problems;
5. Actively work to eliminate enterprise business policies roadblocks to excellent performance;
6. Document discussions with suppliers in order to be able to track the steps the supplier take to improve contract performance;
7. Recognize successful efforts to improve performance

17.2.5.2 Monitor and Control Contract Performance

This is the project management process for planning a set of systematic observation techniques and activities focused on close monitoring and control of the contract performance in order to:

1. Ensure compliance and fulfillment of the contract conditions; and
2. Recommend necessary alterations to the contract objectives/goals.

Monitoring should be commensurate with the criticality of the service or task and the resources available to accomplish the monitoring. A generic form of the "Contract Performance Control" process is shown in Fig. 17.2.

Choose Control Subject

The first step of the "Contract Performance Control Process" is "Choose the Control Subject"—Control subjects are contract performance measures around which the control process is built. They provide information that answers the following question: "Would I do business with this supplier again?"

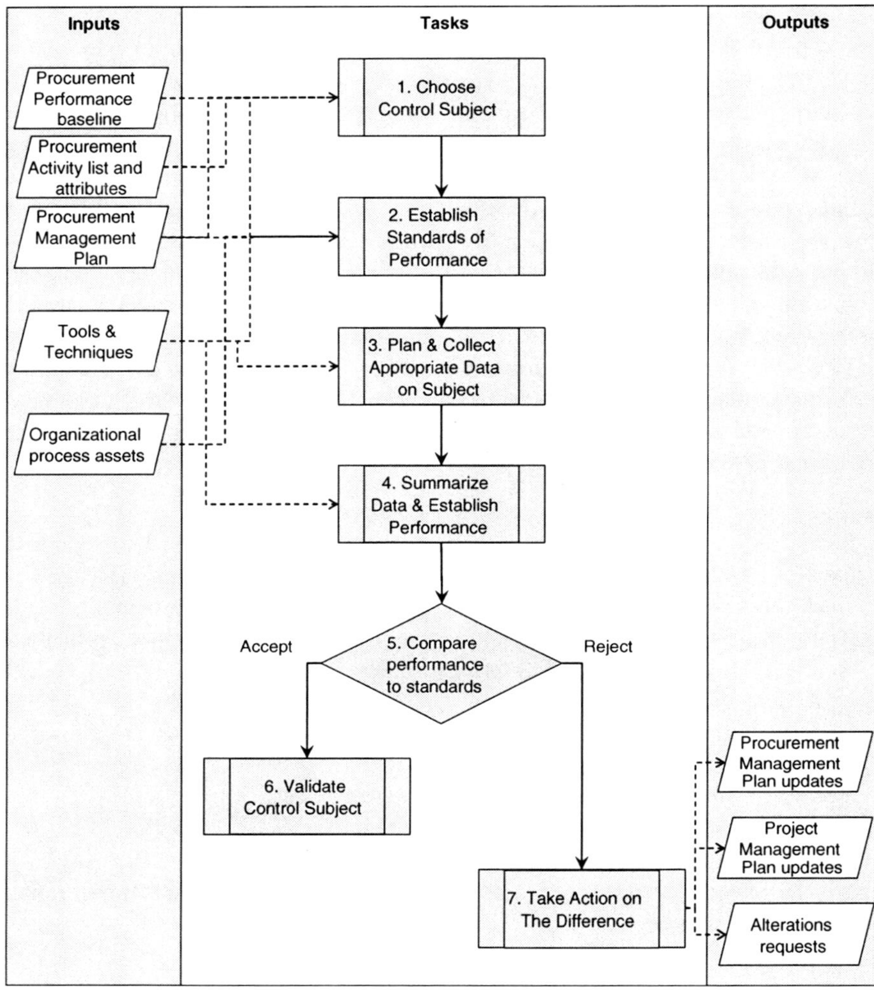

Fig. 17.2 The contract performance control process

Control subjects include the following basic elements, but are not limited to:

1. *Quality performance elements*—as defined in contract standards;
2. *Cost performance elements*—how close to cost estimates;
3. *Schedule performance elements*—timeliness of completion of interim and final milestones.
4. *Business relations*—professional behavior and overall business-like concern for the interests of the customer, including timely completion of all administrative requirements and customer satisfaction.

Establish Standard Performance

The second step of the "Contract Performance Control Process" is "Establish Standard of Performance"—It relates to collecting the standards of quality, cost, schedule and business performance baseline agreed upon in the procurement contract.

Plan and Collect Appropriate Data

The third step of the "Contract Performance Control Process" is "Plan and Collect Appropriate Data" on the chosen "Control subject"—It relates to establishing the means of tracking contract work progress in order to determine the actual performance of the contract work.

Data collection should be accomplished as an ongoing activity over a period of time where data is collected regardless of when performance analyses are carried out. The collected data aims to ensure a clear and concise record of a supplier's performance on the procurement contract, task order or other contractual document, based on a discussion with the supplier about recent performance.

Summarize Data and Establish Actual Performance

The fourth step of the "Contract Performance Control Process" is "Summarize Data and Establish Actual Performance" of the chosen "Control subject"—Performance tracking information is typically summarized weekly for shorter projects and at least monthly for larger projects through performance progress reporting. This information should indicate which deliverables have been completed and which have not.

The content and format of performance progress report is established in accordance with the enterprise business policies and should be tailored to the size, and complexity of the contractual requirements. The performance report should track four basic assessment elements—cost, schedule, technical performance (quality of procurement items delivered) and business relations including customer satisfaction, and could use five basic ratings: exceptional, very good, satisfactory, marginal, and unsatisfactory as indicated below.

1. *Exceptional*—Performance meets contract requirements and significantly exceeds contract requirements to the benefit of the portion of the project that is included within the related contract. For example, the supplier implemented innovative or business process reengineering techniques, which resulted in added value to the portion of the project that is included within the related contract. The contractual performance of the element or sub-element being assessed (i.e., the control subject) was accomplished with few minor problems for which corrective actions taken by the supplier were highly effective.
2. *Very Good*—Performance meets contractual requirements and exceeds some to the benefit of the portion of the project that is included within the related contract. The contractual performance of the element or sub-element being

assessed (i.e., the control subject) was accomplished with some minor problems for which corrective actions taken by the supplier were effective.
3. *Satisfactory*—Performance meets contractual requirements. The contractual performance of the element or sub-element being assessed (i.e., the control subject) contains some minor problems for which proposed corrective actions taken by the supplier appear satisfactory, or completed corrective actions were satisfactory.
4. *Marginal*—Performance does not meet some contractual requirements. The contractual performance of the element or sub-element being assessed (i.e., the control subject) reflects a serious problem for which the supplier has submitted minimal corrective actions, if any. The supplier's proposed actions appear only marginally effective or were not fully implemented.
5. *Unsatisfactory*—Performance does not meet contractual requirements and recovery is not likely in a timely or cost effective manner. The contractual performance of the element or sub-element being assessed contains serious problem(s) for which the supplier's corrective actions appear or were ineffective.

The ratings given by the project manager should reflect how well the supplier met the cost, schedule and performance requirements of the contract and the business relationship. Suppliers are not expected to be perfect in their execution to reach contract requirements.

A critical aspect of the assessment rating system described below is the second sentence of each rating that recognizes the supplier's resourcefulness in overcoming challenges that arise in the context of contract performance. The project manager should look for overall results, not problem free management of the contract.

The following are suggested guidelines often used for assigning ratings on a supplier's compliance with the contract performance, cost, and schedule goals as specified in the Statement of Work. The guidelines for Business Relations are meant to be separate ratings for the areas mentioned. All the areas do not need to fit the rating to give the rating for the category.

Technical Performance (Quality of Product/Service)

1. Exceptional
   - Met all performance requirements, Exceeded 20 % or more
   - Minor problems, Highly effective corrective actions, Improved performance/quality results
2. Very Good
   - Met all performance requirements, Exceeded 5% or more
   - Minor problems, Effective corrective actions
3. Satisfactory
   - Met all performance requirements
   - Minor problems, Satisfactory corrective actions

4. Marginal
   - Some performance requirements not met
   - Performance reflects serious problem, Ineffective corrective actions
5. Unsatisfactory
   - Most performance requirements are not met
   - Recovery not likely

Cost Control

1. Exceptional
   - Significant reductions while meeting all contract requirements
   - Use of value engineering or other innovative management techniques
   - Quickly resolved cost issues, Effective corrective actions facilitated cost reductions
2. Very Good
   - Reduction in overall cost, price while meeting all contract requirements
   - Use of value engineering or other innovative management techniques
   - Quickly resolved cost/price issues, Effective corrective actions to facilitate overall cost/price reductions
3. Satisfactory
   - Met overall cost/price estimates while meeting all contract requirements
4. Marginal
   - Do not meet cost/price estimates
   - Inadequate corrective action plans, No innovative techniques to bring overall expenditures within limits
5. Unsatisfactory
   - Significant cost overruns
   - Not likely to recovery cost control

Schedule (Timeliness)

1. Exceptional
   - Significantly exceeded delivery requirements, All on-time with many early deliveries to the benefit of the portion of the project that is included within the related contract.
   - Quickly resolved delivery issues, Highly effective corrective actions
2. Very Good
   - On-Time deliveries, Some early deliveries to the benefit of the portion of the project that is included within the related contract.
   - Quickly resolved delivery issues, Effective corrective actions
3. Satisfactory
   - On-time deliveries
   - Minor problems, Did not effect delivery schedule
4. Marginal
   - Some late deliveries
   - No corrective actions

5. Unsatisfactory
   - Many late deliveries
   - Negative cost impact, Loss of capability for the portion of the project that is included within the related contract.
   - Ineffective corrective actions, Not likely to recover

Business Relations

1. Exceptional
   - Highly professional, Responsive, Proactive
   - Significantly exceeded expectations
   - High user satisfaction
   - Significantly exceeded contract goals
   - Minor changes implemented without cost impact, Limited change proposals, Timely finalized and approved change proposals
2. Very Good
   - Professional, Responsive
   - Exceeded expectations
   - User satisfaction
   - Exceeded contract goals
   - Limited change proposals, Timely finalized and approved change proposals
3. Satisfactory
   - Professional, Reasonably responsive
   - Met expectations
   - Adequate user satisfaction
   - Met contract goals
   - Reasonable change proposals, Reasonable finalized and approved cycle
4. Marginal
   - Less Professionalism and Responsiveness
   - Low user satisfaction, No attempts to improve relations
   - Unsuccessful in meeting contract goals
   - Unnecessary change proposals, Untimely finalized and approved change proposals
5. Unsatisfactory
   - Delinquent responses, Lack of cooperative spirit
   - Unsatisfied user, Unable to improve relations
   - Significantly under contract goals
   - Excessive unnecessary change proposals to correct poor management
   - Significantly untimely finalized and approved change proposals

Compare Actual Performance to Standard

The fifth step of the "Contract Performance Control Process" is "Compare Actual Performance to Standards"—It includes any or all of the following activities:

1. Compare the actual performance to the baseline goals.
2. Interpret the observed difference; determine if there is conformance to the goals.
3. Decide on the action to be taken.
4. Stimulate corrective action.

Validate Control Subject

The sixth step of the "Contract Performance Control Process" is "Validate Control Subject"—It relates to acceptance decisions from the performance control results, which will indicate how well the chosen "Control subject" has fulfilled the contract objectives.

Take Action on Difference

The last step of the "Contract Performance Control Process" is "Take Action on the Difference." It relates to actuate alterations which restores conformance with contract performance goals.

The decision to issue alterations, i.e., corrective or preventive actions, is to ensure that the observed non-conformance to performance requirements are repaired and brought into compliance with contract performance requirements or specifications. Requested alterations are processed for review and approval by both the project manager and the supplier. Requested alterations can include direction provided by the supplier, or actions taken by the supplier, that the other party considers a constructive change to the contract. Since any of these constructive changes may be disputed by one party and can lead to a claim against the other party, such changes are uniquely identified and documented.

A constructive change to procurement can be an oral or written act, or an omission to act, by someone on the project who has actual or apparent authority to act, which is of such a nature that it can be construed to have the same effect as a written change order. Some of the actions which can result in constructive changes are:

1. Accelerating or delaying the period of the supplier performance;
2. Giving a supplier a specification which contains defects;
3. Changing the specification or statement of work or terms & conditions;
4. Interfering with the performance of the supplier;
5. Rejecting a supplier's deliverables even though they meet the procurement specification;
6. Adding additional and/or excessive testing;
7. Stopping and starting the work of the supplier

In issuing request for alterations, the project manager should, however, remember that "Process improvement" projects having procurements create legal relationships with their suppliers. It is the responsibility of the project manager to define the procured work. If the definition is not adequate, or changes for whatever reason interfere with a supplier's performance, the performing supplier may be entitled to extra compensation for their services.

17.2.6 Close Contracts

"Close Contracts" is the project management process which refers to verification that all administrative matters are concluded on contracts that are otherwise physically complete. A contract is considered physically complete when the supplier has completed performance and the project manager has inspected and accepted the supplies and services.

Just because the supplier has made all deliveries does not necessarily mean that a procurement contract is completed. There are often residual issues which must be addressed. Among them are the orderly close out of each procurement, the storage of all files, and in particular the settlement of all outstanding alterations/changes and residual claims the supplier may have against the project manager. Claims do not settle themselves and the passage of time works primarily in the supplier's favor, not the project manager. The project team may want to go on to exciting new assignments. But the suppliers will want to get paid for everything they did during performance.

Thus, the "Close Contracts" project management process begins when the contract has been physically complete, i.e., all services have been performed and products delivered. Closeout is completed when all administrative actions have been completed, all disputes settled, and final payments have been made. The process can be simple or complex depending on the contract type for cost-reimbursement contracts. This process requires close coordination between the project manager (or qualify designee), the finance office, and the supplier.

The contract audit process also affects contract closeout on cost-reimbursement contracts. Contract audits are required to determine the reasonableness, allowability, and allocability of costs incurred under cost reimbursement contracts. Although there is a pre-award audit of the supplier's proposal, there is a cost-incurred audit of the supplier's claim of incurred costs and a close out audit to reconcile the supplier's final claim under the contract to incurred costs previously audited. When there is a delay in completing the cost-incurred and closeout audits, project managers often cannot complete the closeout process for many cost reimbursement contracts.

The best way for any procurement to end is to have the supplier completely satisfy the statement of work, make all deliveries as specified, and comply with all provisions of the procurement contract. Without question, this is the preferred choice. But sometimes it does not happen that way. There are circumstances in which the project manager and sometimes the supplier may want to end their relationship before completion of the procurement. What are the legal ramifications of such actions? At this point the project will need to receive competent legal advice, coming either from their corporate legal counsel, or more likely from the professional procurement manager assigned to support the project manager.

There are essentially three situations in which the contractual relationships between the project manager and the supplier can be terminated early.

17.2.6.1 Termination for Cause or Default (Actions by the Supplier)

The most common cause of early termination will likely be through the actions of the supplier, which will fall short of fulfilling the critical requirements of their procurement contract. The supplier will breach their contract, which is defined as: *"Failure, without legal excuse, to perform any promise which forms the whole or part of a contract."*

In short, the supplier fails to perform the critical obligations required by their contract, and these actions provide sufficient justification for the project manager to terminate their contract.

The term "material" breach is often used, which means a large or an important breach of contract. The significance of the term "material" in this case is that the action gives the injured party a legitimate excuse to not complete their end of the bargain. However, minor, trivial, or annoying actions on the part of the supplier will not give the project manager a cause to cancel a contract. A breach of contract must be based on a significant event, going to the core of their relationship.

In the world of commerce, often the breach of contract may not have taken place, rather the breach will be anticipated, or highly probable based on conditions surrounding the supplier. Two additional definitions of breach of contract come into play, anticipatory and constructive:

1. *Anticipatory breach of contract.* Such occurs when the "promisor" without justification and before he has committed a breach makes a positive statement to "promisee" indicating he will not or cannot perform his contractual duties.
2. *Constructive breach.* Such breach takes place when the party bound to perform disables himself from performance by some act, or declares, before the time comes, that he will not perform.

The effect of a supplier breach of contract can be costly to the supplier depending on the egregious nature of their actions. In such cases the supplier may not be able to recover all of their costs incurred. The supplier is likely to be entitled to no profit for their work performed. Of greater consequence, the supplier will likely be liable for compensatory damages the project manager may have to incur as they place the same procurement with another supplier in order to complete the project. In this case the supplier may be liable to the project manager for the costs of taking the same work to another firm for performance.

17.2.6.2 Termination for the Convenience (of the Buyer)

If the project manager executes a termination for its convenience the supplier must be notified, and once notified must take positive steps to minimize the incurrence of further liabilities. The project manager must then negotiate with the supplier to make the supplier financially whole, to cover all their reasonable expenses, and pay a reasonable profit for the supplier effort up to the termination.

Terminations for the convenience of a project manager must be done in good faith. There must be a legitimate basis for the termination for convenience. Simply to secure a lower price from a new supplier would likely not be considered a good faith termination by the project manager.

17.2.6.3 Absolute Right to Terminate the Agreement (by Either the Project Manager or the Supplier)

In unusual circumstances, the parties to a contractual relationship will sometimes (rarely) insert a contract provision allowing either of the parties to cancel their contract, by simply giving notice to the other party. This provision sort of negates the very purpose of a contract. Such provisions are used infrequently, sometimes in employment contracts or contracts for professional services. Often the only stipulation is that termination will take place after a specified number of days have passed after notification to terminate.

17.2.6.4 Identifying and Recording Lessons Learned

The last issue to be discussed in the systematic closeout of any procurement contract is something we know we should do, but rarely ever take the time. That is the final position of the project manager, and the project team, describing for the benefit of "future generations" just what went well, and what could have been handled perhaps better on the project being completed. What could we have done better, and should we do differently on the next similar procurement contract. How should we deal with a particular supplier, and perhaps, should we use this seller again on the next project?

A lessons' learned session focuses on identifying procurement successes and failures, and includes recommendations to improve future performance on procurement contracts. During the life cycle of the procurement contract, the project team and key stakeholders identify lessons learned concerning the technical, managerial, and process aspects of the procurement work. The lessons learned are compiled, formalized, and stored through the project's duration.

The focus of lessons learned meetings can vary. In some cases, the focus is on strong technical or product development processes, while in other cases, the focus is on the processes that aided or hindered performance of the work. Teams can gather information more frequently if they feel that the increased quantity of data merits the additional investment of time and money.

Lessons learned provide future procurement teams with the information that can increase effectiveness and efficiency of project management. In addition, phase-end lessons learned sessions provide a good team-building exercise. Project managers have a professional obligation to conduct lessons learned sessions for all procurement contracts with key internal and external stakeholders, particularly if the procurement contract yielded less than desirable results. Some specific results from lessons learned include:

1. Update of the lessons learned knowledge base
2. Input to knowledge management system
3. Updated corporate policies, procedures, and processes
4. Improved business skills
5. Overall product and service improvements
6. Updates to the risk management plan

Develop Communication Management Plan

18

Communication is the activity of conveying information. It requires a sender, a message, and an intended recipient. It also requires that the communicating parties share an area of communicative commonality. The communication process is complete once the receiver has understood the message of the sender. Feedback is an essential part of communication. Communication ranks high among the factors leading to the success of a "process improvement" project. In particular, what is required is constant, effective communication among everyone involved in the project or affected by the "process to be improved." Projects are made up of people getting things done. Getting the right things done in the right way requires communication among all the stakeholders. This chapter presents the project management processes for ensuring that the right people have the right information to make the necessary decisions and carry them out.

18.1 Project Communication

Project communication is the exchange of project-specific information with the emphasis on creating understanding between the sender and the receiver. Effective communication in a "process improvement" project is one of the most important factors contributing to the success of the project. Here, the project team must provide timely and accurate information to all stakeholders. During the course of a project, members of the project team prepare information in a variety of ways to meet the needs of project stakeholders. These stakeholders in return, provide feedback to the project team members.

Project communication includes general communication between team members but is more encompassing. It utilizes the Work Breakdown Structure (WBS) as framework, it is customer focused, it is limited in time, it is product focused with the end in mind, and it involves all levels of the enterprise business. From the process improvement perspective, for each WBS element, there are:

1. Suppliers who provide inputs needed for the WBS element
2. Task managers who are responsible for delivering the WBS element
3. Customers who receive the products of the WBS element

A. van Aartsengel and S. Kurtoglu, *Handbook on Continuous Improvement Transformation*, DOI 10.1007/978-3-642-35901-9_18, © Springer-Verlag Berlin Heidelberg 2013

Table 18.1 The S.I.P.O.C.

Previous WBS Task Manager	Area of communicative commonality	WBS Task Manager	Area of communicative commonality	Next WBS Task Manager
S Suppliers	I Inputs	P Process	O Outputs	C Customers
Suppliers are systems, people, organizations, or other sources of the materials, information, or other resources that are consumed or transformed in the process.	Inputs are materials, information, and other resources provided by the suppliers that are consumed and transformed in the process.	The process is a set of logically related discrete elements (tasks, actions, or steps) taken in order to achieve a particular end.	Outputs are the outcomes (products or services) produced by the process and used by the customers.	Customers are the persons, group of people, companies, systems, and downstream processes, recipients of the process outcomes.

Suppliers must communicate with the task managers, and the task managers must communicate with suppliers and customers. The supplier is often the task manager for an earlier deliverable in the project lifecycle; the customer may be a task manager for a later deliverable. Good project communication practice includes notifying the next task manager in the project delivery chain about when to expect a deliverable. The supplier and customer may also be the functional manager. By considering the process associated with the WBS element, a very effective diagram that depicts this flow of information is the Suppliers-Inputs-Process-Outputs-Customers (S.I.P.O.C.) diagram, illustrated in Table 18.1. Working from the right letter of its acronym, the S.I.P.O.C. identifies the customers, the outcomes or the process, the inputs to the process and the suppliers.

As indicated in a previous section, the S.I.P.O.C. diagram is built from the inside-out, starting at the center with a four to seven steps high-level map of the WBS associated process, followed by an identification of the process outcomes and the associated customers, and finishing with identification of the inputs to the process and the sources of these inputs; i.e., their suppliers. As such, it established the channels of communication between the task manager and supplier and the task manager and customer.

It is important to note the established communication should be reciprocal between the task manager and supplier and the task manager and customer; i.e., although communication is the responsibility of the task manager, the customer/supplier should always validate expected deliverable dates.

The project communication plan is a part of the overall project plan. It builds on the project work plan and the WBS interfaces (areas of communicative commonality), which shows:

1. What will be produced on the project—the deliverable including the WBS
2. Who will produce it
3. When will it be produced

Using the WBS as framework, project communication is the responsibility of everyone on the project team. The project manager, however, is responsible to develop the Project Communication Management Plan with the input from the task managers and stakeholders. A task manager responsible for a deliverable needs to know why the customer wants the deliverable, what features they want, how long it will take, and how they want to receive it. Within the communication process, the task manager must tell their customer exactly when to expect the deliverable. If that deliverable is linked to a WBS element on the critical path, it is even more important that task manager informs their internal customer when the deliverable will arrive. The recipient functional manager must have their staff ready to start work immediately after it arrives.

The task manager must ensure that internal customers know about any changes in the delivery date. This allows the recipient functional manager to schedule their resources accordingly. The task manager must follow up with the customer of each deliverable. The task is not complete merely because the final product is delivered to the customer. The task manager must contact the customer to confirm that the deliverable met his/her needs and expectations. The task manager should enter feedback that others might use in future projects into the enterprise business organizational process assets (lessons learned database and training materials).

18.2 Project Communication Management

Project Communications Management is the project management knowledge area that employs the processes required to ensure timely and appropriate generation, collection, distribution, storage, retrieval, and ultimate disposition of project information. These processes provide the critical links among people and information that are necessary for successful project communications. Project managers use project communication management to:

1. Develop a communication plan for the project;
2. Distribute information via the methods that reach customers most effectively;
3. File data using the enterprise business Project Filing System;
4. Archive records in accordance with the enterprise business Records Retention policies.

In accordance with the Project Management Body of Knowledge guide lines, the constituent processes used during the development of the project communication management plan include the following:

1. Communications Planning—determining the information and communications needs of the project stakeholders.
2. Information Distribution—making needed information available to project stakeholders in a timely manner.

3. Stakeholders Relationship Management—managing communications to satisfy the requirements of and resolve issues with project stakeholders.
4. Performance Reporting—collecting and distributing performance information.

This includes status reporting, progress measurement, and forecasting.

These four constituent processes interact with each other and with the project management processes in the PDSA "Process Groups." Each aspect of executing any of these can involve effort from one or more persons, based on the needs of the project. Each aspect occurs at least once in every "process improvement" project and occurs in one or more project phases.

18.2.1 Communications Planning

This is the project management process for identifying and specifying the areas of communicative commonality (illustrated in Table 18.1) and associated content (including senders and receivers) within these areas of communicative commonality. This communication planning process also specifies suitable means of conveying the identified information.

Communications planning starts with providing answers to the following these questions for each WBS element:

1. Who is involved in the WBS communication process (i.e. senders and receivers)? The identified stakeholders, such as Project Team Members, project sponsors, project management and staff, customer management and staff, and external stakeholders.
2. What is being communicated? The message; the content of the area of communicative commonality or the information being communicated.
3. When is the information communicated? Weekly, monthly, quarterly, as needed, or as identified.
4. How is the information conveyed? In a meeting, a memorandum, an email, a newsletter, a presentation, etc.
5. Who will provide the information being communicated?

The "Communications Planning" process builds on the project WBS list, the project schedule, customer and stakeholder register, the enterprise business environment factors, and the organizational process assets. It should culminate with the release of a formal document called a Project Communication Plan.

A project communications plan describes the information to be disseminated to all project stakeholders to keep them regularly informed of the progress of the project. It is the written strategy for getting the right information to the right people at the right time. The stakeholders identified on the statement of work, the organization chart, and the responsibility matrix are the audience for most project communication. But on every project, stakeholders participate in different ways, so each has different requirements for information.

A clear communications plan is vital to the success of the project, as it helps ensure that all project resources and stakeholders are working towards the same project objectives, and that any hurdles are overcome in a planned and informed manner. The project communication plan must contain the information needed to successfully manage project deliverables. It includes the following:

1. Brief introduction and background, which provide answers to the question, "Why does the project need communication plan?"
2. A list of the project sponsor, project manager, project team members, and other key stakeholders.
3. Methods of communications to be used to convey information, including formal meetings to be held (who, what, when, how).
4. Project reporting information, which provide answers to the question, "How will project performance be collected and distributed to the internal and external project stakeholders?"
5. Stakeholders (internal and external) information requirements analysis, which is designed to help the project team analyze internal and external stakeholder needs by gathering the following information from each stakeholder:
   - Goals for the "process improvement" project. What is each stakeholder's desired outcome for the project? The project manager should ensure at the start that there is a consistent vision for the project.
   - Preferred methods of communication. Project team members will use this information as a means to meet individual communication needs. If the team cannot reasonably communicate through each stakeholder's preferred medium, the team needs to negotiate a method to ensure that each stakeholder receives and understands the project communication.
   - Preferred method for recognizing performance of the team, within the constraints of what is achievable. The project team uses this information to plan appropriate celebrations at the completion of each project component.
6. Communication matrix, which is used to track project performance by project component and WBS element. The WBS product list is the input. It includes the WBS codes, WBS titles, sub-products. To complete the communication matrix, the project team indicates if the sub-product is required, who produces it, who receives it, the method of transmittal, and the date submitted; that is:
   - A schedule of the communication events, methods and release dates;
   - A matrix highlighting the resource involved in each communication event;
   - A clear process for undertaking each communication event within the project.

This plan should be coordinated and endorsed by all key functions and stakeholders supporting the project. The project communication plan is a framework and should be a living, evolving document that can be revised when appropriate. The communication plan is part of the project management plan. There are two key steps to be followed to develop a project communication plan:

1. Identify the communications requirements;
2. Build a communications schedule.

18.2.1.1 Identify the Communications Requirements

The first step towards creating an effective project communications plan is to list the project stakeholders and their interests. For instance, a project sponsor will be

Table 18.2 Stakeholders communications requirements

Stakeholder	Information requirement
Project Sponsor	Project status information (schedule, budget and scope) Understanding of critical project risks and issues Information required to approve each project phase.
Project Review Group	Project status information (schedule, budget and scope) Detailed knowledge of important risks and issues Information regarding proposed project changes (for approval).
Project Manager	Detailed project status information (schedule, budget and scope) Understanding of current project deliverable quality Detailed knowledge of all risks, issues and change requests.
Project Leader	Project activity and task status information Day-to-day knowledge of issues and risks identified.
Project Member	Status of the activities and tasks they are dependent on Awareness of events which may affect their ability to undertake their role.
Quality Manager	Progress of each deliverable against quality standards and criteria set Detailed understanding of all quality issues for resolution.
Procurement Manager	...
...	...
Other Bodies	...

interested in the overall progress of the project, whereas an external body may be concerned with legislative or regulatory compliance. For all stakeholders identified and recorded in the customer and stakeholder register, describe the information required to keep them appropriately informed of the progress of the project. Table 18.2 below provides examples.

18.2.1.2 Build a Communications Schedule

Building a communication schedule is done by describing each communication event, including its purpose, method and frequency by completing Table 18.3. Completion of Table 18.4 should help the project manager identify the people participating in each communication event and their roles. The unique identifier (ID) can be used to link events listed in Table 18.3 to the participating parties listed here. It is also important to list any assumptions made during the communications planning process. For example, it might be assumed that:

1. The communications tools will be provided as required;
2. Adequate communications resources will be available when needed;
3. The communications staff has the required expertise.

Table 18.3 Stakeholders communications schedule

ID	Event	Description	Purpose	Method	Frequency	Date
1.1.	Project Team Meeting	Meeting involving all team members to discuss the work in progress, recently completed, coming up.	Keep the team informed of the project status and ensure that issues, risks or changes are raised.	Verbal	Weekly	d/m/y
1.2.	Quality review meetings	Regular meetings involving the quality manager and quality review staff to ascertain the level of quality of all project deliverables.	To ensure that quality issues are identified early, thereby providing time to enhance the quality of each deliverable and meet the quality criteria.	Verbal	Monthly	d/m/y
1.3.	Stage-gate review meetings	Formal meetings held at the end of each phase to identify the overall status of the project, the quality of the deliverables produced and any outstanding risks, issues or changes	To control the progress of the project through each phase in the project life cycle, thereby enhancing its likelihood of success	Verbal	Weekly	d/m/y
1.4.	Change approval group meetings	Formal meetings held regularly to review change requests	To provide a formal process for the approval of project changes	Verbal	Every two weeks	d/m/y
1.5.	Customer acceptance meetings	Held with the customer to obtain final acceptance of a set of completed project deliverables	Provide a controlled process for the acceptance of deliverables and ensure customer requirements are met	Verbal	Following deliverable's completion	d/m/y
1.6.	Supplier performance meetings	Regular meetings with each supplier to discuss performance issues and product delivery status	To provide a forum within which to assess supplier performance and resolve supplier issues	Verbal	Monthly	d/m/y
1.7.	Status reports	Reports providing the status of the schedule, budget, risks and issues	To keep all key project stakeholders informed of the status of the project	Status report	Weekly	d/m/y
...	...	...	...	...	...	d/m/y

Table 18.4 Stakeholders communications matrix

ID	Project Sponsor	Project Manager	Project Leader	Project Member	Quality Manager	Procurem ent Manager	Comm. Manager	Other Bodies
1.1.	–	A	R	R	R	R	R, M	R
1.2.	–	R	R	–	A	–	M	R
1.3.	A	R	–	–	–	–	M	–
1.4.	A	R	–	–	–	–	M	–
1.5.	R	A	R	–	R	–	M	R
1.6.	–	R	–	–	–	A	M	–
1.7.	R	A	R	–	R	R	M	–
...	...	...	...	...	...			...

Key:
A = Accountable for the communications event, develops and distributes materials and facilitates meetings.
R = Receives communications materials provided, takes part in meetings.
M = Monitors communications process and provides feedback.

In building a communication schedule, the project manager should also list any risks identified during the communications planning process. For example:

1. Key communications staff leaves during the life of the project;
2. The requirements for communication change during the project;
3. Communications are not undertaken effectively.

The project communications plan document is created by collating all of the materials previously listed.

The Communication Plan should be reviewed continuously throughout the project lifecycle to ensure that it remains effective. Periodically, the project manager must solicit feedback from the project stakeholders to ascertain if the project communication is sufficient to suit the stakeholder's needs. It might happen that during certain phases of the project, project stakeholders will need greater detail or more frequent information. During other phases, certain stakeholders may need summary information, or may request notification only if problems arise.

After the project communications plan has been agreed, the communications management process should be invoked to ensure that all communication events are undertaken in a clear and coordinated manner.

18.2.1.3 Information Distribution

The "Information Distribution" project management process focuses on taking the facts and happenings in regards to the "process improvement" project and disseminating this information to all of the relevant parties, with a particular focus on providing information to those who have a financial stake in the ultimate outcomes of the project.

Proper information distribution makes information available to project stakeholders in a timely manner. Following the communication plan ensures that all members of the project team are aware of their responsibilities to communicate with external stakeholders. The more information stakeholders have regarding a project or deliverable, the less likely last minute conflicts, changes, or complaints will affect a project. Team members can improve overall project communication by adhering to the following communication guidelines:

1. Awareness
   - Base communication strategies on stakeholder needs and feedback.
   - Ensure that communication is shared in a timely manner.
2. Content
   - Advocate open, honest, face-to-face, two-way communication.
   - Create an environment where project team members and other stakeholders can constructively challenge behavior and ideas.
3. Context
   - Remember that communication is two-way. Listen as well as deliver the message.
   - Involve senior management when appropriate.
4. Communication flow
   - Coordinate communication with project milestone events, activities, and results.

   - Include key stakeholders in developing an interest-based conflict management process.
5. Effectiveness
   - Conduct regular assessments of the communication plan and process.
   - Communication must focus on the customer.
6. Format and media
   - Take advantage of existing communication vehicles and opportunities.
   - The project team has a variety of methods to share information.

The methods of information dissemination can come in means including regularly scheduled conferences and or meetings, regularly scheduled conference calls in which some or all members of the project team participate, informal written communications such as periodic updates via email and of other short form, less formal means of communications, as well as formal reports that may or may not have been requisite to the completion of the project. Information distribution is essential to assuring that the financial stakeholders are fully aware of the progress throughout as it helps to assure no surprises arise at the time that deliverables are expected to be final.

18.2.1.4 Stakeholders Relationship Management

"Stakeholder Relationship Management" is the project management process which refers to managing communications to satisfy the needs of, and resolve issues with, project stakeholders. As the name suggest, it is essentially stakeholder relationship management as it is the relationship and not the actual stakeholder groups that are managed.

Successful achievement of the project activities inevitably require project stakeholders to contribute to realization of the agreed outcomes. The project stakeholders may be impacted upon by the work to accomplish the project activities or by the project outcomes. The management of their needs, requirements and expectations captured and kept in the customer and stakeholder register must be considered as a critical part, beyond the iron triangle of time, cost and performance, required for a successful realization of a "process improvement" project objectives.

The stakeholder analysis performed at the project initiation is only useful if it is used. Stakeholder management is where analysis and practice meet. It allows using the analysis to help gain support and buy-in for the "process improvement" project.

As indicated in the project initiation section, once identification of stakeholders and their concerns have been performed, the project manager has to respond to their concerns in some way—at least by acknowledging them, whether they can be satisfied or not—and the project manager must find a way to move forward with as much support from stakeholders as he/she can muster.

As we mention already, it is not practical, and usually not necessary, to engage with all stakeholder groups with the same level of intensity all of the time during the course of the project. Being strategic and clear as to whom you are engaging with and why, before jumping in, can help save both time and money. This requires prioritizing the stakeholders and, depending on who they are and what interests they

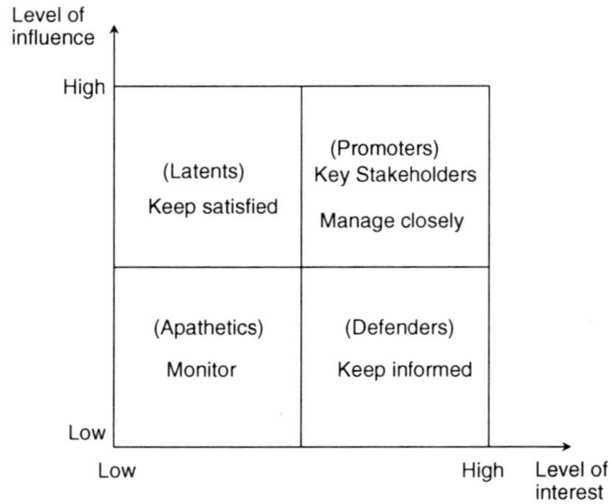

Fig. 18.1 Influence/interest grid for stakeholder prioritization

might have, figuring out the most appropriate ways to engage them. Stakeholder analysis should assist in this prioritization by assessing the significance of the project to each stakeholder group from their perspective, and vice versa. It is important to keep in mind that the situation might be dynamic and that both stakeholders and their interests might change over time, in terms of level of relevance to the project and the need to actively engage at various stages.

Stakeholder analysis (stakeholder mapping) is a way of determining who among stakeholders can have the most positive or negative influence on the "process improvement" project, who is likely to be most affected by the "process improvement" project, and how one should work with stakeholders with different levels of interest and influence.

Most methods of stakeholder analysis or mapping divide stakeholders into one of four groups, each occupying one space in a four-space grid—the influence versus interest grids—as illustrated in Fig. 18.1.

As shown in Fig. 18.1, low to high influence over the "process improvement" project runs along a line from the bottom to the top of the grid, and low to high interest in the "process improvement" project runs along a line from left to right. Both influence and interest can be either positive or negative, depending on the perspectives of the stakeholders in question. The lines describing them are continuous, meaning that people can have any degree of interest from none to as high as possible, including any of the points in between. The "key stakeholders" would generally appear in the upper right quadrant.

The purpose of this diagram is to help in understanding and determining what kind of influence each stakeholder has on the "process improvement" project and its potential success. That knowledge in turn can help deciding on how to manage stakeholders—how to marshal the help of those that support the "process improvement" project, how to involve those who could be helpful, and how to convert—or at least neutralize—those who may adversely affect the "process improvement" objectives.

The first step in developing the stakeholder relationship management is to understand clearly where each stakeholder lies in the influence/interest grid, shown in Fig. 18.1 which was developed at the project "initiation" stage. Someone that has both a major interest in and considerable influence over the enterprise business and/or the "process improvement" project would go in the upper right-hand corner of the upper right quadrant of the stakeholder influence/interest grid. Stakeholders with neither influence nor interest would go in the lower left-hand corner of the lower left quadrant. Those with a reasonable amount of influence and interest would go in the middle of the upper-right quadrant, etc. Eventually, the grid will be filled in with the names of stakeholders occupying various places in each of the quadrants, corresponding to their levels of power and interest.

The next step is to decide who needs the most attention. These are stakeholders who can be most helpful, i.e., those with the most influence. Influential people with the highest interest are most important, followed by those with influence and less interest. Those in the lower right quadrant—high interest, less influence—come next, with those with low interest and low influence coming last.

The names in parentheses in Fig. 18.1 are another way to define the same stakeholder characteristics in terms of how they relate to the effort.

Promoters—Promoters have both great interest in the "process improvement" project and the power to help make it successful (or to derail it). They are the most important amongst stakeholders. They are the ones who can really make the "process improvement" project move forward, and they care about and are invested in the project. If they are positive, they need to be cultivated and involved. The project manager must find work for them (not just tasks) that they will enjoy, and that contribute substantively to the "process improvement" project, so they can feel responsible for part of what is going on. The project manager must attention to their opinions, and accedes to them where it is appropriate. If their ideas are not acted on, the project manager should make sure that they know why and why an alternative seems like the better course. As much as possible, promoters must be considered integral parts of the project team. When stakeholders who could be promoters have an adverse effect on the project objectives, the major task of the project manager is to convert them as they can become the most powerful opponents to "process improvement" project, and could make it impossible to succeed. Thus, they need to be treated as potential allies, and their concerns should be addressed to the extent possible without compromising the effort.

Defenders—Defenders have a vested interest and can voice their support in the "process improvement" project, but have little actual influence on the "process improvement" project in any way.

Latents—high influence/low interest. These are stakeholders largely unaffected by the "process improvement" project that could potentially be extremely helpful, if they could be convinced that the "process improvement" project is important either to their own self-interest or to the greater good. The project manager must approach and inform them, and keep communication with them over time. Offer them opportunities to weigh in on issues relating to the "process improvement" project, and demonstrate to them how the "process improvement" project will have a positive effect on issues they are concerned with. If shifted over to the promoter category, they will become valuable allies. Once again, there is the possibility that these latent stakeholders could be negative and oppositional to the "process improvement" project. If that is the case, it might be best not to stir a sleeping dragon. If they are not particularly affected by or concerned about the "process improvement" project, even if they disapprove of it, the chances are that they will simply leave it and you (as project manager) alone, and it might be best that way. If they begin to voice opposition, then your first attempt might be at conversion or neutralization, rather than battle. If that does not work, then you might have to fight.

Apathetics—those with low interest and low influence. These stakeholders simply do not care about the "process improvement" project one way or the other. They may be stakeholders only through their membership in a group or their position in the enterprise business; the "process improvement" project may in fact have little or no impact on them. As a result, they need little or no management. The project manager should keep them sporadically informed by newsletter or some similar device, and do not offend them, and they will not bother you or get in the way.

Another way to look at stakeholder relationship management—remembering that all the people and groups here are stakeholders, those who can affect and are affected by the "process improvement" project—is that the most important stakeholders are those most dramatically affected. Some of those, at least before the effort begins, may be in the lower left quadrant of the grid. They may be too involved in trying to survive—financially or strategically—from day to day to think about an effort to change their situation.

Therefore, the stakeholder relationship management depends on what the purpose is in involving stakeholders. If the purpose is to marshal support for the "process improvement" project, then each group—each quadrant of the grid—calls for one kind of attention. If the purpose is primarily participatory, then each quadrant calls for another kind of attention.

The methods of communications identified for each stakeholder in the communications management plan are utilized during stakeholder relationship management. Face-to-face meetings are the most effective means for communicating and resolving issues with stakeholders. When face-to-face meetings are not warranted or practical (such as on international projects), telephone calls, electronic mail, and other electronic tools are useful for exchanging information and dialoguing.

18.2.1.5 Performance Reporting

The performance reporting process involves the collection of all baseline data, and distribution of performance information to stakeholders to keep them informed of the many variables that describe how the project is proceeding as compared to the plan and baseline data. Generally, this performance information includes how resources are being used to achieve the project objectives. Performance reporting should generally provide information on scope, schedule, cost, and quality. Many projects also require information on risk and procurement. Performance reports may be prepared comprehensively or on an exception basis using the enterprise business established reporting system.

Performance Report Types

There are five types of performance report that are often used in enterprise businesses:

1. Current period reports
2. Cumulative reports
3. Exception reports
4. Stoplight reports
5. Variance reports
6. A3 Reports

Current Period Reports

These reports cover only the most recently completed period. They report progress on those activities that were open or scheduled for work during the period. Reports might highlight activities completed and variance between scheduled and actual completion dates. If any activities did not progress according to plan, the report should include a discussion of the reasons for the variance and the appropriate corrective measures that will be implemented to correct the schedule slippage.

Cumulative Reports

These reports contain the history of the project from the beginning to the end of the current report period. They are more informative than the current period reports because they show trends in project progress. For example, a schedule variance might be tracked over several successive periods to show improvement. Reports can be at the activity or project level.

Exception Reports

Exception reports report variances from the established plan. These reports are typically designed for senior management to read and interpret quickly. Reports that are produced for senior management merit special consideration. Senior managers do not have a lot of time to read reports that tell them that everything is on schedule and there are no problems serious enough to warrant their attention. In such cases, a one-page, high-level summary report that says everything is okay is usually sufficient. It might also be appropriate to include a more detailed report as an attachment for those who might wish to read more detail. The same might be true of exception reports. That is, the one-page exception report tells senior managers about variances from plan that will be of interest to them, while an attached report provides more details for the interested reader.

Stoplight Reports

Stoplight reports are a variation that can be used on any of the previous report types.

When the project is on schedule and everything seems to be moving as planned, put a green sticker on the top right of the first page of the project status report.

This sticker will signal to senior managers that everything is progressing according to plan, and they need not even read the attached report.

When the project has encountered a problem—schedule slippage, for example—you might put a yellow sticker on the top right of the first page of the project status report. That is a signal to upper management that the project is not moving along as scheduled but that you have a get-well plan in place. A summary of the problem and the get-well plan may appear on the first page, but they can also refer to the details in the attached report. Those details describe the problem, the corrective steps that have been put in place, and some estimate of when the situation will be rectified.

Red stickers placed on the top right of the first page signal that a project is out of control. Red reports are to be avoided at all costs because they mean that the project has encountered a problem, and there is no get-well plan or even a recommendation for upper management available. Senior managers will obviously read these reports because they signal a major problem with the project. On a more positive note, the red condition may be beyond the project manager's control.

Variance Reports

Variances are deviations from plan. They are the algebraic difference between what was planned and what actually occurred. As such, variances can be positives or negatives.

Positive Variances—Positive variances are deviations from plan that indicate that an ahead-of-schedule situation has occurred or that an actual cost was less than a planned cost. This type of variance is good news to the project manager, who would rather hear that the project is ahead of schedule or under budget. Positive variances bring their own set of problems, which can be as serious as negative variances. Positive variances can allow for rescheduling to bring the project to completion early, under budget, or both. Resources can be reallocated from ahead-of-schedule projects to behind-schedule projects.

Not all the news is good news, though. Positive variances also can result from schedule slippage! Consider the allowed funds. Falling short of funds means that not all costs were expended, which may be the direct result of not having completed work that was scheduled for completion during the report period.

On the other hand, if the ahead-of-schedule situation is the result of the project team's finding a better way or a shortcut to completing work, the project manager will be pleased. This situation may be a short-lived benefit, however. Getting ahead of schedule is great, but staying ahead of schedule presents another kind of problem. To stay ahead of schedule, the project manager will have to negotiate alterations/changes to the resource schedule. Given the aggressive project portfolios in place in most companies, there is not much reason to believe that resource schedule changes can be made. In the final analysis, being ahead of schedule may be a myth.

Negative Variances—Negative variances are deviations from plan that indicate that a behind-schedule situation has occurred or that an actual cost was greater than a planned cost.

Being behind schedule or over allocated funds is not what the project manager or his reporting manager wants to hear. Negative variances, just like positive variances, are not necessarily bad news. For example, you might have overspent because you accomplished more work during the report period than was planned. But in overspending during this period, you could have accomplished the work at less cost than was originally planned.

In most cases, negative time variances affect project completion only if they are associated with critical path activities or if the schedule slippage on noncritical path activities exceeds the activity's total float. Variances use up the float time for that activity; more serious ones will cause a change in the critical path.

Negative cost variances can result from uncontrollable factors such as cost increases from suppliers or unexpected equipment malfunctions. Some negative variances can result from inefficiencies or error.

Variance Reports—Variance reports do exactly what their name suggests they report differences between an established plan and the actual performance. The report has three columns:

1. The planned number
2. The actual number
3. The difference, or variance, between the two

A variance report can be provided in one of two formats:

1. The first is numeric and displays a number of rows with each row giving the actual, planned, and variance calculation for those variables in which such numbers are needed. Typical variables that are tracked in a variance report are schedule and cost. For example, the rows might correspond to the activities open for work during the report period and the columns might be the planned cost to date, the actual cost to date, and the difference between the two. The impact of departures from plan is signified by larger values of this difference (the variance).
2. The second format is a graphical representation of the numeric data. It might be formatted so that the plan data is shown for each report period of the project and is denoted with a curve of one color; the actual data is shown for each report period of the project and is denoted by a curve of a different color. The variance need not be graphed at all because it is merely the difference between the two curves at some point in time. One advantage of the graphic version of the variance report is that it can show the variance trend over the report periods of the project, while the numeric report generally shows data only for the current report period.

Typical variance reports are snapshots in time (the current period) of the status of a "control subject" being tracked. Most variance reports do not include data points that report how the project reached that status. Project variance reports can be used to report project as well as activity variances. For the sake of the managers who will have to read these reports, we recommend that one report format be used regardless of the variable being tracked. Top management will quickly become comfortable with a reporting format that is consistent across all projects or activities within a project. It will make life a bit easier for the project manager, too.

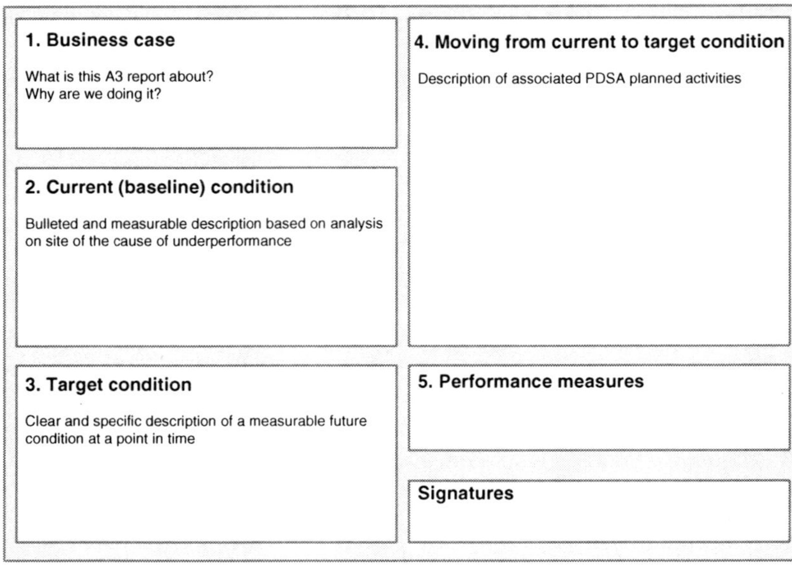

Fig. 18.2 Example of A3 report template

A3 Reports

The A3 Report is a Toyota-pioneered practice of getting the problem, analysis, corrective actions, and action plan reported on a single sheet of large (A3) paper. It summarizes observations or current status of the "process improvement" project. It includes:

1. Target condition
2. Proposals
3. Plans
4. Key points from reflections

The format of the A3 report generally mirrors the steps of the PDSA model within the "process improvement" project. They are written in a succinct, bulleted, and visual style that tells a story with data.

The A3 report of the current status of a "process improvement" project is the "story" itself that is built up and presented on the single page. It provides a way of communicating a proposed idea or a problem on a single A3-sized sheet of paper, encouraging efficiency, clarity, and disciplined thinking. The format of an A3 report varies depending on the purpose and theme. Figure 18.2 presents the typical sections of an A3 report.

Each section of the A3 report builds upon the previous one. The better you define the business case, the better you can assess the current (baseline) condition. The better you assess the current condition, the better you can develop an appropriate target condition. And so on. This sequence of five sections is a tool that gives members of the "process improvement" team a routine and mental pattern for approaching any process or situation, and to help them learn the "process improvement" pattern. The sections distill part of the "process improvement" pattern down to a point where it becomes accessible and usable by anyone.

Although the A3 report is typically on one page, there can be additional pages of backup documentation.

How and What Performance Information to Report

As input to each of these report types, activity managers and the project manager must report the progress made on all of those activities that were open for work (in other words, those that were to have work completed on them during the report period) during the period of time covered by the status report. Recall that the planning estimates of activity duration and cost were based on little or no information. Now that some work on the activity have completed, the project manager should be able to provide a better estimate of the duration and cost exposure. This reflects itself in a re-estimate of the work remaining to complete the activity. That update information should also be provided. The following is a list of what should actually be reported.

Determine a Set Period of Time and Day of Week

The project team will have agreed on the day of the week and time of day by which all updated information is to be submitted. A project administrator or another team member is responsible for seeing that all update information is on file by the report deadline.

Report Actual Work Accomplished During this Period

What was planned to be accomplished and what was actually accomplished are two different things. Rather than disappoint the project manager, activity managers are likely to report that the planned work was actually accomplished. Their hope is to catch up by the next report period. Project managers need to verify the accuracy of the reported data rather than simply accept it as accurate. Spot-checking on a random basis should be sufficient. If the activity was defined according to the completion criteria, as is discussed in a previous section, verification should not be a problem.

Record Historical & Re-estimate Remaining (In-Progress Work Only)

Two kinds of information are reported:

1. All work completed prior to the report deadline is historical information. It will allow variance reports and other tracking data to be presented and analyzed.
2. The other kind of information is futures-oriented. For the most part, this information is re-estimates of duration and cost and estimates to completion (both cost and duration) of the activities still open for work.

Report Start and Finish Dates

These are the actual start and finish dates of activities started or completed during the report period.

Record Days of Duration Accomplished and Remaining

"How many days have been spent so far working on this activity?" is the first number reported. The second number is based on the re-estimated duration as reflected in the time-to-completion number.

Report Resource Effort Spent and Remaining

Whereas the preceding numbers report calendar time, these numbers report labor time over the duration of the activity. Resource effort can be expressed in hours or day spent and remaining effort can be expressed in-progress work only. Thus there are two numbers to consider here. One reports labor completed over the duration accomplished. The other reports labor to be spent over the remaining duration.

Report Percent Complete

Percent complete is the most common method used to record progress because it is the way we tend to think about what has been done in reference to the total job that has to be done. Percent complete is not the best method to report progress, though, because it is a subjective evaluation. When you ask someone "What percent complete are you on this activity?" what goes through his or her mind? The first thing he or she thinks about is most likely "What percent should I be?" followed closely by "What's a number that we can all be happy with?"

To calculate the percent complete for an activity, you need something quantifiable. At least three different approaches have been used to calculate the percent complete of an activity:

1. Duration
2. Resource work
3. Cost

Each of these could result in a different percent complete! So when we say percent complete, what measure are we referring to?

If you focus on duration as the measure of percent complete, where did the duration value come from? The only value you have is the original estimate. You know that original estimates often differ from actual performance. If you were to apply a percent complete to duration, however, the only one you have to work with is the original estimated one. Therefore, percent complete is not a good metric.

Our advice is to never ask for and never accept percent complete as input to project progress. Always allow it to be a calculation. The calculated value that we recommend above all others is one based on the number of tasks actually completed in the activity as a proportion of the number of tasks that currently define the activity. Recall that the task list for an activity is part of the work package description. Here we count only completed tasks. Tasks that are underway but not reported as complete may not be used in this calculation.

Develop Risk Management Plan 19

This chapter is concerned with the subject of risk management and its component parts, risk management planning, risk identification, risk assessment, risk quantification, risk response development, and risk monitoring and control. Risk is present in any situation in which decisions must be made under constraint and uncertainty with imperfect information. To properly develop a project risk management plan, we should first understand the nature of "risk."

19.1 Understanding the Nature of Risk

In order to bring the subject of risk management to life, we will start with Harry Cendrowski's (Cendrowski & Mair, 2009) excellent example on the human mind and car driving.

When operating a car down the road, we are never sure that surrounding drivers will operate their cars in a rational manner. However, we enter such a situation with an a priori belief that other drivers are indeed rational. After all, they must pass a test to obtain a driver's license. Within our car, we constantly identify uncertain events and assess associated risks that may prevent us to reach our destination on time, safe and sound, and at minimum costs (i.e., to succeed in our car driving venture). In fact, our mind constantly enumerates the risks associated with the car driving activity, quantifies the risk, and then compels us to make a decision based on this assessment. Our mind also continually evaluates and updates the a priori belief that other drivers are rational with respect to every car that is within a personal "envelope of concern."

Driving at a steady speed, we are not very concerned with the actions of those cars and associated drivers far behind us. While we can see other cars in the rearview mirror, the likelihood that such a driver's actions impact our own decisions is low. If driver far behind us loses control, it does not impact us, although it could impact a group of drivers behind us. However, we are very concerned with the actions of those cars and associated drivers in front of us—most particularly, those immediately ahead of our own vehicle—and those to our side. If these

A. van Aartsengel and S. Kurtoglu, *Handbook on Continuous Improvement Transformation*, DOI 10.1007/978-3-642-35901-9_19, © Springer-Verlag Berlin Heidelberg 2013

individuals make an error in judgment, the consequences to us could be severe. Our "envelope of concern" is thus concentrated to the front and sides of our vehicle rather than behind it. With this simple example we have introduced five central notions associated with the work "risk":

1. A constraint
2. An uncertain event
3. A likelihood of occurrence of this uncertain event
4. A magnitude of the effects of occurrence of this uncertain event
5. A subjective judgment concerning these likelihood and magnitude.

The constraint that surrounding drivers will operate their cars in a rational manner compels us to enter to drive the car. An uncertain event that could occur is a driver losing control of his/her car. The likelihood that a random driver loses control is identical no matter where this driver is located with respect to us. However, the magnitude of the risk differs based on the location of the driver or based on our "envelop of concern." Our mind subjectively evaluates both magnitude and likelihood when we are assessing risks. This assessment is then used to make decisions based on information we perceive. Whether or not we are conscious of it, our mind quantifies these "risks," and we make decisions based on this quantification. Figure 19.1 illustrates the last three concepts as related to the occurrence and magnitude of an uncertain event within certain constraints.

Figure 19.1 is an illustration of a simple risk matrix, sometimes referred to as a heat map. This is a commonly used method of illustrating risk likelihood and the magnitude of risks should their events materialize. The use of the risk matrix to illustrate risk likelihood and magnitude is a fundamentally important risk management tool. The risk matrix can be used to plot the nature of individual risks, so that the project manager can decide whether the risk is acceptable and within the risk appetite and/or risk capacity of the project. We also use the word magnitude rather than severity, so that the same style of risk map can be used to illustrate risk associated with opportunity events. Severity implies that the risk event is undesirable and is, therefore, related to threat events.

Figure 19.1 also highlight the degree of subjective judgment concerning risk regions rating based on the likelihood of occurrence of an event and the magnitude of the effects of its occurrence, should it occur. There can be little disagreement about the risk regions and their associated level of risk if the likelihood of occurrence of an event and the magnitude of the effects of its occurrence are:

1. A low likelihood of occurrence of an event and a low magnitude of the effects of its occurrence will result in the subjective risk region of the quadrant, and requires additional guidelines to rate the risk level.
2. A high likelihood of occurrence of an event and a high magnitude of the effects of its occurrence will determine the high risk region of the quadrant.
3. A low likelihood of occurrence of an event and a low magnitude of the effects of its occurrence will determine a second low risk region of the quadrant to the overall achievement of desired success criteria.

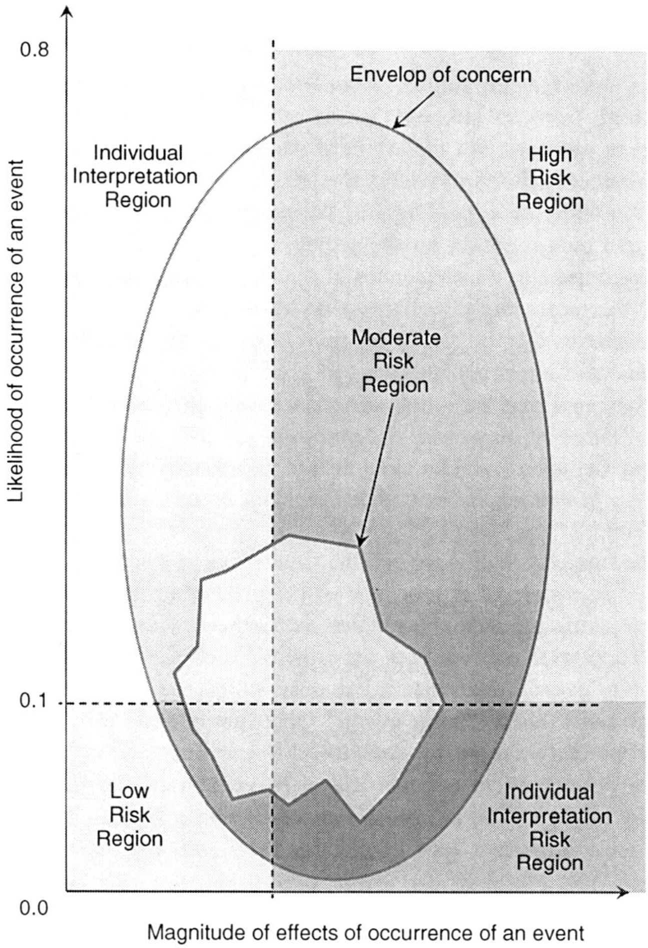

Fig. 19.1 Example of likelihood, magnitude and subjective risk judgment

4. A low likelihood of occurrence of an event and a high magnitude of the effects of its occurrence will result in the subjective risk region of the quadrant, and requires additional guidelines to rate the risk level.

Although risk management might come naturally to our minds, it is not an involuntary process within a project. A project must establish, utilize, and monitor risks within the enterprise business operating environment to effectively perceive changes in the project's environment.

19.2 Characterizing Risk

From the example above, we would say that "risk" is a word that exists only in the future tense, with no past tense, and with only actual occurrence of events that could affect predefined or specified success criteria. We shall define and characterize "risk" as:

"A combination of a constraint and a measurable uncertainty that an event, let call it a 'risk event,' occurs which would affect predefined or specified success criteria."

Here, we think of uncertainty as an outcome subject to an uncontrollable random event stemming from an unknown probability distribution. The measure of the consequence or evaluation of the effect of occurrence of such event on predefined or specified success criteria measures the "risk impact."

Figure 19.1 maps likelihood against the magnitude of an event. However, the more important consideration for the project manager is not the magnitude of the event, but the impact or consequences of its occurrence. For example, a large fire could occur that completely destroys a warehouse of a distribution and logistics enterprise business. Although the magnitude of the event may be large, if the enterprise business has produced plans to cope with such an event, the impact on the overall business may be much less than would otherwise be anticipated. The magnitude of an event may be considered to be the inherent risk level of the event and the impact can be considered to be the risk-managed level. The impact (or consequences) of the occurrence of an event is usually more important than its magnitude.

Within the framework of a project, the occurrence of a risk event may happen to the benefit or detriment of at least one of the project's success criteria that often serve as the determining factors for which risks are worth taking and for which risks are not. If it happens to the benefit of the project's success criteria, it is referred to as an "opportunity event," otherwise it can only inhibit the project's success criteria and it is referred to as a "threat event." Opportunity risks are the risks that are (usually) deliberately sought by the project team. These risks arise because the project team is seeking to enhance the achievement of the project objectives, although they might inhibit the project success criteria if the outcome is adverse. This is the most important type of risk for the future long-term success of any project.

Project success criteria often include the delivering of project outcomes in accordance with agreed-upon schedule, cost, and quality. If the project manager does not know what success criteria are driving the project, he or she cannot hope to identify the project risks that may impede the road to success. Knowledge of the project "success criteria" is also necessary in order to set the important element in the project risks monitoring and control process. As the "success criteria" is expanded and made more complicated, assessing, monitoring and controlling the project risks also becomes complicated.

The constraint associated with a risk, let call it "risk constraint," could include aspects of the project's or enterprise business environment that may contribute to project risks, such as poor project management practices, poor practice of the eight continuous improvement dimensions described in the previous chapter, lack of integrated management systems, concurrent multiple projects, or dependency on external participants who cannot be controlled, etc....

19.3 Characterizing Project Risk Management

Within the framework of a project, risk management is the project management process required to identify, assess, analyze, and respond to project risks should the "risk events" occur, monitor and control project risks, throughout the life cycle of the project, in order to minimize the likelihood and impact of the consequences of adverse events on the achievement of the project objective. Not all projects require a formal risk management approach. In enterprise businesses where success is the norm and failure is a rarity, risk management is relegated to obscurity in the hope that project managers will be able to handle project issues and problems as they occur. Nevertheless, to obtain the maximum benefit, risk management must become a systematic process applied in a disciplined manner throughout the project lifecycle.

A "process improvement" project is intended to break new ground. It will face a very wide range of risks that can impact the project outcomes. The desired overall aim of such project may be stated as objectives in the project scope, but the events that can impact the project may inhibit what it is seeking to achieve, enhance that aim, or create uncertainty about the outcomes. Project risk management offers an integrated approach to the identification, assessment, control and monitoring of these types of risk. For very large "process improvement" projects it might be necessary to appoint a risk manager, who can devote all or most of his or her time to ensuring that a comprehensive risk strategy is put in place and then reviewed from time to time throughout the project to ensure that it remains valid.

From the business point of view, the presence of risk is desirable since a natural balance exists between risk and opportunity. High-risk investments tend to pay proportionately larger premiums and conversely, smaller returns are associated with low-risk investments. Most commercial decisions are made under conditions of uncertainty or risk.

Addressing risks proactively and consistently throughout the project lifecycle will increase the chances of achieving the project objective. Waiting for unfavorable events to occur and then reacting to them can result in panic and costly responses. Risk management may be aided by a range of skills, tools, and techniques used to manage risk when accomplishing specific activity tasks. Some level of risk planning should be done during the early phases of the project life cycle to make sure, for example, that a contractor understands the risks involved with bidding on a proposed project. With knowledge of potential risks, the contractor can include contingency or management reserve amounts in the bid price. On the other hand, if the risks seem too great, the contractor may decide not to bid on a proposed project.

Failure to adequately manage the risks faced by a project can be caused by inadequate risk recognition, insufficient assessment of significant risks and failure to identify suitable risk response activities. Also, failure to set a project risk management strategy and to communicate that strategy and the associated responsibilities may result in inadequate management of risks. It is also possible that the risk management procedures or protocols may be flawed, such that these protocols may

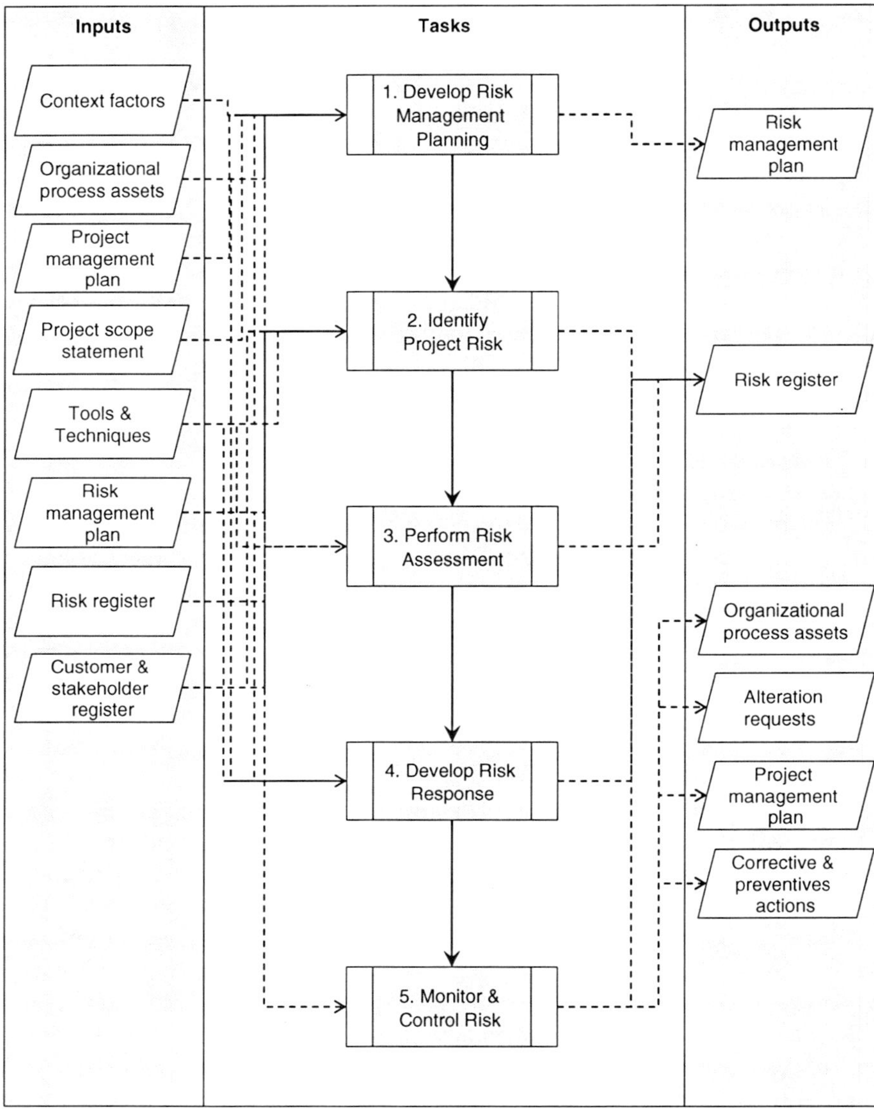

Fig. 19.2 The risk management process

actually be incapable of delivering the required outcomes. The consequences of failure to adequately manage project risk can be disastrous and result in inefficient project work, project not completed on time and agreed project outcomes that are not delivered, or were incorrect in the first place.

Risk management has well-established stages that make up the risk management process, as illustrated in Fig. 19.2, although it is presented in a number of different ways and often uses differing terminologies. These stages build into valuable risk management activities, each of which makes an important contribution. In this handbook, the risk management process is taken as a narrow set of activities; the constituent project management processes of which include the following:

1. Develop Risk Management Planning
2. Identify Project Risks
3. Perform Risk Assessment
4. Develop Risk Response Planning
5. Monitor and Control Risk

These five constituent processes interact with each other and with the project management processes in the PDSA "Process Groups." Table 19.1 gives examples of "risk events" that can occur during a project lifecycle. Each aspect of executing any of these five constituent processes can involve effort from one or more persons, based on the needs of the project. Each aspect occurs at least once in every "process improvement" project and occurs in one or more project phases.

19.4 Develop Risk Management Planning

This is the project management process that compels project managers to devote organized, purposeful thought to project risk management and to provide organizational infrastructure to aid them as they attempt to:

1. Isolate and minimize risk
2. Eliminate risk where possible and practical
3. Develop alternative courses of action
4. Establish time and money reserves to cover risks that cannot be mitigated

The "Develop Risk Management Planning" process builds upon:

1. The environment context within the enterprise business, which reflects the attitudes toward risk and the risk tolerance of the enterprise business and people involved in the project. In every project, there is a risk environment. There are risks that will have to be faced, and there are different ways to deal with them. The risk management planning draws together the risk policies, practices, and procedures of the enterprise business into a cohesive whole that will address the nature of risk peculiar to the project.
2. Risk management policies within the enterprise business will offer insight into the amount of information and risk reporting that is required on projects, as well as general guidance on risk qualification, quantification, and response development. That guidance may include, but is not limited to, organizational definitions and descriptions of approaches to the risk procedure, guidance on risk reserve allocation, explanations of risk probability and impact descriptions, and clarification on proper application of risk response strategies.
3. In some enterprise businesses, risk management is sufficiently well entrenched that there are standard forms and formats for risk management plans available within the organizational process assets. This is more common in organizations where

Table 19.1 Example of risk events and conditions internal to projects

Project Management Plan	**Scope Management Plan**	**Quality Management Plan**
Risk Events • In correct start of plan relative to project life cycle.	Risk Events • Changes in scope to meet project objectives.	Risk Events • Performance failure or environmental impact.
Risk Conditions • Inadequate planning or resource allocation. • Anything which directly or indirectly reduces the chances of project success.	Risk Conditions • Inadequate planning. • Poor definition of scope or work break down. • Inconsistent or incomplete definition of requirements..	Risk Conditions • Inconsistent, incomplete or unclear definition of quality requirements. • Inadequate quality control program.
Time Management Plan	**Cost Management Plan**	**Risk Management Plan**
Risk Events • Specific delays on planned schedule	Risk Events • Impacts of accidents • Unpredictable price changes	Risk Events • Overlooking a risk • Change in work necessary to achieve the objectives
Risk Conditions • Errors in estimating time or resource availability • Poor allocation and management of float • Scope creep without due allowance for time extension.	Risk Conditions • Cost estimating errors. • Lack of investigation of predictable problems • Inadequate productivity cost, change or contingency control.	Risk Conditions • Ignoring risk • Inappropriate or unclear assignment • Inappropriate or unclear contractual assignment of risk
Procurement Plan	**Human Resources Plan**	**Communication Plan**
Risk Events • Contractor insolvency • Claims settlement or litigation	Risk Events • Strikes, terminations, organizational breakdown. • Force majeure events	Risk Events • Inaction or wrong action due to incorrect information or communication failure.
Risk Conditions • Un-enforceable conditions/clauses. • Incompetent or financially unsound contractors. • Adversarial relations. • Inappropriate or unclear contractual assignment of risk.	Risk Conditions • Conflict not managed • Poor organization, definition or allocation of responsibility • Poor use of accountability • Absence of leadership • Consequence of ignoring or avoiding risk.	Risk Conditions • Carelessness in planning or in communicating • Improper handling of complexity. • Lack of adequate consultation with project's customers.

there is a project management office or project support office. These formats encourage consistency and knowledge transfer as risk management history is conveyed continually from project to project and from team to team. These inputs may take some time to collect. Gathering these data is frequently done in concurrence with other project efforts, such as cost estimating and high-level scheduling. Ideally, these efforts would precede the planning steps as the insights from risk management planning may have a significant impact on the outcomes.
4. The project scope statement, which addresses and documents the "process improvement" project and deliverables requirements, the boundaries of the project, the methods of acceptance, and a high-level scope control.
5. The project management plan, which integrates and documents major deliverables, assumptions, and constraints of other project management processes within the PDSA Process Groups.

The risk management planning is not specific to the project risks but instead it addresses the framework in which those risks will be addressed. The project risks are addressed in the subsequent steps of the "Risk Management Plan" process. In developing a risk management planning, team members should work to build a documentation that will encourage consistent adherence to the risk management policy and procedure within the enterprise business and to ensure that there is an unchanging vision as to the levels of risk that are deemed tolerable. They should review all the available inputs and acknowledge (and document) any deviation from enterprise business practices.

The "Develop Risk Management Planning" process should produce an overall risk management plan, a risk approach within which the project will function. This risk management plan includes oversight on operational definitions of risks, practices, timing, metrics, risk thresholds, evaluation, tracking, and roles and responsibilities associated with the risk management effort. A preliminary risk budget may also be developed, although more in-depth documentation and funds support is frequently developed during or after risk quantification.

The Risk Management Planning process should be completed early during project planning as it facilitates risk planning across multiple project management processes (illustrated in Table 19.1) and the PDSA "process Groups," and it is crucial to successfully performing these project management processes.

As an integral part of normal project planning and management, risk planning is sensibly done and repeated and should occur at regular intervals. Some of the more obvious times for evaluating the risk management plan include:

1. In preparation for major decision points and changes
2. In preparation for and immediately following evaluations
3. As significant unplanned change occurs that influences the project

A "process improvement" project is guided by a series of plans through the PDSA model that provide the rationale and intended processes through which the project will be executed. The risk management plan document, outcome of the "Develop Risk Management Planning" process, is recommended as part of this suite of guiding documents. Carl L. Pritchard (2010), in his excellent book on "Risk Management: Concepts and Guidance," provides an approach, illustrated in Table 19.2, to the content of a risk management plan.

Table 19.2 Generic risk management plan content

Project:______________________________	Date:_____________________(original)
Team:_______________________________	______________________(revised)

1. Description
   1.1 Objective (from project charter)
   1.2. Project
      1.2.1. Project description (from the work breakdown structure)
      1.2.2. Key functions (from charter and work breakdown structure)
   1.3. Required operational characteristics
   1.4. Required technical characteristics
   1.5. Required support (from roles/responsibilities)
2. Project summary
   2.1. Summary requirements (V.O.B., V.O.C and V.O.P)
   2.2. Integrated schedule
3. Risk environment
   3.1. Enterprise business risk management policy
   3.2. Stakeholder risk tolerances (from the customer and stakeholder registry)
   3.3. Enterprise business risk management plan template
4. Risk data collection
   4.1. Operational definitions
   4.2. Organizational risk practices
   4.3. Risk reviews and reporting frequency
   4.4. Risk metrics
   4.5. Risk thresholds
   4.6. Implementation
      4.6.1. Evaluation
      4.6.2. Tracking
      4.6.3. Roles/responsibilities
5. Application issues and problems
   5.1. Risk identification
   5.2. Risk qualification
   5.3. Risk quantification
   5.4. Risk response planning
   5.5. Risk monitoring and control
6. Other relevant plans
7. Approach summary
8. References
9. Approvals

19.5 Identify Project Risks

This is the project management process used to establish a base pool of risks by finding and documenting which real risks may affect the project objectives and what their root causes are, if they occur. It is not, however, a process of inventing highly improbable scenarios in an effort to cover every conceivable possibilities. Until project risks are identified and described in an understandable way, project risks cannot be assessed or managed.

Identification of real risks associated with a particular project starts with an intimate knowledge of the enterprise business, the market in which it operates, the legal, social, political and cultural environment in which it exists, as well as the development of a sound understanding of its strategic and operational objectives, including factors critical to its success and the threats and opportunities related to the achievement of these objectives. This should be followed by an understanding of the project itself and an effective description of the project risk events. What is the project scope, what are the project deliverables, and indeed what are the underlying project objectives? The answers to these questions will have a considerable effect on the risk characteristics of the project.

If the project team members are certain that an event will occur, that event is not a risk; it is a certainty. Certain events are not to be handled by the risk management process. Valid information is important in identifying risks and in understanding the likelihood and the consequences of each risk. Existing information sources need to be accessed and, where necessary, new data sources developed. Although it is not always possible to have the best or all information, it should be as relevant, comprehensive, accurate and timely as resources will permit. This means that it is critical to have specialist and experienced staff assist the project team in the risk identification activity.

19.5.1 Project Risks Classification

To the project team, risks are primarily rooted in the "process to be improved" with the requirements to deliver a specified product or service at a specified time for a specified cost. A properly planned "process improvement" project will provide the project with some reserve funds and slack time to work around unanticipated problems and still meet original cost, schedule, and performance goals. But a wide variety of problems can keep the project team from meeting project objectives: the project outcomes may not attain the performance level specified, the actual costs may be too high, or delivery may be too late. There is, of course, a risk that the original cost, schedule, and performance goals were unattainable, unrealistic, or conflicting.

The risks associated with a project such as the internal development or the external procurement of a critical component may be divided into the three traditional areas of project management, often referred to as the project's triple constraint:

1. *The Risks associated with Technical, Quality, or Performance:* The possibility that the outcome of a project activity being developed or procured will not perform to the levels needed by the project. Without question, technical risks are paramount to the success of any project. If a critical activity outcome does not work it will have an adverse impact on the success of any project. Technical risks are most often "show stoppers" and they must be corrected.
2. *The Risks with Schedule Performance:* The possibility that a critical activity outcome needed by the project will not be available in the time-frame needed, and/or that the technical risks will cause an adverse impact on the project schedule. Depending upon the circumstances, schedule risks can be merely an

annoyance, or possibly have a catastrophic impact on the project. Schedule risks are second in criticality, right next to technical performance.

3. *The Risks with Cost Performance:* The possibility that the costs of the critical activity outcomes will exceed that which is has been estimated, funded, or even available to the project, and that the technical and/or the schedule risks will have an adverse impact on the costs of the subproject. Of the three categories, cost risks are typically the least serious. The risks of cost growth are a distant third in the triple constraint.

All three of these risk categories are interrelated such that unfavorable results in any one of these three risk areas will likely have a resulting adverse effect in one or the other two. Technical performance will unquestionably be the primary concern over both schedule and cost risks. However, too tight allocated funds, or too ambitious a schedule, can also have a detrimental effect on the technical and or quality performance factors.

To make it manageable, risk identification should be approached in a methodical way to ensure that all significant risks facing the project have been identified and all the risks flowing from the project activities defined. All associated volatility related to the project activities should be identified and classified or categorized. An appropriate classification should mirror an enterprise business risk needs and enable the project team to better identify the project risk tolerance, risk capacity and total risk exposure in relation to each risk, group of similar risks or generic type of risk. Here, we think of project risk capacity as a measure of the total "risk impact" on the project that the project manager considers the project capable of absorbing, based on its contingency funds.

Project risks can be classified in many different ways and presented in various formats. Risk classifications may be named differently, and certain risks may be included in different categories from one project to another. However, it is most important that the project team ensures that risk has been considered in all areas that are relevant to the development of a comprehensive risk profile and ultimately encompassed in an effective project risk management plan.

Examples of risk classifications that the project team should consider when conducting its risk identification are illustrated in Fig. 19.3 and Table 19.1. This information is intended to provide guidance for discussion and consideration during the risk identification process. It is not intended to be comprehensive or all inclusive, given that each "process improvement" project has unique considerations.

Figure 19.3 and Table 19.1 show that some specific risks can have both external and internal drivers and therefore overlap the two areas. Internal project risks can cover nearly any topic within the project scope and an example includes the following: A project management plan that assumes a staff size of seven, but there are only four resources available. The lack of resources could impact the time required to complete the work and the activities would be late. External drivers can be classified further into types of risk such as strategic, financial, operational, hazard, etc.

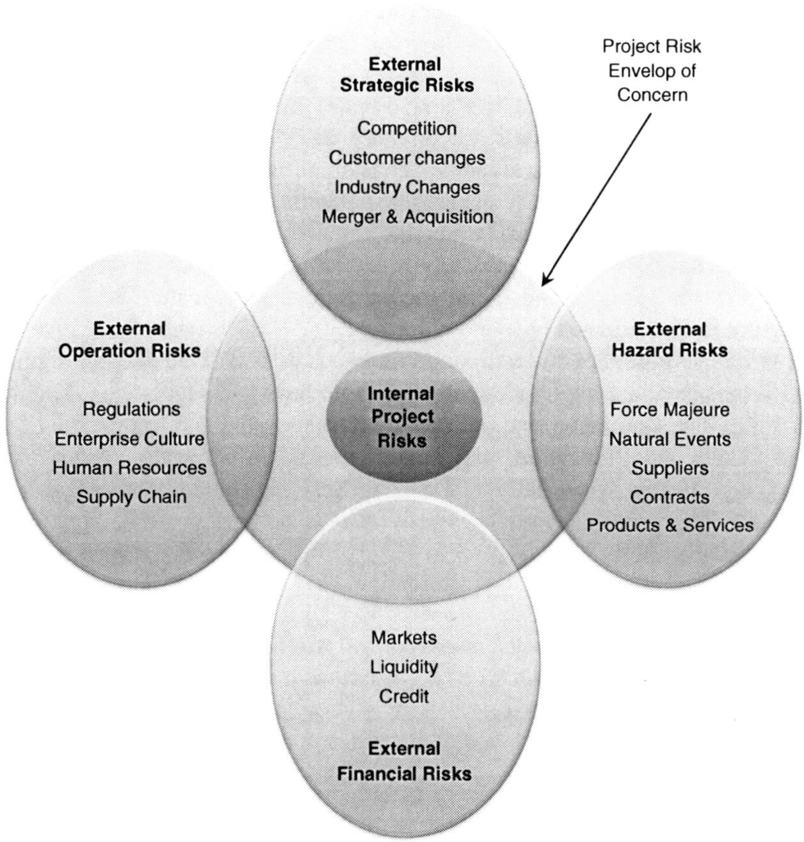

Fig. 19.3 Example of project risks classification

1. *External*. When analyzing external risks, the project team should consider customers, suppliers, and competitors. Risk identification in this area should also include discussions regarding risk to the enterprise business brand and reputation as well as risks associated with new competition, outsourcing, suppliers, partners, and financial or other crisis or disasters.
2. *Financial*. Considerations in this area include risks associated with credit/cash management; financial markets, such as interest rate fluctuations, debt and equity structure; and financial reporting, including the production of accurate, timely financial statements and appropriate disclosures that may have impacts on the "process improvement" project. Risk identification could include discussions regarding processes, controls, and potential deficiencies related to the financial close and financial statement preparation as well as risks to the achievement of all documented financial reporting objectives of the project.
3. *Operational*. Big-picture operational risks focuses on the people (human resources), process (product development, marketing), and physical assets

(property, plant, equipment) that are most important to carry out the "process improvement" project activities. In some cases, these are the items that, typically, ultimately have the greatest impact on cash flow. For example, in many "process improvement" projects, a significant portion of revenue is lost due to errors or issues with process and/or those individuals involved in those processes. In a number of the situations considered in this area, the probability of incident occurrence is low. However, the potential consequences are substantial.
4. *Strategic.* Assessment of strategic risk requires consideration as to whether outlined strategies are appropriately reflected in the objective of the "process improvement" project and if the project supports the enterprise business in meeting its documented business objectives.
5. *Regulatory.* Overall consideration in this area is on risk of meeting all regulatory requirements and compliance with applicable laws and regulations. This focus includes financial, labor, and policy such Sarbanes-Oxley, Occupational Safety and Health Administration, and environmental, respectively. Securities and Exchange Commission, Internal Revenue Service, Department of Labor, and industry regulations should be considered as part of the risk identification process.
6. *Information.* When identifying information, the project team should concentrate on risks related to intellectual property as well as information technology that support processes, operations, and reporting within the "process improvement" project. This includes hardware, software, and network support. Discussion should center on whether information systems are reliable, secure, and adequately support the project and if data/information is relevant, reliable, and timely.

There is no universal classification system that fulfils the requirements of all any project. It is likely that each risk will need to be classified in several ways in order to clearly understand its potential impact on the project success criteria.

Although it is not a formalized system, the classification of risks into short, medium and long term helps to identify risks as being related (primarily) to operations, tactics and strategy, respectively. This distinction is not clear-cut, but it can assist with further classification of project risks.

A short-term risk has the ability to impact the project success criteria, key dependencies and core processes, with the impact being immediate. These risks can cause disruption to normal efficient execution of project activities immediately at the time that the risk event occurs. Short-term risks are predominantly threat risks, although this is not always the case. These risks are normally associated with unplanned disruptive events, but may also be associated with cost control in the project. Short-term risks usually impact the ability of the project to maintain efficient core processes that are concerned with the continuity and monitoring of routine project activity works. "Process improvement" projects that require new approaches or new tools may suffer in the short term yet may have higher productivity and performance levels in the medium or long terms.

A medium-term risk has the ability to impact the project success criteria following a (short) delay after the risk event occurs. Typically, the impact of a medium-term risk on the project success criteria would not be apparent immediately, but would be apparent within months, or at most a year after the risk event. Medium-term risks usually impact the ability of the project to maintain effective core processes that are concerned with the management of project. These medium-term risks are characteristics of projects improvement, product developments, product launch and the like. For example, if a new computer software system for use in the project is to be installed, then the choice of computer system is a long-term or strategic decision. However, decisions regarding the activity to implement the new software will be medium-term decisions with medium-term risk attached.

In general terms, long-term risks will impact several years, perhaps up to 5 years, after occurrence of the risk event or the taken decision. Long-term project risks therefore relate to strategic decisions. When a decision is taken to launch a new product by improving an existing process, the impact of that decision (and the success of the product itself) may not be fully apparent for some time. This perspective of launching a new product might include, among other things, introducing new engineering issues related to project support and production into the "process to be improved" earlier in the project. Although it is desirable to make decisions based on long-term implications, it is not always feasible. The project manager is often forced to act on medium or short-term risk considerations.

One reason for this is a change in personnel. Ideally, the same team member will stay with a project from the earliest phases through closeout. However, because ideal conditions rarely exist, a given project will likely employ several management and staff teams. As a result, the transition in project management personnel often creates voids in the risk management process. These voids, in turn, create knowledge gaps, whence valuable information collected earlier in the project could be lost. Precious time must therefore be spent becoming familiar with the project, often at the sacrifice of long-term planning and risk management.

Another reason for acting on short-term risk considerations is project advocacy. Sudden shifts in organizational priorities can wreak havoc on long-term plans (a risk area in itself). This results in short-term actions to adjust to new priorities. Often, these decisions are made before long-term effects can be thoroughly evaluated. And lastly, in some instances, long-term effects are not always apparent at the time a decision must be made.

19.5.2 Risk Description

Risk events are most effective when they are described clearly and in depth. A high-quality risk event description will describe the potential occurrence and how it would influence the project. The objective of risk description is to display the identified risks in a structured format, for example, by using a table—the "Risk Register." The purpose of the risk register is to form an agreed record of the significant risks that have been identified. Also, the risk register will serve as a record of the control activities that are currently undertaken and details of intended additional controls. It will also be a record of the additional actions that are proposed to improve the control of the particular risk. It is important that the risk register should not become a static document. It should be treated as a dynamic element and considered to be the risk action plan for project.

At its most simple, the risk register can be stored as a document held on computer. However, there are many more sophisticated forms of risk registers, including records of significant risks held on databases. Where quantification of exposure is required, then a simple risk register held as a document is unlikely to be sufficient. This is true of systems for recording not only project risks but also operational risks within the enterprise business, where quantification of risk exposure is required.

The information set out in the risk register should be very carefully considered and constructed. For example, the risks set out in the register need to be precisely defined so that the cause, source, event, magnitude and impact of any risk event can be clearly identified. Also, the existing control activities, together with any additional controls that are proposed, must be described in precise terms and accurately recorded. An example of risk register is provided in Table 19.3. Risk control activities should be described in sufficient detail for the controls to be auditable. This is especially important when the risk register relates to the routine activity works undertaken within the project. Risk registers should also be produced for projects and to support strategic decisions.

A project risk register has to be a very dynamic document. Details of the risks faced by the project, as recorded in the risk register, should be discussed at every project review meeting. As well as risk registers being relevant to project review meetings, they should also support business decisions. In this case, the precise format of a risk register may be less formal.

The risk description table, illustrated in Table 19.4, can be used to facilitate the description and assessment of risks. The use of a well designed structure is necessary to ensure a comprehensive risk identification, description and assessment process. By considering the consequence and likelihood of each of the risks set out in the table, the project team should be able to prioritize the key risks that need to be analyzed in more detail. Identification of the risks associated with business activities and decision making may be categorized as strategic, tactical, and operational.

19.5.3 Project Risks Data Collection

Gathering project risks data is one of the greatest challenges in project risk management as there is a propensity for risk identification and risk information gathering to become highly subjective. The tools and techniques that are applied in risk identification are as varied as the projects they serve. However, some groups of tool and technique types are most commonly applied. They included, but are not limited to:

1. Analogous and Lessons Learned Techniques
2. Brainstorming Technique

Table 19.3 Risk register

RISK REGISTER
Project Name:
Project Manager:

Summary				Description					Preventive Action			Contingency Action		
ID	Date Identified	Identified by	Received by	Description of Risk	Description of Impact	Likelihood Rating	Impact Rating	Priority Rating	Preventive Actions	Action Resource	Action Date	Contingency Actions	Action Resource	Action Date

3. Expert Interviews Technique
4. Delphi Technique
5. Project Network Analysis Techniques
6. The Program Evaluation and Review Technique (PERT)

Table 19.4 Risk description

1. Name of Risk	Name and/or identifier of risk
2. Scope of Risk	Qualitative description of the events, their size, type, number and dependencies
3. Risk Category	According to classification; e.g. strategic, operational, financial, hazard, internal to project, etc...
4. Risk Time Frame	Time frame is the beginning and end dates of when a risk may occur
5. Risk Root Causes	These are the fundamental conditions or events that may give rise to the identified risk. It sharpens the definition of the risk and allows grouping risks by causes. Effective risk responses can be developed if the root cause of the risk is addressed.
6. Stakeholders	Stakeholders and their expectations as described in the "Customer and stakeholder" registry.
7. Quantification of Risk	Significance and Likelihood
8. Risk Tolerance	Loss or gain potential and financial impact of risk Value at risk Likelihood and size of potential losses/gains Objective(s) for control of the risk and desired level of performance
9. Risk Treatment & Control Mechanisms	Primary means by which the risk is currently managed Levels of confidence in existing control Identification of protocols for monitoring and review
10. Potential Action for Improvement	Recommendations to reduce risk

19.5.3.1 Analogy and Lessons Learned Techniques

The analogy and lessons-learned techniques for risk identification are based on the idea that any project—no matter how advanced or unique it is—does not represent a totally new system. Most projects originated or evolved from existing projects or simply represent a new combination of existing components or subsystems. A logical extension of this premise is that the project manager can gain valuable insights concerning various aspects of a current project's risks by examining the lessons learned; i.e., the documented successes, failures, problems, and solutions of similar existing or past projects. The experience and knowledge gained or lessons learned can be applied to the task of identifying potential risk in a project and developing a strategy to handle that risk.

The analogy comparison and lessons-learned techniques provide a sense of enterprise business history and experience. They involve identifying past or existing activities similar to those in the current project effort and reviewing and using risks data from these past or existing activities as initial entries into the risk register for the current project. The term "similar" refers to the commonality of various characteristics that define a project. The analogy may be similar in technology, function, contract strategy, "process to be improved," or other area used in the project. The key is to understand the relationships among the project characteristics and the particular aspects of the project being examined.

Project managers can apply lessons learned or compare the commonality of various characteristics of existing projects to various characteristics of new projects in all phases and aspects of a project whenever historical data are useful.

The analogy and lessons-learned techniques are especially valuable when the "process to be improved" is primarily a new combination of existing sub-processes. The value increases significantly when recent and complete historical project risk data are available. When properly done and documented, analogy comparison provides a good understanding of how project characteristics affect identified risks.

The analogy and lessons-learned techniques build on three types of data:

1. Description and project characteristics of the "process to be improved" and its discrete components (or sub-processes).
2. Description and project characteristics of the existing or past projects activities and their components (or tasks).
3. Detailed data (cost, schedule, and performance) for the past projects activities being reviewed.

The description and project characteristics are needed to draw valid analogies between the current and past projects activities. Detailed data (cost, schedule, and performance) are required to evaluate and understand project risks and their potential effect on the current project.

There are two limitations to the analogy comparisons and lessons learned techniques.

1. *Availability of data*—If common project characteristics cannot be found or if detailed data are missing from past activities, the data collected will have limited utility.
2. *Accuracy of the analogy drawn*—Past activities may be somewhat similar, but rapid changes in technology, manufacturing, methodology, and so on, may make comparisons inappropriate.

Within a project management framework, the analogy and lessons-learned techniques have several applications:

1. Too often, enterprise businesses fail to scrutinize the failings of past designs, only to learn later that the project at hand is failing for the same reasons as a project just a year or two before. Analogy comparisons will provide a sense of corporate history and experience.
2. For project status reporting, the analogy comparison technique can serve only to ascertain certain numbers that may have been used to establish the baseline for the project.
3. Major planning decisions should rely very heavily on an enterprise business lessons learned documented in the enterprise organizational process assets. History is an excellent teacher, and using the enterprise business historical

experience with similar projects activities can prove invaluable. If certain approaches have been attempted to carry out certain activities, it is vital to find out whether they succeeded or failed.

4. For planning decisions in procurement management, the issue of contract strategy selection can be developed using analogy comparison techniques. If work with a similar client, similar project, or similar resources has failed in part due to using one contract strategy, it is worthwhile to consider alternate strategies. Furthermore, terms such as "past performance," "performance history," and "preferred vendor" all reflect some analysis of analogous projects activities. These are valuable analyses because enterprise businesses should not repeat the mistake of dealing with a less-than-acceptable vendor.

19.5.3.2 Brainstorming Technique

With the brainstorming approach, the project team, through ideas generation under the leadership of a facilitator and judgment of multidisciplinary experts, identifies a comprehensive list of project risks as initial entries into the risk register for the current project. A brainstorm is more than a basic core dump of information. It is the expression of ideas that then feeds other ideas and concepts in a cascade of data. It encourages team members to build on one another's concepts and perceptions. It circumvents conventions by encouraging the free flow of information.

As indicated in the V.O.C. data collection, the brainstorming technique is a facilitated sharing of information, without criticism, on a topic of the facilitator's choosing. It brings forth information from participants without evaluation, drawing out as many answers as possible and documenting them. There are no limits to the information flow or direction during a brainstorming session. Brainstorming is designed to encourage thinking outside of conventional boundaries so as to generate new insights and possibilities.

For risk identification, as an example, the facilitator might ask "For the 'process to be improved' sub-process, what are the risks? What adverse effects could happen if the sub-process is altered?" Participants can then fuel their imagination with ideas as the facilitator documents or catalogs each new suggestion.

Although the brainstorming technique may not be the most efficient tool or the most thorough technique, its familiarity and broad acceptance make it the tool of choice for many risk analysts. The technique requires limited facilitation skills and familiarity with any premise being presented to the group for clarification purposes. And whereas it may be viewed as a generic tool, the fact that most participants are aware of the process and the tool's nuances make it desirable in a variety of risk management settings. Because risk is a phenomenon that exists only in the future, and everyone has the ability to intuit some aspect of the future, brainstorming as an idea generation tool is a logical application.

Brainstorming can be used in a variety of risk management practices, including efforts to identify risks, establish qualification schemes, clarify quantification assumptions, and generate potential risk responses. It can draw on project team members, management, customers, and vendors. Virtually any stakeholder can contribute information.

This technique is applicable in virtually every step of the risk management process. Its broad utility makes it appealing in a variety of settings:

1. During risk identification to establish a base pool of risks
2. During qualification to work toward terms and terminology as to what constitute high, moderate, and low risks or impacts in the various categories of risk.
3. During the risk qualification step to capture environmental assumptions and potential data sources.
4. During the response development step to generate risk approaches and to examine the implications thereof.

The brainstorming technique is effective when directed at a clear, easily discernible goal, which is crucial. Without an objective for the outcomes, risk brainstorms can easily deteriorate into complaint sessions. Brainstorms are virtually without equal in environments where quick analysis is needed and individuals with a willingness to participate are available. For risk identification, qualification scheme discussions, and risk response development, the brainstorming technique can produce volumes of valuable information from which the best available responses can be derived. Brainstorms afford new perspectives, and new perspectives are essential to the success of any risk management effort because risk management is a foray into the unknown.

19.5.3.3 Expert Interviews Technique

With the interview approach, the project team, through interviews of experienced project participants, stakeholders, and subject matter experts can identify risks, identifies a comprehensive list of project risks as initial entries into the risk register for the current project. Obtaining accurate judgments from subject matter experts is one of the most critical elements in both risk identification and risk analysis because:

1. The information obtained identifies areas that are perceived as risky.
2. The interviews provide the basis for taking qualitative information and transforming it into quantitative risk estimates.

Using expert interviews required reliance on technical expertise. Because every project is a temporary effort undertaken to create a unique product, service, or result, all information necessary for an accurate risk analysis cannot usually be derived from previous projects activities data. However, obtaining the information from experts can be frustrating and can often lead to less than optimal results.

Nearly all risk analysis techniques require some expert judgment. However, it can sometimes be difficult to distinguish between good and bad judgment, and this aspect makes the approach and documentation even more important than usual. The project team performing the interview task is likely to receive divergent opinions from many experts, and as a result, the project manager must be able to defend the ultimate position he or she takes.

The expert interviews technique is relatively simple. Basically, it consists of identifying appropriate experts and then methodically questioning them about risks in their areas of expertise as related to the project. The technique can be used with individuals or groups of experts. The process normally obtains information on risk associated with all three facets of the triple constraint: schedule, cost, and quality.

This technique is recommended for all projects. Expert interviews focus on extracting information about what risks exist and how severe they may be. Interviews are most useful in risk identification but may apply in other processes as well. When questioning experts about risks on a project, it is logical to pursue potential risk responses and alternatives, as well as information pertaining to likelihood of occurrence of risk events and potential impact on the project success criteria.

When conducted properly, expert interviews provide very reliable qualitative information. Transforming qualitative information into quantitative distributions or other measures depends on the skill of the interviewer. Moreover, the technique is not without problems. Those problems include:

1. Wrong expert identified
2. Poor quality information obtained
3. Expert's unwillingness to share information
4. Changing opinions
5. Conflicting judgments

The expert interviews technique has the advantage of being applicable in a wide variety of situations:

1. *Applying expert interviews technique in milestone preparation is direct and important.* Because the objectives are to ensure that planning has been comprehensive and the project is ready to move forward into its next phase, in-depth consultation with both internal and external customers is vital.
2. *Design guidance is frequently a function of expert interviewing.* The expert interviews are useful for making decisions ranging from considering technology alternatives for the "process to be improved" to choosing technology components. To understand how uncertainties relate to one another and how the alternatives compare, expert interviews are often used in the data gathering stage.
3. *Source selection is a prime application for expert interviews.* In many cases, interviews determine which candidates to eliminate for a subcontract or consulting position. In addition, if the expert interview is conducted properly during source selection, it can open new avenues for later negotiation with the source.
4. *Expert interviews also serve other applications.* They can be used to establish the enterprise business risk tolerances and thresholds, as well as the general culture for risk responses. The interviews can be used to explore specific risk events or general risk strategies. As a tool, interviews have perhaps the greatest breadth of any of the basic risk management tools.

19.5.3.4 Delphi Technique

Although people with experience of particular subject matter are a key resource for expert interviews, they are not always readily available for such interviews and, in many instances, prefer not to make the time to participate in the data gathering process. The Delphi technique works to address that situation by affording an alternative means of educing information from experts in a fashion that neither pressures them nor forces to leave the comfort of their own environs.

The Delphi technique has the advantage of drawing information directly from experts without impinging on their busy schedules. It also allows for directed follow-up from the experts after their peers have been consulted.

As indicated in the V.O.C. data collection chapter, the Delphi technique (created by the Rand Corporation in the 1960s) derives its name from the oracle at Delphi. In Greek mythology, the oracle god Apollo foretold the future through a priestess who, after being question, channeled all knowledge from the gods, which an interpreter then catalogued and translated. In the modem world, the project manager or facilitator takes on the role of the interpreter, translating the insights of experts into common terms and allowing for his or her review assessment. The cycle of question, response, and reiteration is repeated several times to ensure that the highest quality of information is extracted from the experts.

This technique is recommended when the project's experts cannot coordinate their schedules or when geographic distance separates them. The technique is also appropriate when bringing experts together to a common venue may generate excess friction.

The inputs for the Delphi technique are questions or questionnaires. The questionnaire addresses the risk area(s) of concern, allowing for progressive refinement of the answers provided until general consensus is achieved. The questionnaire should allow for sufficient focus on the areas of concern without directing the experts to specific responses.

Outputs from the process are progressively detailed because all iterations should draw the experts involved closer to consensus. The initial responses to the questionnaire will generally reflect the most intense biases of the experts. Through the iterations, the facilitator will attempt to define common ground within their responses, refining the responses until consensus is achieved.

The Delphi technique relies heavily on the facilitator's ability both to generate the original questions to submit to the experts and to distill the information from the experts as it is received. The process is simple but is potentially time-consuming. Its steps are as follow:

1. *Identify experts and ensure their participation.* The experts need not be individuals who have already done the work or dealt with the risks under consideration, but they should be individuals who are attuned to the enterprise business, the customer, and their mutual concerns. Experts can be defined as anyone who has an informed stake in the project and its processes. Commitments for participation should come from the experts, their direct superiors, or both.
2. *Create the Delphi instrument.* Questions asked under the Delphi technique must be sufficiently specific to draw out information of value but also sufficiently general to allow for creative interpretation. Because risk management is inherently an inexact science, attempts to generate excessive precision may lead to false assumptions. The Delphi questions should avoid cultural and organizational bias and should not be directive, unless there is a need to evaluate risk issues in a niche rather than across the entire project spectrum.

3. *Have the experts respond to the instrument.* Classically, this is done remotely, allowing the experts sufficient time to reflect over their responses. However, some enterprise businesses have supported encouraging questionnaire completion en masse during meetings to expedite the process. No matter the approach, the idea is to pursue all the key insights of the experts. The approach (e-mail, social networks, and meetings) for gathering the experts' observations will largely determine the timing for the process as a whole.
4. *Review and restate the responses.* The facilitator will carefully review the responses, attempting to identify common areas, issues, and concerns. These will be documented and returned to the experts for their assessment and review. Again, this may happen by mail or in a meeting, although the classic approach is to conduct the Delphi method remotely.
5. *Gather the experts' opinions and repeat.* The process is repeated as many times as the facilitator deems appropriate in order to draw out the responses necessary to move forward. Three process cycles are considered a minimum to allow for thoughtful review and reassessment.
6. *Distribute and apply the data.* Once sufficient cycles have been completed, the facilitator should issue the final version of the documentation and explain how, when, and where it will be applied. This is important so that the experts can observe how their contributions will serve the project's needs and where their issues fit in the grander scheme of risks and risk issues up for discussion.

The Delphi technique is frequently used when there are only a handful of experts who have an understanding of the project. It is also used when certain experts have insights about a particular aspect of the project that cannot be ignored. Although some other risk identification, assessment, and response development tools have broad application, the Delphi technique is a more exacting tool, drawing out only the responses or types of responses desired. The information acquired from the Delphi technique can be used to support risk identification, qualification, quantification, or response development.

The technique generates relatively reliable data (for a qualitative analysis) because multiple experts subject the information to at least three iterations of reviews. The iterative nature of the process and the requisite reviews tend to enhance accuracy, although the use of inappropriate experts or the development of poorly couched questions may produce less than optimal results. Still, because there are multiple reviewers, some built-in safeguards ensure a measure of reliability.

The Delphi technique has broad utility because of its use of the experts' skills and insights. The applicability of the technique is assessed on a relative scale of high, moderate, and low.

1. Process improvement and design guidance are prime applications for the Delphi technique. They are creative endeavor requiring multiple perspectives. As such, the Delphi technique is a classic tool for bringing different approaches to the fore and selecting the best possible approach.
2. Project status reporting is an area where the Delphi technique can provide more balanced insight than other tools can. Some projects falter because there is not a

common understanding of the work accomplished, but the Delphi technique by its nature can reorient a team. Since the tool draws out consensus of the experts, it can facilitate in-depth analyses of project status. The tool's value here is moderate.
3. Because the experts in an organization tend to make major planning decisions, the Delphi technique can be seen as viable here. Particularly in situations where there is significant conflict over planning decisions, the Delphi technique has high applicability due to its capacity to get a common vision from a group of experts.
4. Contract strategy selection in procurement management is an area where experts are frequently called upon to make decisions, and likewise, conflict can be significant. As with planning decisions, the Delphi technique can serve extremely well in these situations.
5. Source selection in procurement management may be an application of the Delphi technique. If the experts in the technique are familiar with the needs of the procurement and if they are attuned to the enterprise business limitations, the Delphi technique may be appropriate. However, the tool's utility here is moderate at best.

The Delphi technique is peerless in allowing for thoughtful review of the subject matter experts' insights. As such, enterprise businesses may be able to use this technique to establish risk responses, to identify risks, or to assess risk performance to date. However, the drawbacks associated with the timing of the process tend to limit its utility. When time is not of the essence, the Delphi technique can create some of the most thorough qualitative analyses available to the project manager.

The outputs of the Delphi technique are sets of modified responses to the questionnaire, from which the project team identifies a comprehensive list of project risks as initial entries into the risk register for the current project. Although participants generate those responses, the facilitator has the ultimate responsibility to produce final outputs based on an amalgam of responses from subject matter experts to each question or issue.

19.5.3.5 Project Network Analysis Techniques

As indicated in a previous section, project network schedules formalize the project's internal functions and processes and result in graphics that depict the project's activities and their relationships (predecessors, successors, and parallel tasks). Network diagrams are valuable tools because they:

1. Establish project completion dates based on performance rather than arbitrary deadlines.
2. Provide a sense of resource requirements over time, particularly when multiple resources will be deployed on multiple tasks simultaneously.
3. Highlight activities that drive the end date of the project.

Significant outputs of a network analysis are identifying the critical path, which consist of those activities that must be finished on time or the project will be delayed. Activities in the critical path compose the longest single path through the network. Their total duration represents the project duration. Most modern project management software highlight critical path activities so that they can be recognized for their importance. While these tools help identify some potentially higher-risk activities, they also identify those activities with free time or slack. Activities not on the critical path can afford some modest schedule slippage without affecting the overall project schedule.

A key issue in network development is selecting the appropriate level of detail. As with most project work, it is accepted practice to establish general process flows before working at the work package level. By their very nature, high-level networks embed significantly greater uncertainty. Detailed networks require a higher level of effort to generate but minimize the uncertainty associated with the relationships in the project. Realistically, as project requirements and information become more readily available, network models evolve to greater levels of detail.

Networks are formulated based on project activities, interrelationships among activities, and constraints, such as time, money, human resources, technology, and so on. Because all projects have these characteristics, network analysis applies universally. Using the technique is easier if network-based project schedules already exist because the project team can then make logic modifications so that network data can be incorporated into risk management plan as appropriate. If a network does not already exist, one must be created to apply this technique. The time saved by transforming an existing network rather than creating one provides a strong argument for network-based project scheduling from the beginning of the project.

Network analyses are critical to risk identification, given their role in ensuring that schedule objectives are met. These analyses focus attention on the relationships of activities and the interrelationships of risk among those activities. Although the network analysis models sometimes fail to give cost risk its due, they are invaluable early in the project when schedule risk is at its greatest. As with most tools, these are not the only tools required to evaluate or mitigate risk comprehensively. However, when used with other tools and techniques, network analyses are invaluable to the project team.

19.5.3.6 The Program Evaluation and Review Technique (PERT)

The Program Evaluation and Review Technique (PERT), described in a previous section of the "Time Management Process," was the first significant project-oriented risk analysis tool. As indicated already, the PERT objectives included managing schedule risk by establishing the shortest development schedule, monitoring project progress, and funding or applying necessary resources to maintain the schedule.

As projects have more work packages, the PERT becomes more reliable technique for risk identification. A project of 10 or 15 work packages will still have high levels of schedule variability even if the PERT is applied. However, if a project has more work packages, including many occurring concurrently, the PERT will balance out some of the natural incongruities and inaccuracy. This technique is also perceived as being more reliable when the standard deviations are calculated and then applied as schedule targets.

The PERT has broad utility because it affords clarity on probability of meeting deadlines. Its overall utility of PERT is high in that it provides the means to establish a fair, reasonable schedule with risk factored in and with a nominal level of additional effort.

1. The technique supports project progress reporting because it can provide a sense of the likelihood of achieving schedule activities. Since many project progress reports include requests for information on the probability of schedule success and estimated time to complete, PERT has a high level of utility here.
2. PERT also supports major planning decisions for many of the same reasons. Planning decisions and approaches are frequently resolved by opting for the approach that best meets customer requirements and schedule deadlines. Since PERT affords clarity on probability of meeting deadlines, it can plan a major role in planning decisions.
3. The technique can be indispensable in milestone preparation since milestones are a function of the schedule (or vice versa). PERT is easily applied to determine the likelihood of achieving certain milestones or to determine milestone realism.

19.5.3.7 Recording All Identified Risks

The identified project risks are typically recorded in a document called a project "risk register," the entries of which are described in Table 19.3. It ultimately contains the outcomes of the other risk management processes as they are conducted throughout the project lifecycle. The preparation of the risk register begins in the "Identifying Project Risks" process with the following information, and then becomes available to other project management processes within the PDSA Process Groups.

19.6 Perform Risk Assessment

The third step of the project "Risk Management Process" is "Perform Risk Assessment"—It relates to judging the likelihood of occurrence and the impact of each identified risk event, quantifying the importance of identified risk events, and allocating ownership. Frequently, no further analysis needs to be done. Although risk assessment is vitally important, it is only useful if the conclusions of the assessment are used to inform decisions and/or to identify the appropriate risk responses for the type of risk under consideration.

19.6.1 Likelihood of Occurrence of a Risk Event

As we have indicated in a previous section, "risk" is a word that exists only in the future tense, with no past tense, and with only actual occurrence of events that could affect predefined or specified objectives. It is therefore not possible to measure any characteristic of a risk event in the present. Consequently, during this step, the project team members will subjectively estimate a measure of the likelihood of occurrence of each risk event based on historical information and knowledge, that has been accumulated on past similar projects and operations work activities.

The likelihood of occurrence of each risk event can be determined on an inherent basis for any particular risk, or can be determined at the current level of risk, paying regard to the control measures that are in place.

For threat events, previous historical information and knowledge may be a good indication of how likely the risk event is to occur. For a fleet of motor vehicles in the automobile industry for example, there is certain to be a history of motor breakdowns. Controls will be in place to reduce the likelihood of these events. In this case, an enterprise business assesses the likelihood of vehicle breakdowns on an inherent basis and also on the basis of current controls. There are, however, difficulties in assessing the inherent likelihood of motor breakdowns, because certain assumptions would have to be taken about what effect the removal of controls would have on the likelihood of motor breakdowns.

Even if an assessment of the breakdown likelihood at the inherent level cannot be undertaken, the enterprise business will still need to determine the importance of the vehicle maintenance programme in preventing vehicle motor breakdowns and whether the maintenance activities provide value for money. In relation to vehicle motor breakdowns, the company may have driver training processes in place and, again, the effectiveness of these processes can be determined by evaluating inherent and current levels of risk. Whether levels of risk are evaluated at inherent or at current level, there is no doubt that benchmarking the performance of the fleet against the average performance of the industry will be a useful exercise.

A number of techniques have been developed to assist project teams in estimating a measure of the likelihood of occurrence of risk events by providing values against which the likelihood of the risk event occurring can be compared, asking whether the probability of the selected risk event is more, or less, or the same as the value being presented. The aim of these techniques is to adjust the comparator until the assessor cannot distinguish between the probability that the selected risk event occurs and the subjective value being presented. This subjective value is then taken as the best estimate of a measure of the likelihood of occurrence of the selected risk event. There are different ways of presenting probabilities against which estimates of measures of the likelihood of occurrence of risk events can be compared. The project team could also use a relative scale representing estimated values of measure of the likelihood of occurrence of risk events, ranging from 0.05 to 0.8:

1. 0–0.05: Almost impossible the risk event will occur
2. 0.05–0.10: Low likelihood of occurrence of the risk event
3. 0.10–0.20: Moderate likelihood of occurrence of the risk event
4. 0.20–0.40: High likelihood of occurrence of the risk event
5. 0.40–0.80: Very high likelihood of occurrence of the risk event

If the project team members are certain that an event will occur (i.e., scale greater than 0.80), that event is not a risk; it is a certainty. Certain events are not to be handled by the risk management process.

19.6.2 Effect of Occurrence of a Risk Event: Risk Impact

Evaluating the impact or effect of occurrence of a risk event on the project success criteria is also done subjectively by the project team members based on historical information and knowledge, that has been accumulated on past similar projects and operations work activities.

An impact scale, which reflects the significance of impact, either negative for threats or positive for opportunities, on each the project success criteria can be used by the project team. Impact scales are specific to the success criteria potentially impacted, the type and size of the "process improvement" project, the enterprise business intended strategy and financial state, and the enterprise business sensitivity to particular impacts.

Relative scales for impact are also assigned descriptors such as "very low," "low," "moderate," "high," and "very high," reflecting increasingly extreme impacts as defined by the enterprise business.

Negative relative "risk impact" scale:

1. *0–0.05: Very low, negligible "risk impact";* if the risk event occurs, it will have no effect on the project success criteria. All requirements will be met.
2. *0.05–0.10: Low, minor "risk impact";* if the risk event occurs, the project success criteria will be lightly affected. Minimum acceptable requirements will be met. Most secondary requirements will be met.
3. *0.10–0.20: Moderate "risk impact";* if the risk event occurs, the project success criteria will encounter moderate degradations. Minimum acceptable requirements will be met. Some secondary requirements may not be met.
4. *0.20–0.40: High, serious "risk impact";* if the risk event occurs, the project success criteria will encounter major degradations. Minimum acceptable requirements will be met. Secondary requirements may not be met.
5. *0.40–0.80: Very high, critical "risk impact";* if the risk event occurs, the project will fail.

Positive relative "risk impact" scale:

1. *0–0.05: Very low, negligible "risk impact";* if the risk event occurs, it will have no effect on the project success criteria. All requirements will be met.
2. *0.05–0.10: Low, minor "risk impact";* if the risk event occurs, the project success criteria will be lightly affected. All requirements will be met.
3. *0.10–0.20: Moderate "risk impact";* if the risk event occurs, the project success criteria will encounter moderate improvements. All requirements will be met
4. *0.20–0.40: High, positive "risk impact";* if the risk event occurs, the project success criteria will encounter major improvements. All requirements will be met
5. *0.40–0.80: Very high, positive noteworthy "risk impact";* if the risk event occurs, the project will succeed. All requirements will be met.

In accordance with the Project Management Body of Knowledge guidelines, the relative scale above may represent the enterprise business desire to avoid high-impact threats or exploit high-impact opportunities, even if they have relatively low probability.

Table 19.5 Definition of risk impact scale for "Threat Events" on four project success criteria

		Risk Impact Relative scales				
		Very low -0.05	Low -0.10	Moderate -0.20	High -0.40	Very high -0.80
Project Success Criteria	Reduce Cost	Insignificant cost increase	< 10% cost increase	10-20% cost increase	20-40% cost increase	> 40% cost increase
	On Time Delivery	Insignificant time increase	< 5% time increase	5-10% time increase	10-20%time increase	> 20% time increase
	Agreed Upon Scope	Scope decrease barely noticeable	Minor areas of scope affected	Major areas of scope affected	Scope reduction unacceptable	Project outcomes is effectively useless
	Agreed Upon Quality	Quality degradation barely noticeable	Only very demanding applications are affected	Quality reduction requires sponsor approval	Quality reduction unacceptable	Project outcomes is effectively useless

In using these scales, it is important that the project team understands what is meant by the numbers and their relationship to each other, how they were derived, and the effect they may have on the different success criteria of the project.

Tables 19.5 and 19.6, adapted from the Project Management Body of Knowledge guidelines, provide examples of negative and positive impacts definitions that might be used in evaluating risk impacts related to four project success criteria.

19.6.3 Risk Matrix: Importance or Ranking of Risks

When undertaking a risk assessment, it is quite common to identify a hundred or more risks that could impact the project success criteria. This is an unmanageable number of risks and so a means is required to reduce the number that will be considered to be priority issues for management.

The importance of a risk is the level of the risk before any actions have been taken to change the likelihood or impact of the risk event. Although there are advantages in identifying the risk importance, there are practical difficulties in identifying this with certain types of risks. Techniques for ranking risks are well established, but there is also a need to decide what scope exists for further improving control. Consideration of the scope for further cost-effective improvement is an additional consideration that assists the clear identification of the priority significant risks.

There are many different styles of risk matrix. The most common form of a risk matrix is one that demonstrates the relationship between the likelihood of the risk materializing and the impact of the associated event should the risk materialize.

Table 19.6 Definition of risk impact scale for "Opportunity Events" on four project success criteria

Project Success Criteria		Risk Impact Relative scales: Very low +0.05	Low +0.10	Moderate +0.20	High +0.40	Very high +0.80
	Reduce Cost	Insignificant cost reduction	< 10% cost reduction	10-20% cost reduction	20-40% cost reduction	> 40% cost reduction
	On Time Delivery	Insignificant time reduction	< 5% time reduction	5-10% time reduction	10-20% time reduction	> 20% time reduction
	Agreed Upon Scope	Scope increase barely noticeable	Minor areas of scope affected	Major areas of scope affected	Scope increase unacceptable	Project outcomes is effectively useful
	Agreed Upon Quality	Quality improvement barely noticeable	Only very demanding applications are affected	Quality improvement approved by sponsor	Satisfying Quality improvement	Project outcomes is effectively useful

As well as likelihood and impact, other features of the risk can be represented on a risk map, similar to the one illustrated in Fig. 19.1. For example, the scope for achieving further risk improvement is often represented using a risk map. In this case, the risk map will demonstrate the level of risk, in relation to the additional measures that can be taken to improve the management of that risk and thereby set a target level for it.

We would say that a risk is significant if it could have an impact in excess of the benchmark test for significance for that type of risk. Identification of potentially significant risks will be undertaken during a risk ranking exercise. It is necessary to further decide on the:

1. Size of the impact that the event would have on the organization;
2. Scope for further improvement in control.

This will lead to the clear identification of the priority significant risks.

One of the most effective techniques for distilling a listing of identified risks into a ranked listing is by creating a matrix of risk importance which reflects both the degree to which a risk may come to fruition, times the degree to which the risk will have an impact a project, should it happen. Thus, the risk importance is generally represented as:

$$Risk\,Importance = \mu(likelihood\,of\,occurrence) \times Risk\,Impact$$

Where $\mu(likelihood\,of\,occurrence)$, is (an estimate of) a measure of the likelihood of occurrence of the risk event.

Table 19.7 Risk rating matrix

		Risk Impact Relative Scales									
		Threats					Opportunities				
		0.05	0.10	0.20	0.40	0.80	0.80	0.40	0.20	0.10	0.05
Estimate of measure of likelihood	0.80	0.040	0.080	0.160	0.320	0.640	0.640	0.320	0.160	0.080	0.040
	0.40	0.020	0.040	0.080	0.160	0.320	0.320	0.160	0.080	0.040	0.020
	0.20	0.010	0.020	0.040	0.080	0.160	0.160	0.080	0.040	0.020	0.010
	0.10	0.005	0.010	0.020	0.040	0.080	0.080	0.040	0.020	0.010	0.005
	0.05	0.003	0.005	0.010	0.020	0.040	0.040	0.020	0.010	0.005	0.003

Using the scales defined previously on estimates of measures of the likelihood of occurrence of risk events and on risk impacts, the project team should establish an importance matrix, as illustrated on Table 19.7, which leads to rating the identified risks as low, moderate or high.

1. 0–0.02: Low priority risk event; if the risk event occurs, the project success criteria will be lightly affected.
2. 0.02–0.10: Moderate priority risk event; if the risk event occurs, the project success criteria will be affected moderately.
3. 0.10–1.0: High priority risk event; if the risk event occurs, the project success criteria will be highly affected.

The project team can develop importance matrices separately for each of the project success criterion or it can develop one overall importance matrix for the project as weighted sum of individual importance matrix on each of the project success criterion.

19.6.4 Risk Prioritization

Project risk identification can result in a long list of risk events even for so-called smaller or medium-size projects. It is not always possible to track and manage every risk event at the relative expense that could be incurred in terms of cost and time. In order for the project team to concentrate on significant risks, a test for risk significance is required. For project risks that will have a financial or commercial impact, the benchmark test is likely to be based on monetary value. For project risks that could disrupt the infrastructure or activity works associated with the project, a benchmark test based on the risk impact, cost and duration of disruption is appropriate. This may vary according to the nature of the risk and whether it is a financial or non-financial one. In enterprise businesses, identifying a financial test for significance can be undertaken in a number of ways. Many enterprise businesses will have authorization procedures for spending.

The project team could also prioritize the identified risks based on their importance to identify the most significant project risks that should be managed immediately. For example high priority threats, if they occur, require priority action and aggressive response strategies. Moderate priority threats may not require proactive management action beyond being monitored or adding a contingency reserve. In the same vein, high priority opportunities, which offer the greatest benefit, if they occur, should be targeted first by the project team. Low priority opportunities should be monitored.

While most risks should be monitored, those with higher probabilities and impacts will usually require the most concerted attention of the project manager and project team. Also, lower probability and impact risk events should still be tracked and monitored because their probability and impact could change as the project progresses. An initially low probability and impact risk event could ultimately become one with high impact.

The result of risk prioritization is a list of identified risks that can be examined at regular intervals during the course of the project according to the assigned importance. The project manager should maintain a risk watch list, containing a list of the major risks that have been identified, assessed and prioritized for risk response action. For large projects, appropriate managers at each level of management in the project will maintain their own risk watch lists for their areas of responsibility.

As indicated at the beginning of this section, risk assessment relates to subjectively judging the likelihood of occurrence and impact of each identified risk event, ranking identified risk events, and allocating ownership. Frequently, no further analysis needs to be done although some risks might warrant more analysis, including quantitative risk analysis.

19.7 Develop Risk Response Planning

This is the project management process for developing options and actions to enhance the likelihood and impacts of positive events, and to reduce the likelihood and impacts of negative events to the project success criteria. Subsequent to the documentation and assessment of relevant risks, the project team must determine its response to each identified risk. Planning risk response includes the identification and assignment of one or more persons (the "risk response owner") to take responsibility for each agreed-to and funded risk response.

Risk response development is a critical element in the risk management process that determines what action (if any) will be taken to address risk issues evaluated in the identification and assessment efforts. All the information generated to date becomes critical in determining what the project team will do that is in keeping with the risks, the stakeholder's risk tolerance, the project tolerances, and the customer culture.

19.7.1 Planning Responses

A response plan should be developed before project risk event occur. Then, if a project risk event should occur, the project team simply executes the plan already developed. Planning ahead provides the time to carefully analyze the various options and determine the best course of action. That way, the project team is not forced to make a hasty, and perhaps less thoughtful, response to a threatening situation.

All risks have causes; sometimes multiple risks within a given project arise from a common cause. In developing risk responses, the project team should work to identify any common causes as those causes may have common risk responses. As the Project Management Body of Knowledge indicates, the "Risk Response Planning" process addresses the risks by their priority, inserting resources and activities into the allocated funds, schedule, and project management plan, as needed. Planned risk responses must be appropriate to the significance of the risk, cost effective in meeting the challenge, timely, realistic within the project context, agreed upon by all parties involved, and owned by a responsible person. Selecting the best risk response from several options is often required.

There are several risk response strategies available and depending on whether an identified risk event has negative of positive impacts on the project success criteria. The project team should select the strategy or mix of strategies most likely to be effective for each risk. For the selection purpose, risk analysis tools, such as decision tree analysis, can be used to choose the most appropriate responses. Then, specific actions developed to implement the chosen strategy or to develop an alternative plan for implementation if the selected strategy turns out not to be fully effective, or if an accepted risk occurs.

19.7.1.1 Strategies for Threat Events

Three strategies typically deal with threat events or risk events that may have negative impacts on project success criteria if they occur. These strategies are to avoid, transfer, or mitigate. If the project team is unable to identify a response option that will reduce the likelihood and impact to an acceptable level, then avoidance is chosen as the appropriate risk response to that individual risk. Transfer and mitigate responses are selected for implementation when the project team determines that the activity will reduce the residual (i.e., current level of) risk to an acceptable level.

Risk Avoidance

In many situations, a lower risk choice is available from a range of risk alternatives. Selecting a lower risk option or alternative approach represents a risk avoidance decision. Certainly, not all risk can or should be avoided. On occasion, choosing a higher risk can be deemed more appropriate because of design flexibility, enhanced performance, or the capacity for expansion.

Risk avoidance involves changing the project management plan to eliminate the threat posed by an adverse risk, to isolate the project success criteria from the risk's impact, or to relax the project success criterion that is in jeopardy, such as extending the schedule or reducing scope. Some risks that arise early in the project can be avoided by clarifying requirements, obtaining information, improving communication, or acquiring expertise.

Activities associated with a high likelihood of risk occurrence with significant negative financial impact typically evoke a response that results in recommendation of complete avoidance of the activity. Simply stated, these are situations where the consequences or impact are so great that the project team chooses to avoid them completely. Risk avoidance commonly is invoked in cases where the probability and consequence of the risk impact is significant enough to potentially have a severe negative effect on a major project success criterion.

It may be possible to eliminate the source of risk, and therefore prevent it from happening. This may involve an alternative strategy for completing the project. For example, rather than assigning work to a new, less expensive contractor, the project team may choose to reduce the risk of failure by using a known and trusted contractor—even though the cost may be higher. A project team can never avoid all risk, but it can try to eliminate as many sources as possible.

In the area of procurement, many enterprise businesses have adapted a policy of using only experienced, dependable, proven suppliers, in lieu of holding an open public competition and awarding an order to the lowest price. The lowest bid price may not be the best price if the award goes to a seller who has no demonstrated expertise in a given area.

Another risk avoidance technique is to use only pre-qualified suppliers, using a two-step procurement practice. Step one pre-qualifies the prospective suppliers according to established criteria, and eliminates the unqualified suppliers. Step two solicits bids from only qualified sources.

Some enterprise businesses, in an effort to reduce the risks on highly complex procurements, will take two deliberate steps before issuing their Request for Proposal (RFP) to selective suppliers. First, they will take the Request for Proposal package and conduct an in-house "bid-ability review." This process will assign an experienced team to examine the solicitation package and determine whether or not anyone can make sense of it. Often this results in a re-write of the draft Request for Proposal. If prospective suppliers cannot discern what is wanted in the Request for Proposal, the risks of suppliers adding useless contingencies will increase.

Another practice with complex procurements is to take the Request for Proposal package and perform an "independent cost estimate" for the new work. The estimate gives the project an independent benchmark against which the suppliers cost proposals may be compared.

Risk Transfer

Also known as "deflection," risk transfer strategy is the effort to shift responsibility or consequence for a given risk to a third party. It involves moving the negative impact of a threat, along with ownership of the response to a third party, usually for the payment of a risk premium. For example, the project team can avoid the chance of a cost overrun on a specific activity by writing a fixed-price contract. In such a case, the contractor agrees to complete the work for a predetermined (higher) price and assumes the potential consequences of risk events. If the risk is low, the project team could choose to accept the risk and write a cost-plus contract, paying the contractor only the actual costs plus a predetermined profit.

Other examples of risk transference include the purchase of insurance, bonds, guarantees, and warranties. This transfer is often described as risk financing. The fundamental principle of insurance is that the insurance company is contracted to pay a certain sum of money in the event of defined circumstances arising or defined events occurring. Insurance contracts can require the insurance company to pay for losses suffered directly by the insured. This is first-party insurance and includes property damage insurance. Other types of insurance contract the insurance company to pay compensation to other parties if they have been injured or suffer loss because of the activities of the insured. This is third-party insurance and includes motor third-party and public/general liability.

Insurance contracts are contracts of utmost good faith. This means that the insured party is required to disclose all information relevant to the insurance contract. If this information has not been disclosed, the insurance company or underwriter has the right to refuse to continue to provide insurance cover and may refuse to pay any claims that have arisen.

There are advantages and disadvantages associated with the use of insurance as a risk transfer mechanism. The advantages of insurance are that it provides indemnity against an expected loss. Insurance can reduce uncertainty regarding hazard events that may occur. It can provide economic benefits to the insured, because the loss may be greater than the insurance premium. Finally, insurance can provide access to specialist services as part of the insurance premium. These services may include advice on loss control.

The disadvantages of insurance include the delays often experienced in obtaining settlement of an insurance claim and the difficulties that can arise in quantifying the financial costs associated with the loss. There may be disputes regarding the extent of the cover that has been purchased and the exact terms and conditions of the insurance contract. Finally, the insured may have difficulty in deciding the limit of indemnity that is appropriate for liability exposures. This may result in under-insurance and the subsequent failure to have claims paid in full.

There are alternatives to insurance when a project manager wishes to transfer the financial consequences of a threat event. Alternatives to insurance are sometimes referred to as alternative risk transfer or alternative risk financing techniques. The risk financing options available to an organization include:

1. Conventional insurance;
2. Contractual transfer of risk;

3. Captive insurance companies;
4. Pooling of risks in mutual insurance companies;
5. Derivatives and other financial instruments.

Transferring an identified and assessed risk rarely serves to eliminate the risk. Instead, if creates an obligation for mitigation, acceptance, or avoidance on another individual or business function taking the responsibility for its management.

Transferring liability for risk is most effective in dealing with financial risk exposure. Risk transference nearly always involves payment of a risk premium to the party taking on the risk. Transference tools can be quite diverse and include, but are not limited to, the use of insurance, performance bonds, warranties, guarantees, etc. Contracts may be used to transfer liability for specified risks to another party. In many cases, use of a cost-type contract may transfer the cost risk to the buyer, while a fixed-price contract may transfer risk to the seller, if the project's design is stable.

Risk Mitigation

Risk mitigation is the most common of all the risk handling strategies on threats. It is the process of taking specific courses of action to reduce the probability and/or reduce the impact of occurrence of risk events. This often involves using reviews, risk reduction milestones, novel work approaches, and similar management actions. The project manager must develop risk mitigation plans and then track activities based on those plans. All these actions are built into the project plan (cost plans, schedule plans) and ultimately into the work breakdown structure.

Mitigation plans are steps taken to lower the probability of occurrence of the risk event or to reduce its impact should it occur. For example, the project team can reduce the likelihood of a product failure by using proven technology rather than cutting-edge technology. Mitigation costs should be appropriate to the likelihood of the risk event and its potential impact on the project success criteria. Some mitigation strategies may not take a lot of effort, but may have large payoffs in eliminating the potential for disaster. On a project with a tight deadline, the risk of delayed delivery of raw materials may be disastrous. If two vendors can provide materials at essentially the same price, but one has a much larger inventory and a significantly better history of on-time delivery, choosing the vendor with the better track record may be an easy mitigation strategy with a potentially large payoff.

Taking early action to reduce the probability and/or impact of a risk occurring on the project is often more effective than trying to repair the damage after the risk has occurred. Adopting less complex processes, conducting more tests, or choosing a more stable supplier are examples of mitigation actions. Where it is not possible to reduce probability, a mitigation response might address the risk impact by targeting linkages that determine the severity. For example, designing redundancy into a subsystem may reduce the impact from a failure of the original component.

19.7.1.2 Strategies for Opportunity Events

Three strategies typically deal with opportunity events or risk events that may have positive impacts on project success criteria if they occur. These strategies are to exploit, share, or enhance (PMI, 2010).

Exploit Risk

This strategy seeks to eliminate the uncertainty associated with a particular upside risk by making the opportunity definitely happen. Directly exploiting responses include assigning more talented resources to the project to reduce the time to completion, or to provide better quality than originally planned.

Share Risk

Sharing a positive risk involves allocating ownership to a third party who is best able to capture the opportunity for the benefit of the project. Examples of sharing actions include forming risk-sharing partnerships, teams, special-purpose companies, or joint ventures, which can be established with the express purpose of managing opportunities.

Enhance Risk

This strategy modifies the level of an opportunity by influencing, facilitating or strengthening the conditions that create occurrence of the identified risk event to increase its likelihood of occurrence, and by identifying and maximizing key drivers of these positive-impact risks. Impact drivers can also be targeted, seeking to increase the project's susceptibility to the opportunity.

19.7.1.3 Strategies for Both Threat and Opportunity Events

When there is a low likelihood of occurrence of a risk event, when the potential impact on the project success criteria is low, or when the cost of mitigation is high, a satisfactory response may be to accept the risk. For example, say that the economy moves into a recession midway into a project to reengineer an automobile manufacturing plant for increased efficiency and output. The enterprise business may choose to proceed with the project anyway, and accept the risk that lower sales will reduce the return on investment below what was expected. The project team accepts certain risks by virtue of the fact that the enterprise business operates a business.

Acceptance, also known as retention, is the decision to acknowledge and endure the consequences if a risk event occurs. It is broken down into two basic types of acceptance, active and passive.

1. *Passive acceptance* is the acceptance of risk without taking any action to resolve it, cope with it, or otherwise manage it. The only actions required in passive acceptance are documentation of the risk, as well as acknowledgement by management and the team (and the customer, if appropriate) that the risk exists and that the organization is willing to endure its consequences, should the risk occur.
2. *Active acceptance* acknowledges the risk as well, but calls for the development of contingency plans, and in some cases, fallback plans. Contingency plans are implemented to deal with risks only when the risk events come to pass. This may include detailed instructions on how to manage risks retroactively or may be as simple as a contingency reserve budget established for the project.

Contingency reserves are frequently fodder for discussion because some view these reserves as project panaceas and others see them as a crutch for those who cannot manage effectively. These reserves are sometimes referred to as contingency allowances. Enterprise businesses should not establish universal rules for applying contingency, such as flat percentages or fixed monetary (or schedule) amounts. Instead, contingency reserves should reflect the degree of risk acceptance in a project, as well as the overall levels of risk associated with the project.

Enterprise businesses may set contingency values by applying culturally acceptable metrics to the risk models. They may also set contingency reserves through negotiation with the project manager or by using the expected values of the project's assessed risks. Nonetheless, if contingency reserves are to be applied, they must reflect the realities of the project as a unique effort toward a specific objective, thus requiring a specific level of risk support.

Fallback plans are implemented in active acceptance to deal with managing accepted risks if the contingency plans are insufficient. These plans represent the safety net that ensures the entire project will not collapse in failure.

The benefit of undertaking assessment of project risks is that the difference between the current level and the baseline level can be identified. This will give an indication of the importance of the existing control measures and information is used by internal auditors to help identify critical controls and set audit priorities.

19.7.2 Developing a Response Plan

After considering the options of avoiding, transferring, mitigating, exploiting, sharing, enhancing or accepting the risk, the project team may develop a risk management plan, contingency plans, and reserves.

1. *A risk management plan documents the procedures that will be used to manage risk throughout the project lifecycle.* It lists potential risk events, the conditions or signs that may warn of the impending event, and the specific actions to be taken in response.
2. *Contingency plans describe the actions to be taken if a risk event should occur.* Reserves are provisions in the project plan to mitigate the impact of risk events. These are usually in the form of contingency reserves (funds to cover unplanned costs), schedule reserves (extra time to apply to schedule overruns), or management reserves (funds held by general management to be applied to projects that overrun).

After identifying your plans to avoid, transfer, mitigate, exploit, share, enhance or accept the risk, the project team may need to add specific activities to the work breakdown structure and other plans.

Selecting the proper strategy require the project manager and the project team to identify specific strategies for each risk. It may also require that the project manager and its team identify single strategies that may apply to a broader subset of risks or to common causes. A popular tool for identifying such opportunities is the risk response strategy matrix. This matrix encourages the examination of risk responses both in the context of other risks in the project as well as in the context of the other risk responses.

19.7.2.1 Risk Response Matrix

In risk response development, one of the key challenges is finding strategies that will not take longer to implement than the project itself. The risk response matrix addresses that concern by affording individuals and team members the opportunity to analyze and generate strategies that deal with multiple risks and cause the fewest problems in terms of other project risk.

The risk response matrix is created by using the quality function deployment constructs described in the previous chapter. It is concerned with the translation of identified and assessed project risks to substitute risk response measures, language and priorities. The translation is done by building the analogous "House of Quality" associated with the identified and assessed project risks and risk responses.

Figure 19.4 shows a typical "Risk Response Matrix." The matrix structure and visual nature give both discipline and guidance to the conversion process by exploring the information that it contains. The identified and assessed project risks, represented on the left side by WHATs—and the risk responses—represented on the top by HOWs—are the input to the matrix and the foundation for further activities. Next to the identified and assessed project risks is the importance rating of those risks.

The body of the "Risk Response Matrix"; i.e., the "Relationship Matrix" is where the relationships are categorized. This is where identified and assessed project risks "translated" into operation terms. It is also where interactions between relationships between a given WHAT and a given HOW are identified so that the synergistic effect meeting the identified and assessed project risks is seen. In filling in the "Relationship Matrix," a great deal of dialogue will take place between the project team members. Here the team will identify the relationships between the identified and assessed project risks that matter the most and the risk responses. The relationships are defined as strong, medium/some, and weak/possible relationships.

Even though a risk response strategy may have been created primarily to resolve or deal with a single risk, each risk response strategy should be evaluated for its own potential impact on the other risk events listed. Risk response strategies frequently have unforeseen consequences (both favorable and unfavorable) when considered against the project's other risk events.

To document the influence of the risks, a positive strong/medium/weak relationship can indicate when a risk strategy will have a positive strong, medium or weak influence on a risk event (for instance, a +9 next to budget overrun would indicate that the strategy would likely strongly reduce overall cost or minimize the possibility of budget overruns). A negative strong/medium/weak can indicate when a risk strategy might have a negative strong/medium/weak influence on the risk event (for example, a −3 next to schedule delay would indicate that the strategy will likely add to the schedule or increase the probability of delays).

As indicated already, a strong relation is often taken to be equal to 9, a medium equals 3, and a weak equals 1. Symbols are usually used in order to aid in the recognition of patterns. Numbers are substituted in at a later time to calculate weight at the bottom of the matrix. During the dialogues that will take place, the project team members must address both the operation content (especially the

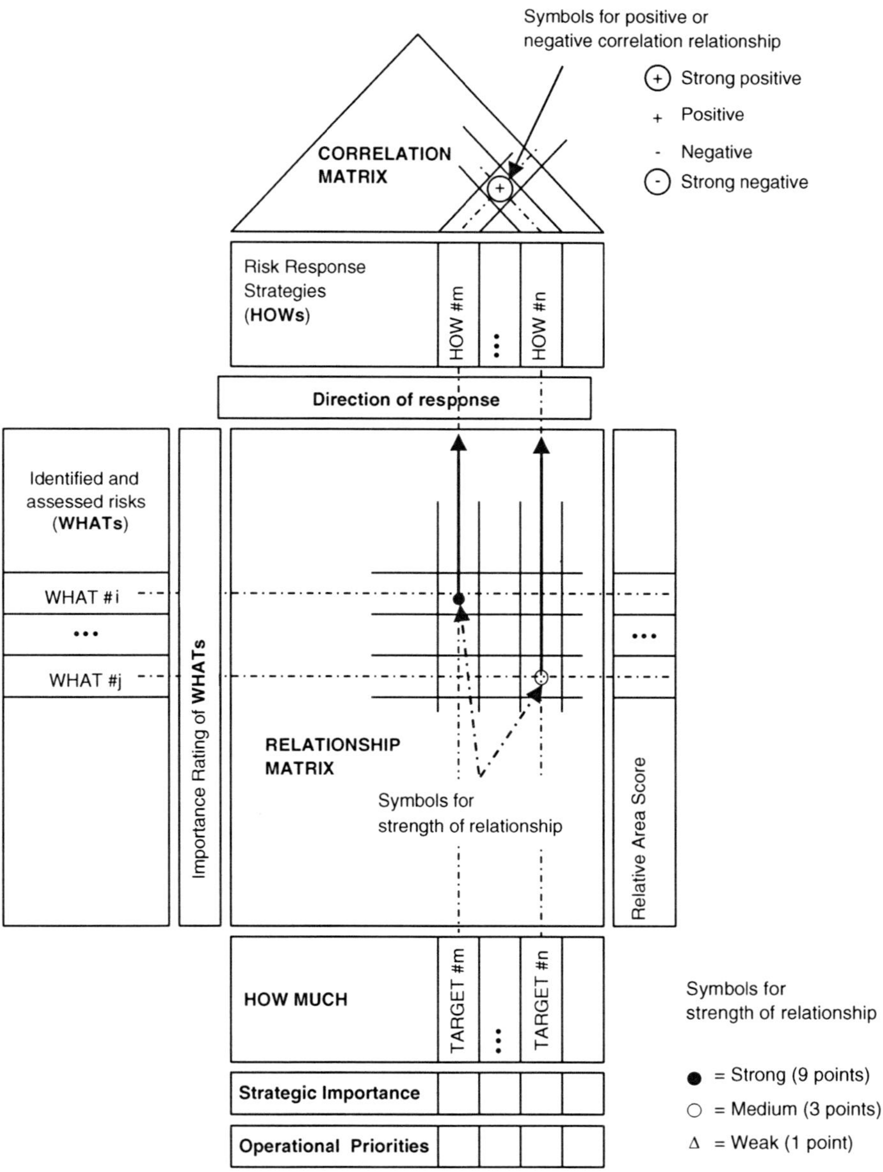

Fig. 19.4 The risk response matrix

validity of the identified and assessed project risks) and the context issues. Specifically, they must address the inevitable concerns (spoken or unspoken) about how this risk response effort will differ from those of the past.

The "Correlation Matrix" ("Roof" of the House) assesses the risk response strategies' impact on other strategies or compares the risk response strategies

(HOWs) to determine if they are in conflict or assisting each other. It identifies positive and negative relationships/correlations, that is, technical trade-offs. This is valuable because in most cases, these risk response trade-offs have not been documented prior to this time. The trade-offs are often the source of compromises because of the limitations of currently available resources or facilities. By identifying them early on, the project team can narrow their efforts. Determining the correlation matrix is often undertaken as a matter of course rather than as a formal step in the process of creating the "Risk Response Matrix."

Below the "Relationship Matrix," the project team sets strategic priorities and expected levels (HOW MUCH) of risk response strategies impacts on the identified and asses project risks that matter the most. These expected levels form the "wish list" that drives the risk response effort and allow determining the risk response strategies with the greatest overall positive impact. Some compromises may be required when all of the wish list items are challenging and cannot be implemented. Although this is a subjective decision, it is tempered by virtue of the tool's indications that some risk strategies have a broader span of influence than others. Thus, by determining which risk strategies in general are the most beneficial and have the least negative influence, it is possible to review options in the context of the project's overall risk environment.

Ideally, the risk response matrix should include the standard risks of cost and schedule. The risk response matrix is applied after that the project team has identified and assessed risks to establish those that are the greatest concerns. It is best applied when the skills and insights of the entire team can be exercised since team members may have widely different perceptions as to what constitutes a corresponding strategy or a conflicting risk approach. The risk response matrix should be applied whenever risk strategies are being evaluated and should be a part of any strategy assessment or major risk reassessment.

The outputs from the "Risk Response Matrix" can be used in basic decision making or to present information to upper-level or executive management to facilitate their decision making. However, the information ultimately needs to be captured, reviewed, and presented to build organizational support and acceptance for the risk management options selected.

Ideally, the project team that has completed risk response planning will have established a contingency reserve for the necessary funds and time to deal with project risk. They will have an adjusted work breakdown structure that reflects issues that surfaced during risk response assessment and incorporates any new activity the strategies require. They also will have communicated the risks, risk strategies, and any residual (or leftover) risks to the enterprise business management team to ensure there is buy-in on the approach. Moreover, they will have contractual agreements to support any risk deflection. As a byproduct, there is also the possibility that new risks will arise as a result of the new strategies. Those new risks should be examined using the same process as the earlier risks-identification, assessment, and response planning-as appropriate.

19.8 Monitor and Control Risk

This is the project management process for planning a set of systematic observation techniques and activities focused on identifying, analyzing, and planning for newly arising project risks, keeping track of the identified risks and those on the risk registry, reanalyzing existing risks, monitoring trigger conditions for contingency plans, monitoring residual risks, and reviewing the execution of risk responses while assessing their effectiveness. It connects risk management to other project management processes. Continuous monitoring and control of risks is an important part of implementation, particularly for large projects or those in dynamic environments.

After project risks are identified, assessed, and clear responses are developed, those findings must be put into action. Risk monitoring and control involves implementing the risk management plan, which should be an integral part of the project plan. Two key challenges are associated with monitoring and control. The first is putting the risk plans into action and ensuring the plans are still valid. The second is generating meaningful documentation to support the process.

Through the "Risk Monitoring and Control" process, illustrated in Fig. 19.5, implementing the risk plans should be a function of putting the project plan into action. If the project plan is in place and the risk strategies have been integrated, then the risk plans should be self-fulfilling. Ensuring that the plans are still valid, however, is not as simple. Risk monitoring involves extensive tracking of the risks and their environment. Have the plans been implemented as proposed? Were the responses as effective as anticipated? Did the project team follow organizational policy and procedure? Are the project assumptions still valid? Have risk triggers occurred? Have new external influences changed the organization's risk exposure? Have new risks surfaced?

Answers to these questions may drive radically different approaches to the project and to its risks. Alternative strategy development, reassessments, reviewing contingency plan implementation, or planning anew may be essential to project survival or success.

19.8.1 Choose Control Subject

The first step of the "Project Risk Monitoring and Control Process" is "Choose the Control Subject"—Each project success criteria in terms of each individual work package objective is a control subject; a center around which the risk monitoring and control process is built.

19.8.2 Establish Standard Performance

The second step of the "Project Risk Monitoring and Control Process" is "Establish Standard of Performance"—It relates to collecting the standards of performance

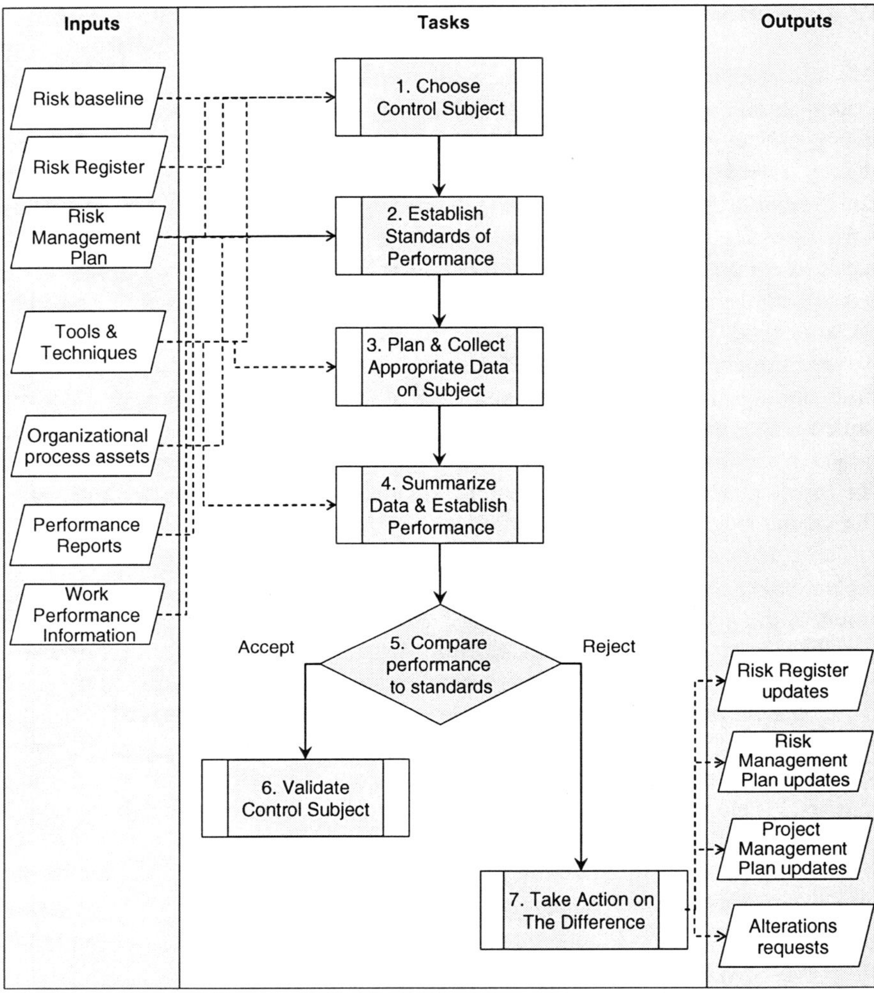

Fig. 19.5 The project risk monitoring and control process

baseline on the success criteria related to each individual work package objective. For each control subject it is necessary to know its standard of performance.

19.8.3 Plan and Collect Appropriate Data

The third step of the "Project Risk Monitoring and Control Process" is "Plan and Collect Appropriate Data" on the chosen "Control subject"—It relates to establishing the means of tracking success criteria related to each individual work package objective in order to determine the actual performance of the project in terms of its earned value, described in a previous section of the "Cost Management" process. Here, the percent complete of in-progress schedule work package activities, their costs and agreed upon quality requirements should be included in the data collection outcomes. To facilitate the periodic reporting of project performance, a template created for consistent use across various project organizational components can be used throughout the project life cycle. The template can be paper-based or electronic.

Tracking of success criteria related to each individual work package objective begin with the collection of information needed to accomplish the prescribed risk assessments. Data collection can be specified to occur at some recurring point in time when data is needed for risk assessment purposes, or it may be accomplished as an ongoing activity over a period of time where data is collected regardless of when risk assessments reviews or risk audits are performed.

As the project progresses, there are risk-specific evaluations to facilitate risk control. Formal risk audits examine the project team's success at identifying risks, assessing probability, and developing appropriate strategies. The frequency of risk audits is largely determined by the duration of the project and the criticality of the deliverables involved. A project with mission-critical deliverables will, by its very nature, undergo more frequent audits than a project developed for a support mission.

Risk reviews, though less formal than risk audits, are vital nonetheless. Risk reviews allow for an examination of the risks, probabilities, impacts, and strategies, largely to determine if supplemental action or review will be required. As with audits, the criticality of the project and its duration determine in large part the frequency of such reviews.

An ongoing data collection approach is recommended, particularly if risk assessments are conducted infrequently, for example, only monthly or quarterly. This removes the burden of trying to capture or recreate past data that may have been replaced by current data. Also, ongoing data collection (even without formal risk assessment) provides indicators of potential project performance issues or problems that would not otherwise surface in a timely manner.

19.8.4 Summarize Data and Establish Actual Performance

The fourth step of the "Project Risk Monitoring and Control Process" is "Summarize Data and Establish Actual Performance" of the chosen "Control subject"—Tracking success criteria related to each individual work package objective is typically summarized weekly for shorter projects and at least monthly for larger projects. To ensure proper risk monitoring and control, the project manager (or a qualified designee) should review and approve all newly identified and assessed project risks. Such approval should not be "rubber stamped." Rather, the approval process should prompt a detailed examination of planned project performance versus acquired performance, in conjunction with verifying success criteria related to each individual work package objective.

19.8.5 Compare Actual Performance to Standard

The fifth step of the "Project Risk Monitoring and Control Process" is "Compare Actual Performance to Standards"—The act of comparing the actual project performance of the chosen "Control subject" to standards is performed by carrying out any or all of the following activities:

1. Compare the actual project performance to the project completion date, cost and quality goals.
2. Interpret the observed difference; determine if there is conformance to the goals.
3. Decide on the action to be taken.
4. Stimulate preventive and/or corrective actions.

During project implementation, one of the key responsibilities of project manager is to measure project performance. This responsibility entails monitoring project performance and project risks to detect and analyze deviation from the established baselines.

19.8.6 Validate Control Subject

The sixth step of the "Project Risk Monitoring and Control Process" is "Validate Control Subject"—It relates to acceptance decisions from the risk monitoring and control results, which will indicate how well the chosen "Control subject" has been absorbed by the project, how much work has been completed, and the extent of residual risks that the project is facing.

The project manager and other team members monitor the "Control Subject" throughout the project its life, looking for triggers and signs that may warn of impending risk events that may affect its performance. When risk events happen, the appropriate corrective action identified in the risk management plan must be executed. When an unplanned risk event occurs, a response must be developed and implemented. After implementation of the response, the risk management plan should be reviewed and updated if necessary. It may also be necessary to adjust other project plans or the basic project objectives. As changes in the project occur, it may be necessary to repeat the steps of identifying, assessing, and planning responses to risk.

19.8.7 Take Action on Difference

The last step of the "Project Risk Monitoring and Control Process" is "Take Action on the Difference." It relates to actuate alterations, through preventive and corrective risk control actions, which restores conformance with the project goals.

19.8.7.1 Preventive Controls

These are the most important type of risk control actions, and all projects will use preventive controls to treat certain types of risks. Prevention or elimination of all risks is not possible on a cost-effective basis.

The advantage of preventive controls is that they reduce threats or they enhance opportunities, so that less consideration of it is required. In reality, this may not be a cost-effective option and may not be possible for operational reasons. The disadvantages of preventive controls are that beneficial project activities may be eliminated and either outsourced or replaced with something less effective and efficient.

19.8.7.2 Corrective Controls

Corrective controls are the next option after it has been decided that preventive controls are not technically feasible, operationally desirable or cost-effective. Corrective controls are capable of producing an entirely satisfactory result, whereby the current level of risk is reduced to within the risk allowance of the project.

The advantage of many corrective controls is that they can be simple and cost-effective. Also, they do not require that existing practices and procedures are eliminated or replaced with alternative methods of work. The controls can be implemented within the framework of existing activities. The disadvantage of some corrective controls is that the marginal benefits that are achieved may be difficult to quantify or confirm as cost-effective.

19.8.7.3 Balancing Controls

It is very easy to get carried away with corrective controls. The more controls that are put in place, the lower the project risk, and the less likely it will be for the project to deviate from its baselines performance. Sometimes, corrective controls are over-engineered and their cost is disproportionate to the benefit that is achieved. It is for the project manager and the team members to identify where expensive and/or ineffective corrective controls have been implemented. Very often, corrective controls are put in place because of regulatory requirements. This may be unsatisfactory from the point of view of the enterprise business and introduce additional costs and/or inefficiency. However, it is for the project manager and the team members to ensure that the appropriate level of corrective control is achieved in order to comply with the minimum requirements of legislation.

Cost aside, there is another impact to consider. Figure 19.6 shows an example of results of analysis of the balance between the cost of controls and a reduction in the potential loss from threat events that implementing these controls would achieve. As this figure illustrates, there is a point of diminishing returns that minimizes the net cost risk for having chosen a particular level of control. That point represents the lowest total cost as a balance between cost of control and the level of potential cost of risk impact. Therefore, when selecting and implementing controls, it is important to ensure that cost-effective controls are selected.

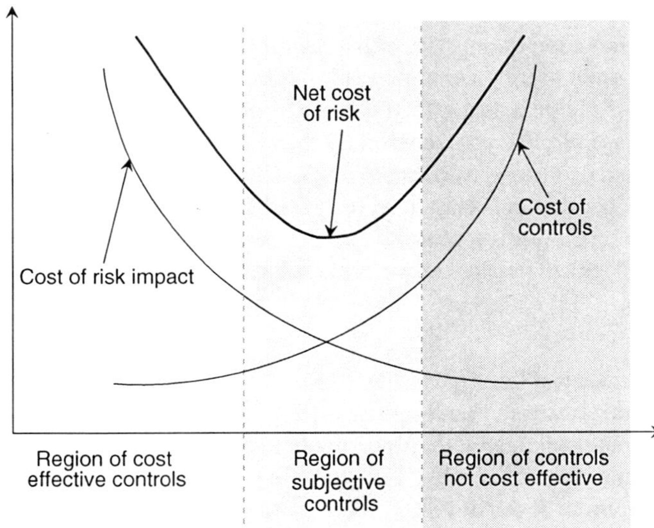

Fig. 19.6 Example of cost effective risk control analysis

The project manager needs to strike a balance between the extent of the corrective controls and the risk of unfavorable outcomes. Just as in the insurance industry, compare the cost of the policy against the dollar value of the loss that will result from the consequences.

It can be seen from Fig. 19.6 that a significant reduction in potential loss is achieved with the introduction of low-cost controls. This section of the diagram is labeled "Region of cost effective controls." The centre section of the diagram illustrates that spending more on controls achieves a reduction in the net cost of risk up to a certain point. In this segment, subjective judgment is required on whether to spend the additional sum on controls. On the right-hand side of the diagram, spending more on controls achieves only a marginal reduction in potential loss. In this segment, further controls are not cost-effective.

As risk control and monitoring are applied, data are generated. Responses succeed and fail. Some risks materialize and some do not. Likelihood of risk occurrence shift and time alters impact values. These changes may drive changes in the project's existing risk identification checklists and should also be captured in a risk register along with any new information. Retention of this information with the project plan significantly increases the probability that others will reuse this information as the project plan is appropriated for use on other, similar efforts. Risk strategies and their outcomes are critical elements of enterprise business organizational assets. Failure to properly store them in an accessible fashion is to diminish the value of the project and the project team in their contributions to technical capital.

Conduct the Project Retrospective 20

This chapter is concerned with the reflection process that must be performed at the end of each significant milestone of the "PDSA Study" project phase and from which the project team reassembles to look back on what results were actually delivered at the milestone and to what extent the team has met the expectations for the considered milestone time period.

20.1 Understanding the Reflection Process

In learning organizations, learning is an integral part of the work been performed and not another activity one has to do—it is seamless to the work itself. Unless the "Plan-Do-Study-Act" improvement cycle informs the very structure of work throughout the enterprise business—forming an interdependent network of feedback processes among all levels—organizational learning will unlikely be realized.

Only in learning organizations are people encouraged to continually challenge the way things are, along with the corresponding beliefs and assumptions. Only in learning organizations do people freely and willingly challenge what they hold in their mind—both individual and collective minds—in performing their work through the process of learning toward improvement.

On the rare occasions when a project team puts effort into capturing lessons, it is usually the last act before the project is formally closed down. It would be just a little unfair to say that this is closing the gate after the project team has bolted, but it goes some way to explaining why many potentially useful reports end up forgotten. In the process of facilitating learning throughout the enterprise business, the project manager or the project team leader should not wait until the end of the "process improvement" project to take stock. He/she should be alert to opportunities for learning lessons at all stages of the project life cycle. In this way, immediate adjustments can be made to the way that the "process improvement" project is planned or performed. The reflection process is just the tool needed to facilitate learning throughout the "process improvement" project life cycle.

A. van Aartsengel and S. Kurtoglu, *Handbook on Continuous Improvement Transformation*,
DOI 10.1007/978-3-642-35901-9_20,

The reflection process integrates or links thought and task execution with reflection. It involves thinking about and critically analyzing one's actions with the goal of improving one's professional practice (Schön, The reflective practitioner, 1983; Schön, 1987). Here, engaging in reflective practice requires individuals to assume the perspective of an external observer in order to identify the assumptions and feelings underlying their practice and then to speculate about how these assumptions and feelings have affected the achievement of a significant milestone objective.

The reflection process is a means of enhancing informal learning among project team members in the workplace. Informal learning is a vitally important aspect of learning within enterprise businesses. It has moved beyond its traditional role as a means of preparing professionals for the workforce and into the function of ongoing competence development through practices such as action learning, which was originally conceived by Reginald W. Revans (1971).

Conducting the project retrospective is performing the reflection process performed at the end of each significant milestone of the "PDSA Study" project phase and from which the project team reassembles to look back on what results were actually delivered at the milestone and to what extent the team has met the expectations for the considered milestone time period. As indicated already, it involves thinking about and critically analyzing one's actions with the goal of improving one's professional practice (Schön, The reflective practitioner, 1983; Schön, 1987).

When the project team members reflect at the end of a significant milestone, they question the assumptions behind their tacit knowledge revealed in the way they carried out tasks and approached problems and think critically about the thoughts that got them into this fix or this opportunity. The project team members may, in the process, restructure strategies of action, understandings of phenomena, or ways of framing problems. Much of a project team's work is focused on problems that occurred while studying deliverables. Problems in this context are simply a gap between what is desired at a milestone and what currently exists. The central elements of problem solving are the following steps, which can be seen both on a macro level (in project plans, for example) or on a micro level (in meeting agendas focused on more specific issues): Identify the problem, collect data about the problem, analyze the data to determine the root causes, develop possible solutions, select the most appropriate solution, implement the solution, and evaluate and monitor the situation after implementation.

20.2 When to Start the Reflection Process?

The reflection process should begin when the application of the project team's know-how does not produce the expected milestone results, the activities it conducted and the project management processes used throughout the milestone time period in the "PDSA Plan" project phase have failed to meet expectations.

The project team may decide to ignore the failure or it may respond to it by reflecting in one of two ways:

1. It may reflect "on action" by allowing its members to step away (i.e. assume the perspective of external observers) from the planning process and thinking back on their experience to understand how part of their tacit knowledge, which was revealed in the way they approach problems and carry out tasks required to reach the milestone considered, contributed to an unexpected outcome.
2. Alternatively, the project team may "reflect in the midst of the planning process without interrupting it."

Thus, the reflection process is the project management process by which the project team "stops any activity and think" about what it does or has done in order to "interpret and give meaning to the unexpected outcome at a significant milestone."

20.3 Layers of Reflection

Based on the object of the reflection process itself, three layers of reflection can be distinguished (Mezirow, 1991): content reflection, process reflection, and premise reflection.

1. *Content Reflection*—Content reflection, involves reviewing how ideas have been applied in solving problems at each stage of the problem-solving process used during the milestone time period. This first level of reflection is typical of a project team retrospective, where the team focuses on reviewing how ideas have been applied in solving problems at each step during the milestone time period.
2. *Process Reflection*—Process reflection, examines the problem-solving process itself, focusing on the procedures and assumptions involved in previous "process improvement" projects. This second level of reflection is also important for project teams, yet it is the type of reflection that is typical of a process manager, whose task is to help project managers improve the processes common to multiple projects.
3. *Premise Reflection*—Premise reflection, goes one step further by uncovering the assumptions that guided the need to address the problem in the first place. This third level of reflection is typical of the type of reflection required at the senior management level, where the primary task is to ensure that the enterprise business is solving the right problems to begin with, and that the right projects and programs have been selected to achieve the enterprise business' intended strategy.

These layers do not imply that the project team cannot or should not utilize premise reflection, for example, or that senior managers need not engage in content reflection. Rather, the three levels of reflection are aimed at helping applicable organizational members reflect in the ways that are most productive for achieving their tasks during the milestone time period. Shareholders do not hold project teams accountable for setting the right strategy. Likewise, senior managers need to empower project teams to deliver project work successfully at each significant milestone. Reflecting at three levels promotes the type of reflection for each constituency that is the best fit with the task at hand during the milestone time period. Its purpose is not to create or embed unnecessary hierarchy, but to facilitate productive reflection among the right people at the right time.

While focusing the right type of reflection at the right level is important, it is also valuable to enable project members at each level to identify problems that affect their performance, even if these problems require deeper or higher levels of reflection.

20.4 Facilitating Learning and Continuous Innovation

Conducting the project retrospective at the end of each significant milestone during the project's life cycle is the primary means for facilitating learning and continuous innovation in an enterprise business. A project retrospective, as indicated here above, enable project teams to systematically learn from experience so that they can improve upon their strategies, reduce the risk of failures and surprises, and deliver high-quality work.

To be effective, a project retrospective should be facilitated by an experienced, trained, objective facilitator from outside the project team who helps draw people out to share their perspectives, promote effective learning and reflection, and create a positive context for "process improvement" rather than one of finger-pointing, defensiveness, avoidance, or blame.

As an experienced hand, the facilitator must work to build the enterprise business' ongoing capability for "Continuous Improvement" maturity at three levels: project, process, and strategy. The facilitator should also work with program managers, project managers, project management office (PMO) personnel, and senior leaders to devise practical ways to integrate action-reflection cycles into the enterprise business' ongoing work routines and project management processes.

At the project management process level, the facilitator should work in liaison with the project management office (PMO), if one exists, to facilitate cross-project improvement. He/she conducts reflection sessions with the "process improvement" project manager to improve the project management processes that are common to multiple "process improvement" projects. PMOs are well positioned to bring innovations from one project team to the next once these improvements are identified. However, many PMOs are not yet fully equipped to play this knowledge brokering role. Instead, despite the best of intentions, they focus on promulgating rules and enforcing standards, often with limited feedback from project managers and project teams. They may neglect to involve others in action-reflection cycles aimed at improving the project management processes they define.

At the project level, the facilitator should team up with the project manager to facilitate regular reflection sessions with the project team. He/she conducts a retrospective, at which the team reviews what it intended to accomplish in that milestone time period, what was actually delivered, the reasons for the results attained, and what can be done to sustain or improve those results for the next time period.

After the retrospective session, the project manager works with the facilitator and the PMO leader to document the results and communicate them to team members, sponsors, and key stakeholders. "Report-out" meetings with senior managers and the PMO may be useful for generating support for the team's improvement actions.

21 Assess Overall Plan and Implementation

This is the phase review process performed at the end of the planning phase. It is a checkpoint to assess the overall plan and implementation proposal and ensure that the project has achieved its stated objectives as planned by refining previously provided answers to the three fundamental questions, which form the basis and the preliminary step of the PDSA model:

1. What is intended to be realized or accomplished by the "process improvement" project?
2. How will the realized or accomplished outcome of the "process improvement" project be recognized as is an improvement?
3. What alterations to the system affected by the "project to be improved" can be made based on the realized or accomplished outcome of the "process improvement" project?

21.1 Perform Planning Phase Review

A phase review form is completed to formally request approval to proceed to the next phase of a project. The phase review form should describe the status of the:

1. Overall project;
2. Project schedule based on the project plan;
3. Project expenses based on the financial plan;
4. Project staffing based on the resource plan;
5. Project deliverables based on the quality plan;
6. Project risks based on the risk register;
7. Project issues based on the issues register.

The review form should be completed by the project manager and approved by the project sponsor. To obtain approval, the project manager will usually present the current status of the project to the project board for consideration. The project board (chaired by the project sponsor) may decide to cancel the project, undertake further work within the existing project phase or grant approval to begin the next phase of the project.

A. van Aartsengel and S. Kurtoglu, *Handbook on Continuous Improvement Transformation*, DOI 10.1007/978-3-642-35901-9_21,

A sample phase review form for the project planning phase is shown in Table 21.1.

21.2 Identify and Document Lessons Learned

The last issue to be discussed at the end of a phase review process performed is something we know we should do, but most project managers rarely ever take the time. That is the final position of the project manager, and the project team, describing for the benefit of "future generations" as well as of next phases of the project just what went well, and what could have been handled perhaps better on the project planning phase. What could have been done better, and should be done differently on the next similar project planning phase?

A lesson learned session focuses on identifying ways of learning that have merit (quality), worth (value), or significance (importance) for the next phases of the "process improvement" project or for future projects within the enterprise business. During the project planning phase, the project team and key stakeholders should identify lessons learned concerning the project management element in which problems arose, how they arose, which positive or negative development was encountered, and what concrete, practical solutions or recommendations was used based on this experience.

The project manager must ask team members, stakeholders, and the project sponsor to help compile the lessons learned document. He/she should ask them what went well during the course of the project planning and what could have gone better. The following is the information you should include in the lessons learned document:

1. How the project management processes were used throughout the project planning and how successful they were in planning and tracking progress.
2. How well the project plan and project schedule reflected the actual work of the project.
3. How well the alteration/change management process worked and what might have worked better.
4. Why corrective actions were taken and whether they were effective.
5. Causes of performance variances and how they could have been avoided.
6. Outcomes of corrective actions.
7. Risk response plans that have been identified and whether they adequately addressed the risk events.
8. Unplanned risk events that occurred at the planning phase.
9. Mistakes that occurred and how they could have been avoided.
10. Team dynamics, including what could have helped the team perform more efficiently.

Lessons learned document should not be limited to only the items on this list above. Anything that worked well, or did not work well, that will help team members perform their next project better or smooth out problems before they get out of hand should be identified and documented here. Lessons learned should include detailed,

Table 21.1 Phase review form for the planning phase

PROJECT DETAILS

Project name: Project manager: Project sponsor:	Report prepared by: Report preparation date: Reporting period:

Project description:
[Summarize the overall project achievements, risks and issues experienced to date.]

OVERALL STATUS

Overall status: [Description]
Project schedule: [Description]
Project expenses: [Description]
Project deliverables: [Description]
Project risks: [Description]
Project issues: [Description]
Project changes: [Description]

REVIEW DETAILS

Review category	Review question	Answer	Variance
Schedule	Was the phase completed to schedule?	[Y/N]	
Expenses	Was the phase completed within budgeted cost?	[Y/N]	
Deliverables:			
Project plan	Was a project plan approved?	[Y/N]	
Resource plan	Was a resource plan approved?	[Y/N]	
Financial plan	Was a financial plan approved?	[Y/N]	
Quality plan	Was a quality plan approved?	[Y/N]	
Risk plan	Was a risk plan approved?	[Y/N]	
Communications plan	Was a communications plan approved?	[Y/N]	
Procurement plan	Was a procurement plan approved?	[Y/N]	
Statement of work	Was a statement of work released?	[Y/N]	
Request for information	Was a request for information released?	[Y/N]	
Request for proposal	Was a request for proposal released?	[Y/N]	
Supplier contract	Was a supplier contract approved?	[Y/N]	
Risks	Are there any outstanding project risks?	[Y/N]	
Issues	Are there any outstanding project issues?	[Y/N]	
Alterations/Changes	Are there any outstanding project alterations/changes?	[Y/N]	

APPROVAL DETAILS

Supporting documentation:
[Reference any supporting documentation used to substantiate the review details above.]

Project sponsor
Signature: Date:

This project is approved to proceed to the"PDSA Do" phase.

specific information about behaviors, attitudes, approaches, forms, resources, or protocols that work to the benefit or detriment of projects. They are crafted in such a way that those who read them will have a clear sense of the context of the lesson learned, how and why it was derived, and how, why, and when it is appropriate for use in other projects. Lessons learned at this stage represent both the mistakes made during the planning phase and the newer "tricks of the trade" identified during a project planning effort.

The content of a lesson learned report should be provided in context, in detail, and with clarity on where and how it may be implemented effectively. Because lessons learned are often maintained in a corporate database, the lesson learned documentation will frequently include searchable keywords appropriate to the project and the lesson.

The process of identifying and documenting lessons learned at this stage of the project lifecycle is particularly useful for projects that failed to pass the phase review, because there are many things that can learn from projects that fail phase reviews that will help prevent next projects from suffering the same fate.

Recording lessons learned information in the organizational process assets is one critical consideration, but equally important is the establishment of protocols to ensure access to the recorded information on a consistent basis. Lessons learned may be captured and logged in depth, but if they are not accessed by project managers and team members within the enterprise business in the future, they do not serve any real function. Access of recorded lessons learned may be encouraged through creative documentation approaches, physical location (hallways and project war rooms), or by including the mandate to access lessons learned as a key component of the performance criteria for project managers and team members.

Conclusion to "PDSA Plan" 22

Throughout the previous chapters, we have illustrated and developed the "PDSA Plan" Process Group processes needed to define and refine the "process improvement" project objectives, and plan the course of action required to attain the objectives and scope that the project is undertaken to address. The described constituent project management processes help gather information from many sources, with each having varying levels of completeness and confidence, establish baseline measurement, and develop the project management plan.

One of the biggest mistakes often made in developing the project management plan for "process improvement" projects is the failure to collect baseline data. Without baseline data, no before-and-after comparisons can be made, and it is impossible to know if there has been any improvement. Furthermore, without baseline data, credible learning measurement targets can not be established at other PDSA project phases. Without such vantage points, process improvement efforts in most enterprise businesses are often designing interventions without knowing how much improvement is wanted and in what areas. This is like "shooting in the dark"—not a very good idea—either with a gun or a process improvement program!

If relevant data is being continuously tracked in the enterprise business, baseline data should be relatively easy to collect. Unfortunately, if data collection is time-consuming, most enterprise business professionals and their clients are reluctant to invest their scarce resources in collecting baseline measurement data. This is a serious mistake that has come back to haunt most business functions that fail to track the effectiveness of their process improvement interventions—and, without baseline data, no meaningful measurement comparisons can be made.

The "PDSA Plan" Process Group processes also help identify, define, and mature the scope, cost, and schedule the "process improvement" project activities. As new project information is discovered, additional dependencies, requirements, risks, opportunities, assumptions, and constraints are identified or resolved. We have illustrated that as more project information or characteristics are gathered and understood, follow-on actions may be required due to the multi-dimensional nature

A. van Aartsengel and S. Kurtoglu, *Handbook on Continuous Improvement Transformation*, DOI 10.1007/978-3-642-35901-9_22, © Springer-Verlag Berlin Heidelberg 2013

of project management processes, which causes repeated feedback loops for additional analysis.

We have shown that within the project management framework, a complete planning of the "process improvement" project is indispensable to enhance the chance of achieving the project objectives. Not only is a complete planning the roadmap to how the work will be performed, but it is also a tool for decision making. It suggests alternative approaches, schedules, and resource requirements from which the project manager can select the best alternative. Furthermore, we have shown through the project management constituents that effective planning of the "process improvement" project within the PDSA framework is based on a foundation of effective data collection system, and almost everything else done during the planning phase is based on that.

Data collected determines what the constituent processes used during the "PDSA Plan" planning do, and data collection works through these constituent processes to touch every part of the "process improvement" project. The right sets of collected data trigger the right planning activities—because they represent factual information from many sources with each having varying levels of completeness and confidence and from which baselines are established. The goals that a "process improvement" project sets will depend on what data the project team collects. These goals, however, are really nothing more than "target values" established on a particular data collection scale. But, the project manager and the project team must first define that scale. The data collection scale can be net profits, customer satisfaction, reduce cost, reduce cycle time, improve productivity, reduce defect rate, decrease potential risk level, stakeholder influence and interest, etc... If the collected data are not reliable, the established project baselines and everything else will be as well. Thus, the data collection system is the engine that drives the "process improvement" project planning.

In order for the full power of data collection system, hence project planning, to be realized, there must be an optimal environment for effective use of collected data, there must be considerable interaction at each use of planning constituent processes leading to new insights about what data to collect, how to collect it, how to establish baselines, and what are the subsequent right planning decisions. Attaining the optimal environment, as we indicated in the previous chapter, requires a specific and intensive set of actions—a transformation process progressing from improving context of measurement, to improving focus, to improving integration, to improving interactivity—the four aspects of paramount importance to making progress on moving the "Process Improvement & Management" initiative from its current maturity stage to "Continuous Improvement" maturity stage. Within the "Process Improvement and Management" dimension of "Continuous Improvement" transformation, the factors which contribute to transform the interactivity of "Process Improvement and Management" include the following:

1. Frequent Interactivity
2. Effective and robust dialogue
3. Collaborative learning
4. Appropriate use of technology

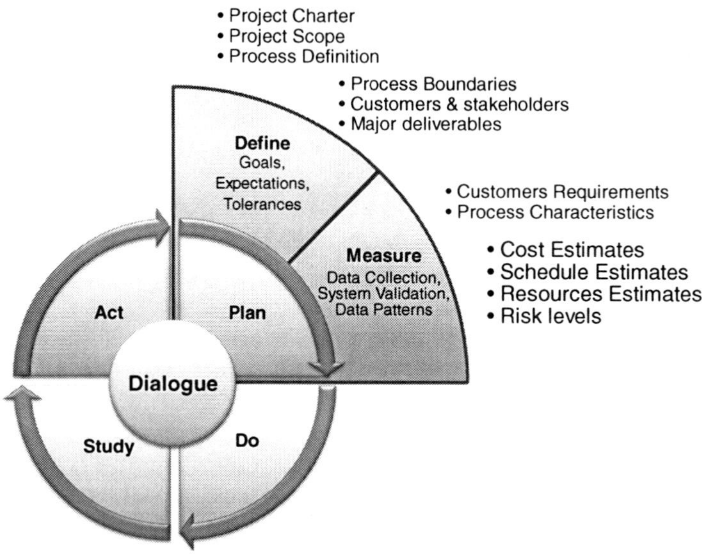

Fig. 22.1 Minimum activities of the "PDSA Plan" phase

Performing the PDSA constituent processes should include highly interactive and iterative (ongoing) discussions, or dialogues, which are the most important aspects of data collection system. As indicated already, these dialogues should be built on the foundation of a positive context of measurement, focus, and integration as indicated in our first book entitled "A Guide to Continuous Improvement Transformation: Concepts, Processes, Implementation." Effective integration and interactivity of the PDSA constituent processes will do more than anything else to break down the silos that are also keeping enterprise businesses from realizing "Process Improvement & Management" transformational potential.

Figure 22.1 shows the minimum activities that are part of the "PDSA Plan" planning phase.

It can be noted that in this figure we use the "Define" and "Measure" nomenclature of the six-sigma literature for convenience and consistency with existing literature. We have placed "Dialogue," which is what enables this continual reassessment, at the very center of the PDSA Cycle in Fig. 22.1. It is in fact the basic unit of a "process improvement" project work. You cannot plan a "process improvement" project well without robust dialogue with customers and stakeholders. How people involved in a "process improvement" project talk to each other, talk to customers and stakeholders, absolutely determines how well the "process improvement" project will progress towards its objectives.

The word "dialogue" should be understood in the sense of "sharing collective meaning" and strongly differentiated from "discussion." The word "discussion" comes from the same root word as percussion and concussion and has to do with beating one thing against another. The word "communication" is a more general term meaning "to make something common." So, communication can be done by discussion or dialogue. When information is made common through discussion, it is often two monologues—an attempt to convey your opinion to another person, and nothing more. Very few people are skilled at dialogue, and very few project team members currently have a strong capacity for dialogue.

A dialogue is a mutual search for shared meaning or understanding. In order to take advantage of this opportunity to dialogue during a "process improvement" project, team members need to consider themselves as equals, each having valuable insights to share on the "process to be improved" being considered. The belief that some are more "expert" than others and that some are "subordinate" to others will undermine dialogue; hence it will undermine the achievement of the development of a credible and robust plan. It will cause some to defer to the others who may have a superior knowledge, or a superior position.

Dialogue in a "process improvement" project thrives on openness, candor, and inviting multiple viewpoints. In dialogue, diversity of perspective is almost always good—whether it be functional, cross-functional, local, global, systemic, or whatever. The more perspectives involved, in theory at least, the richer the dialogue can be, and the much higher levels of knowledge, insight, and wisdom can be generated.

Dialogue as interactivity should incorporate learning, understanding, defining, listening, modeling, hypothesizing, balancing, linking, integrating, etc. It is an important part of the total transformation of data collection system. Transformational and emergent collections of data, especially, require the synergy and support that interactivity around the collected data provides.

As we have indicated in the previous chapter, robust dialogue starts when people go in the planning phase with open minds. They are not trapped by preconceptions or armed with a private agenda. They want to hear new information and choose the best alternatives, so they listen to all sides of the debate and make their own contributions. When people speak candidly, they express their real opinions, not those that will please the power players or maintain harmony.

Think about the planning meetings that you, as enterprise executive, project manager or leader, or team member have attended—those that were a hopeless waste of time and those that produced energy and great results. What was the difference? . . .the difference was in the quality of the dialogue.

Robust dialogue alters the psychology of a project team. It can either expand a project team's capacity in executing tasks or shrink it. It can be energizing or energy-draining. It can create self-confidence and optimism, or it can produce pessimism. It can create unity, or it can create bitter factions. Robust dialogue brings out reality, even when that reality makes people uncomfortable, because it has purpose and meaning.

"PDSA Do" Process Group 23

Having carefully planned the "process improvement" project, the project manager, or the project team leader, and the project team are now ready to start the "PDSA Do" project phase. This phase is typically the longest phase of the project. It is the phase within which:

1. The defined project management plan is carried out;
2. Deliverables are physically built and presented to selected groups of stakeholders and customers;
3. Problems and unexpected observations are documented;
4. Data, resulting from prototyping and piloting a solution, that are useful for answering the questions asked in the project plan and that can be compared to the predictions are collected; and
5. Analysis of these data starts.

Although much of the learning will come from the "PDSA Study" project phase, this "PDSA Do" project phase has its own unique learning opportunities. Documenting problems and unexpected occurrences during the pilot of selected solution(s) will promote learning about aspects of the solution(s) that studying the planned results will not. The information obtained during this "PDSA Do" project phase should prepare for the effective learning in the "PDSA Study" phase.

To ensure that the customer's requirements are met, the project manager monitors and controls the production of each deliverable by executing a suite of planned management processes. After the deliverables have been physically constructed and verified by the stakeholders, a phase review is carried out to determine whether the project is complete and ready for closure.

23.1 The "PDSA Do" Constituent Processes

Figure 23.1 shows the activities to be undertaken during the "PDSA Do" project phase. It illustrates those processes performed to carry out the work defined in the project management plan to accomplish the project's objectives. It involves

A. van Aartsengel and S. Kurtoglu, *Handbook on Continuous Improvement Transformation*, 443
DOI 10.1007/978-3-642-35901-9_23, © Springer-Verlag Berlin Heidelberg 2013

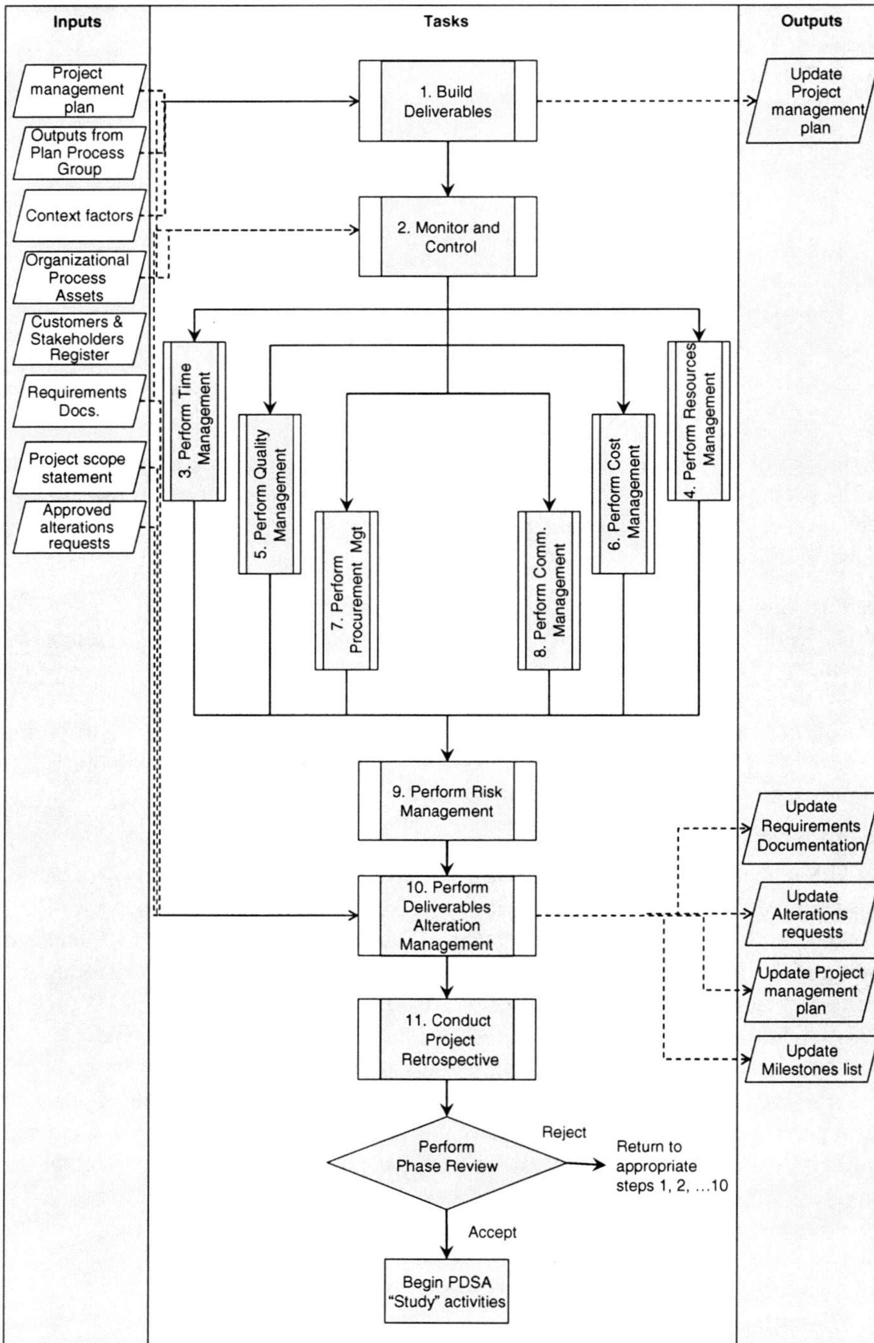

Fig. 23.1 "PDSA Do" process group

coordinating people and resources, as well as integrating and performing the activities of the project in accordance with the project management plan.

The "PDSA Do" Process Group includes the following key processes of the process improvement plan indicated in a previous section:

1. Build Deliverables
2. Monitor and Control Execution
3. Perform Time Management Plan
4. Perform Quality Management Plan
5. Perform Procurement Management Plan
6. Perform Communication Management Plan
7. Perform Cost Management Plan
8. Perform Resources Management Plan
9. Perform Risk Management Plan
10. Perform Deliverables Alteration Management
11. Perform Project Retrospective
12. Perform "PDSA Do" Phase Review
13. Identify and Document Lessons Learned

These constituent processes interact with each other and with the project management processes in the PDSA "Process Groups." Each aspect of executing any of these can involve effort from one or more persons, based on the needs of the project. Each aspect occurs at least once in every "process improvement" project which involves procurement of some of its activities and occurs in one or more project phases.

To successfully deliver the project on time, within budget and to specification the project manager needs to fully implement each of the activities listed in this section. Even though the management processes listed may seem obvious, it is extremely important that the project manager implements each process in its entirety and communicates the process clearly to the project team. A large percentage of projects worldwide have failed because of a lack of formalization of these simple, yet critical project management processes.

While integrating and performing the activities of the project in accordance with the project management plan, deviations from established performance baselines will cause some alterations of the project plan. These alterations can include activity durations, resource productivity and availability and unanticipated risks. Such alterations may or may not affect the project management plan, but can require an analysis. The results of the analysis can trigger an alteration (or a change) request that, if approved, would modify the project management plan and possibly require establishing a new baseline. The vast majority of the project's allocated funds could therefore be expended in performing the "PDSA Do" Process Group processes.

Build Deliverables 24

This is the project management process used to physically construct or build the "process improvement" project deliverables. It is the most time-consuming activity in the project. Whether you are improving a production process, or a customer service offering process, the project will consume the majority of its available resource building the actual deliverables for acceptance by the customer.

The steps undertaken to build each deliverable will vary depending on the type and complexity of the "process improvement" project been undertaken, however its elements can be described here in real details. The key activities required to build each deliverable will be clearly specified within the terms of reference and project plan accordingly.

The "Build Deliverables" Process Group includes the following key processes:

1. Identify and Quantify Assignable Causes of Variations
2. Explore Cause-and-Effect Relationship
3. Verify Identified Assignable Causes
4. Analyze Process Steps And Tasks
5. Generate Improvement Solutions
6. Assess Risk and Pilot Solution(s)

The following section is concerned with the first key process. In the following chapters, we will address the remaining key processes.

24.1 Identify and Quantify Assignable Causes of Variations

This is the "process improvement" project management process used to establish a base pool of all important assignable causes of variations associated with the "process to be improved" under consideration.

When a "process to be improved" is operated unpredictably it is subject to the effects of unknown, dominant assignable causes. Once the problem addressed by the "process improvement" project has been focused through the "PDSA Plan"

A. van Aartsengel and S. Kurtoglu, *Handbook on Continuous Improvement Transformation*, DOI 10.1007/978-3-642-35901-9_24, © Springer-Verlag Berlin Heidelberg 2013

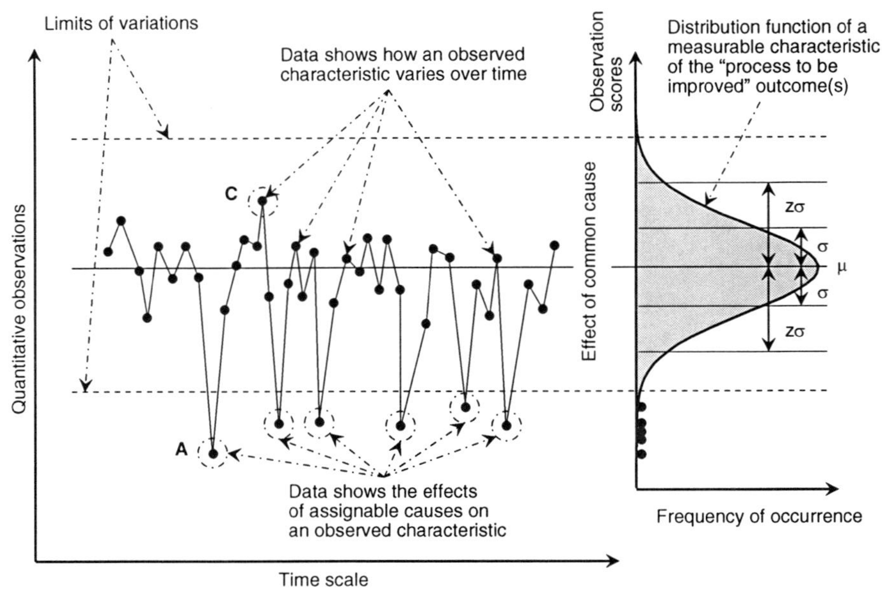

Fig. 24.1 Effects of assignable causes in process outcome over time

phase of the PDSA model, the project team must identify assignable causes of variations using:

1. Process behavior charts, as illustrated in Fig. 24.1, as the unique operational definition of assignable causes.
2. Interviews as the primary method used to establish actual boundary conditions of occurrence of assignable causes. They are the key part of any investigation process.

24.1.1 Process Behavior Charts

In Fig. 24.1 the horizontal scale is time, and the vertical scale, which represents quantitative observations, is quality performance. Thus, the plotted points show quality performance as time progresses.

The chart also exhibits three horizontal lines. The middle line is the average of past performance and is therefore the expected level of performance. The other two lines are statistical limits of variations lines or "limit lines." They are intended to separate assignable causes from common causes.

Point C on the chart differs from the historical average. However, since point C is within the limit lines, this difference could be due to common causes. Hence we assume that there is no special cause. Point A also differs from the historical average, but is outside of the limit lines. Now the odds are heavily against this being due to common causes. Hence we assume that point A is the result of assignable causes.

Assignable causes of variation are those causes that are not intrinsically part of the process been considered but arise because of specific circumstances. They are typically sporadic, and often have their origin in single process input variables. When they occur, they signal a significant occurrence of change in the process and they lead to a statistically significant deviation from the norm. Assignable causes of variation are indicated by a disruption of the stable, repeating pattern of variation. They result in unpredictable process performance and must therefore be identified and systematically removed before taking other steps to improve quality of the "process to be improved" considered.

If the "process to be improved" is to be improved without identifying the assignable causes, the results will be of limited validity and dubious utility. Thus, identifying and quantifying the amount of assignable causes of variations in the "process to be improved" is a critical step towards its improvement.

A process that has only common causes, affecting its outcomes is referred to as a stable process, or one that is in a state of statistical control. In a stable process, the causal system of variation remains essentially constant over time. This does not mean that there is no variation in the outcomes of the process, or that the variation is small, or that outcomes meet the specified requirements. It implies only that the variation is predictable within statistically established limits of variation. In practice, this means that improvement can be achieved only through a fundamental change to the process.

A process whose outcomes are affected by both common and assignable causes of variation is referred to as an unstable process. An unstable process is not necessarily one with large variations. Rather, the magnitude of variations from one period to the next is unpredictable. If assignable causes can be identified and removed, the process becomes stable; its performance becomes predictable. In practical terms, this implies that the system can be put back to an original level of performance by identifying the assignable causes and taking appropriate action. Once a change is made, continuing to plot data over time and observe the patterns helps to determine whether the change has eliminated the assignable cause.

24.1.2 Interviews

This is the second phase of the identification of assignable causes. It includes interviews with appropriate personnel, collecting physical evidence, and conducting other research, such as performing a sequence-of-events analysis, which is needed to provide a clear understanding of the events leading to occurrence of assignable causes.

The interview process for identifying assignable causes is the primary method used to establish actual boundary conditions of occurrence of assignable causes. It is crucial for the project team members conducting interviews to be good listeners with good diplomatic and interviewing skills.

For those occurrences of assignable causes with significant impact on the "process to be improved" outcomes, all key personnel must be interviewed to get a complete picture of the events leading to occurrence of assignable causes.

In addition to those directly involved in the events, individuals having direct or indirect knowledge that could help clarify the events should be interviewed. The following is a partial list of interviewees:

1. All personnel directly involved with the events leading to occurrence of assignable causes (the project team members conducting interviews must review any written witness statements).
2. Supervisors and managers of those involved in the events leading to occurrence of assignable causes (including contractor management).
3. Personnel not directly involved in the events leading to occurrence of assignable causes but who have similar background and experience.
4. Applicable technical experts, training personnel, and equipment vendors, suppliers, or manufacturers.

It is extremely important for the project team members conducting interviews to convey the message that the purpose of these interviews is fact finding not fault finding. The task of the project team members conducting interviews job is simply to find out what actually happened and why it happened. It is important for the project team members conducting interviews to clearly define the reason for the evaluation to the interviewee at the beginning of the interview process. Interviewees must understand and believe that the reason for the evaluation is to identify assignable causes of variation observed from the process behavior charts. If they believe that the interview process is intended to fix blame, little benefit can be derived.

To listen more effectively the team members conducting interviews must be prepared for the interviews, and preparation helps avoid wasting time. Prepared questions or a list of topics to discuss helps keep the interview on track and prevents the interviewers from forgetting to ask questions on key topics. Each interview should be conducted to obtain clear answers to the following questions related to the events leading to occurrence of assignable causes of variations:

1. What happened?
2. Where did it happen?
3. When did it happen?
4. What changed?
5. Who was involved?
6. Why did it happen?
7. What is the impact?
8. Will it happen again?
9. How can recurrence be prevented?

24.1.2.1 What Happened?

Clarifying what actually happened is an essential requirement of identifying assignable causes. Here, the natural tendency is to give perceptions rather than to carefully define the actual event leading to occurrence of assignable causes. The purpose of this first interview question is to clearly and specifically identify and describe a problem detected on the process behavior charts in an effort to focus a root cause analysis and corrective action efforts. A problem is (1) a deviation from a requirement or expectation; (2) when "actual" observation is different from "should" observation; (3) an undesirable event, situation, or performance trend; and/or (4) the primary critical for a situation to occur during execution of the "process to be improved."

When an event leading to occurrence of an assignable cause occurs, the project team may face only suspected problem areas and/or conditions that are not well defined or substantiated by facts. Furthermore, the manner in which the problem is described initially may be very subjective, opinionated, or ambiguous. Therefore, it is important for the project team to include as much detail as the facts and available data permit to focus the scope of the root cause analysis and solution selection.

24.1.2.2 Where Did It Happen?

A clear description of the exact location of the event leading to occurrence of assignable causes helps isolate and resolve the problem. In addition to the location, the team members conducting interviews should determine if the event leading to occurrence of assignable causes also occurred in locations similar to the one where the current "process to be improved" is executed. If similar process inputs variables from which assignable causes have their origin are eliminated, then the event sometimes can be isolated to one, or a series of, forcing function(s) totally unique to the location where the "process to be improved" is executed.

For example, if the process behavior charts from Turbine A indicate occurrence of assignable causes and process behavior charts from Turbines B, C, and D in the same system did not, this indicates that the reason for occurrence of assignable causes is probably unique to Turbine A. If Turbine B, C, and D exhibit similar symptoms, however, it is highly probable that the assignable cause is systemic and widespread to all the turbines.

24.1.2.3 When Did It Happen?

Isolating the specific time that an event occurred greatly improves the project team's ability to determine its source. When the actual time frame of an event leading to occurrence of an assignable cause is known, it is much easier to quantify the process, operations, and other variables that may have contributed to the event.

However, in some cases (e.g., product-quality deviations), it is difficult to accurately fix the beginning and duration of the event. Most plant-monitoring and tracking records do not provide the level of detail required to properly fix the time of this type of incident. In these cases, the project team should evaluate the operating history of the affected process area to determine if a pattern can be found that properly fixes the event's time frame. This type of investigation, in most cases, will isolate the timing to events such as the following:

1. Production of a specific product.
2. Work schedule of a specific operating team.
3. Changes in human resources and intervention.
4. Changes in methods and procedures used in every step of the process.
5. Changes in management systems and methodologies.
6. Changes in ambient environment.

24.1.2.4 What Changed?

Occurrence of assignable causes and major deviations from acceptable performance levels in process outcomes do not just happen. In every case, specific variables, singly or in combination, caused the event leading to occurrence of assignable causes to occur. Therefore, it is essential that any changes in the system that occurred in conjunction with the event be defined.

No matter what the event is (i.e., equipment failure, environmental release, accident, etc.), the project team must quantify all the variables associated with the event leading to occurrence of assignable causes.

24.1.2.5 Who Was Involved?

The project team should identify all personnel involved, directly or indirectly, in the event leading to occurrence of assignable causes. Events leading to occurrence of assignable causes often result from human actions or inadequate skills to execute the process. However, the project team must remember that the purpose of these interviews is to identify and quantify assignable causes that are displayed on the process behavior charts, not to place blame.

All comments or statements derived during this part of the investigation on assignable causes should be impersonal and totally objective. All references to personnel directly involved in the occurrence of assignable causes should be assigned a code number or other identifier. This approach helps reduce fear of punishment for those directly involved in the occurrence of assignable causes of variations. In addition, it reduces prejudice or preconceived opinions about individuals within the enterprise business.

24.1.2.6 Why Did It Happen?

If the preceding questions are fully answered, it may be possible to identify and quantify an assignable cause with no further investigation. However, the project team must exercise caution to ensure that the real cause has been identified. It is too easy to address the symptoms or perceptions without a full analysis. It is also common for people working on improvement efforts to jump to conclusions without studying the real causes, target one possible cause while ignoring others, and take actions aimed at surface symptoms.

Symptoms are the tangible evidence or manifestation(s) indicating the existence or occurrence of something wrong. They are not the cause, but are the manifestation of the problem and are their occurrence is indicated by process behavior charts. Based on these symptoms, an apparent cause (i.e., the immediate or obvious reason for a problem) may be assigned. Corrective action(s) can resolve the apparent cause, but it is only when the real reason the event leading to occurrence of the assignable cause is identified and treated that recurrence can be prevented.

At this point, the project team should generate a list of what may have contributed to the occurrence of assignable causes of variations. The list should include all factors, both real and assumed. This step is critical to the process. In many cases, a number of factors, many of them trivial, combine to cause a serious variation in the process outcomes.

All assumptions included in this list of possible assignable causes should be clearly noted, as should the assignable causes that are proven. A sequence-of-events analysis provides a means for separating fact from fiction during the analysis process.

24.1.2.7 What Is the Impact?

The project team should quantify the impact of the event leading to occurrence of assignable causes before embarking on a full cause-and-effect relationship analysis or a full root cause and failure analysis. Again, not all events, even some that are repetitive, warrant a full analysis.

This part of the identification and quantification process should be as factual as possible. Even though all the details are unavailable at this point, the project team should attempt to assess the real or potential impact of the event. Identified assignable causes must be quantified so that further improvements focus on the deep assignable causes, and not on the original symptoms of "process to be improved" underperformance.

24.1.2.8 Will It Happen Again?

If the preliminary interviews determine that the event leading to occurrence of assignable causes is nonrecurring, the interview process may be discontinued at this point. However, a thorough review of the historical records associated with the process inputs variables from which assignable causes have their origin should be conducted before making this decision. The project team should ensure that it truly is a nonrecurring event before discontinuing the identification.

All reported events should be recorded and the files maintained for future reference. For events leading to occurrence of assignable causes found to be nonrecurring, the project team should establish a file that retains all the data and information developed in the preceding steps. Should the event or a similar one occur again, these records are an invaluable investigative tool.

A full investigation should be conducted on any event leading to occurrence of assignable causes that has a history of periodic recurrence, or a high probability of recurrence, and a significant impact in terms of economics.

24.1.2.9 How Can Recurrence Be Prevented?

Although this is the next logical question to ask, it generally cannot be answered until the entire root cause analysis of assignable causes is completed.

24.1.3 Types of Interviews

One of the questions to answer in preparing for an interview is "What type of interview is needed for this investigation on assignable causes?" Interviews can be grouped into three basic types: one-on-one, two-on-one, and group meetings.

24.1.3.1 One-on-One

The simplest interview to conduct is that where the project team member interviews each person necessary to clarify the event leading to occurrence of assignable causes. This type of interview should be held in a private location with no distractions. In instances where a field walk-down is required, the interview may be held in the employee's work space.

24.1.3.2 Two-on-One

When controversial or complex event leading to occurrence of assignable causes are being investigated, it may be advisable to have two interviewers present when meeting with an individual. With two investigators, one can ask questions while the other records information. The interviewers should coordinate their questioning and avoid overwhelming or intimidating the interviewee.

At the end of the interview, the interviewers should compare their impressions of the interview and reach a consensus on their views. The advantage of the two-on-one interview is that it should eliminate any personal perceptions of a single interviewer from the investigation process.

24.1.3.3 Group Meeting

A group interview is advantageous in some instances. This type of meeting, or group problem-solving exercise, is useful for obtaining an interchange of ideas from several disciplines (i.e., maintenance, production, engineering, etc.). Such an interchange may help resolve an event or problem.

This approach also can be used when the investigator has completed his or her evaluation and wants to review the findings with those involved in the event leading to occurrence of assignable causes. The investigator might consider interviews with key witnesses before the group meeting to verify the sequence of events and the conclusions before presenting them to the larger group. The investigator must act as facilitator in this problem-solving process and use a sequence-of-events diagram as the working tool for the meeting.

Group interviews must not be used in a hostile environment. If the event leading to occurrence of assignable causes is controversial or political, this type of interview process is not beneficial. The personal agendas of the participants generally preclude positive results.

24.2 Conclusion

As indicated early in this chapter, if the project team can identify and remove assignable causes, the process becomes stable; its performance becomes predictable. In practical terms, this implies that the system can be put back to an original level of performance by identifying the assignable causes and taking appropriate action. Once a change is made, quality control through continuous plot of data over time and observation of the patterns helps to determine whether the change has eliminated the assignable cause.

Once a potential assignable cause has been identified, the project team must quantify its effects on the "process to be improved" outcomes in order to target the improvement efforts correctly and thereby avoid wasted resources.

Explore Cause-and-Effect Relationship 25

This is the "process improvement" project management process required to ensure that the project team explores cause-and-effect relationship of assignable causes of variations, associated with the "process to be improved" under consideration, with the observed characteristic of the "process to be improved" outcomes.

Because of the nature of relationships that are not intrinsically part of the "process to be improved" process and their effects on its outcomes, exploring cause-and-effect relationship of assignable causes of variations will sometimes provide insights that are neither available from raw "process to be improved" data nor readily evident.

25.1 Ishikawa's Cause-and-Effect Diagram

Ishikawa's cause-and-effect diagram (also called fishbone diagrams, or herringbone diagrams, cause-and-effect diagrams or Fishikawa) is the commonly used tool to illustrate how various causes and sub-causes create a special effect on a process outcome.

Ishikawa's cause-and-effect diagram, shown in Fig. 25.1, is a causal diagram that shows the causes of a specific event. It was designed to sort the potential causes of a problem while organizing the causal relationships. Professor Kaoru Ishikawa developed this tool in 1943 to explain to a group of engineers at Kawasaki Steel Works how various manufacturing factors could be sorted and interrelated.

Ishikawa's cause-and-effect diagrams are drawn primarily to illustrate the possible causes of a particular problem by sorting and relating them using a classification schema. They build from right to left because the Japanese language of its creator reads from right to left. Each cause or reason for imperfection is a source of variation.

Major causes category branches can be initially identified using the following main process inputs category:

A. van Aartsengel and S. Kurtoglu, *Handbook on Continuous Improvement Transformation*,
DOI 10.1007/978-3-642-35901-9_25,

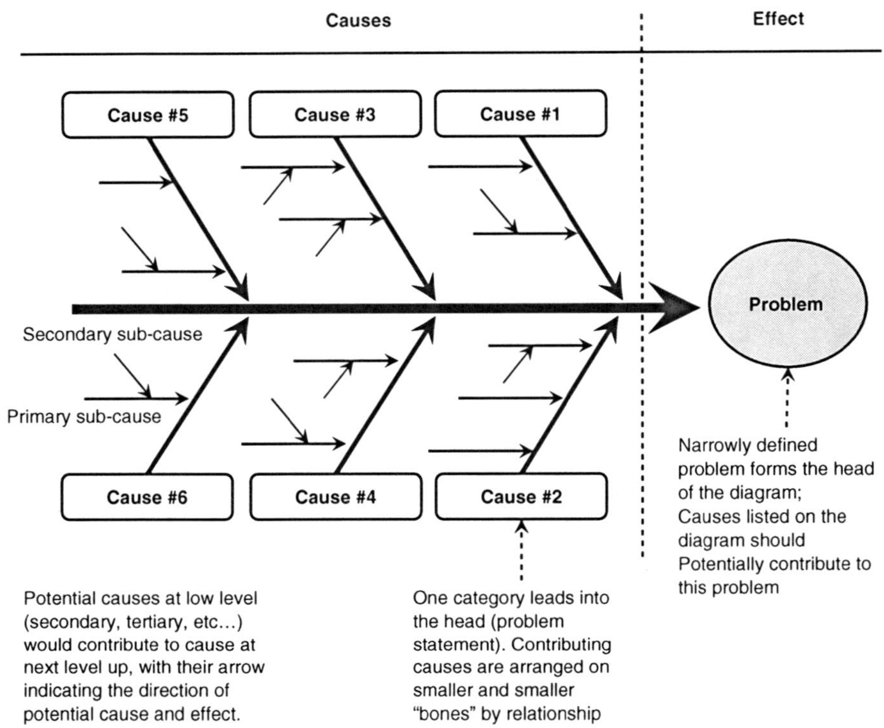

Fig. 25.1 Ishikawa (fishbone) diagram

1. *Man*—This refers to human resources and intervention that is necessary for a process to be completed successfully. From employees to management, this category clarifies the roles and responsibilities of every person involved in the process.
2. *Machines*—The performance of individual machines is important for the assessment of a process and whether any improvement will be necessary. To reduce process variation amongst different machines, it is important to provide regular maintenance and replacement as part of the process.
3. *Methods*—Methods and procedures used in every step of the process are an important component of inputs. To assess process variation from one production unit to another, the project team will need to assess whether production methods are being adhered to or not.
4. *Mother Nature*—While the environment cannot be controlled in many instances, enterprise businesses must assess its impact on processes. The environment, for instance, impacts the availability and transportation of raw materials and products.
5. *Management*—Management systems and methodologies are important inputs in processes. Whether formal or informal, a management system ensures that an enterprise business functions as a single unit with a shared vision.
6. *Materials*—Materials refer to both raw and manufactured elements of process inputs. When making furniture, for example, materials include wood products,

metal screws, paint, paper and labeling products, and many more production materials. The quality, availability, ease of transportation of materials has a strong impact on a process and its success in producing services or products.

7. *Measurement systems*—Every process dictates the type of data collection system that needs to be put in place. Using the right type of data collection system ensures that the appropriate data and information are collected.

Ishikawa's cause-and-effect diagrams depict the general concern associated with a negative outcome and allow for exploration of that concern in the context of its numerous causes (and, in turn, the causes' causes). They also:

1. *Establish the premise for analysis.* In Ishikawa diagrams, it is important to focus on a single issue to be addressed as the net effect of all causes in the cause-and-effect diagram. The broader the premise, the more likely there will be legion fishbones supporting it. Conversely, a narrower premise will yield a more directed analysis of the identified causes.
2. *Build the basic diagram structure.* The basic structure is consistent in most cause-and-effect analyses, similar to the one in Fig. 25.1. The basic structure includes causes related to personnel, equipment, methods, and materials. Although organizations may have broadly different risk issues and concerns, these remain the four classic elements of the structure.
3. *Identify the causes and their causes.* The key in this diagram is to identify root causes for significant concerns. As new causes are identified, the question is asked, "What caused that cause?" This continues until all causes associated with the effect have been exhausted.

With Ishikawa's cause-and-effect diagrams, the key input is the effect that will undergo scrutiny. Then, as the diagram is developed through interviews, brainstorming sessions and 5-why questioning, the inputs become the causes of that effect, and their causes and their causes. The effort continues until all root causes (including some that critics might deem minutiae) are developed. Here, a root cause represents that most basic reason for an undesirable condition or problem which, if eliminated or corrected, would have prevented it from occurring. The root cause may be described in a binary sense (that is, its existence or evidence to the contrary) or in a qualitative sense, meaning that measures intended to preclude its occurrence may be missing or less than adequate.

Root causes usually are defined in terms of specific or systematic factors. Since it is seen as the most basic cause, a root cause is expressed in terms of the least common organizational, personal, or activity denominator. With this view of the meaning of a root cause, the project team must take care to distinguish symptoms clearly from causes, as well as apparent causes from root causes in constructing Ishikawa's cause-and-effect diagram.

25.1.1 Drawing Ishikawa's Cause-and-Effect Diagram

The following are steps to be used in the process of constructing Ishikawa's cause-and-effect diagram.

1. Review the focused problem statement, write it on the right side, and draw an arrow from the left to the right side, as shown in Fig. 25.1. The focused problem statement should be measurable, focused on the gap between what between what is and what should be, which reflects a change or deviation from requirements, norm, standard, or expectation, and it should state the effect as to what is wrong and why it is wrong.
2. Write the main factors that may be causing the problem by drawing major branch arrows to the main arrow. Primary causal factors of the problem can be grouped into items with each forming a major branch, (i.e. cause #i) as shown in Fig. 25.1.
3. For each major branch, detailed causal factors are written as twigs on each major branch of the diagram. On the twigs, still more detailed causal factors are written to make smaller twigs, as shown in Fig. 25.1.
4. Develop other main "bones" by ensuring that all the items that may be causing the problem are included in the diagram.
5. Select possible causes to verify with data.

Outputs of the process of constructing Ishikawa's cause-and-effect diagram are lists of causes linked to the resulting effects that they cause. These outputs should be prioritized based on economic significance or on other criteria of importance. In order to set priorities, the project team should:

1. Review the lists of assignable causes linked to the resulting effects that they cause.
2. Identify which of the assignable causes listed are the most likely contributors to the "process to be improved" underperformance problem.
3. Consider how measurable each of the assignable causes listed are. In general, the project team should focus on the causes that it can easily collect data on. However, some important causes may be hard to measure or observed, and the project team may need to be creative in coming up with ways to measure those causes.
4. Consider which of the assignable causes listed the project team should take prompt corrective action to restore the status quo. In practice many assignable causes identified do not result in corrective action. The usual reason is that the process alterations or changes involving assignable causes are too numerous—the available personnel cannot deal with all of them. Corrective action is taken for the high-priority cases; the rest must wait their turn. A further reason for failure to take corrective action is a persistent confusion between statistical control limits and quality tolerances.

25.2 Fault Tree Diagram (FTD)

A fault tree analysis is an analytical technique, whereby an undesired state of a "process to be improved" and affected system is specified—often a state that is critical from a safety operation standpoint—and the "process to be improved" or the affected system is then analyzed in the context of its environment and operation to find all credible ways in which the undesired event can occur. Fault Tree Analysis methodology is well described in the "Fault Tree Handbook (NUREG-0492)" by the US Nuclear Regulatory Commission (Vesely, Goldberg, Roberts, & Haasl, 1981).

The fault tree itself is a graphic diagram of the various parallel and sequential combinations of faults that will result in the occurrence of the predefined undesired event. The faults can be events that are associated with any component of the system considered, human errors, or any other pertinent events which can lead to the undesired event. Here, the term 'event' denotes a dynamic change of state that occurs in a system element affected by the "process to be improved," which can include hardware, software, human, and environmental factors. A 'fault event' is an abnormal system state. A normal event is expected to occur.

Hence, a fault tree diagram depicts the logical interrelationships of basic events that lead to the undesired event, which is often referred to as the "top event" of the fault tree. It is a logic block diagram that displays the state of a system (top event) in terms of the states of its components (basic events). It is a detailed deductive graphic that usually display considerable information about the system considered.

Fault trees are used to ensure that all critical aspects of a system are identified and controlled. Deduction constitutes reasoning from the general (effect) to the specific (cause). In a deductive analysis of the "process to be improved" and associated system, we postulate that the "process to be improved" itself has failed in a certain way, and we attempt to find out what modes of its discrete elements or system/component behavior contribute to this failure. In other words, we attempt to find out what chain of events caused the failure of the "process to be improved" to produce outputs in conformance with requirements.

A fault tree diagram does not depict all possible failures of the "process to be improved" or affected system. It is tailored to its top event, which corresponds to some particular system failure mode, and the fault tree therefore includes only those faults that contribute to this 'top event.' Furthermore, these faults are not exhaustive as they cover only the most credible faults as assessed by the project team.

In itself, a fault tree diagram is not a quantitative model. It is a qualitative diagram that can be evaluated quantitatively and often is. This qualitative aspect is true of virtually all varieties of system models.

The structure of a fault tree is shown in Fig. 25.2. The undesired event appears as the top event and is linked to more basic fault events by intermediate event statements and logic gates.

25.2.1 Drawing Fault Trees: Gates and Events

A Fault Tree Diagram uses a top-down analytic method and Tree diagram structure to dissect and define the relationship of related causes of a specific problem or problematic outcome, using a unique set of classification symbols. In this technique, some specific state, which is generally a state of failure, of the "process to be improved" and associated system is postulated, and chains of more basic faults contributing to this undesired failure event are built up in a systematic way.

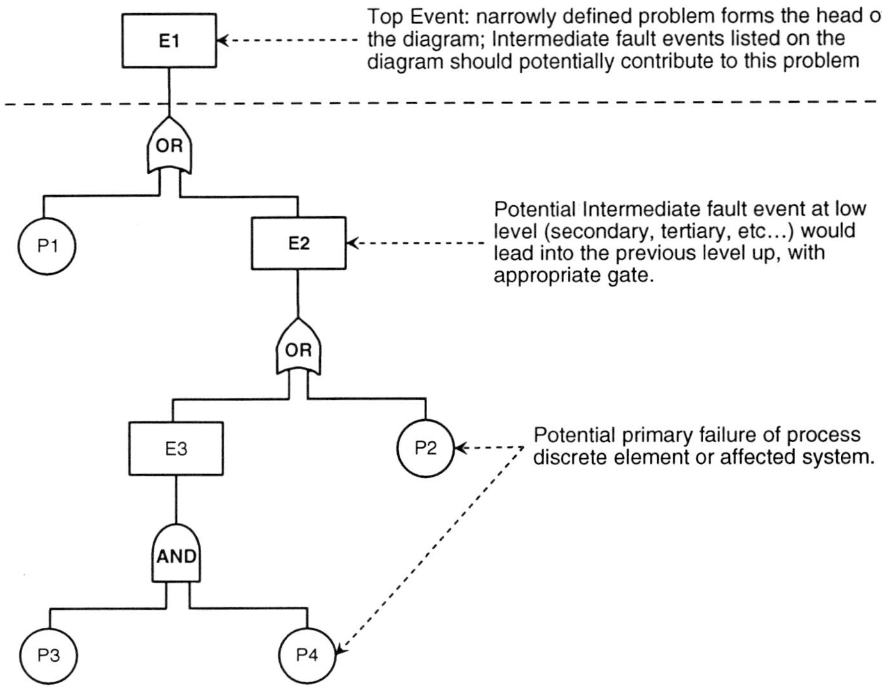

Fig. 25.2 Example of a fault-tree logic tree

This method represents graphically the Boolean logic associated with a particular "process to be improved" and associated system failure. It uses a graphic "model" of the pathways within a system that can lead to a foreseeable, undesirable loss event (or a failure). The pathways interconnect contributory events and conditions, using standard logic symbols (AND, Priority AND, OR, Voting OR, XOR, Inhibit, etc.. . .). The basic constructs in a fault tree diagram are gates (or conditions) and events (blocks). The approach starts with a failure (either potential or existing) and probes backward toward the fundamental events or root causes.

The following are steps to be used in the process of constructing Ishikawa's cause-and-effect diagram.

1. Identify the key failure event to be studied, often called a primary or top event, including the boundaries that limit the analysis.
2. Draw a rectangle at the top center and label it with this key failure event.
3. Examine the "process to be improved" and the system affected by it, and identify elements and events related to the key failure event being analyzed, using the appropriate background information and documentation.
4. Starting with the primary failure event, construct a hierarchical relationship of related events and elements.
   (a) For each event, determine if it is a basic failure, or if can it be analyzed further for its own causes.

Table 25.1 Classic fault tree diagram symbols

Name of Gate	Classic Symbol	Description
Primary events symbols: The primary events of a fault tree are those events, which, for one reason or another, have not been further developed. These are the events for which probabilities will have to be provided if the fault tree is to be used for computing the probability of the top event.		
Basic Event	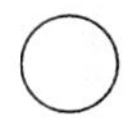	The circle describes a basic initial fault event that requires no further development. In other words, the circle signifies that the appropriate limit of resolution has been reached.
Conditional Event	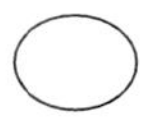	The ellipse is used to record any conditions or restrictions that apply to any logic gate. It is used primarily with the INHIBIT and PRIORITY AND gates.
Undeveloped Event	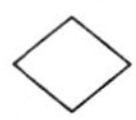	The diamond describes a specific fault event that is not further developed, either because the event is of insufficient or because information relevant to the event is unavailable.
External Event		The house is used to signify an event that is normally expected to occur: e.g., a phase change in a dynamic system. Thus, the house symbol displays events that are not, of themselves, faults.
Intermediate Event	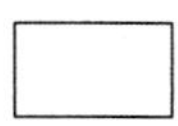	An intermediate event is a fault event, which occurs because of one or more antecedent causes acting through logic gates. All intermediate events are symbolized by rectangles.
Gates symbols: There are two basic types of fault tree gates: the "OR-Gate" and the "AND-Gate." All other gates are really special cases of these two basic types. With one exception, gates are symbolized by a shield with a flat or curved base.		
OR-GATE		The "OR-Gate" is used to show that the output event occurs only if one or more of the input event occurs. There may be any number of input events to an "OR-Gate." Furthermore, causality never passes through an "OR-Gate"; that is, for an "OR-Gate," the input faults are never the causes of the output fault. Inputs to an "OR-Gate" are identical to the output but are more specifically defined as to cause.
AND-GATE		The "AND-Gate" is used to show that the output fault occurs only if all the input faults occur. There may be any number of input faults to an "AND-Gate." In contrast to the "OR-Gate," the "AND-Gate" does specify a causal relationship between the inputs and the outputs; that is, the input faults collectively represent the cause of the output fault. The "AND-Gate" implies nothing whatsoever about the antecedents of the input faults. When describing the events input to an "AND-Gate," any dependencies must be incorporated in the event definitions if the dependencies affect the system logic. Dependencies generally exist when the failure "changes" the system affected by the "process to be improved."
PRIORITY AND-GATE		The "PRIORITY AND-Gate" is a special case of the "AND-Gate" in which the output event occurs if all input events occur in a specified ordered sequence. The sequence is usually shown inside an ellipse drawn to the right of the gate.
EXCLUSIVE OR-GATE		The "EXCLUSIVE OR-Gate" is a special case of the "OR-Gate" in which the output event occurs only if exactly one of the input events occurs. The exclusive OR differs from the usual or inclusive OR in that the situation where both input events occur is precluded.
INHIBIT-GATE		The "INHIBIT-Gate", represented by the hexagon, is a special case of the "AND-Gate." The output is caused by a single input, but some qualifying condition must be satisfied before the input can produce the output. The condition that must exist is the conditional input, spelled out within an ellipse symbol.

   - If it is a basic failure, draw a circle around it.
   - If it is not a basic failure, draw a rectangle around it.
   - If appropriate, use the other symbols available to better define this element, such as an undeveloped event.
   (b) For each event, determine how it is related to the subsequent events that it causes (in the hierarchical flow). Select the appropriate Gate Symbol for each event and related causes.
   - The lower-level events are the input events.
   - Input events cause output events and are placed above in the hierarchy.
   - The input and output events are linked by gates and are placed in between them.
   (c) Repeat Steps a. and b. until all the tree branches depict basic or undeveloped events at its outer-most end.
   (d) (Optional) Determine the probabilities of each basic or un-developed event and mathematically calculate the probabilities of each higher-level event and the top event.
5. Analyze the Fault Tree diagram to understand its multiple relationships and define strategies to prevent potential events that lead to failures.
   (a) Use the Gate Symbols to assist in determining the type of relationship and efficiently determine appropriate preventative strategy.
   (b) Focus first on the most likely to occur events.
6. Document these strategies in a contingency action plan, describing what to do if a failure occurs and who is accountable for taking action.

A typical fault tree is composed of a number of symbols, which are described in Table 25.1 (Vesely et al., 1981).

Verify Identified Assignable Causes

26

This is the "process improvement" project management process required to ensure that the project team builds, as precisely as possible, a factual understanding of existing "process to be improved" assignable causes of underperformance. Its purpose is to get sufficient and accurate information or collect sufficient data to confirm which potential causes actually contribute to the underperformance problem and further focus the "process improvement" effort.

The constituent project management processes, used during the verification of assignable causes, include the following:

1. Plan Assignable Causes Data Capturing
2. Collect Cause-and-Effect Relationship Data
3. Analyze Collected Data

26.1 Plan Assignable Causes Data Capturing

The first step in the process of verifying assignable causes is to "Plan Assignable Causes Data Capturing." In much the same as with the voice of the process, this is the project management process for documenting the actions necessary to define, prepare, integrate, and coordinate into one document all subsidiary data capturing actions for assignable causes that matter the most to the "process to be improved" underperformance. Planning for data collection on assignable causes that matter the most includes, but is not limited to the following steps:

1. Identify assignable causes data and clarify goals
2. Develop operational definitions and procedures
3. Develop sampling strategy
4. Validate data collection system

A. van Aartsengel and S. Kurtoglu, *Handbook on Continuous Improvement Transformation*, DOI 10.1007/978-3-642-35901-9_26, © Springer-Verlag Berlin Heidelberg 2013

26.1.1 Identify Assignable Causes Data and Clarify Goals

The first step in planning for assignable cause data collection, as with any data collection, is to identify the assignable cause data and clarify goals. The purpose here is to ensure that the assignable cause data, which the project team collects, will provide the answers needed to carry on the "process improvement" project successfully. Knowing what type of data the project team will be dealing with also tells which tool should be used to capture it. The right data should:

1. Describe the issue or problem that the "process to be improved" is facing;
2. Describe related conditions that might provide clues about assignable causes of underperformance of the "process to be improved";
3. Lead to analysis in ways that answer the project team questions.

Desired characteristics data for assignable cause that matter the most are: sufficient, relevant, representative, and contextual. As with customers and process requirements, there are two types of data for assignable causes: qualitative and quantitative data.

Qualitative data on assignable causes are obtained from description of observations or measures of characteristics of the process outcomes in terms of words and narratives statements. They can be grouped by highlighting key words, extracting themes, and elaborating concepts.

Quantitative data on assignable causes are obtained from description of observations or measures of characteristics of the process outcomes in terms of measurable quantity and in which a range numerical values are used without implying that a particular numerical value refers to a particular distinct category. As indicated already in a previous section, data originally obtained as qualitative information about description of observations of characteristics of the process outcomes may give rise to quantitative data if they are summarized by means of counts; and conversely, data that are originally quantitative are sometimes grouped into categories to become qualitative data.

As recommended during the processes of collecting customers and process requirements, one of the most important things that the project team should also do in planning for data collection on assignable causes is to draw and label the graph that will communicate the findings before the actual data collection begins. This directs the project team to exactly what assignable cause data is needed.

26.1.2 Develop Operational Definitions and Procedures

For an assignable cause, a process behavior chart provides the unique operational definition. It is a description of term as applied to a specific situation of the "process improvement" project to facilitate the collection of meaningful (standardized) assignable cause data.

When collecting data for an assignable cause, it is important to define terms of process charts very clearly in order to assure that all the appraisers or people collecting and analyzing the data have the same understanding. Any assignable cause data for which an "operational definition" has not been defined often will lead to inconsistencies and erroneous results.

With processes, it is easy to assume that those collecting the data understand what and how to complete the task. However, appraisers or people collecting data have different opinions, views, and working habits and these will affect the data collection. As a result, a process behavior chart for assignable causes should also be very precise and be developed to avoid possible variation in interpretations and to ensure consistent and quality data collection.

The procedures associated with an operational definition for an assignable cause defines exactly how the project team will proceed to collect and record the assignable cause data. The template shown in Fig. 8.2 is equally valid for use for data collection for an assignable cause. During this planning step, the following must also be considered by the project team:

1. Importance of the assignable cause data;
2. Accuracy of the assignable cause data;
3. Completeness of the assignable cause data capturing.

26.1.2.1 Cause-and-Effect Data Collection Sources

Information on the "process to be improved" performance outcomes, hence on identified assignable causes, can only come from observational studies and/or experimental studies. Moreover, assignable causes are typically sporadic, and often have their origin in single process input variables.

Experimental Studies

As indicated in a previous chapter, an experimental study is a methodical procedure carried out with the goal of observing, verifying, explaining, or establishing the validity of a hypothesis. Experimental studies vary greatly in their goal and scale, but always rely on repeatable procedure and logical analysis of the collected results.

Observational Studies

We have also indicated in a previous section that a "process improvement" project is concerned with optimizing a system or getting the "process to be improved" to a higher performance. Therefore, the project team cannot depend solely upon experimental results which are always obtained in a limited context. The project team has to deal with the response variable in the presence of all of the factors that have an impact upon it. The project team cannot simply study some input variables from which assignable causes originate and ignore others. But, of necessity, every experiment will choose some input variables from which assignable causes originate and exclude others. So while the project team may begin with a set of experiments, it needs to remember that limited results and conditional relationships do not tell the whole story. Eventually the project team will need a holistic approach, and this is what observational studies provide.

With an observational study, the data arise as a side effect of some continuing operation or on-going execution of the "process to be improved." It may take longer to discover things with an observational study, but all the possible interactions and all the various factors are present and are allowed to make their contribution to the results of the study. When a factor makes its presence known in an observational study, the project team can be certain that it is a dominant factor.

With an observational study, the clues to the source of any particular behavior will come from the context for each observed event. Here the key to discovery is the connection between context and the observed behavior. The data on assignable causes will have to be interpreted in terms of their context. Moreover, since any of the input variables is not been ignored in an observational study, there is no need for any insurance device, like randomization. In fact, any attempt to impose randomization on an observational study will merely result in confusion. In an observational study some of the most important information may consist of the time order sequence for the data. Therefore, with an observational study, any careful data to be collected must preserve the time-order sequence of the data.

26.1.2.2 Prioritize Cause-and-Effect Data

Since data collection can consume a tremendous amount of time, it is critical that the project team focus on the inputs variables that matter the most and from which assignable causes originate. Two funneling tools can be used for this purpose: a prioritization matrix and a FMEA matrix.

Prioritization matrix—There are two applications for a prioritization matrix: linking response variables to identified assignable causes and linking assignable causes input and process variables to response variables. A prioritizations matrix, as shown in Table 14.1, can be used when the project team has determined that too many assignable causes input variables might have an impact of the response variable and collecting data on all possible variables would cost too much time and resources (including money).

Failure Mode and Effect Analysis (FMEA)—The FMEA is an effective step-by-step approach for focusing the data collection effort on those inputs variables that matter the most (i.e. those related to the identified process critical elements) for the current "process to be improved." It is a structured approach to identify, estimate, prioritize and evaluate risk associated with execution of the identified "process to be improved" critical elements. Failures are unwanted features of a characteristic of a "process to be improved" outcomes; it is any errors or defects, especially ones that affect the customer, and can be potential or actual. "Effects analysis" refers to studying the consequences of those failures.

26.1.3 Develop Sampling Strategy

From all the possible interactions between selected inputs from which assignable causes originate and extraneous factors the totality of all associated responses and about which the data should be collected can still be relatively large and it might not be possible, nor it is necessary, to collect information from the total population considered. It is incumbent on the project team to clearly define the target population. There are no strict rules to follow, and the project team must rely on logic and judgment. The population is defined in keeping with the questions to be answered and the objectives of capturing assignable causes' data.

Sometimes, the entire population will be sufficiently small, and the project team can include the entire population in the study. Collecting the data in this case is called a "census data collection" because data is gathered on every input and associated response of the target population.

Usually, the target population is too large for the project team to attempt to observe and record all data. A small, but carefully chosen sample can be used to represent the population. The sample should reflects the characteristics of the population from which it is drawn and the goal in choosing a sample is to have a picture of the population, disturbed as little as possible by the act of gathering information. The sampling methods described in a previous section can be used to achieve this purpose.

26.1.4 Validate Data Collection System

The "data collection system" consists of data sample, appraisers or people executing the data collection tasks, operational definitions and procedures followed to collect the data. The events associated with any one of these constituents are not conveyed to the other constituents; that is, the constituents of a "data collection system" are statistically independent. Validation of the data collection system follows the procedure described in the validation of the V.O.C. data collection system in a previous chapter.

26.2 Collect Cause-and-Effect Relationship Data

Once the plan for collecting the assignable causes' data is established, the next step is to begin the actual data collection, from the determined data collection sources (i.e. observational studies and/or experimental studies), on those assignable causes inputs variables X that matter the most for the current "process to be improved" and their associates response variable $Y = f(X) + \varepsilon$.

If the "process to be improved" is being operated unpredictably, then there are dominant assignable causes present which the project team has failed to identify in the past. Conducting an experiment on the known cause-and-effect relationships is pointless. It will generally be more profitable to use an observational study to discover and verify these assignable causes of exceptional variation than it will be to study the known factors that are already been controlled.

On the other hand, if the "process to be improved" is being operated predictably, then the project team will have the process behavior chart to give feedback about any disturbances that may occur from time to time, and in the meantime the project team will be operating the process reasonably close to its full potential (given the current "process to be improved" configuration). Here the project team may design and conduct experiments to optimize the process by effectively exploring the cause-and-effect relationship between numerous process input variables, from which potential assignable cause of variation originate, and the process response variable. But the proven ability to operate the process predictably will remove the need for randomization and blocking.

Finally, if the "process to be improved" is unpredictable and no process behavior chart is being used, then the knowledge gained from the experiment is likely to be lost within few weeks due to the confusion and chaos that typically surround business line operations. Without the operational discipline required to operate the process predictably, any knowledge gained will be gained in vain.

26.2.1 Design and Conduct Experiments

Designing and conducting experiments to collect data is an approach based on experimental studies for effectively exploring the cause-and-effect relationship between numerous process input variables, from which potential assignable cause of variation originate, and the process response variable. It is a methodical procedure carried out with the goal of observing, verifying, explaining, or establishing the validity of a hypothesis on the cause-and-effect relationship. Design of experiments or simply experimental studies, vary greatly in their goal and scale, but always rely on repeatable procedure and logical analysis of the collected results. They are a component of the learning process. We design and conduct experiment to learn. Learning through experimentation is a complex mechanism combining one's hopes, needs, knowledge, and resources.

Designing and conducting experiments to collect data has been described in a previous chapter.

26.2.2 Design and Conduct Observational Studies

Statistically designed experiments are essential in R&D. Randomization and blocking are powerful tools in experimental studies for increasing the sensitivity of the analysis in the face of sources of variation that the project team cannot do anything about. But in the industrial context, where assignable causes of exceptional variation must be identified in order to remove their effects from production or service process, and where there is the luxury of successive iterations to confirm hypotheses, an observational approach will complement and complete any program of experimentation.

As indicated earlier, if the "process to be improved" is being operated unpredictably, then there are dominant assignable causes present which the project team has failed to identify in the past. Conducting an experiment on the known cause-and-effect relationships is pointless. It will generally be more profitable to use an observational study to discover and verify these assignable causes of exceptional variation than it will be to study the known factors that are already been controlled.

Furthermore, the project team cannot depend solely upon experimental results which are always obtained in a limited context. The project team has to deal with the response variable in the presence of all of the factors that have an impact upon it. The project team cannot simply study some factors and ignore others. But, of necessity, every experiment will choose some factors and exclude other factors. So while the project team may begin with a set of experiments, it needs to remember that limited results and conditional relationships do not tell the whole story. Eventually the project team will need a holistic approach, and this is what observational studies provide.

With an observational study, the data arise as a side effect of some continuing operation or on-going execution of the "process to be improved." It may take longer to discover things with an observational study, but all the possible interactions and all the various factors are present and are allowed to make their contribution to the results of the study. When a factor makes its presence known in an observational study, the project team can be certain that it is a dominant factor.

With an observational study the clues to the source of any particular behavior will come from the context for each observed event. Here the key to discovery is the connection between context and the observed behavior. The V.O.P. data collected through observational studies will have to be interpreted in terms of their context. Moreover, since none of the input variables, from which potential assignable causes of variation may originate, are been ignored in an observational study, there is no need for any insurance device, like randomization. In fact, any attempt to impose randomization on an observational study will merely result in confusion. In an observational study some of the most important information may consist of the time order sequence for the data. Therefore, with an observational study, any careful data to be collected must preserve the time-order sequence of the data.

26.3 Analyze Collected Data

The purpose of analyzing the results collected from the designed and conducted experiments is to take the experiments results and piece together the cause-and-effect relationship between the selected subset X of potential input variables, from which assignable causes of variation may originate, and the response Y by a relation of the form:

$$Y = f(X) + \varepsilon$$

How much effect does the potential input variables X, from which assignable causes of variation may originate, have on the response Y? What mathematical form does this relationship take on? What is the main effect or quantitative influence a single experiment factor has on the response Y, as there will be a main effect for each factor in the experiment? What is the interaction effect of one potential input variables, from which assignable causes of variation may originate, interacting with another? Even though the project team can calculate all the main and interaction effects of the input variables, from which assignable causes of variation may originate, are they all significant? Are they all necessary? Which effects are more significant? The Pareto Principle tells that a relatively small subset of all the possible effects explains the vast majority of the output response. So how do the project team determine which effects to hold on and which to cast aside? These are the questions that the analysis performed by the project team will answer.

If the factors selected for the designed and conducted experiments have no impact on the response Y, the calculated main and interaction effects will just be random—normally distributed and centered around zero. But if any one of the effects is significant, it will depart from the random cluster of the rest.

26.3.1 Summarize Data & Display Patterns

The major purposes of summarizing the collected data, on those assignable causes inputs variables that matter the most for the current "process to be improved" and their associates "process to be improved" outcomes, and displaying their patterns are:

1. To help get the "process to be improved" into a "satisfactory state," which one might then be content to monitor if not persuaded by arguments for the need of improvement.
2. To provide a route to investigate further what can be accomplished by removing assignable causes that matter the most and by operating the current "process to be improved" up to its full potential. To get the "process to be improved" to operate up to its full potential it is necessary to operate the process predictably. To operate the process predictably is to operate it with minimum variance. Unpredictable operation will inevitably increase the variation, which will lower the capability indexes and increase the effective cost of production and use of the process outcome.

Process behavior charts, described in a previous section, will help to measure what the "process to be improved" is doing and to determine when the "process to be improved" is not operating up to its full potential. A very useful graph to summarize collected data, which the project team can use, is a scatter plot. Its purpose here is to provide a visual indication of the relationship between the input variables from which assignable causes originate and the observed characteristic of the "process to be improved" outcomes.

26.3.1.1 Patterns of Data in Scatter Plots

The project team can use scatter plots to analyze patterns in bivariate data. These patterns are described in terms of linearity, slope, and strength, as illustrated in Fig. 26.1.

On these patterns, linearity refers to whether a data pattern is linear (straight) or nonlinear (curved). Slope refers to the direction of change in the score of the response variable Y when the value of the input variable X, from which assignable causes originate, increases. If the score of the response variable Y also increases, the

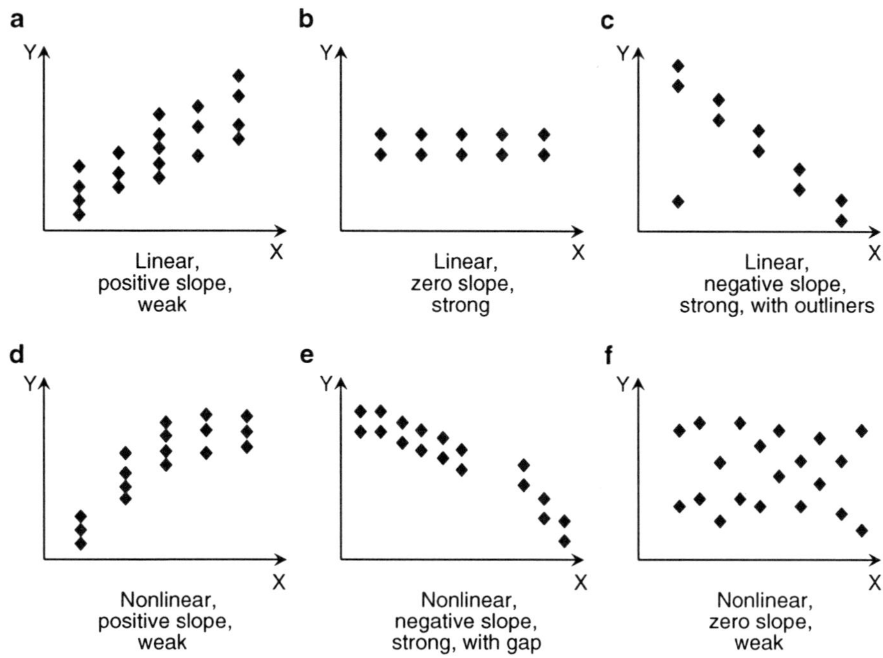

Fig. 26.1 Scatter plot patterns

slope is positive; but if the score of the response variable *Y* decreases, the slope is negative. Strength refers to the degree of "scatter" in the plot. If the dots are widely spread, the relationship between variables is weak. If the dots are concentrated around a line, the relationship is strong. Additionally, scatter plots can reveal unusual features in data sets, such as clusters, gaps, and outliers.

26.3.2 Analyze Cause-and-Effect Relationship Data

This is the project management process used to verify, based on collected data, whether the cause-and-effect relationship truly exists or whether the identified assignable causes that matter the most and the observed characteristic of the "process to be improved" outcomes are related.

26.3.2.1 Correlation

One of the most basic measures of the association between the input variables from which assignable causes originate and the observed characteristic of the "process to be improved" outcomes is the correlation coefficient. Although there are a number of different types of correlation coefficients, the most commonly used is the Pearson product-moment correlation coefficient. Three other types of correlation coefficients are: the point-biserial coefficient, the Spearman rho coefficient, and the phi coefficient.

For a Pearson product-moment correlation, both of the variables, the input variable X from which assignable causes originate and the response variable Y, must be measured on an interval or ratio scale and are known as continuous variables.

There are two fundamental characteristics of correlation coefficients that the project team should carefully consider. The first of these is the direction of the correlation coefficient and the second is the strength or magnitude of the relationship.

Direction of the Correlation Coefficient

Correlation coefficients can be either positive or negative (as indicated by the slope of the corresponding scatter plots).

A positive correlation, as illustrated in Fig. 26.1a, d, indicates that the values on the two variables being analyzed, the input variable X from which assignable causes originate and the response variable Y, move in the same direction. That is, as scores on one variable go up, scores on the other variable go up as well (on average). Similarly, on average, as scores on one variable go down, scores on the other variable go down. Here, we say "on average" because it is important to note that the presence of a correlation between two variables does not mean that this relationship holds true for each member of the sample or population. Rather, it means that, in general, there is a relationship of a given direction and strength between two variables in the sample or population.

A negative correlation, as illustrated in Fig. 26.1c, e, indicates that the values on the two variables being analyzed move in opposite directions. That is, as scores on one variable go up, scores on the other variable go down, and vice versa (on average).

Strength or Magnitude of the Relationship

The second fundamental characteristic of correlation coefficients is the strength or magnitude of the relationship. Correlation coefficients range in strength from -1.00 to $+1.00$. A correlation coefficient of 0.00 indicates that there is no relationship between the two variables being examined, as illustrated in Fig. 26.1b, f. That is, scores on one of the variables are not related in any meaningful way to scores on the second variable. The closer the correlation coefficient is to either -1.00 or $+1.00$, the stronger the relationship is between the two variables. A perfect negative correlation of -1.00 indicates that for every member of the sample or population considered, a higher score on one variable is related to a lower score on the other variable. A perfect positive correlation of $+1.00$ reveals that for every member of the sample or population considered, a higher score on one variable is related to a higher score on the other variable.

Perfect correlations are never found in actual business applications. Generally, correlation coefficients stay between -0.70 and $+0.70$. Some textbook authors suggest that correlation coefficients between -0.20 and $+0.20$ indicate a weak relation between two variables, those between 0.20 and 0.50 (either positive or negative) represent a moderate relationship, and those larger than 0.50 (either positive or negative) represent a strong relationship. These general rules of thumb for judging the relevance of correlation coefficients must be taken with a grain of salt.

Calculating the Correlation Coefficient

Pearson's correlation coefficient between two variables is defined as the covariance of the two variables divided by the product of their standard deviations:

$$\rho_{XY} = \frac{\mathrm{cov}(X, Y)}{\sigma_X \sigma_Y}$$

The above formula defines the population correlation coefficient, commonly represented by the Greek letter ρ (rho). The covariance of any two variables, X and Y, is $Cov(X, Y) = E((X - \mu_X)(Y - \mu_Y))$, where $\mu_X = E(X)$ is the mathematical expectation of the variable X. The covariance is the average cross-product of X, deviated from its mean, times Y, deviated from its mean. The covariance measures the extent to which two variables vary together. Positive covariances suggest that higher values of X are associated with higher values of Y. Negative covariances, on the other hand, indicate that higher values of X are associated with lower values of Y. The value of the covariance depends on the units of measurement of X and Y.

The correlation is a "standardized" covariance, constructed so as to fall in the interval $[-1, +1]$, with an absolute value of 1 indicating perfect correlation and a value of zero indicating no correlation. Another way to view the correlation is that it is the covariance between two standardized variables. Covariance and correlation are designed to detect linear association. Hence, either may be zero if two variables vary together in a systematic but nonlinear fashion.

Substituting estimates of the covariance and variances based on a sample gives the sample correlation coefficient, commonly denoted r:

$$r = \frac{1}{n-1} \sum_{i=1}^{n} \left(\frac{X_i - \bar{X}}{s_X}\right) \left(\frac{Y_i - \bar{Y}}{s_Y}\right),$$

Where, $(X_i - \bar{X})/s_X$, $\bar{X}$, and s_X are the standard score, sample mean, and sample standard deviation, respectively.

In this formula, notice what is happening. First, we are multiplying the paired standard scores together. When we do this, notice that if an individual case in the sample has scores above the mean on each of the two variables being examined, the two standard scores being multiplied will both be positive, and the resulting cross product will also be positive. Similarly, if an individual case has scores below the mean on each of the two variables, the standard scores being multiplied will both be negative, and the cross product will again be positive. Therefore, if we have a sample where low scores on one variable tend to be associated with low scores on the other variable, and high scores on one variable tend to be associated with high scores on the second variable, then when we add up the products from our multiplications, we will end up with a positive number. This is how we get a positive correlation coefficient.

Now consider what happens when high scores on one variable are associated with low scores on the second variable. If an individual case in a sample has a score that is higher than the average on the first variable (i.e., a positive standard score) and a score that is below the mean on the second variable (i.e., a negative standard score), when these two standard scores are multiplied together, they will produce a negative product. If, for most of the cases in the sample, high scores on one variable are associated with low scores on the second variable, the sum of the products of the standard scores will be negative. This is how we get a negative correlation coefficient.

What the Correlation Coefficient Does, and Does Not, Tell Us

Correlation coefficients such as the Pearson are very prevailing statistics. They allow the project team to determine whether, on average, the values on one variable are associated with the values on a second variable. Before moving on, there is one common misconception between the concepts of correlation and causation that should be addressed here. Correlation between two variables does not imply that there is a cause-and-effect relationship. Correlation (co-relation) simply means that variation in the scores on one variable correspond with variation in the scores on a second variable. Causation means that variation in the scores on one variable causes or creates variation in the scores on a second variable. Correlations only describe the relationship; they do not prove cause and effect. Correlation is a necessary, but not a sufficient condition for determining causality. There are three requirements needed to infer a causal relationship

1. A statistically significant relationship between the variables
2. The causal variable occurred prior to the other variable
3. There are no other factors that could account for the cause

Correlation studies do not meet the last requirement and may not meet the second requirement.

When we make the leap from correlation to causation, we may be wrong. Indeed, evidence of a relationship between two variables (i.e., a correlation) does not necessarily mean that there is a causal relationship between the two variables. However, it should also be noted that a correlation between two variables is a necessary ingredient of any argument that the two variables are causally related.

In addition to the correlation-causation issue, there are a few other important features of correlations worth noting. First, simple Pearson correlations are designed to examine linear relations among variables. In other words, they describe average straight relations among variables. For example, if the project team finds a positive correlation between two variables, it can predict how much the scores in one variable will increase with each corresponding increase in the second variable.

Statistically Significant Correlations

The project team calculates correlation coefficients to know whether a correlation found in sample data represents the existence of a relationship between two variables in the population from which the sample was selected: the input variable X from which assignable causes originate and the response variable Y. In other words, the project team wants to test whether the correlation coefficient is statistically significant.

To test whether a correlation coefficient is statistically significant, the project team should begin with the null hypothesis that there is absolutely no relationship between the two variables in the population, or that the correlation coefficient in the population equals zero. The alternative hypothesis is that there is, in fact, a statistical relationship between the two variables in the population and that the population correlation coefficient is not equal to zero. So what the project team should be testing here is whether the correlation coefficient is statistically significantly different from 0. These two competing hypotheses can be expressed with symbols as follows:

$$H_0 : \rho = 0$$

$$H_0 : \rho \neq 0$$

Where ρ, the population correlation coefficient.

Student's t-distribution is used to test whether a correlation coefficient is statistically significant. Therefore, the project team must conduct a t-Test. As with all t-Tests, the t-Test used for correlation coefficients involves a ratio, or fraction. The numerator of the fraction is the difference between two values. The denominator is the standard error. When the project team wants to see whether a sample correlation coefficient is significantly different from zero, the numerator of the t-Test formula will be the sample correlation coefficient, r, minus the hypothesized value of the population correlation coefficient ρ, which in the null hypothesis is zero. The denominator will be the standard error of the sample correlation coefficient:

$$t = \frac{r - \rho}{s_r}$$

Where r is the sample correlation coefficient, ρ is the population correlation coefficient, and s_r is the standard error of the sample correlation coefficient.

Fortunately, with the help of a little algebra, the project team does not need to calculate s_r to calculate the t-value for correlation coefficients. Indeed, for the sake of knowledge, the formula for s_r is given to be:

$$s_r = \sqrt{(1 - r^2) + (n - 2)}$$

Where r^2 is the correlation coefficient squared and n is the number of cases in the sample.

The formula for calculating the t value is:

$$t = r \cdot \sqrt{\frac{n - 2}{1 - r^2}}$$

Where the degrees of freedom is the number of cases in the sample minus two (i.e., $n - 2$).

With $100(1 - \alpha)\%$ confidence, the project team should reject the null hypothesis if the calculated absolute value of t is greater or equal than the percentile $t_{\alpha/2}$; that is, $|t| \geq t_{\alpha/2}$.

The Coefficient of Determination

Although correlation coefficients give an idea of the strength of the relationship between two variables, the input variable X from which assignable causes originate and the response variable Y, they often seem a bit nebulous. If the project team gets a correlation coefficient of 0.40, is that a strong relationship? Fortunately, correlation coefficients can be used to obtain a seemingly more concrete statistic: the coefficient of determination. Even better, it is easy to calculate.

When we want to know if the input variable X from which assignable causes originate and the response variable Y are related to each other, we are really asking a somewhat more complex question: Are the variations in the scores on the input variable X from which assignable causes originate somehow associated with the variations in the scores on the response variable Y? Put another way, a correlation coefficient tells us whether we can know anything about the scores on the response variable Y if we already know the scores on the input variable X from which assignable causes originate. In common statistical language, what we want to be able to do with a measure of association, like a correlation coefficient, is be able to explain some of the variance in the scores on the response variable Y based on our knowledge of the scores on the input variable X from which assignable causes originate. The coefficient of determination tells us how much of the variance in the scores of one the response variable Y can be understood, or explained, by the scores on the input variable X from which assignable causes originate.

One way to conceptualize explained variance is to understand that when two variables are correlated with each other, they share a certain percentage of their variance. The stronger the correlation, the greater the amount of shared variance, and the more variance the project team can explain in one variable by knowing the scores on the second variable. The precise percentage of shared, or explained, variance can be determined by squaring the correlation coefficient. This squared correlation coefficient is known as the coefficient of determination. The stronger the correlation, the greater the amount of shared variance, and the higher the coefficient of determination. Even though the coefficient of determination is used to tell us how much of the variance in one variable can be explained by the variance in a second variable, coefficients of determination do not necessarily indicate a causal relationship between the two variables.

26.3.2.2 Regression

Correlation involves a measure of the degree to which two variables are related to each other. A closely related concept, coefficient of determination, provides a measure of the strength of the association between two variables in terms of percentage of variance explained. Both of these concepts basically just confirm that there is a linear relationship between two variables and it quantifies how linear that relationship is. What these two correlation concepts do not tell us is how much a given change in one variable will change a related variable. To get that type of information, the project team needs to become acquainted with predictive concepts; namely, regression: a step beyond correlation.

Regression is a very common versatile statistic in which a random response, or outcome, variable Y, is posited to be a function of a set of input, or explanatory variables, denoted by $X = (X_1, X_2, \ldots, X_m)$. The regression function can be linear, in the sense that Y is defined to be a weighted sum of constants times explanatory variables $X = (X_1, X_2, \ldots, X_m)$, or it can be non-linear. Regression, particularly multiple regressions, allows examining the nature and strength of the relations between the variables, the relative predictive power of several independent variables on a dependent variable, and the unique contribution of one or more independent variables when controlling for one or more covariates.

Simple regression analysis involves a single input, or independent, or predictor variable and a single response, or dependent, or outcome variable. The difference between a Pearson correlation coefficient and a simple regression analysis is that whereas the correlation does not distinguish between input and response variables, in a regression analysis there is always an input variable and a designated response variable. That is because the purpose of regression analysis is to make predictions about the value of the response variable given certain values of the input variable. This is a simple extension of a correlation analysis. Of course, the accuracy of the predictions will only be as good as the correlation will allow, with stronger correlations leading to more accurate predictions. Therefore, simple linear regression is not really a more powerful tool than simple correlation analysis. But it does give the project team another way of conceptualizing the relation between two variables: the input variable X from which assignable causes originate and the response variable Y.

The real power of regression analysis can be found in multiple regression, which involves models that have two or more input variables and a single response variable.

Variables Used in Regression

As with correlation analysis, in regression the input and response variables need to be measured on an interval or ratio scale. Dichotomous (i.e., categorical variables with two categories) input variables can also be used. There is a special form of regression analysis, logit regression, which allows examining dichotomous dependent variables. In this section, we limit our consideration to those types of regression to that involve a continuous dependent variable and either continuous or dichotomous predictor variables.

Regression Analysis

The modeling of the relationship between a response variable and a set of explanatory variables is one of the most widely used of all statistical techniques. We refer to this type of modeling as regression analysis. A regression model provides the user with a functional relationship, between the response variable and explanatory variables, that allows the user to determine which of the explanatory variables have an effect on the response. The regression model allows the user to explore what happens to the response variable for specified changes in the explanatory variables.

The basic idea of regression analysis is to obtain a model for the functional relationship between a response variable (often referred to as the dependent variable) and one or more explanatory variables (often referred to as the independent variables).

Regression analysis, particularly simple linear regression, is a statistical technique that is very closely related to correlations. In fact, when examining the relationship between two continuous (i.e., measured on an interval or ratio scale) variables, either a correlation coefficient or a regression equation can be used. Indeed, the Pearson correlation coefficient is nothing more than a simple linear regression coefficient that has been standardized. The benefits of conducting a regression analysis rather than a correlation analysis are:

1. Regression analysis yields more information, particularly when conducted with one of the common statistical software packages, and
2. The regression equation allows the project team to think about the relation between the two variables of interest in a more intuitive way.

Whereas the correlation coefficient provides a single number (e.g., $r = 0.40$), which the project team can then try to interpret, the regression analysis yields a formula for calculating the predicted value of one variable when we know the actual value of the second variable. In simple linear regression, we begin with the assumption that the two variables are linearly related. In other words, if the two variables are actually related to each other, we assume that every time there is an increase of a given size in value on the X variable (called the input or independent variable), there is a corresponding increase (if there is a positive correlation) or decrease (if there is a negative correlation) of a given size in the Y variable (called the response, or dependent, or outcome, or criterion variable). Thus, the key to understanding regression is to understand the formula for the regression equation.

The simplest form of the regression equation, the linear regression, which expresses the impact of the selected subset of input variables X from which assignable causes originate upon the response Y is given by the relation:

$$Y = b \cdot X + a + \varepsilon$$

Where b represents the non-standardized regression coefficient, or the slope, a represents the intercept (i.e., the point where the regression line intercepts the Y axis), and ε represents a normally distributed uncertainty or disturbance of mathematical expectation equal to zero in depending upon the selected assignable causes input variables X and the "process to be improved" to actually produce the desired output Y.

A fundamental requirement for a causal interpretation to be given to the regression coefficient b in $Y = b \cdot X + a + \varepsilon$ is that the covariance $Cov(X, \varepsilon)$ be equal to zero, or that the equation disturbance, ε, is uncorrelated with the causal variable X. This is often referred to as the pseudo-isolation assumption, the causal assumption, or the orthogonality condition.

The regression equation allows the project team to do two things. First, it lets the project team find predicted values for the response variable Y for any given value of the input variable X. Second, it allows the project team to produce the regression line. The regression line is the basis for linear regression and can help the project team to understand how regression works.

Taking the mathematical expectation of both sides of the simplest form of the regression equation, we have

$$E(Y) = b \cdot X + a$$

Note that a, b, and X are constants at any particular X, and the mean of a constant is just that constant. In other words, according to the model $E(Y)$, the mean of the response variable Y (more properly, the conditional mean of the response variable Y, since this mean is conditional on the value of X) is a linear function of X. That is, the mean of the response variable Y at any given X is simply a point on the regression line. Hence, using regression modeling assumes, at the outset, that the means of the dependent variable at each X -value lie on a straight line. To understand how to interpret the coefficient a and b, we manipulate the mathematical expectation equation above so as to isolate either parameter. For example, if $X = 0$, we have

$$E(Y|X = 0) = a$$

Hence, the regression coefficient a, the intercept, is the mean of Y when X equals zero. To isolate the regression coefficient b, we consider the difference in the mean of Y for observations that are 1 unit apart on X. That is, we evaluate the difference $E(Y|X = x + 1) - E(Y|X = x)$. Notice that this is the difference in $E(Y)$ for a unit difference in X, regardless of the level of X at which that unit difference occurs. We have

$$E(Y|X = x + 1) - E(Y|X = x) = b$$

In other words, the regression coefficient b represents the difference in the mean of the response variable Y in the population for those who are a unit apart on an input variable X from which an assignable cause originates. If X is presumed to have a causal effect on Y, then the coefficient b might be interpreted as the expected change in an observed characteristic of the "process to be improved" outcome, Y, for a unit increase in an input variable X from which an assignable cause originates. In either case, the project team could refer to the difference $E(Y|X = x + 1) - E(Y|X = x)$ as the unit impact of X in the model. The unit impact of X in any regression model—whether linear or not—can always be found by computing the difference $E(Y|X = x + 1) - E(Y|X = x)$ according to the model.

The interpretation of the coefficient b can also be elucidated by taking the first derivative of the mathematical expectation $E(Y)$ with respect to X:

$$\frac{dE(Y)}{dX} = b$$

Thus, we see that the coefficient b is also the first derivative of the mathematical expectation $E(Y)$ with respect to X. Recall that the first derivative, also called the slope of the mathematical expectation $E(Y)$ with respect to X, represents the instantaneous rate of change in $E(Y)$ with an increase in X at point $X = x$. It also represents the slope of the line tangent to the curve relating the response variable Y to the input variable X at point x from which an assignable cause originates. Because the curve in this case is really a straight line, the slope of the tangent line is just the slope of the regression line itself (there is no tangent line in this case, since it is impossible for a line to touch a straight line at only one point). Hence, the instantaneous rate of change in the mathematical expectation $E(Y)$ with increase in X is the same as the change in $E(Y)$ per unit increase in X. In other words, in regression models in which $E(Y)$ is a linear function of X, the first derivative and the unit impact are identical quantities. This is no longer the case for models in which $E(Y)$ is modeled as a nonlinear function of X.

Estimation of Regression Parameters

Estimation of the regression equation and associated parameters using sample data is most often accomplished by employing ordinary least squares estimation (OLS). The requirements enumerated above for the disturbance and for a causal interpretation to be given to the regression coefficient b are largely assumptions required for unbiased estimation of the regression parameters using OLS. Let's review the assumptions. They are:

1. Y is a linear function of X; that is, $Y = b \cdot X + a + \varepsilon$
2. The observations are sampled independently.
3. X and Y are approximately continuous variables.
4. The X -values are fixed over repeated sampling and measured with only negligible error.
5. $E(\varepsilon) = 0$
6. $Var(\varepsilon) = \sigma^2$
7. ε is normally distributed

Assumption 5 also ensures the orthogonality condition that $Cov(X, \varepsilon) = 0$. The reason for this is straightforward. If there were a linear relationship between ε and X, reflected by a nonzero covariance, it would take the form $E(\varepsilon) = \gamma_0 + \gamma_1 X$. That is, the mean of the errors would be a linear function of X. If the mean of the errors is the same (in particular, it is zero) at X, this implies that there is no linear relationship between the error and X. This, in turn, implies that $Cov(X, \varepsilon)$ is zero.

To develop the rationale for employing ordinary least squares estimation (OLS), we consider the measuring the prediction error $\varepsilon = Y - E(Y)$ and determine

"How well does this regression equation line fit the data?" One way to tell is to examine the total prediction error made in using this equation to predict Y. The first impulse might simply be to sum the prediction errors for the n considered cases of the collected sample data to assess total error. However, that does not work well, since large positive errors and large negative errors tend to cancel each other out. The project team might therefore end up with a small total error even with a poorly fitting line. Instead, using the ordinary least squares estimation, the errors are first squared and then summed. The resulting quantity is called (fittingly) the sum of squared errors and denoted SSE:

$$SSE = \sum_{i=1}^{n} (Y_i - E(Y))^2$$

The idea behind ordinary least squares is to choose as the regression parameters estimates of the population regression line that sample line that minimizes SSE; hence the resulting coefficients a and b are called the ordinary least squares estimates. Finding the coefficients a and b that minimize SSE is a minimization problem in two variables, which is readily solved using the techniques of differential calculus.

$$a = \bar{Y} - b\bar{X}$$

$$b = \frac{Cov(X, Y)}{s_X^2} = r \cdot \frac{s_Y}{s_X}$$

Notice that the regression coefficient is simply the correlation coefficient times the ratio of the standard deviations for the two variables involved. When we multiply the correlation coefficient by this ratio of standard deviations, we are roughly transforming the correlation coefficient into the scales of measurement used for the two variables. Notice that there is a smaller range, or less variety, of scores on our 7 variable that there is on our X variable in this example. This is reflected in the ratio of standard deviations used to calculate b.

It is important for the project team to remember that when the regression equation is used to find predicted values of an observed characteristic of the "process to be improved" outcome, Y, for different values of input variables X from which assignable causes originate, the actual value of Y is not been calculated. Only predictions about the value of Y are been made. Whenever predictions are been made, the results will sometimes be incorrect. Therefore, there is bound to be some error ε in the predictions about the values of Y at given values of X. The stronger the relationship (i.e., correlation) between X and Y variables, the less error there will be in my predictions. The error is the difference between the actual, or observed, value of Y and the predicted value of Y.

It is also important for the project team to remember that regression analysis is based on correlations. Just as correlations should not be mistaken for proof of causal relationships between variables, regression analyses cannot prove that one variable, or set of variables, causes variation in another variable. Regression analyses can reveal how sets of variables are related to each other, but cannot prove causal relations among variables.

Two other quantities connected to the linear regression model are of interest to estimate. The first is σ^2, the variance of the error terms at each X value. The second is denoted ρ^2 (pronounced "rho-squared") and is called the coefficient of determination of the regression equation. The latter is the primary index of discriminatory power for a regression model. To develop estimators of these quantities, let's consider partitioning the variability in Y, based on the assumption that it is a linear function of X in the population. Since $Cov(X, \varepsilon) = 0$ by assumption (and, of course, $Cov(a, \varepsilon) = 0$, since the covariance of a constant with a variable is always zero), the following holds true:

$$1 = \rho^2 + \frac{\sigma^2}{Var(Y)}$$

This relation shows that the total variability in the response variable Y consists of two proportions: ρ^2, which can also be written as $1 - \sigma^2/Var(Y)$, is the proportion of variability in Y accounted for by the linear regression on the "process to be improved" input variable from which an assignable cause originates, X (i.e., by the structural part of the model); and, $\sigma^2/Var(Y)$ is the proportion accounted for by error. Thus, ρ^2 reflects the ability to account for variation in Y using a linear function of X, and as such, is the ideal measure of discriminatory power for linear regression. The value of ρ^2 ranges from 0, for the case in which X has absolutely no ability to account for Y, to 1.0, for the case in which Y is perfectly determined by X; i.e., all the points lie exactly on the regression line and there is no error. Typically, ρ^2 will range somewhere between these two extremes.

Analyze Process Steps and Tasks 27

We have defined a process as "a set of logically related discrete elements (tasks, actions, or steps) taken in order to achieve a particular end." In this definition, a discrete element, the performance of which is measurable, is meant to be the smallest identifiable and essential piece of activity that serves both as a unit of work and as a means of differentiating between the various aspects of a project or an operation work. Each discrete element is designed to create unique outcomes by ensuring proper control, acting on and adding value to the resources that support the work being completed.

From the perspective of this definition, the discrete elements making up the process can be represented in terms of hierarchies of goals and sub-goals, using work instructions to show when sub-goals need to be carried out. A work instruction describes the statement of the conditions under which each of a set of sub-goals is undertaken to achieve their common super-ordinate goals. To achieve the purpose of the overall process, these discrete elements need to interact. This interaction occurs through discrete elements having inputs and outputs.

"Analyze Process Steps and Tasks" is the project management process used to examine how the "process to be improved" discrete elements are accomplished. It explores the "process to be improved" discrete elements through a hierarchy of goals indicating what a task owner is expected to do, and work instructions indicating the conditions when subordinate goals should be carried out. It collects different sorts of information about the process discrete elements and their context in order to reach satisfactory outcomes where potential problems have been identified and where potential solutions have been proffered. It includes a detailed description of both manual and mental activities, tasks and element durations, tasks frequencies, tasks allocations, tasks complexities, environmental conditions, and any other unique factors involved in or required for one or more people to execute the "process to be improved" successfully.

"Analyze Process Steps and Tasks" entails the project team working through a cycle of decisions which include: making judgments concerning where the team should focus; examining operations in greater detail in order to identify problems

A. van Aartsengel and S. Kurtoglu, *Handbook on Continuous Improvement Transformation*, DOI 10.1007/978-3-642-35901-9_27, © Springer-Verlag Berlin Heidelberg 2013

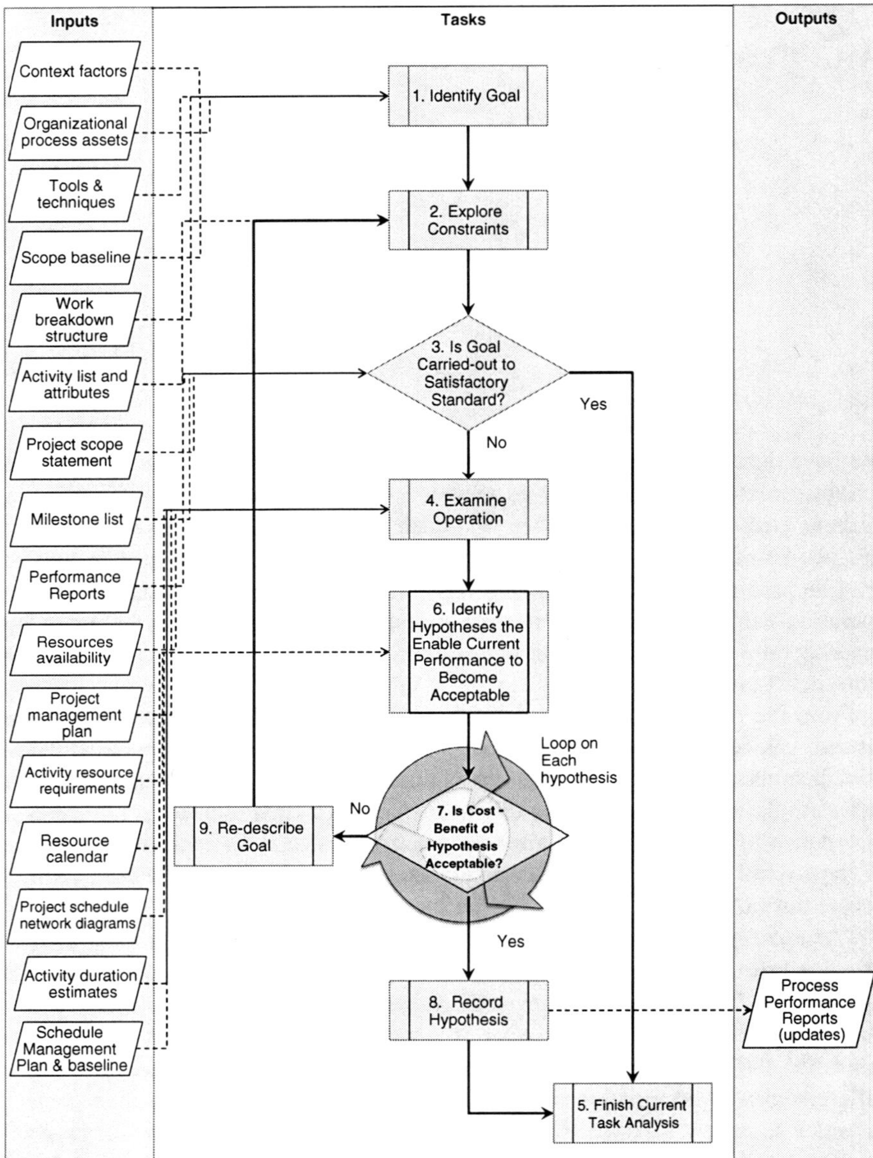

Fig. 27.1 The cycle of task analysis decisions

and solutions; and finding ways to increase the grain of analysis such that closer scrutiny can be given to problem areas.

The constituent project management processes used during the analysis of each "process to be improved" discrete element, illustrated in Fig. 27.1, includes the following:

1. Identify Goal
2. Explore Constraints
3. Is Goal Carried Out To a Satisfactory Standard?
4. Examine Operation (Resources-Task Interaction)
5. Examine Goals by Re-description
6. Record Cost-Benefit Assumptions

These six constituent processes interact with each other and with the project management processes in the PDSA "Process Groups." Each aspect of executing any of these can involve effort from one or more persons, based on the needs of the project. Each aspect occurs at least once in every "process improvement" project which involves procurement of some of its activities and occurs in one or more project phases.

27.1 Identify Goal

The first step in analyzing a process discrete element (task, action, or step) is to identify, state, and focus effort upon the main goal of the discrete element in order to be effective and economical. At this first step, the project team must focus upon areas that have been documented during the project planning as being of concern for assignable causes of variation and it must expresses a suitable working goal to provide focus for the subsequent analysis. If there is no concern for the manner in which the work associated with the discrete element is carried out, then further action is unnecessary. If there is concern, then the project team can first examine the resources-task interaction and the operations underpinning performance of the discrete element, in order to establish whether these problems can be identified or whether improved solutions can be proffered. If no hypotheses are forthcoming, usually because the goal of the process discrete element considered is still at too coarse a level of description for sufficient insight, then the project team should attempt a re-description, into sub-goals and their work instructions.

It is often helpful for the project team first to consider the inputs and outputs to the resource-task system being examined in the process discrete element. This means understanding what information and materials flow elsewhere in the enterprise business and beyond, and what information and materials the current process discrete element relies upon. The outputs will help the project team appreciate the importance of the goal under investigation to the wider enterprise business. This can help later on in understanding both the consequences of occurrence of assignable causes and the "process to be improved" underperformance. It can also account for problems experienced elsewhere. Understanding inputs sets out the resources, information and materials upon which the present process discrete element relies. It is a good analytical practice for the project team to start goal identification by becoming acquainted with the wider system containing the process discrete element considered. Then, if problems arise in the analysis which might be difficult to resolve, alternatives might exist elsewhere in the system which either allow the cause of "process to be improved" underperformance to be minimized or dealt with in a different way.

27.2 Explore Constraints

The second step in analyzing a process discrete element (task, action, or step) is to explore constraints on how responses to inputs to the process discrete element considered are made and on the options that the project team can choose in making recommendations. As goals are discussed, so constraints associated with their attainment or their solutions are encountered. Constraints are particularly important in practical projects as they affect and limit the options that might be adopted and pursued to realize goals. Constraints include detail about the work environment that is assumed to influence performance. They also include limitations on preferred solutions imposed by management and staff.

During this constraints exploration step, the project team may also become aware of factors which are not strictly rational, yet still need to be observed. Exploring constraints tends to be informal and intuitive and either kept in the back of the project team's mind or recorded in a separate note. Sometimes constraints are not obvious at a particular level of description and are only recognized when the goal is explored further.

27.3 Is Goal Carried Out to a Satisfactory Standard?

The third step in analyzing a process discrete element (task, action, or step) is to assess whether the goal will be met to an acceptable standard given prevailing circumstances

Since analysis of a process discrete element (task, action, or step) is a practical activity, there are often time constraints on how it is carried out. For example, it is often essential that a process discrete element analysis is completed in good time so that decisions can be acted upon. Also, there may be a limit on how much access the project team can have to personnel and the workplace. For these reasons effort in analysis should be directed where it is most essential. Thus, before examining a goal, the project team should assess whether the effort is necessary.

Examining the goal of a process discrete element is pointless if the goal is not important. When analysis commences, the main goal to be analyzed invariably warrants examination, otherwise the project would not have been initiated. However, when sub-goals are examined the project team needs to focus on the most critical areas. Judging whether a sub-goal is worthy of further investigation entails a form of cost-benefit analysis. It is concerned with the risk entailed in the discrete element operator carrying out the operation. This involves the likelihood of the operator committing an error and the consequences that would arise when an error occurs. It is often expressed as the $P \times C$ rule, where P stands for the probability of inadequate performance and C stands for the criticality or cost of inadequate performance of the discrete element considered. C is the total cost of corrective actions (rework, scrapping and conformance use) also known as the Cost of Quality (CoQ). It is a measure of the costs specifically associated with achievement or non achievement of an element quality—including all elements requirements established by the business and its contracts with its customers.

Generally, P and C are estimates made by the project team in conjunction with the customer of the outcomes of the process discrete element considered, and often the judgment is made intuitively by the project team in order to make progress. It is difficult for the probability of inadequate performance and for the criticality or cost of inadequate performance to be precisely quantified, though in some circumstances they could be. In a repetitive automobile assembly or inspection task, for example, data could be collected that would show the frequency of error and the costs of rework or the costs of disposal of unsatisfactory items or the costs of replacing faulty goods that had been sold.

Generally, though, the project team relies on estimates. Their product $P \times C$ is simply a convenient shorthand for combining these two factors. Thus, if the estimate of C tends to zero, the product $P \times C$ will tend to zero. In some enterprise business workplaces, for example, staff is required to keep logs as a method of prompting them to keep an eye open for assignable causes of variation of unusual events occurring during production. It may be the case that log entries are never checked and poor log entries go unnoticed. Therefore, C would be low, approaching zero, so P could take any value and still the product $P \times C$ would tend towards zero. In other situations keeping log entries is important, because the data contained in written logs are crucial to evaluating production problems, so this conclusion would not apply. Therefore, despite the extremely low value of P brought about through well-engineered process and good process execution training, the unacceptably high value of C means that there is no case for complacency.

If the value of $P \times C$ is considered unacceptable as the project team delves deeper into the process discrete element considered—that is, the risk of leaving things as they are is unacceptable—then the main goal to be analyzed invariably warrants examination.

Alternatively, the project team can also use the Discrete Element Yield is a criterion used to control the considered discrete element performance. As indicated in a previous section, we can think of it as a percentage of a process discrete element outcomes passing the compliance check (their key parameters fall within certain range of tolerance), in other words these outcomes will not be rejected as defective ones, so additional costs for repairing or scrapping the defective discrete element outcomes will not be incurred to the enterprise business. A process discrete element yield uses the concepts of upper and lower specification limits (and a target limit between them) which are boundaries defining acceptable performance level.

All the outcomes of the process discrete element suiting the range between the upper and lower specification limits or precisely meeting a target limit will make up a discrete element yield rate (the fluctuation of characteristics of these outcomes can be depicted on control charts). Yield loss (quality gaps) is caused by certain faults in the process discrete element, entailing different deficiencies or shortcomings in the process discrete element key parameters. Yield loss rate can be classified by deficiency or defect types and this helps to pinpoint the problematic areas of the process discrete element considered.

If the value of the yield loss associated with the process discrete element is considered unacceptable as the project team delves deeper into the process discrete element considered—that is, the risk of leaving things as they are is unacceptable—then the main goal to be analyzed invariably warrants examination.

27.4 Examine Operation

The fourth step in analyzing a process discrete element (task, action, or step) is to examine the resources-task interaction and the operations underpinning performance of the process discrete element considered, in order to establish whether assignable causes of variation can be identified or whether improvement solutions can be proffered.

Where performance warrants attention, examination should first consider the operation and the resources-task interaction with a view to generating an improvement hypothesis that will overcome the performance deficiency or help make a judgment concerning the cause of this weakness in performance. This phase is central to the "Analyze Process Steps and Tasks" project management process. It may involve attempting to understand information processing, cognition, attitudes, etc. or it may be done informally by the project team relying on experience or human factors knowledge.

During this step, every question which the project team may ask becomes twofold. As the project team delves deeper into the process discrete element considered, it should ask first "What is happening?" And to the answer of that question it should ask, "Why?" Why this is been done? Is it necessary? Can it be eliminated? If there is not a good reason for doing it, recommend that it be eliminated. This is the question that produces the most cost effective changes and should always be asked first. There is no sense in asking more detailed questions about a work step that should not be done at all. When steps of work are eliminated there is little or no implementation cost and the benefit equals the full cost of the performing that step. Many work steps that once served a valuable purpose (possibly years ago) can be eliminated when that purpose no longer exists.

If there is a good reason for performing the work then the project team should ask: "Where is it done and why is it done there?"—"When is it done and why is it done at that time?"—"Who does it and why does that operator do it?" These questions lead to changes in location, timing, and the operator doing the work without changing the process discrete element task itself and therefore they are also highly cost effective. Equipment is relocated closer to the people who use it. Schedules are revised to fit with previous and following portions of the process to produce smoother flow. Tasks are shifted to people better able to perform them. Tasks are combined, eliminating the transports and delays between them that occurred as the work flowed between locations and/or people.

Only after these questions have been asked and answered should the final question be addressed: "How is it done and why is it done that way?" While this question can lead to excellent benefits, it also incurs the greatest costs because changing how a task is done generally requires introducing new technology with new equipment, programming and significant amounts of training. Of course, new technology is important. It is very important, and it should be pursued. But, if the enterprise business wants to maximize profits, it will hold off on changing how the process discrete elements considered are performed until after the previous questions have been properly dealt with. Unfortunately, new technology is so enticing that enterprise businesses often leap into it before asking the earlier questions. They miss easy to install, high payoff opportunities and sometimes wind up having spent a lot of money to automate activities that should not be done at all.

27.4.1 Generate Hypotheses

If the current performance of the process discrete element considered is judged to be unsatisfactory the project team may examine the operator-system interaction to establish a solution to the problem. Different process improvement project aims may require hypotheses to be stated in terms of potential causes of underperformance or design suggestions to overcome the underperformance problem. Thus, the project team should try to identify ways in which human performance might lead to occurrence of assignable causes of variation in the outcomes of the process discrete element considered.

Central to generating hypotheses are the tasks performed within the process discrete elements to fulfill organizational objectives: producing products or services for customers. Tasks may be hypothesized to be value-added, to cycle fast enough to generate entities to meet downstream demand, to control production flow, and to be flexible.

27.4.1.1 Value-Added Tasks

Value is added to outcome of the process discrete element considered when it shows changes in form, changes in fit and changes in function.

To meet the demand of "value", the outcome of the process discrete element considered must fulfilled the following three requirements:

1. *It is customer defined;*
2. *The customer is willing to pay for it;*
3. *It is produced correctly the first time.*

Value-added tasks include only production activities. Non-value-added tasks increase the cost of a product or service but do not increase its value to customers. A memory-jogger acronym often used to remind people of the different types of non-value-added tasks or waste is known as DOWNTIME, which stands for:

1. **D**efect or Rework—Sorting, repetition or making scrap.
2. **O**verproduction—Producing too much, too early and/or too fast.
3. **W**aiting time—People or parts waiting for a work cycle to finish.

4. Non utilized people's intellect—Failure when it comes to exploiting the knowledge and talent of the employees.
5. Transportation—Unnecessary movement of people or parts.
6. Inventory—Materials parked and not having value added to them.
7. Motion—Unnecessary movement of people or parts within a process.
8. Excess processing—Processing beyond the demand from the customers.

Classifying waste as non-value-adding is not always a clear-cut. An enterprise business understands that some activities are necessary non-value-adding—though they are non-value-adding through the eyes of the customer, they are essential to properly operate the business. Activities that are necessary to meet regulatory requirements and accreditation standards fall into this category, as do many activities within support departments that do not provide direct value to the customer, such as human resources, information technology, finance, legal, etc. In these areas, the project team must generate hypotheses to reduce the effort required to assure full compliance and proper operation of the business.

27.4.1.2 Cycle Time, Lead Time and Takt Time

Cycle Time

Processes discrete elements are required to cycle at a rate fast enough to generate entities at a pace to meet customer (or market) demand. The actual rate of processing process discrete element entities is the process discrete element Cycle Time. It is the amount of time taken or allocated to complete the goal of process discrete element entity. It provides a starting point for improvement in costs, quality and inventory.

Lead Time

A lead time is the amount of time, specified by the enterprise business, as process outcome vendor, that elapses from the time a customer expresses an order to buy a process outcome until that order is satisfied. It is the amount of time that the customer waits for the enterprise business, as process outcome vendor, to respond by fulfilling the order. It is the traditional measure used to track the ability, or the inability, of an enterprise business—as a vendor—to react to a customer demand or order.

Lead Time and cycle Time are two different entities. They are not synonymous and they are not interchangeable. To do so is similar to using the terms cost and price interchangeably. They are driven by different factors and used for different purposes.

Cycle Time, in much the same as Cost, is the result of activities. Each activity adds an amount of time taken or allocated to complete the goal, and it also adds a monetary unit amount to a process discrete element entity, to define the total time as well as the total cost. Allocated times and costs make up a part of the total but each element of time or cost is identifiable. The market place sets the selling price. The difference between cost and price is profit. One may address cost elements to increase the profit margin or to lower costs to meet a market supportable price, but the costs do not always drive the price. While cost models add all elements, time models count only the longest of simultaneous time elements together. From this, we can see that a process total cycle time (often referred to as total work content) is the sum of the sequence of process discrete element entities cycle times and the longest component enterprise business lead time as vendor.

The cycle time is process driven, while the lead time is business driven. When a cycle time is changed or reduced, the customer does not necessary see the change or reduction. The customer always sees a change or reduction in lead time. Thus, it is important for the "process improvement" project team to capture the between cycle time and lead time. The lead time must be accurate, adhered to, and timely. The purpose of Cycle Time is to define a benchmark, a starting point, from which to make improvements in costs and quality of processes. Just as we continually challenge costs, so as we continually challenge Cycle Times. The purposes of Lead Time values, in a planning system, is to define when an action needs to be taken so that a result will occur when desired.

Takt Time

Takt (a real word, not an abbreviation) is the German word for rhythm or cadence. A common mistake here is to confuse it with TACT (Total Activity Cycle Time) or similar, which is an entirely different thing. The Takt Time is defined as "The rate at which the end product or service must be produced and delivered in order to satisfy a defined customer demand within a given period of time." Simply put, it is the drumbeat of the customer (or market) demand based on working hours in period. The Takt Time is calculated as:

$$Takt\,Time = \frac{Available\,Work\,Time\,in\,Period}{Available\,Demand\,in\,Period}$$

If a process (or a process discrete element) is perfectly balanced with the market demand, then for every Takt Time increment an entity is processed and used by the customer (or market). For example, if a process discrete element runs 24 h a day and the customer (or market) demand is 240 entities (or units) per day then the Takt Time is given as:

$$Takt\,Time = (24 \times 60)\ \min/240\,units = 6\ \min/unit$$

The process (or a process discrete element) considered would need to produce one unit every 6 min. If an entity is not processed (on average) each and every 6 min, then the process falls behind Customer demand. Thus, if the processing time is known (also known as the Cycle Time) is above the Takt Time, the process discrete element considered falls behind Customer demand. Likewise if the Cycle Time is less than the Takt Time of, for example, 5 min, then the process discrete element is cycling faster than Customer demand and building inventory or spending 1 min in every six waiting, to avoid creating unused inventory.

If the cycle time of the process discrete element is longer than the pace of customer (or market) demand (i.e. Takt Time), then the process discrete element is not cycling quickly enough and inevitably falls behind.

Clearly if the work period (shift time) is less than 24 h per day, then the process discrete element has to go proportionately faster during those times that work is actually done to meet the daily customer demand. For example, if the work period is 12 h per day in the preceding example, then the new Takt Time is now 3 min, because there is only half the available work time and the process has to cycle twice as fast. So, if the Cycle Time is not 3 min, the process discrete element is not balanced with the customer demand.

While the benefits of producing at a rate, equal to customer (or market) demand, are obvious, there are additional hypotheses that need to be generated regarding the applicability of the Takt Time concept in the specific situation of the current process discrete element considered: hypothesis on how balance is the actual demand, i.e. how much does actual demand vary, period to period; and hypothesis on how does the customer want to take delivery (also consider any shipping constraints).

Let's take these hypotheses one at a time: the Takt Time implies a reasonably balance level rate of customer (or market) demand. In some government contract businesses, this is indeed the case. However, in many industries, the actual demand varies significantly period to period.

Knowing that average customer (or market) demand is 100 units/day is of little value when actual customer (or market) demand regularly varies from zero/day to 300/day! If the process (discrete element) considered were to produce at an average rate, while consistently meeting customer (or market) demand, the enterprise business would likely have to carry some finished goods inventory. This strategy has all of the typical "wastes" associated with inventory: Double handling, tracking, risk of obsolescence, hidden quality defects, tied up money and space, etc.

By considering the second hypothesis on how does the customer want to take delivery, it is needless to say, producing at a nice balance rate of customer (or market) demand, all month long, only to ship once/month, defeats much of the benefit of producing at a rate. All month long the enterprise business will be building and storing inventory, double handling (place it into storage, then take it back out of storage to ship it), with all of the wastes associated with inventory.

One of the biggest benefits implied by a process (or a process discrete element) perfectly balanced with the customer (or market) demand is being able to produce linearly, at Takt Time; hence the ability to produce essentially directly to the shipping process (or associated process discrete element) and to transfer the product or service immediately to the customer. That is, the enterprise business wants to produce and ship at the Takt Time rate.

In the circumstance where the customer (or market) does NOT want daily linear deliveries, it often makes more sense to produce and ship Just-In-Time (JIT) according to the customer's shipping constraints. That is, instead of producing linearly and accumulating the product or service to ship, it may make more sense to produce the shipping quantity, at a considerably higher production rate, and then immediately load and ship the product or service.

In the above example, assume that the average demand is 100/day, but the customer (or market) only wants delivery once/week. The hypothesis generated here could be that the enterprise business might choose to produce at a Takt Time of 0.8 min/unit (500 shipping quantity, all produced in 1 day) and to pack and ship the shipment directly off the end of the production line.

Another typical extenuating circumstance is logistical. If the logistical costs of daily shipments are prohibitive, then some of the benefits of Takt Time are lost. The process (or a process discrete element) which is perfectly balanced with the market demand would produce nice and linearly, but then the enterprise business will have to accumulate and store the product or service waiting for an "efficient" transportation batch.

Another limitation to the Takt Time usefulness is the amount of value-adding time required to produce the unit. Let's go back to our example, this time with no constraints on the shipping rate: i.e. the customer (or market) is willing to take daily delivery, and the transportation costs are not prohibitive. Our computed Takt Time is 6 min/unit. But what if the entire value-add time is less than 1 min/unit? That is, the process (or a process discrete element) considered can produce a unit in less than 1 min. Does it make sense to spread out the work beyond 1 min? Probably not!

The Takt Time applicability demands that the total value-add time is substantial enough to justify producing at a rate of customer (or market) demand. So what are some reasonable alternatives hypotheses?

27.4.1.3 Product & Production Leveling: Heijunka

Through the numerous discrete elements of a "process to be improved"; key factors such as time and supply of required inputs, demand for the supply, lack of communication and disorganization can result in one of the most common problems in supply chain management. This common problem is known as the bullwhip effect; also sometimes the whiplash effect. This effect can be explained as an occurrence detected by the supply chain where demands sent to a selected process discrete element create larger variance than its output to the succeeding element in the "process to be improved." These irregular demands in the lower discrete elements of the "process to be improved" often develop to be more distinct higher up in the "process to be improved." This variation can interrupt the smoothness of the supply chain process as each link in the "process to be improved" supply chain will over or underestimate the product or service demand resulting in exaggerated fluctuations. There are many factors that contribute to this bullwhip effect in supply chains, some of which include:

1. Disorganization between each process discrete element supply chain link; with demanding larger or smaller amounts of an input or output than is needed due to an over or under reaction to the supply chain beforehand.
2. Lack of communication between each process discrete element in the supply chain makes it difficult for these elements to run smoothly.
3. Order batching; a process discrete element may not immediately place a demand of inputs to its previous supplier elements; often accumulating the demand first. It may demand inputs daily, weekly or even monthly. This creates variability in the demand as there may for instance be a surge in demand at some stage followed by no demand after.

To prevent such variations in the "process to be improved" final outcomes, it is important to minimize demands and output variations in its constituent discrete elements. Production leveling, also known as production smoothing or by its Japanese original term—heijunka—is a technique for leveling the workload of each process discrete elements for the sake of continuity and consistency regardless to demands fluctuations.

Ohno (1988) illustrates this technique in his teaching with the tale of "the slower but consistent tortoise" versus the "dashing of the hare."

"The slower but consistent tortoise causes less waste and is much more desirable than the speedy hare that races ahead and then stops occasionally to doze. The Toyota Production System can be realized only when all the workers become tortoises."

Heijunka is a two step process of dampening variations from the production schedule:

1. Leveling the workload of each process discrete element over a defined period of time (mostly daily)—smoothing out variations in the overall takt time.
2. Leveling the outcomes (products or services) mix of each process discrete elements within the defined period of time—smoothing out variations in the demand from upstream discrete process elements.

The objective of generating heijunka hypotheses is to induce continuity and consistency in effort, in innovation and improvement on the "process to be improved." Large variations in the "process to be improved" outcome volumes and mix create peaks and valleys in the process. Handling peaks requires additional resources while idling is a direct result of valleys. Peaks need additional resources which include human, machine and time. This may add extra stress on the machinery, people and systems considered. On the other hand when the "process to be improved" outcome volumes are low, the resources would be idling. This obviously is a loss to the enterprise business.

27.4.1.4 Kanban Limits

Kanban is the Japanese word for "signboard" or card. It has become synonymous with demand scheduling. Kanban traces its roots to the early days of the Toyota Production System (TPS). In the late 1940s and early 1950s, Taiichi Ohno developed kanbans to control production between processes and to implement "Just in Time (JIT)" manufacturing at Toyota manufacturing plants in Japan. These ideas did not gain worldwide acceptance until the global recession in the 1970s. By using kanbans, he minimized the work in process (WIP) between processes and reduced the cost associated with holding inventory (Ohno 1988).

Originally, Toyota used kanban to reduce costs and manage machine utilization. However, today Toyota continues to use the system not only to manage cost and flow, but also to identify impediments to flow and opportunities for continuous improvement. Interestingly, Mr. Ohno modeled many of the control points after U.S. supermarkets—hence the term kanban supermarkets.

Through the production facility of most manufacturing enterprise businesses, kanbans are used to control the flow of process outcome; they are typically used when the system affected by the process considered operates on a just-in-time (JIT) or pull-production philosophy. The goal of a system operating with a just-in-time (JIT) philosophy is to produce the appropriate items in the necessary quantity at the right time. The just-in-time (JIT) operating philosophy is an integral part of the Toyota Production System (TPS), which strives to achieve an all around improvement in the economics of manufacturing operations by eliminating wastes of all forms, inventory included.

In a push production operation philosophy, parts of the process outcome are released by process discrete elements to their immediate succeeding process discrete element, which on completion pushes its outcome to the next process discrete element. Here, planning is carried out with the assumption that the demand rate is know and does not show any variability. Furthermore, the operation is based on the assumption that it is better to anticipate future production requirements and plan for them. Products are pushed through the system and are stored in anticipation of demand, which often results in overproduction because anticipated demand may not materialize. Also, there are costs associated with having inventories of products sitting in storage and waiting for consumption. Changes in the demand rate call for duplication in planning and lead to a waste of time and resources, increase lead time and the generation of inventory.

With a JIT or pull-production philosophy, only the last process discrete element interfaces with a demand pattern, and information flows back through the process. In a pull environment, each operation required to produce a product is considered to be the customer of the preceding operation. The kanban, or pull signal, is treated like the customer order. In a pull system environment, items are not processed without a "customer order", i.e. a pull signal. Items are therefore made and/or moved "just in time". Here, the preceding process discrete element(s) must produce the exact quantity of process outcome part withdrawn by the subsequent process discrete element. To achieve this goal, the flow of information is controlled by the flow of Kanbans connecting two adjacent process discrete elements. A Kanban is a tag-like card authorization for the previous process discrete element to produce a specified quantity of units. A Kanban is released when the succeeding process discrete element withdraws units from the output of the preceding process discrete element(s).

There are two type of kanban mainly used on most production floors: the withdrawal Kanban and the production ordering Kanban. A withdrawal Kanban conveys information about the quantity of units that the succeeding process discrete element should withdraw from the associated buffer storage, while a production ordering Kanban specifies the quantity of units that the preceding process discrete element must produce. Kanban cards are therefore used as production triggers. They are attached to process outcomes or to parts of process outcomes that flow through the system affected by the process considered.

A kanban may contain information identifying a process outcome part, a process discrete element, and a production demand order. By keeping track of kanbans, the system affected by the process considered can keep track of the work in process (WIP). In kanban-based production systems, a Kanban card represents a type of signal that indicates when to begin production of a process outcome part. Without this signal, production cannot occur. Many things such as cards, containers, tape spaces, etc..., can be used as kanban signals.

The number of Kanban cards thus limits the amount of work-in-process inventory that can be accumulated between process discrete elements or within a process discrete element. Thus it prevents overproduction, reduce multi-tasking, reduce process outcome lead time and maximize throughput. However, it reveals bottlenecks dynamically so that the project team can address them before they get out of hand. It can also be used to limit the amount of raw material used as input to the process and finished process outcomes. Hence, less cash is tied up, less space, less handling, less handling damage, etc.

What is counter-intuitive about this is that a process throughput increases because of these kanban limits. The kanban limits surface problems (create bottlenecks within the process considered) and force the project team to set or generate hypotheses on them. Because we usually do not like to face up to problems it is very tempting to lift the limits or to dismiss the approach altogether. Raising a kanban limit too high will create too much turbulence and confusion and slow production down. Lowering a kanban limit too far will hold production back and hence slow production down.

It is tempting to say that kanban limits are priorities when the system affected by the process considered operates on a Just-in-Time (JIT) or pull-production philosophy. But it is only all of them together that influence priorities—none of them individually represents a priority, so raising and lowering them becomes meaningless in this analogy. Nor is a limit a warning signal. A limit is a hard limit, and a strict command that no more may enter this process discrete element. That is more than just a warning. But saying that it is a command is also insufficient, as that does not help us understand the issues around making sure the adjustment to be made to the "process to be improved" is just so. Accordingly, the hypotheses to be set on kanban limits by the project team depend on the complexity of the process considered and its outcome(s).

27.4.1.5 Flexibility

Where the above constraints are not an issue, the project team can go ahead and generate a hypothesis on the process (or a process discrete element) considered in order to produce at a rate of customer (or market) demand. However, the project team should generate hypothesis to build in the flexibility, via overtime, flexible workforce, additional work stations, extra raw material, etc. to allow for the actual production rate to vary as required to meet, exactly, the customer (or market) demand.

It is imperative, for any world class operation, that it fulfills its customer commitments with religious precision. The project team can then work with its identified customers to generate hypotheses, identify, and reduce the reasons for a process (or a process discrete element) not to be balanced with the customer (or market) demand. All too often, these "hypotheses" are self imposed: column pricing (providing a unit price discount for the customer when he/she orders a large batch size), in-efficient supply chain (batched deliveries), and false economies of production: "lot size optimization," etc.

The project team may use operational task taxonomies or guidelines to identify improvement hypotheses without requiring any detailed modeling of behavior. In other process improvement projects where suitable design hypotheses are not obviously forthcoming, the project team should try to model behavior as a precursor to suggesting a solution to overcome performance weaknesses. Cognitive task analysis methods, described in a later section below, may be used to serve this end. In practice, a variety of such strategies may be entirely appropriate.

An important aspect of generating hypotheses is observing constraints. Constraints are identified during the second step of the "Analyze Process Steps and Tasks" project management process. It is quite possible that no hypotheses consistent with these constraints are forthcoming. This will warrant a re-description or challenging the constraints.

27.4.2 Examine Resources-Task Interaction

Examining the resources-task interaction under the generated hypotheses requires the project team to focus upon a particular goal (or sub-goal). The project team has considerable freedom to choose how this is done. Indeed, this is where the project team members are invited to draw on other task analysis methods and use their own experience. There are many different methods and perspectives that can be used here, although, in many situations the project team will be content to make an intuitive or expert judgment. Five common resource factors strategies may be employed in order to:

1. Examine customs and manner underpinning the workplace where operation of the "process to be improved" is carried out.
2. Model the psychological processes underpinning the operation of the "process to be improved" discrete element considered;
3. Take advantage of the common similarities that exist between actual operations even from widely differing contexts;
4. Treat each operation of concern to a systematic appraisal using, for example, a checklist;
5. Subject the operation to a process of further data collection using a specialist method.

27.4.2.1 Examining the Workplace Customs and Manners: 5S

Examining customs and manner underpinning the workplace where operation of the "process to be improved" is carried out, is performed using the "5S" concept. The "5S" is defined by Japanese specialists as a set of good customs and manners, deriving from the traditional manner of behavior in house and school. The determination "5S" dates from Japanese words. It is a system to reduce waste and optimize productivity through maintaining an orderly workplace customs and manners and using visual cues to achieve more consistent operational results.

A typical "5S" implementation would result in significant reductions in the square footage of space needed for existing operations. It also would result in the organization of tools and materials into labeled and color coded storage locations, as well as "kits" that contain just what is needed to perform a task. The "5S" concept provides the foundation on which Lean methods, Total Productive Maintenance, Cellular Manufacturing, and Just-in-Time production can be introduced.

The "5S" pillars, Sort (Seiri), Set in Order (Seiton), Shine (Seiso), Standardize (Seiketsu), and Sustain (Shitsuke), provide a methodology for organizing, cleaning, developing, and sustaining a productive workplace where operation of the "process to be improved" is carried out. In the daily work of an enterprise business, routines that maintain organization and orderliness are essential to a smooth and efficient flow of activities. This "5S" concept encourages employees affected by or operating the "process to be improved" to improve their workplace and helps them to learn to reduce waste, unplanned downtime, and in-process inventory.

Seiri

Seiri-(selection, or sort); proper (suitable) preparation of a workplace, manner and instrument of work; with the elimination of everything useless—The first S, focuses on eliminating unnecessary items from the workplace that are not needed for current "process to be improved" operations. An effective visual method to identify these unneeded items is called "red tagging", which involves evaluating the necessity of each item in a work area and dealing with it appropriately. A red tag is placed on all items that are not important for operations or that are not in the proper location or quantity. Once the red tag items are identified, these items are then moved to a central holding area for subsequent disposal, recycling, or reassignment. Enterprise businesses often find that selection enables them to reclaim valuable floor space and eliminate such things as broken tools, scrap, and excess raw material.

Seito

Seito-order (systemic); tidiness in a workplace and preparation of every required tool in the manner enabling simple and quickly utilization. The second S, focuses on creating efficient and effective storage methods to arrange items so that they are easy to use and to label them so that they are easy to find and put away. This second S can only be implemented once the first pillar, "Selection," has cleared the work area of unneeded items. Strategies for effective "Set In Order" in a manufacturing environment include painting floors, affixing labels and placards to designate proper storage locations and methods, outlining work areas and locations, and installing modular shelving and cabinets.

Seiso

Seiso-clearness (cleaning); order in a workplace allowing on increase of safety of workplace, control of equipment and responsibility for the means of production. Once the clutter that has been clogging the work areas is eliminated and remaining items are organized, the next step is to thoroughly clean the workplace. Daily follow-up cleaning is necessary to sustain this improvement. Working in a clean workplace enables workers to notice malfunctions in equipment such as leaks, vibrations, breakages, and misalignments. These changes, if left unattended, could lead to equipment failure and loss of production. Organizations often establish Shine targets, assignments, methods, and tools before beginning the shine pillar.

Seiketsu

Seiketsu-consolidation (standardization); reminding employees about their duties in the aspect of care of used tools and equipment and about keeping the workplace order. Once the first three 5S's have been implemented, the next pillar is to standardize the best practices in the workplace. Standardize, the method to maintain the first three pillars, creates a consistent approach with which tasks and procedures associated with the "process to be improved" are performed.

The three steps in this process are assigning 5S (Select, Set in Order, Shine) job responsibilities, integrating 5S duties into regular work duties, and checking on the maintenance of 5S. Some of the tools used in standardizing the 5S procedures are: job cycle charts, visual cues (e.g., signs, placards, display scoreboards), scheduling of "five-minute" 5S periods, and check lists. The second part of Standardize is prevention—preventing accumulation of unneeded items, preventing procedures from breaking down, and preventing equipment and materials from getting dirty.

Shitsuke

Shitsuke-discipline (self-discipline); adaptation of employees to the principles accepted by the organization, independent elimination of bad custom, training. Sustain, making a habit of properly maintaining correct procedures, is often the most difficult S to implement and achieve. Changing entrenched behaviors can be difficult, and the tendency is often to return to the status quo and the comfort zone of the "old way" of doing things. Sustain focuses on defining a new status quo and standard of work place organization. Without the Sustain pillar the achievements of the other pillars will not last long.

Tools for sustaining 5S include signs and posters, newsletters, pocket manuals, team and management check-ins, performance reviews, and department tours. Organizations typically seek to reinforce 5S messages in multiple formats until it becomes "the way things are done."

27.4.2.2 Examining Equipment Maintenance: Total Productive Maintenance (TPM)

Examining maintenance activities of equipment utilized through the "process to be improved" can alleviate production losses caused by machine breakdowns and supporting just-in-time production policies. The objectives here are to:

1. Examine equipment effectiveness;
2. Examine maintenance efficiency and effectiveness;
3. Examine early equipment management and maintenance prevention;
4. Establish training needs to improve the skills of all people involved;
5. Involve operators in routine maintenance.

Examine Equipment Effectiveness

Examining equipment effectiveness insures that the equipment performs to design specifications. The focus of the project team must be that the enterprise business asset produces the best possible outcomes than the competition can produce. The equipment utilized throughout the "process to be improved" must operate at its design speed, produce at the design rate, and produce a quality product at these speeds and rates. Any inefficiency detected can lead to additional capital investment in equipment to meet the required production output.

Examine Maintenance Efficiency and Effectiveness

Through this objective, the project team must focus on insuring that maintenance activities that are carried out on the equipment are performed in a way that is cost effective. Studies have shown that nearly one-third of all maintenance activities do not add any value to the process outcomes. Hence, it is important to lower the cost of maintenance. The employee involved must understand the basic maintenance planning and scheduling that are crucial to achieving low-cost maintenance.

The project team should also focus on insuring that the equipment maintenance activities are carried out in such a way that they have minimal impact on the up time or unavailability of the equipment. Planning, scheduling, and backlog control are all important if unnecessary maintenance downtime is to be avoided. Thus at this stage, maintenance and operations personnel must have excellent communication in order to avoid down-time due to misunderstandings.

Examine Early Equipment Management and Maintenance Prevention

Through this objective, the project team should focus on insuring that the amount of maintenance required by the equipment is reduced.

Establish Training Needs to Improve the Skills of All People Involved

Through this objective, the project team must focus on insuring that employees have the skills and knowledge necessary to contribute in the desired "improved process" environment. Providing the proper level of training insures that the overall equipment effectiveness is not negatively impacted by any employee who did not have the knowledge or skill necessary to perform the required tasks.

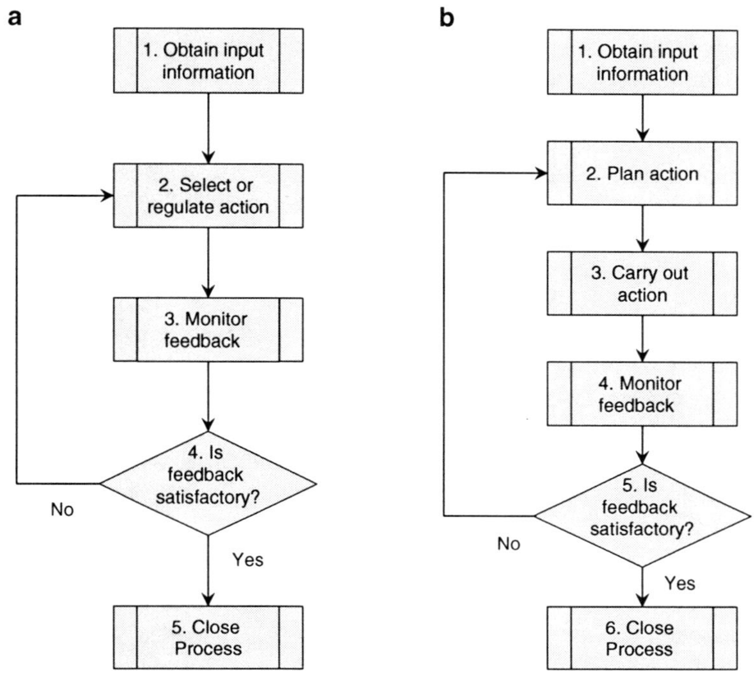

Fig. 27.2 A simple information-model of an operation. (**a**) Shows how the operation is represented using input, action and feedback. (**b**) Substitutes 'action' with planning for an action and executing the action

Involve Operators in Routine Maintenance

Through this objective, the project team should focus on finding maintenance tasks related to the equipment that the operators can take ownership of and perform. These tasks may amount to anywhere from 10 to 40 % of the routine maintenance tasks performed on the equipment. Maintenance resources previously engaged in these activities can be redeployed in other maintenance activities.

27.4.2.3 Modeling Behavior

Modeling behavior entails trying to understand how people accomplish the process discrete element goals by making reference to models of human performance. Here, an operation (i.e. what the operator does) on the process discrete element should be thought of in terms of 'input', 'action' and 'feedback'. That is, competence at an operation implies an ability to collect information (input) pertinent to the execution of the process discrete element; an ability to carry out the action selected in order to move towards the goal stated; and the ability to monitor appropriate feedback to determine whether the action is being executed correctly and is appropriate for dealing with the goal in question.

This is illustrated in Fig. 27.2a. There is no explicit decision-making component in this process, but use of feedback to regulate action to meet the goal implies the necessary planning and decision making skills. This sort of modeling behavior process enables the project team to consider, systematically, the likely sources of human error in the conduct of an operation.

If an input or a feedback weakness is suspected, the project team would be directed towards considering the display of information to the operator, or training in the discrimination, categorization and interpretation of signals. If an action weakness is suspected, the project team could be directed to consider equipment redesign or training.

A modification to this modeling behavior process is to distinguish explicitly between the planning and decision making components and the execution of the action itself—see Fig. 27.2b. This could help the project team focus on very different aspects of performance of the process discrete element considered.

Problems with planning and decision making are very much concerned with operator strategy and knowledge, whereas problems with action could be concerned with motor skills, physical fitness and inappropriately designed controls. Indeed, identifying an assignable cause of variation or a problem as one that is concerned with planning and decision making can open up other alternatives for representing cognition.

27.4.2.4 Identifying Similarities Between Operations and Goals

Another method that the project team can adopt is to try to identify characteristics of a current operation of the process discrete element considered and relate this operation to others encountered—insights or solutions that were seen to be appropriate elsewhere might be promising on this occasion.

Undoubtedly, similarities do exist between operations from different domains. Recognizing and exploiting these is one of the key elements of experience of process improvement teams. For example, many operations of process discrete elements rely on an operator monitoring a system to detect if and when its conditions go out of specification. Examples include people operating automated industrial plant, supervising transportation systems and nursing in intensive care. Each of these domains is very different. However, in all cases, monitoring requires the operator to know the parameters to monitor, knowing target values and knowing the tolerances outside of which an observed parameter must not be allowed to go. These operations also require the operator to be conscientious and monitor routinely and reliably. Knowing these facts about monitoring can alert the project team to a number of potential practical issues. In this way the project team might quickly pinpoint a source of difficulty or an assignable cause of variation.

Similarities between operations may be exploited in formal classification schemes or they might simply be the result of the project team members' experience. For example, operations of process discrete elements concerned with dealing with complex systems, where operators must monitor and maintain system status, all entail combinations of the following standard operations:

1. Monitor for problem
2. Detect potential problem
3. Diagnose problem

4. Make system safe
5. Compensate for problem
6. Rectify problem
7. Recover from problem

These similarities can help the project team see how to re-describe such operations of process discrete elements or they can help pinpoint concerns.

Checklist Approaches

By 'checklist approach' is meant the approach where the project team subjects operations of concern to systematic scrutiny using a list to guide these considerations. Checklists are classic tools of examining resources-task interaction, drawing on the experience of other project managers, operation managers, and past projects to ensure a level of consistency in early resources-task interaction analysis. They consist of simple lists of questions or statements based on lessons learned from earlier projects, which allow the project manager to build early operations of concern lists that reflect operations of concern faced on previous projects. A good example would be where the project team was concerned to ensure that the environmental ergonomics of a workplace were satisfactory. For instance, the project team might wish to establish whether the operator was subject to any extremes of heat, light, sound or draught that might adversely affect performance.

Operations of potential concern would be systematically subjected to a battery of environmental measures and these would be recorded against the operation identified through the "Analyze Process Steps and Tasks" project management process. There are many approaches such as this where the project team might wish to be systematic in recording data as part of a wider comprehensive audit, for example.

The use of checklist technique is recommended for all projects in enterprise businesses where checklists have been developed. The technique is normally applied early in execution of a process discrete element considered, although checklists can also be used at midterm and final process discrete element evaluations.

The inputs used to build the checklists are past experience of project teams and clear documentation of their experiences. Once the checklists have been created, however, the inputs to applying checklists are nothing more than the checklists themselves. The project manager and the project team should take the checklist and openly, honestly discuss the issues and concerns addressed by the tool.

Depending on the construction of the tool, the checklist may do little more than generate red flags to warn of categories of concern or specific operations of concern. If the tool is software-driven and more complex, it may also provide a list of recommended basic actions to guide the project manager and the team toward best-practice experience in handling any of the operations of concern identified in the tool.

Operating under the assumption that a checklist has already been created, the process associated with checklists is among the simplest of all the process tasks analysis tools. Its major steps include de following:

1. *Review the operations of concern checklist.* Ensure that the project team is working with a checklist that is appropriate to the environment, the culture, and the project in question. Because some operations of concern checklists are designed to address issues within a given enterprise business, within a given project type, or within a given process discrete element, it is important to work with a tool that is appropriate to the project at hand.
2. *Answer the questions or check the appropriate boxes on the checklist.* Checklists normally come with guidance to direct the user on appropriate application. Such applications are simple question and answer sessions or rating schemes to assess the likelihood of encountering some common operations of concern.
3. *Review and communicate the guidance provided.* Even though checklists normally include some direction on how to fill them out, they also include guidance on how to apply the findings. In some cases, those findings may represent nothing more than a list of commonly identified operations of concern for the process discrete element considered. However, some of the more advanced checklists will also embed suggestions on standard internal practice and procedure for resolving or managing the operations of concern identified. Guidance of any nature should be communicated to the team.

Enterprise businesses looking to build their internal risk practice can frequently develop that practice by generating checklists. Checklists are often among the first steps that the operator of a process discrete element takes to build a broader understanding of the depth of operations of concern within the process and the support that they can provide in ameliorating some of those operations of concern.

Because checklists are first applied early in the execution of a process discrete element, outputs can be used to provide a general understanding of the nature of the concerns in the process discrete element in a nonthreatening fashion. Data from such checklists tend to cause less anxiety since the questions asked (or statements made) are applied equitably to all operation processes, and the outputs are normally familiar to the enterprise business. Outputs at the end of the process discrete element should be used in any reevaluation of the checklists for additions or deletions.

The reliability of the process of using checklist pivots on the quality of the checklist. A sound checklist built to reflect the enterprise business' culture, nature, and "process to be improved" history will build an excellent set of initial checklists. A checklist that a single individual crafts after a single process improvement project without considering the organizational culture will have limited reliability. The best checklists are those that capture experience from a variety of process improvement projects and a variety of project teams. Answered candidly, checklists of that caliber can generate extremely positive and reliable results.

Other Methods of Examining Resources-Task Interaction

In several cases, the project team might wish to explore the detail of an operation further by using an appropriate method of data collection. This could provide insights where less stringent methods employed within the resources-task interaction analysis have failed.

One such example would be the collection of verbal protocols. The project team might encourage the process discrete element operator to verbalize his or her strategy in carrying out a difficult task, record this speech, and then examine it later on. The verbal protocol could be recorded concurrently as the task is carried out, or it could be recorded afterwards, relying on the operator's memory or allowing the operator to follow a video recording of what took place. In this way, the project team may gain useful insight into the operator's strategy, motivation and justification for action. This process can help identify useful task knowledge that could be used to train people, or it could provide evidence of the need to modify the information available to operators during the execution of the work associated with the process discrete element. One important outcome is that it could show the project team how further re-description of the work could be accomplished.

Verbal protocol analysis is but one formal method that can be employed in order to collect data about operations. Other common techniques include "link analysis" where the project team records the extent to which the operator makes use of different artifacts in the executing a process discrete element, including instrumentation, communications devices, other people and records. Identifying common links can point to ways of reorganizing the workplace. Identifying common patterns of using links can point to the skills and strategies that people use and, so, provide insights for further re-description of the work associated with the process discrete element.

Another useful method is the "withheld information" technique. This simple, but powerful method is helpful in understanding how people diagnose situations in the face of uncertainty. The project team must prepare an information sheet containing the set of information that the operator of a process discrete element considered could use, but then withhold this information until the operator explicitly asks for an information item. By recording the order in which information is requested in reaching a solution to the problem, the project team gains insight into the operator's strategy, the information upon which the operator depends and the types of error the operator makes.

27.4.3 Analyze Cognition Within the Discrete Element Context

Cognition is a crucial aspect of behavior in executing the work associated with a process discrete element especially in connection with those associated with supervision and control of complex automated systems and with human-computer interaction. Cognitive task analysis aims to explore the ways in which people deploy cognitive processes when they carry out tasks. Cognitive task analysis methods include methods for examining behavior and for representing the way task knowledge is organized and the processes through which operators deploy knowledge and skill to demonstrate their expertise.

Some process discrete element warrant a cognitive approach for analysis while other should utilize a non-cognitive approach. Part of the difficulty in distinguishing between cognitive task analysis methods and other forms of task analysis is that the project team is required to make a prior judgment on whether a task is "cognitive" or whether it is not. All tasks rely on processes of cognition to ensure they are carried out effectively. They require information to be monitored, an appropriate response to be selected and the consequences of actions to be evaluated to ensure that actions are adapted and controlled appropriately.

While there are neither wholly cognitive nor wholly non-cognitive tasks, we might refer to those tasks where the operator must interpret masses of information in order to monitor system health, diagnose system states, or work instructions, as 'cognitively loaded tasks'. Even for such tasks, an evenhanded strategy should be adopted in which the need to examine cognition emerges as part of a general analysis process. Otherwise, the project team may be engaged in substantial futile work. For example, a task involving human computer interaction may be extremely difficult, but the difficulties may be resolved by rescheduling work demands, or providing a more supportive learning environment, without undertaking an extensive and expensive examination of the behavior of the operator trying to work in a poorly designed task.

It is also important for the project team to remember that the "Analyze Process Steps and Tasks" project management process is undertaken as part of an applied intervention. This emphasizes two important considerations. First, cognition is affected by the context in which it is deployed. Unless context is understood, the ensuing analysis may be ill-informed because the performance may be affected most critically when certain factors come into play and may not be affected when the context is benign. Thus, dealing with a diagnostic problem may be observed to be straightforward when other aspects of the workplace are well under control; problems may only arise when staff is stressed by other tasks. The second practical consideration is criticality. A task element within a process discrete element may be difficult to execute, but if the consequences of assignable cause of variation are trivial, then time and other resources assigned to analyze that element are not justified.

The project team should also remember that, strategies for examining and representing cognition are not themselves particularly valid in the sense of leading to reliable and meaningful results which apply to all people involved in execution of the "process to be improved." Many of the methods used to collect data in cognitive task analysis, such as observing performance, analyzing verbal protocols, and examining this data from a cognitive perspective, are subjective and do not necessarily provide an account of cognition with proven reliability. But they are useful as methods for engaging the project team with the task. Their role is to help with the process of generating improvement hypotheses.

27.4.4 Estimate Cost-Benefits of Hypotheses: Value Added

One common outcome which follows the examination of a resources-task interaction is that the project team proposes a means by which performance of a process discrete element could be improved. There are often many ways of improving problems associated with the performance of a process discrete element (task, action, or step), with some more expensive than others. An important aspect of the analysis, then, is cost-benefit analysis applied to the hypotheses under consideration. A cost benefit analysis is done to determine how well, or how poorly, a stated hypothesis will turn out. Although a cost benefit analysis can be used for almost anything, it is most commonly done on financial questions. Since the cost benefit analysis relies on the addition of positive factors and the subtraction of negative ones to determine a net result, it is also known as running the numbers.

A cost benefit analysis is a technique that compares the monetary value of benefits with the monetary value of costs in order to evaluate and prioritize issues or hypotheses. The effect of time (i.e. the time it takes for the benefits of a change to repay its costs) is taken into consideration by calculating a payback period. In its simple form, cost benefit analysis uses only financial costs and financial benefits. A cost benefit analysis finds, quantifies, and adds all the positive factors. These are the benefits. Then it identifies, quantifies, and subtracts all the negatives, the costs. The difference between the two indicates whether the planned action is advisable. If the costs of an innovation hypothesis exceed its benefits within a process discrete element (task, action, or step) then, that innovation hypothesis is not worth pursuing. Equally, if several possible, equally valid, hypotheses are considered, then the least expensive should probably be preferred. Again, this is a judgment that is made as a routine within the cycle of activities in a process discrete element analysis.

Cost-benefit analysis becomes increasingly important as technologies for controlling systems or dealing with human factors solutions become more expensive. Training simulation is a good illustration because the technology required to obtain high fidelity simulation in many domains is very expensive and may not appear to be justified, even in terms of the costly events it may avert. It is, however, important to recognize that as analysis of a process discrete element progresses and further opportunities to use the same equipment or technology present themselves, benefits may start to overhaul costs. Therefore, it is important that no potentially useful hypotheses are totally discarded on cost grounds, but kept alive to be reviewed later.

It will be noted that issues of cost-benefit analysis are related to issues of stopping analysis with regard to the $P \times C$ rule described above. A proper cost-benefit analysis would compare regimes and not simply deal with the costs and benefits of improvement innovations in isolation. Calculating the "cost-benefit" factor for any innovation hypothesis entails costing the innovation fully and identifying benefits broadly. Costs include capital and recurrent costs associated with the innovation and the costs that will be incurred as a consequence of risks that will still prevail. Any innovation hypothesis will merely serve to reduce risks;

eliminating risk entirely is fanciful. Moreover, benefits can include improved productivity, but they will also include hidden benefits such as the additional expertise the enterprise business has now gained as a result of the innovation in the process discrete element considered. Introducing virtual reality, for example, may not be justified in terms of the cost-benefits of a single project but may make sense by taking a longer term perspective.

If the cost-benefit of a hypothesis is judged acceptable, then the project team can cease at that point and record the hypothesis.

27.5 Examine Goals by Re-description

The fifth step in analyzing a process discrete element (task, action, or step) is to examine improvement hypotheses, consistent with cost-benefit criteria, that failed to emerge from examination of the resources-task interaction and the operations underpinning performance of the process discrete element considered.

Where the project team has been unable to generate a suitable hypothesis, a re-description of the goal into its sub-goals and their work instructions are warranted. Design hypotheses, consistent with cost-benefit criteria, may fail to emerge from examination of the resources-task interaction. Moreover, the task may be judged to be too complex for fruitful application of a formal method of modeling. In these cases the project team should try to examine the process discrete element in terms of its sub-elements. Exploring sub-elements is achieved through re-description by stating a set of subordinate operations and a "work instruction" specifying the conditions when each is appropriate.

If no hypotheses can be established, the project team may need to challenge the constraints. At the outset of the analysis of the process discrete element, management might have ruled out investing in new technology on cost grounds, thereby limiting the options that may be pursued in the "Analyze Process Steps and Tasks" project management process.

However, a suitable design hypothesis may not be forthcoming. The following example refers to a case where management sought improvements to the skills of controlling an ageing batch process plant in order to improve productivity. A common requirement in process control is that a particular parameter is required to be maintained at a particular value by adjusting an associated parameter. For example, the formulation of a liquid in a vessel may drift off target over time and must be adjusted by the appropriate addition of a particular feed-stock. In older generations of plant there are still many instances where human operators are expected to carry out this sort of adjustment manually.

The task analysis carried out within the constraint of current technology had to focus on training solutions. The skills identified included taking samples, calculating the adjustments to formulation, making the adjustment, then waiting to re-sample and further adjust if necessary. Despite the preferences of management, analysis of these process discrete element elements showed that there were no suitable training solutions to meet the standards required, because plant dynamics were not predictable and additional work demands required staff to direct their attention elsewhere during crucial stages. Therefore, the previously stated constraint on changing technology had to be challenged if any productivity improvements were to be forthcoming. To do this, the project team moved up the hierarchy to a suitable level of description, challenged constraints, and sought different methods of operation that involved automation of control loops.

Sometimes re-description may prove impossible for the project team. This may be because no way of re-describing can be seen by the project team or it may mean that no way can be seen within the given set of resource constraints. To resolve the problem the project team should seek advice. Such advice may provide help in suggesting a method of re-description, or it may provide a more acute examination of the resources-task interaction, leading to an improvement hypothesis. Re-describing operations or goals is a skill that develops with experience.

27.6 Summarize Data & Display Value Stream Diagram

The sixth step in analyzing a process discrete element (task, action, or step) is to summarize data and keep record to show what progress has been made, how the considered process discrete element has been represented and what improvement hypotheses have been proposed. Maintaining good records is essential both to manage the analysis of the process discrete elements and to justify, at a later date, the decisions made. Effective recording methods must account for the complexity of the task in a number of ways and still be relatively easy to follow.

It may also be useful for the project team to record the hypotheses that have been rejected on cost grounds, since these decisions may be revised as new issues emerge. The outcomes of this analysis process are recorded through both value stream diagrams and tables. Diagrams are probably the clearest way to communicate analyses since it shows how subordinate goals and their work instructions are organized to represent the goal they are re-describing. They are useful in showing task structures but do not show the greater detail of information collected and insights gained.

27.6.1 Summarize Data

Careful tasks representation and recording is necessary in communicating results to others. As analysis of the process discrete elements (task, action, or step) progresses, the project team works with one or more task experts to obtain information and represent it in an appropriate way. There will be occasions when input from other people is sought either to confirm results so far obtained or to fill in detail that has been missed out. It is necessary, therefore, to represent any progress to date so that the task expert or informant may provide further information consistent with what has already been collected.

When a first draft of the analysis has been completed a good record is necessary so that the work can be reviewed and validated. The overall content of the analysis must be checked for accuracy, completeness, and to determine whether it satisfies the requirements of the customer of the process discrete element (task, action, or step) considered. To accomplish this requires that the analysis is critically read by the customer. It is also important to ensure that the work can be used by others, for example, people designing training or interfaces or safety analysts.

Analysis of a process discrete element (task, action, or step) should also provide a record to be kept for future reference within the enterprise business. In many instances, analysis needs to be reviewed from time to time. New equipment, staffing changes and new legislation all have consequences for how a process discrete element (task, action, or step) work is carried out. A previous analysis can be reviewed and modified accordingly. If there are incidents which require investigation, having a good analysis on record will enable a swifter response or provide a justification for a decision that subsequently proved problematic. Equally, decisions taken about the process discrete element (task, action, or step) during earlier analyses may need to be reviewed.

Where process discrete element (task, action, or step) are complex, analyses can become large and work may need to continue over days, weeks or even months. Where work on analysis of a process discrete element (task, action, or step) is interrupted for even a few hours, the project team may lose the thread and needs help to recommence the work on the later occasion. Even where analytical work is continuous, some parts of the task will, inevitably, be put aside to concentrate elsewhere. The project team is continually returning to points in the analysis which were previously left. It is vital, therefore, that progress is properly recorded to ensure that previous work is not misinterpreted.

Finally, it must be noted that goals, operations and work instructions cannot, in themselves, provide a comprehensive account of all that has been done. Part of the analysis process is concerned with justifying where the analysis has stopped. Reasons for stopping should be recorded, along with assumptions about the process discrete element context which have been made in deciding when to stop analysis. Sometimes, a level of detail may be felt to be sufficient because training or other improvement suggestions have been made to support the appropriate operator behavior. Unless these are summarized and recorded clearly, justifications made at the time will be forgotten. Hence, there will be insufficient guidance for people wishing to use the analysis for improvement and there will be no clear statement for anyone challenging assumptions made during the analysis.

27.6.2 Display Maps and Flowcharts Diagrams

A process map is detailed flow diagram of the process using standardized icons that drill further into the high level map generated on the SIPOC. The detailed flow diagram provides a detailed picture of a process by mapping all of the steps and activities that occur in the process. This type of flowchart indicates the steps or

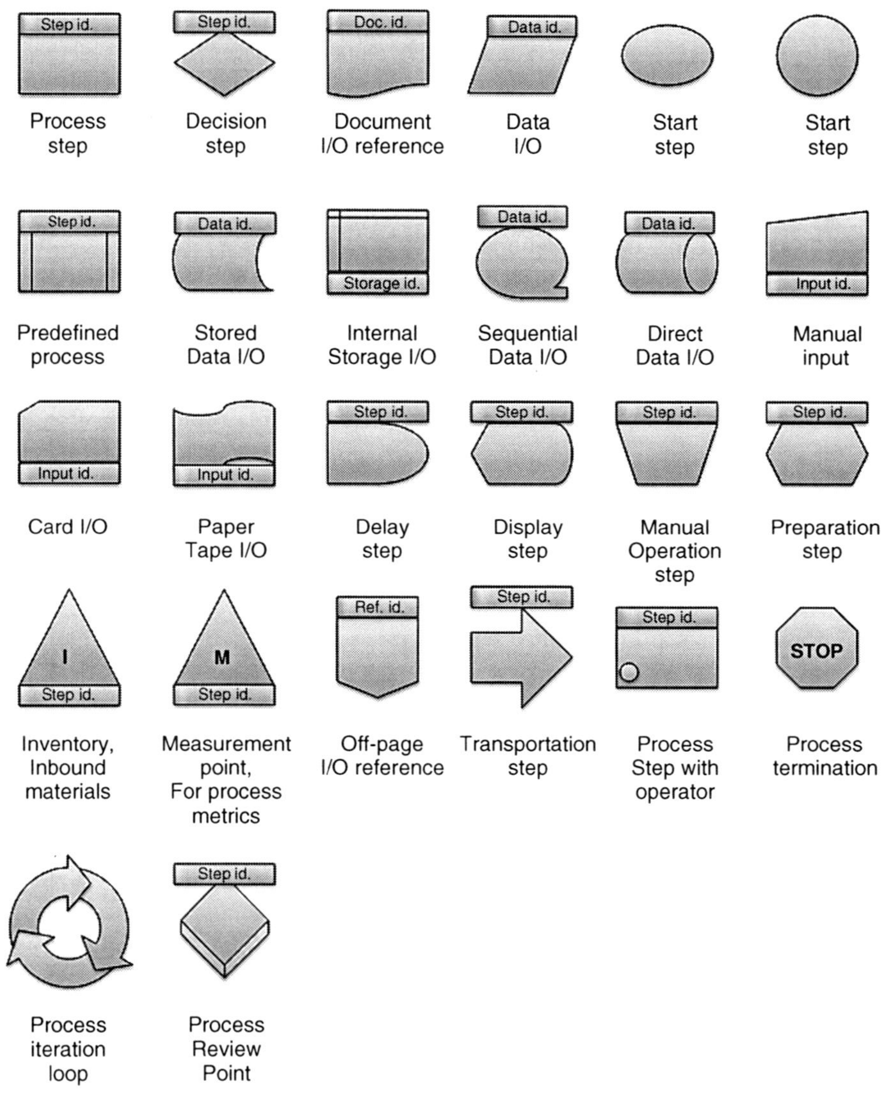

Fig. 27.3 Basic flowchart symbols

activities of a process and includes such things as decision points, waiting periods, tasks that frequently must be redone (rework), and feedback loops. They maps and flowcharts diagrams help make discrete element work visible. Increased visibility improves communication and understanding, and provides a common frame of reference for those involved with the work process.

Process maps are often used to show how the work currently gets done. When used in this way, they represent a snapshot in time that shows the specific combination of the functions, steps, inputs, and outputs that the process uses to provide value to its customers. Thus, process maps and flowcharts help summarizing and documenting the current pathway to customer satisfaction.

Process maps and flowcharts diagrams are usually drawn using some standard symbols; however, some special symbols can also be developed when required. Some standard flow chart symbols are shown in Fig. 27.3.

It is not strictly necessary to use boxes, circles, diamonds or other such symbols to construct a flow chart, but these do help to describe the types of events in the chart more clearly.

Described below are a set of standard symbols which are applicable to most situations without being overly complex.

1. *Rounded box*—use it to represent an event which occurs automatically. Such an event will trigger a subsequent action.
2. *Rectangle or box*—use it to represent an event which is controlled within the process. Typically this will be a step or action which is taken. In most flowcharts this will be the most frequently used symbol.
3. Diamond—use it to represent a decision point in the process. Typically, the statement in the symbol will require a "yes" or "no" response and branch to different parts of the flowchart accordingly.
4. Circle—use it to represent a point at which the flowchart connects with another process. The name or reference for the other process should appear within the symbol.

27.6.2.1 Displaying Basic Information About the Process Current State

When mapping the process current state, the project team must be able to display on a diagram the collected and document information on what is going on within the process as it occurs. The project team should paint a picture of what it looks like at the time it was documented, not what people said or though the process should be doing. This picture, or snapshot in time, is truly the power behind process maps and flowcharts diagrams. What is most important is being able to explain graphically to employees all of the following:

1. What they do
2. How they do it
3. How they interact with the employees in the process
4. How the entire process flows

With this basic information in hand, the mapping team has the necessary background information to start mapping the process. Although this data will provide insight to certain issues and conditions that may exist within the process flow, the process mapping team members must remember to not only look for situations that support what they now know, but also to keep an open mind and map what they have seen during the process analysis walk and map the process. In the event that there are strong preconceived ideas around what is wrong with the process flow, the mapping team might want to conduct a simple brainstorming session prior to actually mapping the current state. By allowing all team members to brainstorm the perceived pain, problems, and issues, any preconceived notions will come to the surface, and discussion will help the team to pinpoint these issues.

However, the project team should not stop with this brainstorming list and discussion. The project manager should conduct a follow-up exercise by having each team member independently work through the brainstorming list to categorize each problem into the generated hypotheses (value-added tasks and cycle times). Each team member should write down which of the generated hypotheses apply to each problem listed. Once each team member has had the opportunity to categorize each problem into the generated hypotheses, the project manager can then total up the number of times each generated hypothesis appears, as well as the number of total generated hypotheses that appear for each problem. From this exercise the project team can oftentimes eliminate these preconceived ideas and refocus on what the team as a whole believes is the biggest problem and which generated hypotheses should receive the most focus.

Determining the proper level of detail to be included in the process map diagram is critical to explaining the opportunities in the process. Although many process and quality improvement professionals have been led to believe that process maps only provides a high-level look at processes, the reality is that this could not be further from the truth. The power of a process map lies in the detail. The level of detail should be determined by the problems or issues being addressed and the audience to which the map set will be presented.

A primary goal of any process current state map should be to capture and display graphically how a process actually operates: in practice, not theory. Another equally important goal of a process map is that it be drawn in such a way as to be understood by anyone. Because a process current state map is a snapshot in time, the viewer should be able to look at the map, and within a relatively short time frame with minimal explanation, understand the process. Because of this goal to have everyone understand the process maps produced, there are very few rules that exist with this discipline.

Therefore, when the project team is mapping the "process to be improved," it should be creative and to use the process map tool in such a way that it is possible to clearly communicate to all operators, management, suppliers, and customers. This does not, however, mean that the project team can draw anything and any way it like. The process map tool itself allows for flexibility to work within any setting, and yet, boundaries still exist. The rules that do exist focus on the following:

1. Standardization of icons use, as much as possible
2. The basic layout of the map
3. Creation of a structured method of documentation and presentation to make the results clearer to the audience

Capturing the main "process to be improved" flow in an organized manner is critical to success in the mapping exercise. After having determined what to map, the process mapping team must now get down on paper what the process flow actually looks like. To do this, it should start with the main flow through the identified value stream, walk the process and map exactly what it sees. The process mapping team should conduct its first-pass walkthrough of the process map by walking the process in reverse order; from delivery to the customer to receipt of process inputs using pencil and paper.

Even though there are many software packages now available to assist with process mapping, using paper and pencil still provides the fastest way to capture what is observed during this initial pass. The following are steps to be followed to identify and map the main "process to be improved" flow of material, information and work performed.

1. Identify and display each process step by using process step boxes.
2. Map how the product or service moves from one step to the next.
3. Map where the inventory is.
4. Map where the operators are located.
5. Map subtasks and parallel flows.
6. Line up process steps and put it all together.

Identify and Display Each Process Step by Using Process Step Boxes

The process boxes, shown in Fig. 27.3, are basic icons, also used in traditional flowcharting, which are used to show where flow starts and stops within the process.

When the process mapping team is walking the process and observing how the process actually operates, it should ignore traditional departments and boundaries within the enterprise business, and focus instead on where flow occurs. Flow typically starts at the point where labor (value-added or non-value-added) is applied to product or service or where machine time is initiated, excluding material handling. As indicated in the "generate hypotheses" sub-section above, a non-valued-added task is anything the customer is not willing to pay for. In contrast, value-added tasks add market form or function to the product or service; simply put, they are what the customer is willing to pay for. Flow stops at the point where the product or service comes to rest. The easiest way to see product "at rest" is to look for piles of inventory on workstations, in baskets, on pallets, etc. Operations that use batch and queue production methods will usually have significant amounts of product "at rest."

As the process mapping team is mapping and moving from one process step to another, there are three other critical items it should look for: how the "process to be improved" product or service is moved, where inventory is, and where the employees are.

Map How the Product or Service Moves From One Step to the Next

The first thing the process mapping team should look for is how the product or service moves from one step to another. Is it moved on to the next step in the process without thought or consideration as to whether the next step is ready and waiting for it to arrive? When work is moved in this manner, it is being "pushed." Materials might also be moved using a FIFO lane, where the material is moved and consumed in a "first-in–first-out" methodology.

Map Where the Inventory Is

The second thing that the process mapping team should be looking for is inventory. The inventory process box is represented by a triangle with an "I" in the middle. This symbol is very appropriate because the triangle, often called a "delta," represents change in many settings. Because the process current state maps are snapshots in time, there is no better way to acknowledge that the inventory level observed and documented on the map will change.

Inventory is what manufacturers live and die for; it is the heart and soul of a production process. If there is too little inventory during execution of a production process, parts shortages begin to occur. When this happens, the process can come to a grinding halt. If there is no inventory, then there is no product to complete, and no finished goods to deliver to the customer. To avoid this situation, manufacturers bring in large quantities of raw material and/or component parts inventory. However, because each and every material supplier or vendor is not linked together to coordinate delivery dates and quantities, once a material supplier or vendor misses a delivery date or ships too small of a quantity, the manufacturer suffers from a similar problem. There is plenty of raw material, but not the right materials. The result is piles of inventory.

By showing on the process map where the inventory is located and how much exists, the process mapping team has the ability to tell a story about how inbound material flows through the process. If multiple stacks of inbound material exist between two process steps, the process mapping team may choose to show it as two separate inventory icons. By quickly counting the amount of inventory that exists at a specific location, the process mapping team can label the inventory icon with a quantity while mapping and then convert it to lead time after the mapping on the floor is complete. The thing to remember when mapping the current state is that, although it is not essential to get the inventory count 100 % accurate, it is important to show what the process mapping team actually saw when it was walking and mapping the process.

Map Where the Process Operators Are Located

Finally, the process mapping team should be looking for and documenting where the operators are located within the process flow. The operator icon, represented by a circle in the left corner within a process step box, is used to show where the operators are actually located when the process is mapped. The process mapping team should not place operator icons where someone said the operators are supposed to be, or usually are, or even where they are budgeted or assigned. Instead, mapping the operators should be based on an actual observation: once again, a true picture of what was observed.

When multiple operators are seen within one process step, the actual number of operators should be placed next to the symbol, and the symbol should be placed inside the process step box.

Map Subtasks and Parallel Flows

As the process mapping team follows the main flow of the "process to be improved" to create a current state map, it may often find forks in the road; i.e., places where two or more process flows come together or move apart. Where these branches in the flow occur, it may find subtasks or parallel flows in the value stream:

1. *Subtasks* occur where subassemblies or other feeder processes exist to provide parts or components to the main flow of the "process to be improved" map. Subtasks typically start from a unique point in the process map and feed into the main flow. When mapping the current state, the process mapping team will usually find where subtasks join the main flow when team members ask where parts come from to get to a process step that they are documenting. When operators explain that subassemblies or components are manufactured elsewhere on-site and join up with the flow at this step, the process mapping team should recognize the additional path and prepare to come back and capture this information. However, on occasion, it may overlook a subassembly path on its first pass through the merge point and not discover it until it gets to the start of the process or later. Either way, once the process mapping team document the main flow, it should use the same technique of walking the process backwards to map this additional flow.
2. *Parallel flows* exist where there are multiple options in the process flow that occur through different parallel paths or when various tasks are completed simultaneously through different paths and then come back together at a single point in the main flow. Parallel or alternate paths through a process flow represent variation and decision making within the process. When the process mapping team is mapping subtasks or parallel flows, the challenge is in knowing when to follow their paths. This requires abandoning the main flow in pursuit of the newly found path. Although it is possible to follow and map this newly found path, and then return to the main flow once the subtask or parallel flow has been documented and displayed, it is usually easier to denote the existence of the path directly on the mapping pad at the point in the process flowchart where it was identified, and then return and capture it on the map once the main flow has been completed.

Line Up Process Steps and Put It All Together

In order to provide a current state map that workers can accept, the picture that the process mapping team paints must be clear and concise. The easiest and most effective way to achieve this is by proper alignment and flow of the process steps on the map. When the process mapping team ensures that the top level of the process flow is the main flow through the process, it sets the stage for gaining this acceptance. Subtasks, parallel paths, rework, etc., must be relegated to the area beneath the main flow.

In order to clearly show the main flow and to make it easier to understand the relationship of the subtasks and other flows within the map, the process mapping team should align the process boxes both horizontally and vertically. If it fails to adhere to this basic alignment concept, workers within the process will more often than not fail to "see the picture."

27.6.2.2 Displaying Basic Information About the Process Future State

Once the process mapping team has documented and displayed the process current state map, it can now create a potential future state map of the "process to be improved" to guide positive change in the days and weeks ahead.

Whereas the process current state map provides the insight into the underperformance problems that exist in the "process to be improved" flow, it is a future state that determines the potential goal for the project team. A process future state map is a projection of how a process flow should look in the short terms future, generally 6–12 months. It is defined by incorporating into the process current state map, generated hypotheses, gap analysis of the "process to be improved" steps and tasks that identifies the gaps (holes) between the current and desired state, industry benchmarks and management input to create a desired future state.

A process future state map summarized and displays improvements suggestions to be made to the process flow that will shorten the overall lead time, reduce non-value added tasks, and eliminated all identified bottlenecks in the process current state. Here, bottlenecks identified, during the analysis of the resources-task interaction under the generated hypotheses, include:

1. Any resource whose capacity limits the amount of information or material that flows through the process.
2. Any resource whose capacity is equal to or less than the demand placed upon it.

Generate Improvement Solutions 28

Once assignable causes of variations have been identified, the associated cause-and-effect relationships explored, the identified assignable causes verified, and the process steps and tasks analyzed and summarized, the project team can begin with generating improvement solutions to guide positive change in the days and weeks ahead.

28.1 Brainstorm Using Available Data Generated So Far

To start generating improvement solutions, the project team must review all generated data about the "process to be improved" and the verified assignable causes of variation, and engage members and selected employees to improve the process. It is critical to remember that employee buy-in is essential to success in improving the "process to be improved."

Employee buy-in is important to gaining agreement and acceptance of the current state of the "process to be improved," and it is vital to the generated solutions to the "process to be improved" underperformance problems if the change process is to move forward. Employees not only want to be a part of the process and have input into what is being reported out, but they also want to know that the issues important to them within the "process to be improved" are being addressed.

By including many of these employees in the brainstorming sessions, the project team can show employees that they are in tune with the "pain" felt working within the "process to be improved." Beginning with a simple and limited duration brainstorming session, team members and selected employees can jump-start the entire process of envisioning the improved process.

As project manager, you must gather the team and selected employees around available data generated about the "process to be improved" and the verified assignable causes of variation to begin this session. Allow each team member and employee to suggest possible solutions to the identified underperformance problems and opportunities on the process current state map. To truly use the concept of brainstorming, do not set any ground rules other than that ideas must

A. van Aartsengel and S. Kurtoglu, *Handbook on Continuous Improvement Transformation*,
DOI 10.1007/978-3-642-35901-9_28, © Springer-Verlag Berlin Heidelberg 2013

be focused on the available data generated so far and the verified assignable causes of variation. As possible solutions are stated, first document them on a flipchart or whiteboard. To this end, it may be beneficial to assign the duty of scribe to one team member to document potential solutions. If a scribe is assigned, ensure that the chosen person understands that any idea that he or she thinks of can also be written on the list.

Employees often have answers to the underperformance problems associated with the "process to be improved," they just don't always know it. During this brainstorming session, it is critically important to bring the workforce into the observation and documentation process of the current "process to be improved." As all generated data about the "process to be improved" and the verified assignable causes of variation are been reviewed, employees should be sought out to bring their pain and frustration to the exercise. Without this input, generating potential solutions could take considerably longer. By including employees from within the "process to be improved" on the project team, you bring an incredible resource to your fingertips for finding solutions. The secret is in getting these team members to open up. They must be willing to give suggestions, and they must be willing to provide feedback on various ideas presented by other team members.

To achieve this, you must ask leading questions. When searching for potential solutions to the underperformance problems associated with the current "process to be improved," ask team members who work within the "process to be improved" leading questions. By asking, "What would you do to fix the problem?" or, "If you had a magic stick and could change one thing about the "process to be improved," what would it be?" the team can quickly understand what is important to the employees. With a good understanding of the "pain" working with the current "process to be improved" and some basic knowledge about all generated data about the "process to be improved" and the verified assignable causes of variation, employees can provide some very insightful solutions. They may not know how to implement, but they have ideas that may be transformed into excellent solutions. They are not only assisting the team in finding the solution, but they are also gaining ownership of the future "improved process."

This format for finding solutions is not new. The concept of brainstorming as a continuous improvement tool is rooted in using the employees who are experiencing the pain to pinpoint the problem(s) and finding all possible solutions in a rapid-fire setting. This concept is laid out in a SIPOC chart. The power of the SIPOC chart is that the supplier of the solutions is focused on the needs of the customer, because they are one and the same. This can create nearly instantaneous ownership in the changes to be implemented.

It may also be of some use to repeat this exact same exercise with other employees working within the current "process to be improved" that are not on the project team. By proactively widening the input base for potential solutions, you can gain more acceptance and ownership faster, instead of waiting on a presentation meeting to start achieving this critical buy-in.

As the brainstorming session continues, there will be potential solutions presented that are met with resistance. When this occurs, stop and discuss the idea as a team. The role of the project facilitator or project manager should be to work through issues such as these, looking for consensus.

As project manager, you should remember here that consensus is a group process where the input of everyone is carefully considered and an outcome is crafted that best meets the needs of the group in order to achieve the project objectives. It is a process of synthesizing the wisdom of all the team members and participant employees into the best decision possible at the time. The root of consensus is the word consent, which means to give permission to. When you consent to a decision, you are giving your permission to the team to go ahead with the decision. You may disagree with the decision, but based on listening to everyone else's input, all the individuals agree to let the decision go forward, because the decision is the best one the entire team can achieve at the current time.

The heart of consensus is a cooperative intent, where team members are willing to work together to find the solution that meets the needs of the project. The cooperative nature of consensus is different mindset from the competitive nature of majority voting. In a consensus process the members come together to find or create the best potential solutions by working together. Key attributes to successfully participation include humility, willingness to listen to others and see their perspectives, and willingness to share own ideas but not insist they are the best ones.

Consensus is not what everyone agrees to, nor is it the preference of the majority; consensus is not:

1. A unanimous vote
2. Having everyone completely satisfied with the outcome
3. Necessarily anyone's first choice
4. Everyone getting everything they want
5. Everyone finally coming around to the "right" opinion.

As project manager, you do not need to have unanimous agreement on potential solutions. The number of potential solutions generated can be used as a benchmark on how long to run a session, but we recommend that these brainstorming meetings last no longer than an hour.

As the brainstorming session progresses, the amount of information drawn on the generated data about the "process to be improved" and the verified assignable causes of variation may become excessively cluttered or confusing. This is the time to start fresh with a clean sheet of paper. Occasionally, as you brainstorm ideas for potential solutions, you or the project team will determine that there is nothing worth saving on the current state of the "process to be improved" and decide that it is a better decision to simply start from scratch on a new sheet. When situations like this occur, you must ensure that the current state of the "process to be improved" remains visible and that the focus continues to be centered on reviewing the opportunities and issues in the current state and identifying potential solutions for improvement. Sticking to this basic rule will produce cleaner, faster, better focused "improved process" to guide implementation efforts.

Once you, the project team members and selected employees have agreed on a list of potential solutions to the "process to be improved" underperformance problems, you must prioritize these potential solutions and perform their cost-benefit analysis.

Table 28.1 Prioritization matrix template

Selection Criteria		C_1	C_2	...	C_j	...	C_M	Total
Weights		w_1	w_2	...	w_j	...	w_M	
Potential Solutions	S_1	p_{11}			p_{1j}			
	S_2							
	...							
	S_i	p_{i1}			p_{ij}			$\sum_{j=1}^{M} w_j * p_{ij}$
	...							
	S_N							

28.2 Prioritize Potential Solutions

At this stage, the project team has invested a lot of emotional energy into the project; however, the merits of the list of potential solutions that it has selected may not be obvious to those outside the team. Furthermore, the project team may be faced with an uncertain or risk-filled pattern of future events associated with those potential solutions or it might have selected potential solutions that do not meet the requirements of the business.

Deciding what is really important from a list of potential solutions to the "process to be improved" underperformance problems can be very difficult, especially if the project team members and the selected employees involved have a difference of opinion about which potential solution should be acted upon first. In such cases, a prioritizations matrix, as shown in Table 28.1, can be used by the project team to identify the best solutions from among a range of possibilities.

Too often, project teams "just discuss" various options and make a choice without using any tool or particularly structured objective evaluation techniques. Project team members can even decide to "go against" the technique's answer and support another alternative as long as they understand why they are doing so. When there is data available to help score criteria and potential solutions, the project team should use a prioritization matrix, rather than simple voting, when the extra effort that is required to find a more confident solution overcomes the risk of making a faulty decision.

A prioritization matrix is a decision-support tool that allows decision makers as project managers to structure and then solve their problem by:

1. Specifying and prioritizing their needs with a list a criteria; then
2. Evaluating, rating, and comparing the different solutions; and
3. Selecting the best matching solution.

It uses a combination of tree and matrix diagramming techniques to do a pair-wise comparison of items and to narrow down options to the most desired or most effective.

To construct a prioritization matrix, the team should first develop a hierarchy of decision criteria, also known as decision model for the selection: whether they are the project defined CTXs quantitative requirements (where X represents Quality, Cost, or Schedule on the "process to be improved"), ease of use, ease of implementation, impact level on the project outcomes, expected monetary value, cost-benefit projections, and the level of acceptance by internal and external stakeholders.

A good criterion reflects key project goals and enables objective measurements to be made. Thus "resource cost" is measurable and reflects a project goal, while "user-friendly" may not reflect any goals and will be difficult to score. CTXs are the measurable product or service characteristics that the customer considers important, and whose performance standards or specification limits must be met to satisfy customer requirements. They usually have four components: characteristic, measure, target, and specification limits.

Secondly, the project team can use a "nominal group technique" to determine the weighting of the criteria. Each team members should weight the criteria from 1 to 5 for example, and the relative weights assigned to selection criteria using ordinal averages from the team members. These are not scientific formulas that produce the "right" answer. Instead, they are tools to help the project manager facilitates discussion around which solution is "best."

Thirdly, the project team should effectively construct the prioritization matrix.

The following are steps to be followed to construct a prioritization matrix:

1. List all selection criteria, as shown in Table 28.1
2. Rank and assign weights to the selection criteria
3. List all potential solutions
4. Evaluate the strength ρ of the relationships between selection criteria and potential solution
5. Cross multiply weight and strength of relationships. The combinations with the highest total are the potential solutions on which the project team needs to focus the improvement efforts on.
6. Highlight the critical few potential solutions that matter the most from the computed totals

28.3 Develop Prototype, Assess Risk & Pilot Solution(s)

Regardless of how good the hierarchy of decision criteria used to select potential solutions is and how much employees, customers and stakeholders feedback is solicited, some employees, customers and stakeholders will always surprise the team with how they approach and use the outcomes of a solution. As such, there is no substitute for putting a solution out in front of a broad array of selected employees, customers and stakeholders to observe and measure their experience. Commonly called piloting a solution, this process involves deploying to production (resp. to service) or to a production-like (resp. to service-like) environment (e.g., staging environment) a limited release of an improved process to validate that a solution works as expected under live conditions with "real" users.

At this stage, the project team should now be ready to develop a prototype of the potential "improved process" resulting from the analysis put forward in the previous section and evaluate solution(s) targeted at the verified assignable causes of underperformance of the "process to be improved." Within a project management context, which includes the actions necessary to define, analyze, develop, implement, assess and formulate all subsidiary plans for developing a prototype and piloting potential solutions, the purpose of the "Develop Prototype, Assess Risk and Pilot Solution(s)" project management process is to obtain early feedback on requirements by providing a working model of the expected "process to be improved" outcome before actually building it.

Since prototypes are tangible, it allows customers and stakeholders to experiment with a model of their final process outcome more quickly and less expensively than only discussing abstract representations of their requirements. Prototypes support the concept of progressive elaboration because they are used in iterative cycles of mock-up creation, user experimentation, feedback generation, and prototype revision. When enough feedback cycles have been performed, the requirements obtained from the prototype are sufficiently complete to move to a design or build phase.

28.3.1 Develop Prototype

Prototype development is an important and vital part of a process improvement project. It should be undertaken by the project team as a sub-project within a project management context, which includes the actions necessary to define, analyze, develop, implement, assess and formulate all subsidiary plans for developing a prototype and piloting potential solutions.

The project team should look at a prototype as the first original approximation of a potential "improved process" outcomes (products or services) in some form that has been or will be copied or developed for a definite purpose in its implementation. It is an original model, a preliminary version of the potential "improved process" outcomes, or a small scale mock-up of a product, or a test run of a service, created to explore its viability as part of a process improvement activity.

Prototyping is the process of realizing a prototype. It is an evolutionary process that can range from just a virtual prototype, which refers to prototypes that are non-tangibles and are usually build for study and analysis, to a physical prototype that refer to the tangible manifestation of the potential "improved process" outcomes and are usually build for testing and experimentation. Its degree of approximation can vary from a rough representation to an exact replication of the desired process outcomes (products or services).

The implementation aspect of the prototype should cover the range of prototyping the complete products (or services) to prototyping part of, or a subassembly or a component of the potential "improved process" outcomes. The complete prototype, as its name suggests, should model most, if not all, the required characteristics of the potential "improved process" outcomes.

28.3.2 Pilot Solution(s)

Once a prototype solution has been developed, it should be further explored and refined by piloting it. Piloting a prototype solution is a way of trying the prototype solution on a temporary basis and learning about its potential impact on the enterprise business. The idea of piloting a prototype solution does not seem to come naturally. People tend to want to make a potential "improved process" part of the day-to-day operations immediately. Being successful at making changes requires a very different approach. Tests should be designed so that as little time, money, and risk as possible are invested while at the same time enough is learned to move toward full-scale implementation of the prototype solution.

In much the same vein as with the development of a prototype solution, piloting solution(s) and assessing risk is a project within a project. It needs a plan that details what, when, why, who, and how the pilot is to be conducted. It needs to be well planned, executed, and concluded. It involves piloting other solution elements to assess support and training of employees affected by the potential "improved process" as well as end-user communications (e.g., launch preparedness announcements).

In many enterprise businesses, the continuation of process improvement projects involving prototypes development often hinges on the success of the developed prototypes to provide impetus to management to forge ahead with it. Having completed the development of the prototype, the project team will be raring to go and have hands-on experience with all the new features to be implemented. How should the potential improvement from the prototype be evaluated before the potential "improved process" is implemented? How can we accelerate the learning as we test the developed prototype? How can the risk of making a change be minimized? The answer is piloting of the prototype to assess its suitability and other criteria.

Once the prototype has been developed, it is important to distinguish between piloting and implementing it. Piloting a prototype is used to evaluate the potential "improved process" on a temporary basis. Implementing the potential "improved process" means making it part of the day-to-day operations or incorporating it into the next version of a product or service. An important practical consequence of piloting a prototype solution before implementing it is that some pilots are expected to fail, and the project team can learn from those failures. This is why piloting a prototype solution on a (very) small scale to build knowledge while minimizing risk is so important. Once a prototype is implemented, the project team should expect very few failures.

The process of piloting a prototype solution is one in which the project team defines and refines how to operate the process with minimum variance. When a process is operated on target with minimum variance it is operating up to its full potential. When that is not enough, the project team places the prototype out in front of a broad array of selected employees, customers and stakeholders to observe and measure their experience. It involves deploying to production (resp. to service) or to a production-like (resp. to service-like) environment (e.g., staging environment) a limited release of an improved process to validate that a solution works as expected under live conditions with "real" users. Thus, this course of action is one that will guide the "process improvement" project team through the complexities of improving the existing "process to be improved."

The goals of a pilot are to stabilize a prototype further using feedback from a broad representation of users and to reach consensus with the stakeholders and employees affected that a solution satisfies their needs and meets their expectations. A pilot should be a well-orchestrated review of selected prototype solution(s) to accomplish specific goals and objectives not a random walk by affected employees through a prototype. Conversely, it should not be a tightly controlled guided tour through a prototype solution either. The project team should make the purpose of the pilot and the measuring of feedback as well as the collection of the V.O.B., V.O.C. and V.O.P. data clear but let employees affected by the potential "improved process" go where they may within the context of what needs to be reviewed this is necessary to surface any unexpected behaviors.

The primary purposes of piloting a prototype are to demonstrate that the selected solution(s), i.e. potential "improved process," works as expected under live conditions and that it meets the project requirements. Conducting a pilot helps to:

1. Adjust a predictable potential "improved process" in the "Threshold State";
2. Tweak a predictable potential "improved process" in the process "Ideal State";
3. Understand and reduce the enterprise business' risk of encountering problems during full-scale deployment of the potential "improved process."

28.3.2.1 Applying the Science of Improvement to Pilot(s)

Throughout this handbook, we have adopted the system theory to explain the dynamics of enterprise businesses, hence processes used within enterprise businesses. An enterprise, by its most basic definition, refers to an assembly of people working together to achieve common objectives through a division of labor. An enterprise provides a means of using individual strengths within a group to achieve more than can be accomplished by the aggregate efforts of group members working individually. The systems approach offers a way of understanding enterprise businesses and their ability to make improvements. Enterprise businesses and particular organizational situations can be analyzed in terms of the various interacting systems they comprise. These may be social, organizational, or technological.

The ability to make improvements on the piloted prototype solution is enhanced by combining subject matter knowledge on the "process to be improved" under improvement and 'profound knowledge' in creative ways. Deming describes 'profound knowledge' in four parts, all related to each other:

1. *Appreciation for a system*—Because most process outcomes (products and services) result from a complex system of interaction among people, procedures, and equipment, it is vital to understand the properties of a system under consideration. We may think of a system as an interdependent assembly of items, people, or processes working together toward a common purpose. The common purpose aligns the parts of the system, while interdependence considers the relationships and interactions among them.

Enterprise businesses are made up of departments, people, equipment, facilities, and functions. If each part of a system, considered separately, decides to operate as efficiently as possible, then the system as a whole will not operate to maximum effectiveness. The management's task is to optimize the system, that is, orchestrate the efforts of all components toward achievement of the stated purpose.

2. *Understanding variation*—As we have indicated in previous chapters, we can think of variation as change or slight difference in condition, amount, or level from the expected occurrence, typically within certain limits, as shown in Fig. 8.4. Variation has two broad causes that have an impact on data collected: common (also called random, chance, or unknown) causes and special (also called assignable) causes.

   Common causes of variation are inherent and an integral part in the process been considered. They can be though of as the "natural pulse of the process been considered" and they are indicated by a stable, repeating pattern of variation.

   Assignable causes of variation are those causes that are not intrinsically part of the process been considered but arise because of specific circumstances. When they occur, they signal a significant occurrence of change in the process and they lead to a statistically significant deviation from the norm. Assignable causes of variation are indicated by a disruption of the stable, repeating pattern of variation. They result in unpredictable process performance and must therefore be identified and systematically removed before taking other steps to improve quality of the system been considered.

   Knowledge about separating variation of outcomes of a process or system into common and special causes helps to decide appropriate actions for that process or system. Inappropriate action may make things worse.

3. *Building knowledge*—In the context of process improvement, analysis, development of prototype solution(s) and piloting of a new "improved process" is a prediction. Prediction that one of several alternatives will be superior to others in the future. The choice may encompass different concepts for a product, or materials, or conditions for operating a process.

   Prediction is essential for effective management. In fact, as Deming said, "Management is prediction." One of the most important functions of process improvement is to enable prediction. Whether we are consciously aware of it, almost everything we do involves some form of prediction, according to some "theory" of action. A theory is nothing more than a cause-and-effect prediction about how planned actions lead to expected outcomes: If I do X, Y will be the result. Why is a particular practice linked to performance improvement—what is the logic?

   In a learning enterprise business, theory-based predictions are systematically being tested and the theory is being revised—using both single-loop and double-loop learning. Without theory, we have nothing to revise, and nothing to learn and build knowledge. We learn by comparing predictions from the theory with actual data.

As part of the analysis of the data from a "PDSA Do" project phase, it is generally useful to compare performance of the alternatives under various conditions. The ultimate aim is prediction of performance of the new "improved process" in the future. The improvement of quality is almost solely focused on the results of the "PDSA Do" project phase.

If the new "improved process" is implemented, improvement will result. The more knowledge one has about how the particular system under consideration functions or could function, the better the prediction and the greater the likelihood the implementation of a new "improved process" will result in improvement. Comparing predictions to results is a key source of learning.

Rational prediction requires theory. A theory represents our current knowledge about how some aspect of the system of interest works. Participants in an improvement effort articulate the basis of their predictions by making their theories (or generated hypotheses) explicit. Stating theories or assumptions helps design pilots to validate these theories; ideas for implementation of new processes can then be improved on the basis of the results of the pilots. If a new "improved process" piloted do not lead to the improvement predicted, the circumstances present must be identified and the understanding gained used to further refine the theory (or generated hypotheses).

Theory-building, process improvement frameworks and organizational learning involve "system thinking." The behavior of a system is not a function of what each part is doing in isolation, but on how the parts interact. In order to understand a system, we need to understand how it fits into the larger system. As understanding increases, so do the ability of individuals—and the enterprise business as a whole—to predict.

In addition to the idea of prediction in learning and improvement, the concept of operational definitions is an important contribution to building knowledge. Operational definitions are used to put communicable meaning to a concept. To develop an operational definition, consideration needs to be given to: a method of measurement or test and a set of criteria for judgment.

Skillfully building knowledge by implementing new "improved processes" and observing or measuring the results is the foundation of improvement. By repeating learning cycles, most circumstances for applying the theory can eventually be categorized, making the theory (or generated hypotheses) useful for predictions in future situations.

The emphasis in most organizations on "doing" actually inhibits learning because managers and employees are not required to take the time to properly predict (develop a theory) and learn (systematically test that theory). As a result, every action or process improvement initiative is viewed as an independent activity, and whether it will succeed or fail depends more on chance. Without good theory-based predictive process improvement and learning, it will be impossible to determine what is working, or not working, and why. Under such circumstances, experience and resources are often wasted.

4. *Human side of change*—Most implementation of new "improved processes" will not happen without the support of people. Focusing only on the implementation

of a new "improved process" and not on its effect on people will doom improvement efforts. Because most improvement efforts involve an informal or formal improvement team, members of a team (or at the least the team leader) should have some knowledge of running effective meetings, active listening, and resolving conflict. Adhering to simple principles to run good meetings such as having an agenda, designating roles (minute taker and the like), agreeing on how decisions will be made, documenting action items, and ensuring all members are heard can have a positive effect on progress. People will usually have some reaction to change. This reaction can range from total commitment to open hostility.

Knowledge of the human side of change helps understand how people, as individuals, interact with each other and with a system. It helps predict how people will react to a specific change and how to gain commitment. It helps understand the motivations of people and their behavior.

Although these four components of Deming's 'profound knowledge' can be addressed separately, their importance in improvement is derived mainly from their interaction. Focusing on appreciation for a system without considering the impact that variation is having on the system will not produce effective ideas for improvement. Similarly, the interplay of the human side of change and the building of knowledge, as seen in areas of study such as cognitive psychology, is critical for growing people's knowledge about implementing new processes that result in improvement.

The use of Deming's 'profound knowledge' along with subject matter knowledge, about the "process to be improved," as an outside view or lens offers important guidance in piloting a prototype solution. Some insights for piloting a prototype solution include, but are not limited to:

1. Understanding interdependencies in the components of the system affected by the potential "improved process";
2. Understanding the relationship between prediction of the potential "improved process" outcomes and knowledge of the system being affected and how these predictions build knowledge;
3. Understanding the temporal effect of the potential "improved process" in the system;
4. Understanding how separating variation of outcomes of a process or system into common and special causes helps to decide appropriate actions;
5. Understanding how to integrate the potential "improved process" in the social system considered within the enterprise business, especially when planning for implementing a prototype solution.

28.3.2.2 Learning Through Pilot(s)

Piloting a prototype solution increases learning and builds knowledge about the causal mechanisms at work in the system affected by the potential "improved process." There are two types of learning that should take place: single-loop and double-loop learning.

Single-loop learning occurs when there is a single feedback loop, with the project team modifying its actions according to the difference between expected and obtained outcomes of the potential "improved process." This is an ongoing error-correction process that continues until an acceptable level of knowledge or action is achieved. With single-loop learning, practice, feedback, error correction, and repetition will increase the project team's ability at improving the prototype solution further.

In double-loop learning, the project team questions the very content of the learning, assesses those beliefs it has taken for granted, and challenges the expectations, values, and assumptions that led its members to adopt the knowledge or engage in the process improvement effort in the first place! The process of building knowledge emphasizes the importance of rational prediction of the outcomes of the potential "improved process."

If during piloting a prototype solution a prediction is incorrect, the theory and hypotheses that were used to generate the potential "improved process" must be questioned and modified. If the project team is able to modify expectations, then a second-order, or double-loop, of learning has occurred. As such, double-loop learning can be thought of as "learning about single-loop learning."

Double-loop learning is important if the project team wants to achieve more than increased efficiency in doing the executing the new "improved process." As the project team conducts pilots of the prototype solution, double-loop learning is taking place. Innovation and transformation are double-loop processes. After double-loop learning has occurred, then single loop learning can take over again, and we can improve the use of the new method or measure.

When considering making predictions about the outcomes of a potential "improved process," it is important to recognize that a very limited set of conditions will be present during piloting. Circumstances unforeseen or not present at the time of piloting a prototype solution may arise in the future. Will the potential "improved process" still result in an improvement under these new conditions? Determining whether the potential "improved process" actually results in improvement during the piloting a prototype solution is important, but this determination is usually much less difficult than considering the effect of the potential "improved process" in the future.

The formulation of a scientific basis for prediction of the outcomes of a potential "improved process" has its beginnings with W. A. Shewhart. Shewhart's concept of "degree of belief" presents a way to think about and assess the depth of an improvement team or individual's knowledge about a potential "improved process." In making any prediction about the outcomes of a potential "improved process," one has some degree of belief (high, medium, or low) that the prediction is correct. One's degree of belief in a prediction about the outcomes of a potential "improved process" depends on two considerations:

1. The extent to which the prediction can be supported by evidence, and
2. The similarity between the conditions under which the evidence was obtained and the conditions to which the prediction applies.

How could the project team quantify the degree of belief? Unlike a probability, confidence level, or statistical significance level, the degree of belief is a concept, not a calculated value. The belief is about a prediction, not a past occurrence. There is not a proven theory to make quantitative statements about the future. The degree of belief is increased as pilots of a prototype solution are conducted and predictions about the outcomes of a potential "improved process" begin to agree with the results of the pilots. If a prediction is incorrect, the hypotheses and analysis performed must be modified and hence learning takes place.

As the scope and scale of pilot(s) gradually increase, the rigor for learning must be increased, as well as the rate of learning. Teams often spend too much time thinking about all of the possible options, ramifications, and implementation issues before proceeding with piloting a prototype solution. Can one learn more by diagnosing the current "process to be improved" or affected system, or by changing something? Improvement efforts are frequently stuck in the diagnostic journey (analysis paralysis). The alternative is to very quickly pilot a prototype solution. Experience has shown this latter approach leads to accelerated learning and improvement. However, it does require a structured approach to piloting a prototype solution and learning.

28.3.2.3 Some Principles for Piloting a Prototype Solution

Anyone who has piloted a prototype solution has probably pondered whether the same results will be obtained when the prototype solution is implemented in the future. Considering these basic principles helps reduce this uncertainty:

1. Pilot on a small scale and build knowledge sequentially.
2. Include a wide range of conditions in the sequence of pilots.
3. Collect In-Process Data over Time—V.O.B., V.O.C., & V.O.P.
4. Determine the process defect rate, capability & performance indices.

Pilot on a Small Scale and Build Knowledge Sequentially

Knowledge is built iteratively by making predictions, about the outcomes of a potential "improved process," that are based on the current hypotheses and analysis, testing the predictions with data, improving the hypotheses and analysis according to the results, making predictions on the basis of the revised hypotheses and analysis, and so forth. The building of knowledge in a series of tests is illustrated in Fig. 28.1.

It is important to minimize the negative impact that can result from a potential "improved process" that does not result in improvement. Table 28.2 summarizes the appropriate scale of piloting a prototype solution for a number of situations. Small-scale pilots are needed if the degree of belief is low and the consequences of failure are large. Planning one large cycle in an attempt to get all of the answers with one pilot should always be avoided. Moving to implementation of a prototype solution should be considered only if:

1. The project team has a high degree of belief that implementation of the potential "improved process" will result in improvement, and
2. The cost of failure is small (losses from a failed test are not significant), and
3. The enterprise business is ready to accept the potential "improved process."

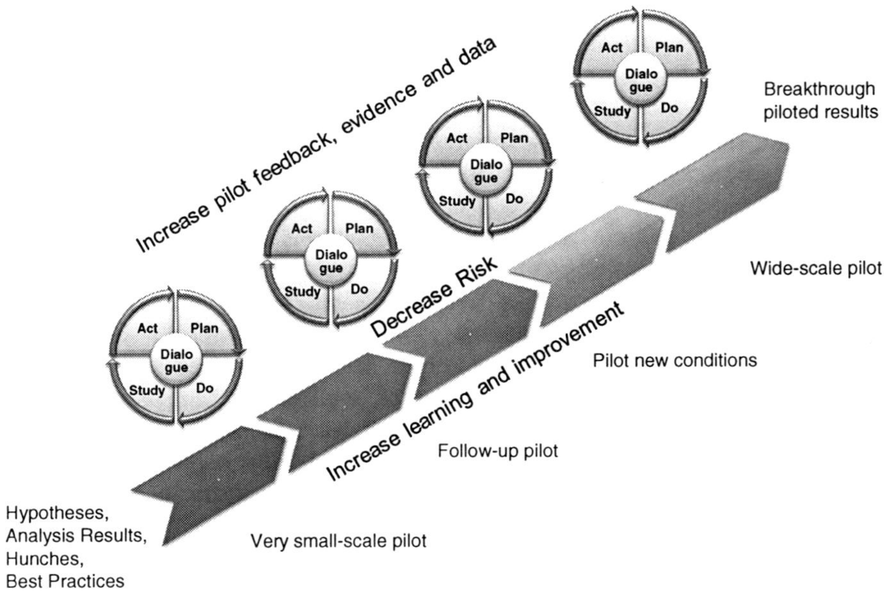

Fig. 28.1 Sequential building of knowledge with piloting through PDSA

Table 28.2 Deciding on the scale of the pilot

		Level of commitment within the enterprise business		
Current belief within the enterprise business	Cost of failure	No commitment	Some commitment	Strong commitment
Low degree of belief that the potential "improved process" will lead to Improvement	Large cost of failure	Very small-scale	Very small-scale	Very small-scale
	Small cost of failure	Very small-scale	Very small-scale	Small-scale
High degree of belief that the potential "improved process" will lead to Improvement	Large cost of failure	Very small-scale	Small-scale	Large-scale
	Small cost of failure	Small-scale	Large-scale	Implement

Piloting a prototype solution on a small scale is an important way of reducing people's fear of accepting the potential "improved process." When (very) small-scale pilots are not considered, people procrastinate. They try to develop the perfect "improved process" because of the potential consequences of a failed pilot. This approach might be particularly prevalent in some big corporations or government agencies where any change to programs or policies is usually scrutinized. When planning a cycle to pilot a prototype solution, much thought should be given to developing ways of building knowledge through small-scale tests.

One should never move quickly to implementation after one successful small-scale test. In most situations, additional cycles for piloting a prototype solution are needed. As the degree of belief in the success of piloting a prototype solution is increased, the scale of the pilot can be increased with less risk.

Similarly, one should never move quickly to abandonment of piloting a prototype solution after one "failed" small-scale pilot. One must first understand why the prediction about the outcomes of a potential "improved process" is incorrect. Is the observed outcome an anomaly? There is much to learn from failure. The whole point of performing a small-scale pilot is to minimize the risk from failed pilots and maximize the learning.

Include a Wide Range of Conditions in the Sequence of Pilots

Including data spread over time is also a convenient way to include a range of conditions in the cycle, the third principle for piloting a prototype solution. How can the project team increase their degree of belief that the potential "improved process" will be effective in the future? Including a wide and varied set of conditions in piloting a prototype solution is the best way to increase degree of belief.

Too often, pilots of a prototype solution are not conducted over a broad range of conditions. Some reasons given for limiting the conditions are limited resources, time constraints, difficulty in analysis of the data, lack of knowledge of how to efficiently include different conditions, and too many possible conditions to consider. The degree of belief in the results of piloting a prototype solution increases as the same conclusions are drawn for a variety of pilot conditions. The time required to increase the degree of belief that the potential "improved process" will result in improvement and persist is a matter of judgment.

If a particular supplier's material proves to be the best in various environmental conditions and on different days and shifts, one feels much safer to use the results to select a supplier than if the pilot was run on one day under constant environmental conditions. The experimenters might also consider running the pilot using material from more than one lot provided by the supplier. Incorporating some or all of these conditions will increase the degree of belief in the results if similar conclusions are seen for all conditions.

Collect In-Process Data Over Time: V.O.B., V.O.C., & V.O.P.

During the different pilots of the prototype solution, the project team should continually collect, compile, and evaluate feedback and operations, and document the potential "improved process" features and functions needed to achieve the desired results identified in the "PDSA Plan" project phase, especially with respect to the critical success factors. From the quality perspective, this includes:

1. *The business needs and expectations* (*Voice of the Business—V.O.B.*). This is the voice of profit and return on investment. Every "process improvement" project has to enable the enterprise business sustainability and meet the needs of the employees and shareholders.
2. *The customers and the stakeholder's needs and expectations* (*Voice of the Customer—V.O.C.*). This is the voice calling back at the "improved process" from beyond its outcomes that offer compensation in return for satisfaction of the customers and stakeholders needs and wants. This voice represents the stated and unstated needs, wants, and desires of the customers and stakeholders, referred to as the customers and stakeholders' requirements.

   Collecting the customers and stakeholders' requirements is as much about defining and managing customers' and stakeholders' expectations as any other key project deliverables that has been built and will be the very foundation of customer acceptance and completing to "process improvement" project. It is also about refining the improvement effort by gathering information on the current situation. Its purpose is to build, as precisely as possible, a factual understanding of existing potential "improved process" conditions and problems or causes of underperformance that may occur.

   The constituent project management processes, used during the capturing of the voice of the customer have been described during the "PDSA Plan" project phase. They include the following:
   - Plan V.O.C. Capturing
   - Collect and Organize Data
   - Analyze Data and Validate CTXs Predictions
3. *The "improved process" needs and expectations* (*Voice of the Process—V.O.P.*). The "improved process" must meet the requirements of the customers and stakeholders, and the ability of this process to meet these requirements is called Voice of the Process. It is a construct for examining what the "improved process" is telling about its inputs and outputs and the resources required to transform the inputs into outputs.

   The constituent project management processes, used during the capturing of the voice of the process have been described during the "PDSA Plan" project phase. They include the following:
   - Plan V.O.P. Data Capturing
   - Collect Data
   - Display Data and Patterns
   - Establish Process Performance
   - Validate/Refine Process Quality Targets

Just because the different pilots of the prototype solution have been carefully designed to meet the learning and performance requirements does not mean that it will be as effective as anticipated once it is deployed.

Incorporating time into the pilots is a more important consideration than the sample size. There is almost always more information in small samples selected over a long time period than a larger one collected over a relatively short period of time. A large sample taken in the winter to determine the results of a pilot may not increase degree of belief about the effect of the potential "improved process" in the summer.

The purpose of collecting these data over time is to get sufficient and accurate information to complete improvement of the "improved process" set forth. Most importantly, the purpose is to get accurate and sufficient data to derive complete functional requirements for the "improved process" outcomes. Although the different pilots of the prototype solution will vary in complexity based on the extent of the changes to be made to the "process to be improved," the "improved process" should be well documented at this point, and the captured feedback and the V.O.C., V.O.B. and V.O.P. data analyzed. The project's success is directly influenced by the care taken in capturing and managing these requirements.

The frequency of in-process data collected is dependent on the criticality of the "improved process" on the enterprise business. This data should provide timely feedback on how well the prototype solution is working, on possible problems that might require some corrective action, opportunities to further enhance the prototype solution, hence the new "improved process," or, in rare instances, discontinue it.

In-Process data collection is high-leverage because it typically requires little effort and can provide extremely valuable, ongoing feedback. Traditional data collection typically occurs too late for such decisions and remedial action to be taken. In order to maximize effectiveness, data collection should occur throughout the prototype solution lifecycle—from initial conception to the end of deployment. In order to maximize effectiveness, the prototype solution should be measured throughout its lifecycle—from initial concept through to full deployment. Remember: Without measurement, it is impossible to manage anything and the earlier the project team starts measuring, the more leverage it can get from the measurement.

As indicated during the "PDSA Plan" project phase, the most strident needs and requirements are those related to the customers. Indeed, the bottom line for the potential "improved process" is the value of its outcomes (products and services) in the eyes of potential customers. Without continuing enthusiasm from customers, sustainability of these outcomes may not last. It is customers' opinions that will determine the value of the "improved process" outcomes. Customers' opinions of the value of the process outcomes determine the "customer value" of these outcomes. The customer value of a process outcome consists of key factors that determine how well customers will appreciate this outcome. For a given process, the customer value may change over time, and a new process outcome that better fits the changing customer value could be a breakthrough product or service.

Determine the Process Defect Rate, Capability & Performance Indices

In conducting pilots a prototype solution through PDSA Cycles, there will be variation in the measures of the characteristics of potential "improved process" outcomes due to causes and conditions unrelated to the prototype solution being piloted.

As we pointed out in Chap. 2, in business applications which operate at a performance permissible limit of variations of z standard deviations, every process outcome within those business applications is intended to add value to the enterprise (businesses & customers) as a whole. It has a set of requirements or descriptions of what an element needs to add value to the enterprise. When a particular element meets those requirements, it is said that it has achieved quality, provided that the requirements accurately describe what the businesses and the customers actually need. Those process outcomes whose characteristics are falling beyond z standard deviations of the expected central tendency are often regarded as flaw, defective, unacceptable, or in non conformance quality. They will undergo more or less corrective actions: rework, scrapping (of whatever can not be reworked) and conformance use.

Having collected in-process data, the project team should determine the new "improved process" rolled throughput yield. Establishing the rate at which defects occur on a characteristic of the "process to be improved" outcomes with respect to the number of "process to be improved" outcomes inspected is complementary to establishing the process yield.

The effect of the prototype solution must be distinguished from these uncontrolled or extraneous conditions. Using control charts to view the patterns of the data over time can assist in this distinction. It is also important for the project team to distinguish between the new "improved process" in a state of statistical control and the new "improved process" that is meeting specifications.

A state of statistical control does not necessarily mean that the outcomes from the new "improved process" conform to specifications. Statistical control limits on sample averages cannot be compared directly with specification limits because the specification limits refer to individual units. For some processes that are not in control, the specifications are being met, and no action is required; other processes are in control, but the specifications are not being met and action is needed. In summary, the project team will need to ensure that the new "improved process" is both stable (in statistical control) and capable (meet product specifications) states.

For a process that is in a state of statistical control, the process capability is a measurable property of the process and it summarizes how much variation there is in the process relative to a set of customer and business specifications. It also allows different processes to be compared with respect to how well an enterprise business controls them. Therefore, the process capability represents the capability of the process to meet its purpose as defined by the enterprise business intended strategy and process definition structures. If a process is out of control and the causes cannot be eliminated economically, the standard deviation and process capability limits nevertheless can be computed (with the out-of-control points included). These limits will be inflated because the process will not be operating at its best. In addition, the instability of the process means that the prediction is approximate.

In most processes, not only are there departures from a state of statistical control but the process is not necessarily being operated to secure optimal process yields; e.g., the average of the characteristic of the process outcome considered is not centered between the upper and lower tolerance limits. To allow for these realities, it is convenient to try to select processes with the 6σ process capability well within the specification range.

Under the normality assumption on the observed characteristic of the new "improved process" outcomes, the project team should arrange the collected in-process data are into subgroups of specific period of time. If the upper and lower specification limits of the process are USL and LSL, the target process mean is T, the estimated expectation of the observed characteristic of the "improved process" is $\hat{\mu}$, the estimated variability of the process (expressed as a standard deviation) within a subgroup is $\hat{s}$, and the estimated overall variability of the process (expressed as a overall standard deviation) is $\hat{\sigma}$, then commonly-accepted estimates of process capability indices within subgroups and overall process performance indices are given in Tables 14.5 and 14.6.

The project team must also analyze in-process data over time to determine whether the potential new "improved process" is in the "Ideal State," toward which every process aspires to. As pointed out already, a process in the "Ideal State" is predictable and all its outcomes are in full conformance. The predictability of the new "improved process" will be the result of purposeful continuous efforts on the part of the enterprise business and the personnel who operate the process. A predictable process is an achievement, requiring constancy of purpose and the effective use of process behavior charts. The conformity of the process outcomes will be the result of having natural process limits that fall inside the specification limits. When the process is operating in the "Ideal State," its centered capability index estimate $\bar{C}_{pk}$ will be close to, or greater than, 1.00. A process in the "Ideal State" satisfies four conditions:

1. The process must be inherently predictable over time.
2. The enterprise business personnel must operate the process in a predictable and consistent manner. The operating conditions cannot be selected or changed arbitrarily.
3. The process central tendency must be set at the proper level.
4. The natural process limits must fall inside the specification limits for its outcomes.

Whenever one of the four conditions above is not satisfied, the possibility of producing non-conforming outcomes exists. When a process satisfies these four conditions, the enterprise business can be confident that nothing but conforming products or services are being produced. Furthermore, the conformity of the process outcomes should continue as long as the process behavior remains predictable.

Therefore, a process that is in the "Ideal State" does not need further improvement. Since the process outcomes stream for a predictable process can be thought of as being homogeneous, the measurements taken to maintain the process behavior chart will also serve to characterize the process outcomes produced by the predictable process.

28.3.3 Assess and Reduce Risk

To further minimize risk, the project team could conduct multiple pilots consisting of separate pilots for different process outcomes characteristics, starting on small scope and gradually increase the scope of successive pilots, as indicated in Fig. 28.1. Effective handling of risk increases the likelihood of success in a project by minimizing the potential for failure and maximizing the potential to use risk for gain. Effective handling of risk involves having a good approach (i.e., risk management process described during the development of the "PDSA Plan" Process Group) and accomplished execution of that approach (i.e., risk management discipline).

Using the Failure Mode and Effect Analysis (FMEA), focusing the data collection effort on those inputs variables from which assignable causes of variation may originate, the project team should identify, estimate, prioritize and evaluate risk associated with the characteristics of the potential "improved process." As indicated already, failures are unwanted features of a characteristic of a "process to be improved" outcome; it is any errors or defects, especially ones that affect the customer, and can be potential or actual. "Effects analysis" refers to studying the consequences of those failures.

Piloting a prototype solution is usually carried out in sub-phases, with each subphase growing in number of employees and customers affected by the potential "improved process" and the distribution of users throughout the enterprise business. The whole purpose of such sub-phases is to slowly roll out to employees and customers throughout the enterprise business and externally to it to validate that the prototype solution and pilot assumptions are accurate and that they can be successful in the production (resp. to service) or to a production-like (resp. to service-like) environment. Typically, each successive sub-phase broadens user involvement (e.g., pilot first to a team, then to a department, and then to a division).

As mentioned already, piloting a prototype solution is also a time when thee project team builds the knowledge base of improved products or services that occur during the roll-out of the "improved process" so that if or when problems occur again (possibly in the full rollout phase of the improved products or services), lessons have been learned and workarounds already created to resolve stumbling blocks. The pilot participants provide feedback about how well the potential "improved process" is performing, and whether it is meeting their expectations and requirements. The project team should use this feedback to resolve issues that arise or to create a contingency plan.

In addition to expanding the scope of piloting a prototype solution by sheer quantity, selecting users, who have different usage requirements of the potential "improved process," can provide an additional level of complexity in piloting selected solution(s) on the developed prototype. Regardless of how much planning and testing are conducted, problems always arise in piloting a prototype solution. It is important for the project team to have the prototype solution still intact so that any outstanding problems can be re-created, tested, and resolved to be tested in the pilot production phase again.

28.3.3.1 Perform Mistake-Proofing: Poka-Yoke

Mistake-Proofing might also be used of as an extension of the FMEA. While a FMEA helps in the prediction and prevention of problems, mistake-proofing emphasizes the detection and correction of mistakes in the potential "improved process" outcomes before they become defects that may subsequently be delivered to either the end customer or the next-in-line customer.

The originator of this idea was Shigeo Shingo from Japan (Shingo, Zero Quality Control: Source Inspection and the Poka-Yoke System, 1986). The term Poka-Yoke comes from anglicizing the Japanese words "poka" (inadvertent mistake) and "yoke" (prevent). The underlying philosophy of Mistake-Proofing explicitly recognizes that:

1. People forget and make errors
2. Machines and processes fail and make errors
3. The use of simple mistake-proofing ideas and methods in product design and process design can eliminate both human and mechanical errors

For a given process, mistake-proofing is very easy to understand as it is grounded in common sense. Its essence is to design both the process outcome and the process itself so that mistakes are either impossible to make or, at the least, they are easy to detect and correct. At the heart of mistake-proofing is simply paying careful attention to every activity in the process considered and then placing appropriate checks and problem prevention facilitators at each step in the process. It is simply a matter of constant data feedback, similar to that required to maintaining your balance while riding a bicycle. Mistake-Proofing is achieved, in its simplest form, by taking the following three sequential steps:

1. *Identify possible errors that might still occur in spite of preventive actions.* At each step of the process considered, simply ask the question "What possible human error or equipment malfunction could take place at this step?" E.g., an apparently symmetrically shaped part could inadvertently be installed backward. This could be an area where a truly negative or paranoid person within the enterprise business becomes an asset.
2. *Determine a way to detect that an error or malfunction either is taking place, or is about to take place.* A guide pin might be added to prevent the incorrect part installation sited in #1 above. The project team should not just rely on people to simply catch their own errors all the time.
3. *Identify and select the specific action to be taken when an error is detected.* There are three basic actions. Listed in their order of preference, they are:
   - *Control.* An action that self-corrects the process error, e.g., a spell-checker/corrector.
   - *Shutdown.* A procedure that blocks or shuts down the process when an error occurs, e.g., a lockout switch.
   - *Warning.* Alert the person involved that something is going wrong. The primary weakness with warnings is the fact that they are frequently ignored, especially if they occur too frequently. Therefore, controls and shutdowns are generally preferred over simple warnings.

Mistake-Proofing during pilot of the prototype solution facilitates understanding of how defects originate and then helps to focus attention on simple devices/methods that can be used to eliminate defects. The real challenge for the project team is to come up with specific methods to detect, self-correct, block/shut down, or warn of a problem occurring in the potential "improved process." This can sometimes require real imagination and creativity, but the emphasis is usually on inexpensive solutions.

Mistake-Proofing can be achieved by 100% inspection while the work is in process, not by the use of quality inspectors between work areas. The key to this inspection is the fact that it is accomplished as an integral part of the work process either by the worker or, better yet, automatically, not by an "inspector."

From an organizational viewpoint, one of the most positive results of implementing mistake-proofing on the potential "improved process" is the fact that it enable people, at all levels and across all functions affected by the potential "improved process", to begin to think in a "preventive mode" rather than in an "after the fact" detection mode relative to design and process errors.

28.3.4 Conclude the Pilot

Once the pilot completion criteria have been satisfied, the project team should conclude the pilot phase and regroup to address all outstanding issues and to refine solution(s) and supporting material based on the results of the pilot phase. The project team must review the parts of the solutions that were successful and the types of problems that were reported so that it can revise and improve the project plan. Based on the captured feedback and the generated V.O.P data, the project team must determine if the piloted solution(s) on the developed prototype meets the success criteria defined in the project plan document. Preferably, the potential new "improved process" should be in the "Ideal State," toward which every process aspires to, to conclude the pilot.

During and at completion of this pilot phase, it is important to document the results. Even with the extensive discovery and improvement work, as well as the prototype testing and pilot sub-phases that have taken place, problems may reoccur in the post-pilot phases, and any documented information on how problems were resolved or configurations made to resolve problems in the pilot phase will help simplify the resolution in future phases.

Ultimately, the pilot phase leads to a decision to proceed with a full implementation and deployment or to slow down so that the project team can resolve problems that could jeopardize deployment of the potential "improved process."

28.3.5 Develop Implementation Plan

In its broadest sense, implementation has been used to mean the execution of a project plan for process improvement. From this broad perspective, the term would include developing, piloting, and activities to sustain a potential "improved process." However in this section, we use the term "implementation" in a narrower sense. Specifically, in developing an implementation plan for a prototype solution our focus is on the activities one takes after piloting has shown the potential solution is positive and leads to improvement.

The aim of this narrower scope of implementation is to make sure that the infrastructure is in place to make the potential "improved process" long-lasting and successful. This includes planning issues such as training, documentation, standardization, adequate resourcing, and social considerations. To summarize, piloting is about learning if the developed prototype solution will result in an improvement, and implementation is about how to make the developed prototype solution an integral part of the system considered.

The PDSA Cycle creates the structure for both piloting and implementing a developed prototype solution. Besides the similarities between piloting and implementation (making predictions, collecting data, and documenting things that go wrong) that result from use of the cycle, there are also important differences, as we summarize below:

Piloting a prototype solution:

1. The prototype solution piloted is not permanent and therefore does not need supporting processes to maintain it beyond a brief period.
2. The opportunities for learning about many aspects of the prototype solution piloted are expected to be significant, including learning from failures. Some percentage of tests—perhaps 25 to 50 %—is expected to result in no improvement, to "fail," but to result in substantial learning nevertheless.
3. The number of people affected by a pilot is usually smaller than the number that would be affected if the prototype solution was implemented. Thus, the awareness of and reaction to a pilot of a prototype solution is often much less.

Implementing a prototype solution:

1. A prototype solution that is implemented is expected to become part of the routine operation of the system under consideration. Therefore, supporting processes to maintain the potential "improved process" will usually need to be designed or redesigned. Supporting processes include feedback and data collection systems, job descriptions, procedures, new employee training, and so on.
2. Because learning can occur anytime action is taken, implementation should be carried out as part of a cycle. However, assuming that piloting has been effective, implementations are expected to result in no improvement. The increased permanence of an "improved process" that is a result of moving from piloting an associated prototype solution to implementation is usually accompanied by increased awareness of and reaction to introduction of the "improved process."
3. Implementation cycles generally require more time than pilot cycles.
4. Normally, the same team that piloted the prototype solution will be involved in its implementation. However, the implementation team often has to be supplemented with others needed to support the more permanent nature of the effort.

It is a common mistake to go straight to implementation and skip piloting. This is the reason why so many implementation efforts in process improvement fail or create innumerable problems. The learning that occurs in pilot cycles is vital to successful implementation. A developed prototype solution can be piloted under a variety of conditions to raise one's degree of belief that improvement can be sustained in the future. Also, the learning from pilot cycles that did not go as planned is very important and differs from the learning from successful cycles. Enterprise businesses cannot afford to learn from failed implementation cycles.

Why is it so hard to implement a potential "improved process"? During the early phases of developing and piloting a prototype solution associated with a potential "improved process," the existing system remains in place. Though these early investigative stages may arouse people's interest, the fact remains that nothing has yet been permanently altered. Once the development and piloting of a prototype solution are finished, it is time to develop an implementation plan on the basis of what was learned. To many people, this implementation would appear to be a matter of simply "installing" what was developed and piloted. If implementation did not involve people, then the physical, emotional, and logical challenges that hinder most planned process improvement efforts might not be an issue. However, most implementation of "improved process" in enterprise businesses have a social component. The social challenges that usually accompany the implementation of an "improved process" (that is, when there is a need for people to change behavior permanently) can surprise the sponsors of the improvement effort.

Implementing an "improved process" can be very challenging to many enterprise businesses, especially when the scope of the "improved process" is broad. Implementation of a simple "improved process" can be made with little interdependence on other processes, people, procedures, or structures. It can be performed little formality, and the steps for implementing are usually readily apparent to the person or persons developing the "improved process." For example, improving procedures for reordering supplies might involve simply posting a reorder list on the bulletin board. To implement a complex "improved process," however, it is generally useful, if not necessary, to develop a formal plan involving procedures and training. These ideas have been found to be important in planning implementation of complex prototype solution associated to "improved process" effectively:

1. Planning implementation as a series of cycles;
2. Planning the provision of support during and after the implementation to ensure that improvement is achieved and maintained;
3. Planning to recognize and address the social aspects of implementing the "improved process."

28.3.5.1 Planning Implementation as a Series of Cycles

For some "improved processes," people do not even have to be told about their implementation. For example, the IT department of an automobile manufacturing plant added additional memory chips to all of the computers in the office over the weekend. On Monday, all of the employees noticed that their applications performed much faster. This change resulted in an increase in productivity of the office.

Depending on the complexity and the risks involved, implementation can be conducted in a number of ways. Three approaches are often considered (Langley et al., 2009), all relying on the use of the PDSA Cycle:

1. *The "Just do it!" or "Cold Turkey" approach*. Many times, implementing a simple change is a matter of doing it—for example, following a flow diagram for a new process. After a successful pilot on a relatively low-risk "improved process," implementation can often be accomplished by running one more cycle to ensure that the predicted results are achieved and the changes are made so as to be irreversible. The effect of the "improved process" on the people involved should also be considered. If unforeseen negative consequences occur, the "Just do it!" approach will maximize their negative impact. If the change is complex and the system is large, one of two types of a phased—in approach should be considered; the parallel approach or the sequential approach.
2. *The parallel approach, which plans implementation of the "improved process" while the old "process to be improved" is still in place*. Sometimes implementation of an "improved process" must be phased in by operating it parallel to the existing "process to be improved." Business cannot just stop during implementation. Implementation of an "improved process" must be accomplished while the business is running so that customers will be satisfied while the "improved process" is being implemented. Implementing complex "improved process" while trying to satisfy normal business demands has been compared to changing the fan belt on a car while the motor is running. Planning and phasing in the implementation of an "improved process" parallel with the existing "process to be improved" should reduce some of the risks. This type of implementation will take a bit longer than the "Just do it!" approach, but it is usually less risky. If the implementation of an "improved process" is planned and implemented properly, then the implementation will produce the expected results.
3. *The sequential approach by time*. Often the implementation of an "improved process" comprises multiple components. The third approach is to plan implementation of the components of an "improved process" sequentially over time. For example, if a medical practice was implementing a system to improve access to primary care (based on the results of several months of piloting, of course), they might plan to first reduce the appointment types and then work down the backlog, before opening up the scheduling to all same—day appointments. These components could be implemented sequentially. After the first few cycles, all the components of the "improved process" may not be implemented, but there is no risk of 100 % failure. When determining whether to use a sequential approach, the project team should consider the following:
   - *Identify people and circumstances that will adopt the "improved process."* What strategy will best use the skills and capabilities of the people involved, consider the environment (other improvement efforts that are going on and the will and support that exists for the "improved process"), and minimize geographical issues?
   - *The impact*. Will a sequential approach result in improvements early in the implementation process?
   - *The potential learning*. Will a sequential approach permit learning that can then be used in the next phase of implementation?

- *Resources*. Will a sequential approach allow the best scheduling and use of available resources?
- *Interdependence*. A sequential approach should not be used if the "improved process" cannot work without all of its components.

28.3.5.2 Planning Resources for Implementation

Implementation of an "improved process" often requires new forms, training, a piece of equipment, or something else that requires resources to be allocated. Needed resources may not be thought through and included in the implementation plan. This is one of the areas where the shift from piloting a prototype solution to implementing it is not fully appreciated. The piloting is done on a small-scale, where resourcing is not an issue. Therefore, PDSA Cycles to learn about the required resources to maintain the change should be planned into the series of implementation cycles.

Understanding how a proposed "improved process" will be maintained should be part of the implementation plan. This approach should apply to all changes. Even small changes can sometimes cause big effects, some desirable and some quite undesirable. The following general points should be considered in planning the development of a system to maintain an "improved processes" implemented in the enterprise business:

1. A process should be developed to capture all important "improved processes" in the enterprise business:
   - In deciding if the "improved process" needs to be documented, the project team should consider if the "improved process" will have a small (localized) or large (interdependent) effect on the system considered.
   - Who has the authority to implement an "improved process"? What will his or her responsibility be in seeing that the "improved process" is properly implemented and documented? Who will be the process owner?
2. How will the implementation of the "improved process" be communicated to those affected inside or outside the defined system?
   - How will improvements and learning be shared with interested people in other departments, divisions of the same company, suppliers, and customers?
   - Will training and education be required to implement the "improved process"?
3. What will the process be for updating flow diagrams, best practices, measurements, and other important process, product or service information?
   - Information technology offers an opportunity for real-time updates to documentation of procedures to all parts of the organization.

28.3.5.3 Planning the Social Aspects of Implementation

Dealing with implementation of "improved process" in many enterprise businesses has become an everyday challenge. There are constantly new methods, tools, products, and sometimes a meeting of very different cultures as our world becomes smaller. How can the enterprise business deals with these changes successfully?

Not too many years ago, it was common for people to raise an issue about the monotony and sameness of the work world. Many people considered it their responsibility to maintain the status quo. Now more and more thoughtful people are agreeing that managing the implementation of "improved processes" is one of our more important tasks. Despite this, there remains much that is misunderstood about how people and enterprise businesses undergo implementation of "improved processes."

After the project team has developed and piloted a prototype solution that it is convinced will lead to improvement, it will expect people to accept the implementation of the "improved process." It is only natural, however, that people will seek to maintain control of their environment. Some sort of reaction should be expected when implementation of an "improved process" is announced. This is the stage at which the project team and sponsor must explain the why and how of the "improved process."

Properly planning to address concerns and questions helps people commit to the change. The behavior of the people affected by the "improved process" might range from open resistance to commitment, depending on how implementation of the "improved process" is communicated and what the circumstances are surrounding it (that is, the degree and relevance of the "improved process," the current situation in the enterprise business, the credibility of management, how "improved processes" were handled in the past, the style of leadership, and the enterprise business' culture). A number of behaviors may be observed:

1. *Resistance*: responding with emotions or behaviors meant to impede an "improved processes" that is perceived as threatening;
2. *Apathy*: feeling or showing little or no interest in the "improved processes";
3. *Compliance*: publicly acting in accord while privately disagreeing with the "improved processes";
4. *Conformance*: changing behavior as a result of real or imagined group pressure;
5. *Commitment*: becoming bound emotionally or intellectually to the "improved processes."

People have a need to understand the physical implications of the "improved processes" ("How much smaller will my new office be?"), the logical implications ("Why is this "improved processes" necessary?"), and the emotional aspect of implementation of the "improved processes" ("How do I feel about this "improved processes"?"). The project team and sponsor of the process improvement project should not view people's initial reactions as negative resistance; however, if these reactions are not properly dealt with, they can develop into full-blown resistance.

Monitor and Control Execution 29

A key part of effective project management is to skillfully manage the actual execution of the project and ensure that it stays on track according to the plan. While the project team is physically constructing each deliverable, the project manager must undertake a series of management processes to ensure that the deliverables building activities are progressing as scheduled and that alterations to the original plan are properly implemented.

The "Monitor and Control Execution" is the project management process necessary to execute a set of systematic observation techniques and activities focused on collecting, measuring, and disseminating performance information, and assessing measurement and trends to affect process improvement. Its purpose is to provide an understanding of the project's execution progress so that appropriate corrective actions can be taken when the project's performance deviates significantly from the plan. When actual status deviates significantly from the expected values, corrective actions are taken as appropriate. These actions may require making alteration to the project deliverables, which may include revising the original plan, establishing new agreements, or including additional mitigation activities within the current plan.

Monitoring and controlling execution of the process improvement project is a very intensive process. It takes place throughout the course of the project. Although the project manager is ultimately responsible to a proper execution of the deliverables building activities, he/she depends on the "eyes and ears" of the project team to make sure that information is captured and acted on. The "Monitor and Control Execution" project management process interacts with every single project management monitoring and control process described in the PDSA Plan process group; namely:

1. Schedule Control Plan;
2. Quality Control Plan;
3. Control Spending Plan;
4. Control Contract Performance;
5. Monitor and Control Risk;

A. van Aartsengel and S. Kurtoglu, *Handbook on Continuous Improvement Transformation*, DOI 10.1007/978-3-642-35901-9_29, © Springer-Verlag Berlin Heidelberg 2013

The "Monitor and Control Execution" project management process also builds on the:

1. Project Management Plan
2. Performance Reports
3. Enterprise Environmental Factors
4. Organizational Process Assets

29.1 Perform Time Management

This is the project management process focused on collecting, measuring, and disseminating schedule performance information required to execute the defined schedule control plan and by which time spent by staff undertaking project tasks to build the required deliverables is recorded against the project. Recording the actual time spent by staff on a project has various purposes. It is used to:

1. Calculate the total time spent undertaking each task as well as the total staff cost of undertaking each task in the project;
2. Enable the project manager to control the level of resource allocated to each task;
3. Identify the percentage of each task completed as well as the amount of outstanding work required to complete each task in its entirety.

The "Perform Time Management" process is undertaken through the completion and approval of timesheets. A timesheet is a document which records an allocation of time against a set of project activities listed on the project plan. Timesheets are typically completed weekly, by all members of the project. This includes project staff, contractors and often suppliers. If timesheets are not recorded, then it may be difficult to accurately assess the amount of time spent undertaking project activities, and therefore become impossible to manage the project constraints of time, cost and quality.

Although the "Perform Time Management" process is usually initiated after the project plan has been formally documented and the project is under way (in other words, during the execution phase of the project), timesheets may be completed at any phase of the project if requested by the project manager. For instance, it may be necessary to record timesheets throughout the entire project to ensure that the full costs of the project are captured.

The following are sub-processes used to document, approve and register timesheets within the project.

1. Document Timesheet
2. Approve Timesheet
3. Register Timesheet

Document Timesheet—This process involves the capture of information related to the time spent undertaking each task on the project. Time spent undertaking each task must be recorded for the entire duration of the completion of the task. Time should be recorded against all project tasks for the entire project execution phase.

From the moment time is spent undertaking a project task, it should be recorded using a timesheet. Timesheets exist in various forms, including paper-based, spreadsheet and software-based formats. The most accurate method of capturing timesheet information is to request that all project staff record time in timesheets as they undertake each task, as opposed to waiting till the end of the reporting period before capturing the information.

Approve timesheet—Once documented, timesheets should be submitted by each member of the project team to the project manager, for approval on a regular (for example, weekly) basis. Following the receipt of a timesheet, the project manager will:

1. Confirm that the tasks undertaken were valid tasks as listed in the project plan;
2. Confirm that the staff member was in fact a resource allocated to complete the task;
3. Decide if the outcome of the time spent is reasonable.

Based on the above information, the project manager will approve the timesheet, request further information from the staff member regarding the time spent, or decline the timesheet and raise a staff issue.

Register Timesheet—The details of all approved timesheets are formally recorded in a timesheet register, enabling:

1. The project plan to be updated with a summary of the time recorded against each task;
2. The cost of each staff member to be calculated and monitored throughout the project;
3. The identification of overtime for the project.

On a regular basis, summarized timesheet information should be extracted from the timesheet register and entered into the project plan. This enables the project manager to:

1. Produce a view of the overall progress of the project to date;
2. Forecast task slippage (that is, identify tasks that might not be completed by the due date);
3. Identify any exceptions (for example, instances where tasks have been completed using more time than had been allocated).

The project manager should then take action based on the extent of the deviation from plan and as a result of execution of the project schedule control plan. Examples of actions taken include:

1. Changing the individual amount of resource allocated to the task;
2. Allocating additional funds to complete the task;
3. Requesting assistance from an external supplier to complete the task;
4. Raising a project issue for action by the project board/sponsor.

Once tasks have been completed, they are marked in the timesheet register and project plan as fully or 100 % completed. After a task has been marked as fully or 100 % completed, no further time can be allocated against it for the duration of the project.

29.2 Perform Resource Management

This is the project management process focused on collecting, measuring, and disseminating resource utilization and performance information required to execute the defined resource management plan. It also tracks team member performance, providing feedback, resolving issues, assessing and improving the competencies and interaction of team members, and coordinating changes to enhance project performance. Objectives include:

1. Improve skills of team members in order to increase their ability to complete project activities;
2. Improve feelings of trust and cohesiveness among team members in order to raise productivity through greater teamwork.

As team development efforts such as training, team building, and co-location are assessed and improved, the project team makes informal or formal assessments of the project team's effectiveness. Effective team development strategies and activities are expected to increase the team's performance, which increases the likelihood of meeting project objectives. The evaluation of the team's effectiveness can include indicators such as:

1. Improvements in skills that allow an individual to perform assigned activities more effectively;
2. Improvements in competencies and sentiments that help the team perform better as a team;
3. Reduced staff turnover rate.

If project team members lack necessary management or technical skills, such skills can be developed as part of the project work or through training. Scheduled training takes place as stated in the staffing management plan. Training includes all activities designed to enhance the competencies of the project team members. Training can be formal or informal. Examples of training methods include classroom, online, computer-based, on-the-job training from another project team member, mentoring, and coaching. Unplanned training takes place as a result of observation, conversation, and project performance appraisals conducted during the controlling process of managing the project team.

Observation and conversation are used to stay in touch with the work and attitudes of project team members. The project team monitors indicators such as progress toward project deliverables, accomplishments that are a source of pride for team members, and interpersonal issues.

Objectives for conducting performance appraisals during the course of a project can include re-clarification of roles and responsibilities, structured time to ensure team members receive positive feedback in what might otherwise be a hectic environment, discovery of unknown or unresolved issues, development of individual training plans, and the establishment of specific goals for future time periods. The need for formal or informal project performance appraisals depends on the length of the project, complexity of the project, organizational policy, labor contract requirements, and the amount and quality of regular communication.

29.3 Perform Quality Management

This is the project management process for performing a set of systematic observation techniques and activities focused on outcomes of the project (i.e., project deliverables and project management processes used to produce the outcomes), to monitor and record results of executing the quality assurance plan in order to:

1. Assess performance of the "process improvement" project and "process to be improved" outcomes; and
2. Recommend necessary alterations to the project objectives and/or "process to be improved" goals.

This project management process focused on collecting, measuring, and disseminating performance information on the quality of the deliverables and management processes required to build, assured and control the quality of the produced deliverables. It process involves undertaking a variety of reviews to assess and improve the level of quality of project deliverables and processes. More specifically, performing the quality management process involves:

1. Listing the quality targets to be achieved from the quality assurance plan;
2. Identifying the types of quality data collection techniques to be undertaken;
3. Implementing quality assurance and quality control techniques;
4. Taking action to enhance the level of deliverable and process quality;
5. Reporting the level of quality attained.

The quality management process is performed by the Quality manager and the Quality reviewer during the execution phases ("PDSA Do" and "PDSA Study") of the project. Although quality assurance methods are initiated prior to this phase, quality control techniques are implemented during the actual construction of each physical deliverable. Without a formal quality management process in place, the basic premise of delivering the project to meet "time, cost and quality" targets may be compromised. The quality management process is terminated only when all of the deliverables and management processes have been completed.

At this phase of the project, the quality manager ensures that the project produces a set of deliverables which attain a specified level of quality as agreed with the customer and reported in the quality assurance plan. The quality manager is responsible for:

1. Reviewing the quality of deliverables produced and management processes undertaken;
2. Ensuring that comprehensive quality targets reported in the quality assurance plan are met for each deliverable;
3. Implementing quality assurance methods to assure the quality of deliverables produced by the project;
4. Implementing quality control techniques to control the quality of the deliverables currently being produced by the project;
5. Recording the level of quality achieved in the quality register;
6. Identifying quality deviations and improvement actions for implementation;
7. Reporting the quality status to the project manager.

With a clear understanding of the quality targets to be achieved, it is time to execute quality assurance and quality control techniques to assure and control the level of quality of each deliverable constructed. Figure 13.3 describes the processes and procedures to be undertaken to assure and control the quality of deliverables and processes within the project.

29.4 Perform Cost Management

This is the project management process for performing a set of systematic observation techniques and activities, and by which costs/expenses incurred on the project, are formally identified, approved and paid and recorded in order to:

1. Assess cost performance of the "process improvement" project; and
2. Recommend necessary alterations to the project objectives and/or "process to be improved" goals.

A generic form of the "Control Cost Management" process is shown in Fig. 16.1. The purpose of this project management process is to accurately record the actual costs/expenses which accumulate during the project life cycle. Cost/expenses incurred are formally documented and recorded through the completion and approval of expense forms. An expense form is a document that is completed by a team member to request the payment of an expense which has already been incurred, or is about to be incurred, on the project. A single expense form may be completed for multiple expenses in the project. Regardless of the number of expenses incurred, payment should not be made to the payee until a completed expense form has been approved by the project manager. Each expense form must specify, but is not limited to, the following items:

1. A detailed description of the expense;
2. The amount of the expense claimed;
3. The payee to whom payment should be made;
4. The invoice number related to the expense (if applicable);
5. The date on which the expense occurred;
6. The activity and tasks listed in the project plan against which the expense occurred;
7. The type of expense (for example labor, equipment, materials, administration);

Expense forms should be completed weekly and provided to the project manager for approval for all project expenses, including contractor, supplier, equipment, materials and administration expenses. Upon approval, the expense information should be recorded into an expense register to enable the project manager to track and record the physical costs of the project. Summarized expense information should also be entered into the project plan to document and record the actual spend against the planned spend. Although expense forms are typically completed during this "PDSA Do" phase of the project, it may be requested that they be completed during any PDSA project phase to ensure that the full costs of the project are captured.

The following are sub-processes used to document, approve and register expense forms within the project.

1. Document expense
2. Approve expense
3. Register expense

Document Expense—This first step involves the capture of information relating to an expense incurred on the project. Expenses are incurred on the project when undertaking project activities and tasks. It is therefore important to identify the project activity and task related to each expense incurred so that the total cost of undertaking project activities and tasks on the project can be calculated. Expense forms should be completed regularly by:

1. Members of the project who have had to incur expenses;
2. Project administrators, on behalf of external suppliers who have issued invoices for goods and services rendered;
3. Contractors allocated to the project for services provided.

Approve Expense—Completed expense forms should be forwarded for review and approval to the project manager who will consider whether:

1. The tasks for which the expense occurred are valid, as listed in the project plan;
2. The expense was originally budgeted, as defined in the project plan;
3. Any unbudgeted expenditure is fair, reasonable and affordable.

The project manager may have authority to approve only budgeted expenditure. Unbudgeted expenditure over a certain limit may require the approval of the project board or sponsor. The project manager may then either:

1. Approve the expense and forward it to the project administrator for payment;
3. Request further information from the person submitting the form;
3. Decline the expense and raise an issue with the person submitting the form.

Following formal approval of the expense by the project manager, payment will be scheduled. It is typical to pay expenses in batches to reduce the administrative workload in making expense payments and manage project cash-flow more effectively.

Register Expense—After the payment has been scheduled, the project administrator should update the expense register to ensure that an accurate record of the approval and payment is kept. Although the register must be updated after the expense has been approved, the register should be updated throughout the process to ensure that the project manager is kept informed of the expense status at all phases in the expense approval cycle. The expense register records the full details of all expense forms submitted, thereby enabling:

1. The project plan to be updated with the expenses recorded against each task;
2. The cost of each staff member to be calculated and monitored throughout the project;
3. The project manager to identify the actual versus budgeted expenditure throughout the project.

On a regular basis, the project administrator should update the project plan with the total expenditure against each task, as listed within the expense register. This enables the project administrator to produce a view of the overall cost of the project to date, and identify any exceptions (such as instances where the actual expenditure exceeds the planned expenditure). The project administrator then provides the project manager with a copy of the updated project plan and identifies any expenditure deviations noted to date. It is then up to the project manager to take action, based on the extent of the deviation from plan. Examples of actions taken could include:

1. Changing the individual/amount of resource allocated to the task;
2. Allocating additional funds to complete the task;
3. Requesting assistance from an external supplier to complete the task;
4. Raising a project issue for action by the project board/sponsor.

Once each task is completed, it is marked as fully complete in the project plan and no further expenditure may be allocated to the task for the duration of the project.

29.5 Perform Procurement Management

As indicated during the development of the project procurement plan, project procurement is the process of obtaining or procuring materials, products, services, or results needed from outside the project boundaries to perform the project work. It commonly involves purchase planning, standards determination, specifications development, supplier research and selection, value analysis, financing, price negotiation, making the purchase, supply contract administration, inventory control and stores, and disposals and other related functions.

"Perform Procurement Management" is the project management process for performing a set of systematic observation techniques and activities focused on close monitoring and control of the in-process contracts performance in order to:

1. Ensure compliance and fulfillment of the contract conditions; and
2. Recommend necessary alterations to the contract objectives/goals.

It involves contract administration performed by the project manager (or a qualified designee) after a contract has been awarded to determine how well the portion of the project that is included within the related contract is been implemented and how well the supplier(s) performs to meet the requirements of the contract. It encompasses all dealings between the project manager and the supplier(s) from the time the contract is awarded until the work has been completed and accepted or the contract terminated, payment has been made, and disputes have been resolved.

In performing this process, the focus is on obtaining procurement items, of requisite quality, on time, and within budget. While the legal requirements of the contract are determinative of the proper course of action of the project manager in administering a contract, the exercise of skill and judgment is often required in order to protect effectively the interests of both the enterprise business and the supplier(s).

How well the project manager administers in-process contracts and discusses with suppliers their current performance determines to a large extent how well the portion of the project that is included within the related contract will be implemented and provide value to the enterprise business. By increasing attention to supplier performance on in-process contracts, project managers are reaping a key benefit: better current performance because of the active dialog between the supplier and the project manager.

Monitoring should be commensurate with the criticality of the service or task and the resources available to accomplish the monitoring. A generic form of the "Contract Performance Control" process is shown in Fig. 17.2.

29.6 Perform Communication Management

As indicated during the development of the project procurement plan, communication is the activity of conveying information. It requires a sender, a message, and an intended recipient. It also requires that the communicating parties share an area of communicative commonality. The communication process is complete once the receiver has understood the message of the sender. Feedback is an essential part of communication.

Project communication is the exchange of project-specific information with the emphasis on creating understanding between the sender and the receiver. Effective communication in a "process improvement" project is one of the most important factors contributing to the success of the project. Here, the project team must provide timely and accurate information to all stakeholders. During the course of a project, members of the project team prepare information in a variety of ways to meet the needs of project stakeholders. These stakeholders in return, provide feedback to the project team members.

Project communication includes general communication between team members but is more encompassing. It utilizes the Work Breakdown Structure (WBS) as framework, it is customer focused, it is limited in time, it is product focused with the end in mind, and it involves all levels of the enterprise business. From the process improvement perspective, for each WBS element, there are:

1. Suppliers who provide inputs needed for the WBS element
2. Task managers who are responsible for delivering the WBS element
3. Customers who receive the products of the WBS element

Suppliers must communicate with the task managers, and the task managers must communicate with suppliers and customers. The supplier is often the task manager for an earlier deliverable in the project lifecycle; the customer may be a task manager for a later deliverable. Good project communication practice includes notifying the next task manager in the project delivery chain about when to expect a deliverable. The supplier and customer may also be the functional manager. By considering the process associated with the WBS element, a very effective diagram that depicts this flow of information is the Suppliers-Inputs-Process-Outputs-Customers (S.I.P.O.C.) diagram, illustrated in Table 18.1. Working from the right letter of its acronym, the S.I.P.O.C. identifies the customers, the outcomes or the process, the inputs to the process and the suppliers.

"Perform Communications Management" is the project management knowledge area that performs the processes required to ensure timely and appropriate generation, collection, distribution, storage, retrieval, and ultimate disposition of project information as defined in the project communication schedule. The project communication schedule developed during the planning phase describes each communication event, including its purpose, method and frequency as indicated in Tables 18.2, 18.3, and 18.4. It provides the critical links among people and information that are necessary for successful project communications. Clear project communication therefore ensures that the correct stakeholders have the right information, at the right time, with which to make well-informed decisions.

Various types of formal communication may be undertaken in a project as indicated in Table 18.4. Examples are releasing regular project status or performance reports, communicating project risks, issues and changes, and summarizing project information in weekly newsletters. Regardless of the type of communication to be undertaken, the method for undertaking the communication will always remain the same:

1. Identify the message content, audience, timing and format.
2. Create the message to be sent.
3. Review the message prior to distribution.
4. Communicate the message to the recipients.

These four processes should be applied to any type of formal communication on the project, including the distribution of:

1. Regular project status reports;
2. Results of phase review meetings;
3. Quality review reports documented;
4. Minutes of all project team meetings;
5. Newsletters and other general communication items.

Although the communications process is typically undertaken after the communications plan has been documented, communications will take place during all phases of the project. This process therefore applies to all formal communications undertaken during the life of the project. Without a formal communications management process in place, it will be difficult to ensure that project stakeholders receive the right information at the right time.

29.7 Perform Risk Management

"Perform Risk Management" is the project management process by which risks to the project are formally identified, quantified and managed during the "PDSA Do" phase of the project. The process entails completing a number of actions to reduce the likelihood of occurrence and the severity of impact of each risk. The risk management process developed during the project planning phase is used to ensure that every risk is formally identified, quantified, monitored, avoided, transferred and/or mitigated.

"Perform Risk Management" has well-established stages that make up the risk management process, as illustrated in Fig. 19.2, although it is presented in a number of different ways and often uses differing terminologies. These stages build into valuable risk management activities, each of which makes an important contribution. In this handbook, the risk management process is taken as a narrow set of activities; the constituent project management processes of which include the following:

1. Identify Project Risks
2. Perform Risk Assessment
3. Develop Risk Response Planning
4. Monitor and Control Risk

These four constituent processes, described in a previous section of the project planning phase, interact with each other and with the project management processes in the PDSA "Process Groups." Each aspect of executing any of these four constituent processes can involve effort from one or more persons, based on the needs of the project. Each aspect occurs at least once in every "process improvement" project and occurs in one or more project phases.

Although the risk management process is undertaken during the "PDSA Do" phase of the project, risks may be identified at any stage of the project life cycle. In theory, any risk identified during the life of the project will need to be formally managed as part of the risk management process. Without a risk management process in place, unforeseen risks may impact the ability of the project to meet its objectives.

29.8 Perform Deliverable Alteration Management

This is the project management process by which alterations or changes to the project scope, deliverables, timescales or resources are identified, evaluated and approved prior to implementation. The process entails completing a variety of control procedures to ensure that if implemented, the alteration or change will cause minimal impact to the project.

This process is undertaken during the "PDSA Do" and "PDSA Study" phases of the "process improvement" project, once the project has been formally defined and planned. In theory, any change to the project during the execution and study phases will need to be formally managed as part of the deliverable alteration or change process. Without a formal deliverable alteration process in place, the ability of the project manager to effectively manage the scope of the project may be compromised. The deliverable alteration management process is terminated only when the execution and study phases of the project are completed.

Figure 29.1 shows the process to be undertaken to initiate, implement and review alterations or changes within the project. Where applicable, alteration roles have also been identified.

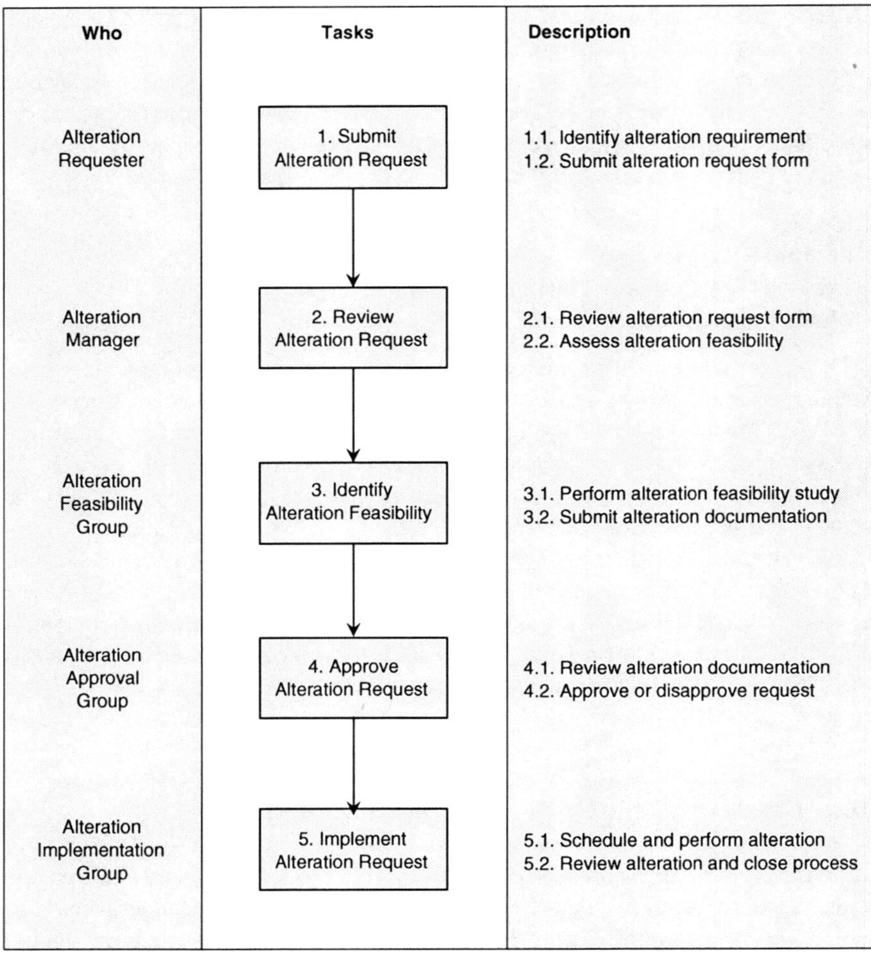

Fig. 29.1 Deliverable alteration management process

29.8.1 Submit Alteration Request

To initiate an alteration to the project deliverable(s), the project manager should first allow any member of the project team to submit a request for an alteration to the project deliverable. The person raising the alteration is called the "alteration requester." The alteration requester initially recognizes a need for alteration to the project deliverables and formally communicates this requirement to the alteration manager. The alteration requester is responsible for:

1. Identifying the need to make an alteration to the project deliverables;
2. Documenting the need for alteration by completing an alteration request form (ARF);
3. Submitting the alteration request form (ARF) to the alteration manager for review.

The alteration requester will document the requirement for alteration to the project deliverable by completing an alteration request form (ARF), summarizing the alteration description, benefits, costs, impact and approvals required.

29.8.2 Review Alteration Request

The alteration manager reviews the request form and determines whether or not a feasibility study is required for the alteration approval group to assess the full impact of the alteration to the project deliverables. The decision will be based on the size and complexity of the proposed alteration. The alteration manager receives, logs, monitors and controls the progress of all alterations within a project. The alteration manager will record the alteration request form (ARF) details in the alteration or change register. Ultimately, the alteration manager is responsible for:

1. Receiving all ARFs and logging them in the alteration or change register;
2. Categorizing and prioritizing all alteration requests;
3. Reviewing all ARFs to determine whether additional information is required;
4. Determining whether or not a formal alteration feasibility study is required;
5. Forwarding the ARF to the alteration approval group for approval;
6. Escalating all ARF issues and risks to the alteration approval group;
7. Reporting and communicating all decisions made by the alteration approval group.

29.8.3 Identify Alteration Feasibility

If deemed necessary, an alteration feasibility study is completed to determine the extent to which the requested alteration to the project deliverables is actually feasible. The alteration feasibility group complete feasibility studies for ARFs issued by the alteration manager. The alteration feasibility group is responsible for:

1. Undertaking investigation to determine the likely options for alterations, costs, benefits and impacts of requested alterations;
2. Documenting all findings within a feasibility study report;
3. Forwarding the feasibility study report to the alteration manager for alteration approval group submission.

The alteration feasibility study will define in detail the alteration requirements, options, costs, benefits, risks, issues, impact, recommendations and plan. All alteration documentation is then collated by the alteration manager and submitted to the alteration approval group for final review. This includes the original alteration request form (ARF), the approved alteration feasibility study report and any supporting documentation.

29.8.4 Approve Alteration Request

A formal review of the alteration request form (ARF) is undertaken by the alteration approval group. The alteration approval group is the principal authority for all

ARFs forwarded by the alteration manager. The alteration approval group is responsible for:

1. Reviewing all ARFs forwarded by the alteration manager;
2. Considering all supporting documentation;
3. Approving or rejecting each ARF based on its relevant merits;
4. Resolving alteration conflict, where two or more alterations overlap;
5. Identifying the implementation timetable for approved alterations.

This group will either, reject the proposed alteration to the project deliverables, request more information related to the alteration, approve the alteration as requested, or approve the alteration subject to specified conditions. Their decision will be based on the level of risk and impact to the project successful achievement resulting from both implementing and not implementing the requested alteration.

29.8.5 Implement Alteration Request

Approved alterations are then implemented. This involves:

1. Identifying a date for implementation of the alteration;
2. Implementing the requested alteration;
3. Reviewing and communicating the success of the implementation;
4. Recording all alteration actions in the change register.

The alteration implementation group will schedule and implement all approved alterations. It is responsible for:

1. Scheduling all alterations within the timeframes provided by the alteration approval group;
2. Testing all alterations to the project deliverables, prior to implementation;
3. Implementing all alterations within the project;
4. Reviewing the success of each alteration, following implementation;
5. Requesting that the alteration manager close the requested alteration in the alteration or change register.

29.9 Conduct the Project Retrospective

This is the reflection process performed at the end of each significant milestone of the "PDSA Study" project phase and from which the project team reassembles to look back on what results were actually delivered at the milestone and to what extent the team has met the expectations for the considered milestone time period. The reflection process integrates or links thought and task execution with reflection. As indicated already, it involves thinking about and critically analyzing one's actions with the goal of improving one's professional practice (Schön, The reflective practitioner, 1983; Schön, 1987). Here, engaging in reflective practice requires individuals to assume the perspective of an external observer in order to identify the assumptions and feelings underlying their practice and then to speculate about how these assumptions and feelings have affected the achievement of a significant milestone objective.

When the project team members reflect at the end of a significant milestone, they question the assumptions behind their tacit knowledge revealed in the way they carried out tasks and approached problems and think critically about the thoughts that got them into this fix or this opportunity. The project team members may, in the process, restructure strategies of action, understandings of phenomena, or ways of framing problems. Much of a project team's work is focused on problems that occurred while studying deliverables.

The reflection process should begin when the application of the project team's know-how to build deliverables does not produce the expected milestone results, the activities it conducted and the project management processes used throughout the milestone time period in the "PDSA Do" project phase have failed to meet expectations. As mentioned in a previous section, the project team may decide to ignore the failure or it may respond to it by reflecting in one of two ways:

1. It may reflect "on action" by allowing its members to step away (i.e. assume the perspective of external observers) from the planning process and thinking back on their experience to understand how part of their tacit knowledge, which was revealed in the way they approach problems and carry out tasks required to reach the milestone considered, contributed to an unexpected outcome.
2. Alternatively, the project team may "reflect in the midst of the planning process without interrupting it."

Conducting a project retrospective at the end of each significant milestone during the project's life cycle is the primary means for facilitating learning and continuous innovation in an enterprise business. To be effective, a project retrospective should be facilitated by an experienced, trained, objective facilitator from outside the project team who helps draw people out to share their perspectives, promote effective learning and reflection, and create a positive context for "process improvement" rather than one of finger- pointing, defensiveness, avoidance, or blame.

After the retrospective session, the project manager works with the facilitator and the project management office (PMO) leader to document the results and communicate them to team members, sponsors, and key stakeholders. "Report-out" meetings with senior managers and the PMO may be useful for generating support for the team's improvement actions.

29.10 Perform Phase Review

This is the phase review process performed at the end of the "PDSA Do" phase to ensure that the project has achieved its stated objectives as planned by refining previously provided answers to the three fundamental questions, which form the basis and the preliminary step of the PDSA model:

1. What is intended to be realized or accomplished by the "process improvement" project?
2. How will the realized or accomplished outcome of the "process improvement" project be recognized as is an improvement?
3. What alterations to the system affected by the "process to be improved" can be made based on the realized or accomplished outcome of the "process improvement" project?

A phase review form is completed to formally request approval to proceed to the next phase of a project. The phase review form should describe the status of the:

1. Overall project;
2. Project schedule based on the project plan;
3. Project expenses based on the financial plan;
4. Project staffing based on the resource plan;
5. Project deliverables based on the quality plan;
6. Project risks based on the risk register;
7. Project issues based on the issues register.

The review form should be completed by the project manager and approved by the project sponsor. To obtain approval, the project manager will usually present the current status of the project to the project board for consideration. The project board (chaired by the project sponsor) may decide to cancel the project, undertake further work within the existing project phase or grant approval to begin the next phase of the project. A sample phase review form for the "PDSA Do" project phase is shown in Table 29.1.

29.11 Identify and Document Lessons Learned

As indicated already, the last issue to be discussed at the end of a phase review process performed is something we know we should do, but most project managers rarely ever take the time. That is the final position of the project manager, and the project team, describing for the benefit of "future generations" as well as of next phases of the project just what went well, and what could have been handled perhaps better on the project "PDSA Do" phase. What could have been done better, and should be done differently on the next similar project "PDSA Do" project phase?

A lesson learned session focuses on identifying ways of learning that have merit (quality), worth (value), or significance (importance) for the next phases of the "process improvement" project or for future projects within the enterprise business. During the "PDSA Do" project phase, the project team and key stakeholders should identify lessons learned concerning the project management element in which problems arose, how they arose, which positive or negative development was encountered, and what concrete, practical solutions or recommendations was used based on this experience.

The project manager must ask team members, stakeholders, and the project sponsor to help compile the lessons learned document. He/she should them what went well during the development of the project planning and what could have gone better. The following is the information you should include in the lessons learned document:

Table 29.1 Phase review form for the "PDSA Do" project phase

PROJECT DETAILS			
Project name: Project manager: Project sponsor:	Report prepared by: Report preparation date: Reporting period:		
Project description: [Summarize the overall project achievements, risks and issues experienced to date.]			
OVERALL STATUS			
Overall status: [Description] Project schedule: [Description] Project expenses: [Description] Project deliverables: [Description] Project risks: [Description] Project issues: [Description] Project changes: [Description]			
REVIEW DETAILS			
Review category	**Review question**	**Answer**	**Variance**
Schedule	Was the phase completed to schedule?	[Y/N]	
Expenses	Was the phase completed within budgeted cost?	[Y/N]	
Deliverables:	Deliverables:		
Deliverable #1	Was the deliverable #1 completed and approved?	[Y/N]	
Deliverable #2	Was the deliverable #2 completed and approved?	[Y/N]	
…	…	…	
Deliverable #n	Was the deliverable #n completed and approved?	[Y/N]	
Risks	Are there any outstanding project risks?	[Y/N]	
Issues	Are there any outstanding project issues?	[Y/N]	
Alterations/Changes	Are there any outstanding project alterations/changes?	[Y/N]	
APPROVAL DETAILS			
Supporting documentation: [Reference any supporting documentation used to substantiate the review details above.] **Project sponsor** Signature: Date: This project is approved to proceed to the"PDSA Study" phase.			

1. How the project management processes were used throughout the "PDSA Do" project phase and how successful they were in building deliverables and tracking progress.
2. How well the project plan and project schedule reflected the actual work carried out during the "PDSA Do" phase of the project.
3. How well the alteration/change management process worked and what might have worked better.
4. Why corrective actions were taken and whether they were effective.
5. Causes of performance variances and how they could have been avoided.
6. Outcomes of corrective actions.
7. Risk response plans that have been identified and whether they adequately addressed the risk events.
8. Unplanned risk events that occurred at the "PDSA Do" project phase.
9. Mistakes that occurred and how they could have been avoided.
10. Team dynamics, including what could have helped the team perform more efficiently.

As indicated at the closure of the "PDSA Plan" phase, lessons learned document should not be limited to only the items on this list above. Anything that worked well, or did not work well, that will help team members perform their next project better or smooth out problems before they get out of hand should be identified and documented here. Lessons learned should include detailed, specific information about behaviors, attitudes, approaches, forms, resources, or protocols that work to the benefit or detriment of projects. They are crafted in such a way that those who read them will have a clear sense of the context of the lesson learned, how and why it was derived, and how, why, and when it is appropriate for use in other projects. Lessons learned at this stage represent both the mistakes made during the "PDSA Do" project phase and the newer "tricks of the trade" identified during a project "PDSA Do" effort.

The content of a lesson learned report should be provided in context, in detail, and with clarity on where and how it may be implemented effectively. Because lessons learned are often maintained in a corporate database, the lesson learned documentation will frequently include searchable keywords appropriate to the project and the lesson.

The process of identifying and documenting lessons learned at this stage of the project lifecycle is particularly useful for projects that failed to pass the phase review, because there are many things that can learn from projects that fail phase reviews that will help prevent next projects from suffering the same fate.

Recording lessons learned information in the organizational process assets is one critical consideration, but equally important is the establishment of protocols to ensure access to the recorded information on a consistent basis. Lessons learned may be captured and logged in depth, but if they are not accessed by project managers and team members within the enterprise business in the future, they do not serve any real function. Access of recorded lessons learned may be encouraged through creative documentation approaches, physical location (hallways and project war rooms), or by including the mandate to access lessons learned as a key component of the performance criteria for project managers and team members.

Conclusion to "PDSA Do" 30

Throughout the previous chapters related to the "PDSA Do" Process Group, we have illustrated and developed the "PDSA Do" constituent processes needed to build the project deliverables and perform the course of action required to attain the objectives and scope that the project is undertaken to address. The described constituent processes help perform the established project management plan.

We have shown that within the project management framework, performing the defined "process improvement" project plan is indispensable to enhance the chance of achieving the project objectives. Furthermore, we have shown through the project management constituents that effective building of the "process improvement" deliverables within the PDSA framework is based on a foundation of effective analysis of the root of assignable causes of variations, effectively exploring the cause-and-effect relationship between numerous process input variables from which assignable cause of variation originate and the process response variable, effective piloting of prototype solution, and almost everything else done during the execution phase is based on that.

Analysis of the root of assignable causes of variations determines what the constituent processes used during the "PDSA Do" project phase do, and this analysis works through these constituent processes to touch every part of the "process improvement" project. The right sets of identified input variables from which assignable causes of variation originate trigger the right analysis activities—because the collected data associated with these input variables represent factual information from many sources with each having varying levels of completeness and confidence and from which baselines were established during the "PDSA Plan" planning phase.

In order for the full power of analysis of the root of assignable causes of variations, hence piloting of prototype solution, to be realized, there must be an optimal environment for effective identification of assignable causes of variation through interviews with appropriate personnel, collecting physical evidence, conducting other research, such as performing a sequence-of-events analysis, which is needed to provide a clear understanding of the events leading to occurrence of assignable causes, and conducting pilots of prototype solution. Thus, there

A. van Aartsengel and S. Kurtoglu, *Handbook on Continuous Improvement Transformation*,
DOI 10.1007/978-3-642-35901-9_30, © Springer-Verlag Berlin Heidelberg 2013

must be considerable interaction at each use of "PDSA Do" constituent processes leading to new insights about the potential assignable causes of variation and what are the subsequent right analysis decisions, and new knowledge base of improved products or services that occur during the roll-out of the "improved process."

Attaining the optimal environment, as we indicated in the previous chapter, requires a specific and intensive set of actions—a transformation process progressing from improving context of identification of assignable causes of variation, to improving focus, to improving integration, to improving interactivity—the four aspects of paramount importance to making progress on moving the "Process Improvement & Management" initiative from its current maturity stage to "Continuous Improvement" maturity stage as we have indicated already. Within the "Process Improvement and Management" dimension of "Continuous Improvement" transformation, the factors which contribute to transform the interactivity of "Process Improvement and Management" include the following:

1. Frequent Interactivity
2. Effective and robust dialogue
3. Collaborative learning
4. Appropriate use of technology

Performing the "PDSA Do" constituent processes should include highly interactive and iterative (ongoing) discussions, or dialogues, which are also the most important aspects of root-cause analysis. As indicated already, these dialogues should be built on the foundation of a positive context of identification of assignable causes of variation, focus, and integration. As was the case with the "PDSA Plan" constituent processes, effective integration and interactivity of the "PDSA Do" constituent processes will also do more than anything else to break down the silos that are also keeping enterprise businesses from realizing "Process Improvement & Management" transformational potential.

Figure 30.1 shows the minimum activities that are part of the "PDSA Do" project execution phase, in addition to the already listed "PDSA Plan" activities.

It can be noted that in this figure we use the "Analyze" and "Improve" nomenclature of the six-sigma literature for convenience and consistency with existing literature. We have placed "Dialogue," which is what enables this continual reassessment, at the very center of the PDSA Cycle in Fig. 30.1. It is in fact the basic unit of a "process improvement" project work. You cannot plan and execute a "process improvement" project well without robust dialogue with customers and stakeholders. How people involved in a "process improvement" project talk to each other, talk to customers and stakeholders, absolutely determines how well the "process improvement" project will progress towards its objectives.

As indicated already, the word "dialogue" should be understood in the sense of "sharing collective meaning" and strongly differentiated from "discussion." The word "discussion" comes from the same root word as percussion and concussion and has to do with beating one thing against another. The word "communication" is

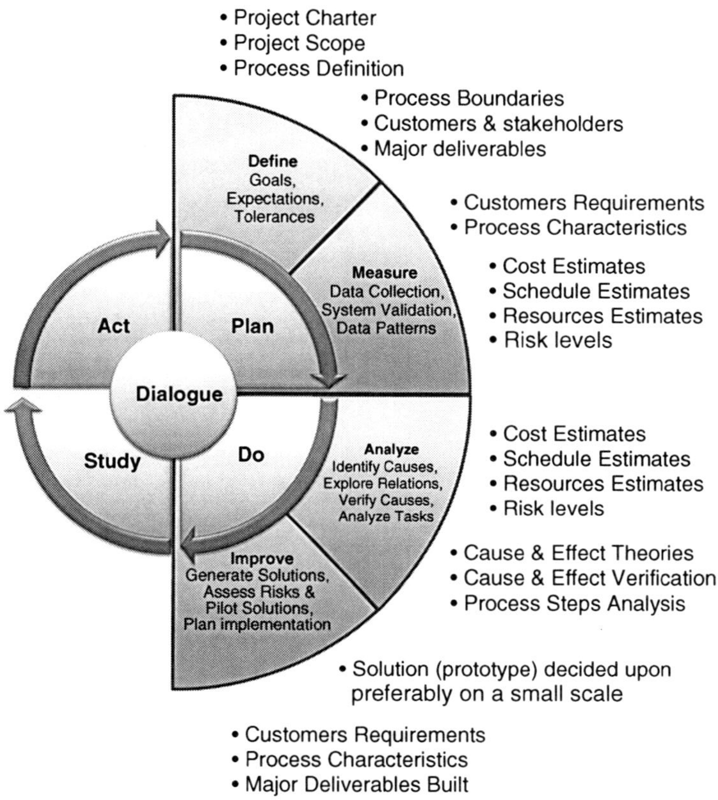

Fig. 30.1 Minimum activities of the "PDSA Do" phase

a more general term meaning "to make something common." So, communication can be done by discussion or dialogue. When information is made common through discussion, it is often two monologues—an attempt to convey your opinion to another person, and nothing more. Very few people are skilled at dialogue, and very few project team members currently have a strong capacity for dialogue.

"PDSA Study" Process Group 31

The purpose of this project phase is to sustain the deliverables built over the long term and build new knowledge through learning from the deliverables built in the "PDSA Do" project phase. It is not enough to determine that development and piloting of a prototype solution associated with an "improved process" resulted in improvement during particular pilots. As the project team builds knowledge about the new "improved process," it will need to predict whether the change introduced by the new "improved process" will result in improvement under the diverse conditions it will face in the future.

Thus, this third step, the "PDSA Study" project phase, brings together the baseline data collected in the "PDSA Plan" project phase and the in-process data resulting from the pilots conducted in the "PDSA Do" project phase. This synthesis is done by comparing the results of the V.O.B, V.O.C., and V.O.P. data analysis to the established baseline data and to the predicted results. If the results of the pilots match the predictions made in the "PDSA Plan" project phase, the project team's degree of belief about their knowledge is increased. If the predictions do not match the data, there is an opportunity to advance their knowledge through understanding why the prediction was not accurate.

This third step is also the phase within which the built deliverables are sustained and presented to the customer for acceptance. To ensure that the customer's requirements are met, the project manager keeps monitoring and controlling the qualification, validation and revalidation of each deliverable by executing a suite of planned management processes. After the deliverables have been physically qualified, validated and accepted by the customer, a phase review is carried out to determine whether the project is complete and ready for closure.

31.1 The "PDSA Study" Constituent Processes

Figure 31.1 shows the activities undertaken during the "PDSA Study" project phase. It illustrates those processes performed to qualify and validate the work defined in the project management plan to accomplish the project's objectives.

A. van Aartsengel and S. Kurtoglu, *Handbook on Continuous Improvement Transformation*, DOI 10.1007/978-3-642-35901-9_31, © Springer-Verlag Berlin Heidelberg 2013

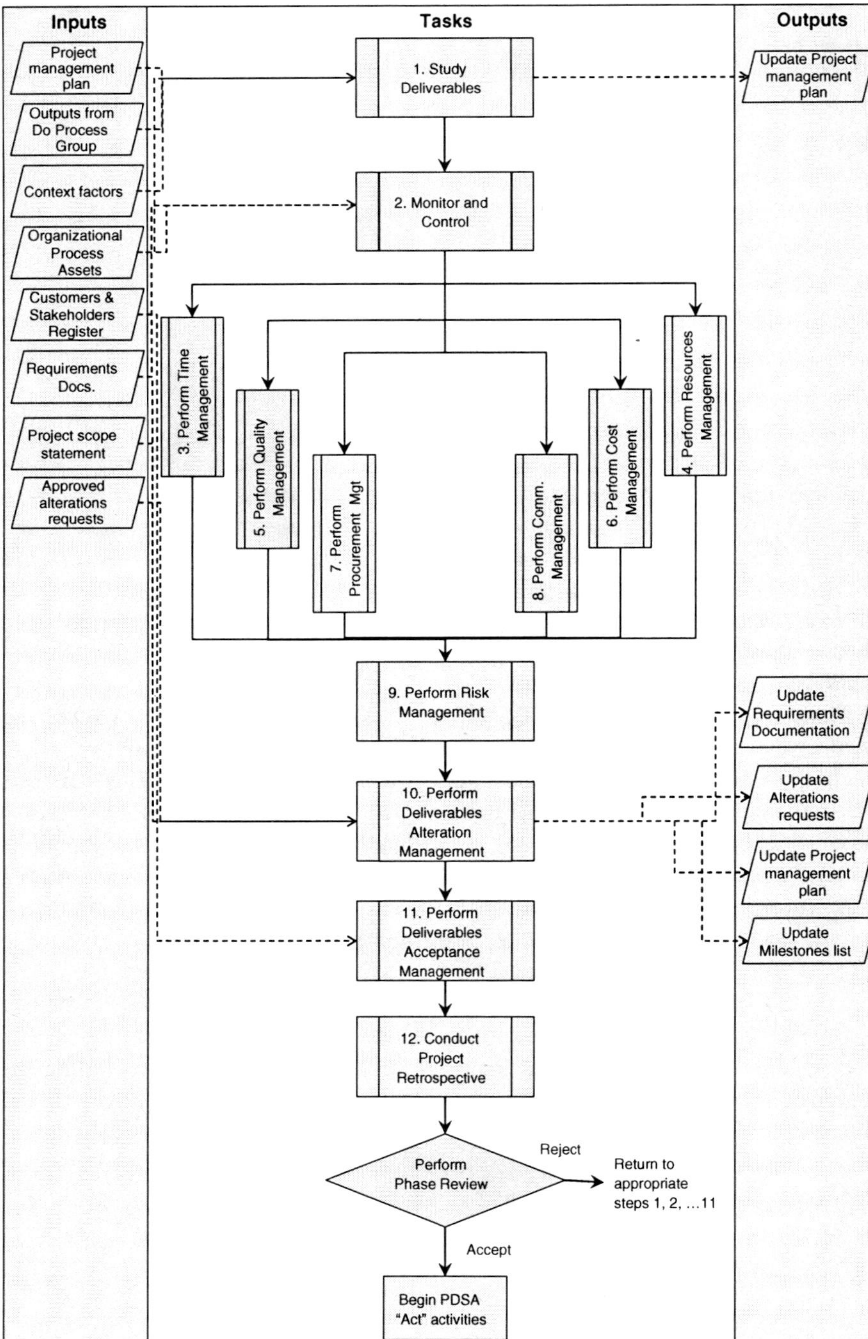

Fig. 31.1 "PDSA Study" process group

It involves coordinating people and resources, as well as integrating and performing the activities of the project in accordance with the project management plan.

The "PDSA Study" Process Group includes the following key processes of the process improvement plan indicated in a previous section:

1. Study Deliverables
2. Monitor and Control Execution
3. Perform Time Management Plan
4. Perform Quality Management Plan
5. Perform Procurement Management Plan
6. Perform Communication Management Plan
7. Perform Cost Management Plan
8. Perform Resources Management Plan
9. Perform Risk Management Plan
10. Perform Deliverables Alteration Management
11. Perform Deliverables Acceptance Management
12. Conduct Project Retrospective
13. Perform "PDSA Study" Phase Review
14. Identify and Document Lessons Learned

To successfully deliver the project on time, within budget and to specification the project manager needs to fully implement each of the activities listed above. Even though the management processes listed may seem obvious, it is extremely important that the project manager implements each process in its entirety and communicates the process clearly to the project team.

While integrating and performing the activities of the project in accordance with the project management plan, deviations from established performance baselines will cause some alterations of the project plan. These alterations can include activity durations, resource productivity and availability and unanticipated risks. Such alterations may or may not affect the project management plan, but can require an analysis. The results of the analysis can trigger an alteration (or a change) request that, if approved, would modify the project management plan and possibly require establishing a new baseline. The vast majority of the project's allocated funds could therefore be expended in performing the "PDSA Do" Process Group processes.

Study Deliverables

32

This chapter is concerned with the project management process necessary to study the project deliverables built over long term. Let recall that implementation of a new "improved process" is about "change in form, quality, or state, over time" from the original "process to be improved." An important consideration here is the notion of "time." As we have indicated in the previous chapter, time is the "ether" of change and we judge that change has occurred against the background of time.

Of course, not all changes result in improvements, thus we use metrics on the background of time for assessing when changes occur, the rate of change, and the extent of change, and also to establish the opposite of change, stability. It is the focus on change and an understanding of the basic principles of the science of improvement outlined in the previous section that leads to efficient and effective improvement efforts. Improvement has meaning only in terms of observation based on given performance measure. Thus, the objectives of the "Study Deliverables" project management process are to collect retrospective data over time, complete the analysis of the data started in the "PDSA Do" project phase, compare data to established baselines, and summarize what was learned.

Retrospective data is the final collected data that occurs at the end of the pilot of the prototype solution deployment. This does not necessarily mean that piloting of the prototype solution is terminated and full implementation is launched, but that it has reached a point at which the prototype solution is sufficiently mature that data can be retrospectively collected. It can also be considered the last in-process data point. The primary purpose of collecting data at this juncture is to make final judgments about the prototype solution, hence the "improved process," including the calculation of Return on Investment (ROI), if desired. However, Retrospective Data is typically too late to inform the most important decisions on the current "improved process"—although these retrospective data can be used to inform decisions about future projects.

The project team should continually collect retrospective data over time, compile, and evaluate feedback and operations, and document the potential "improved process" features and functions needed to achieve the desired results identified in

A. van Aartsengel and S. Kurtoglu, *Handbook on Continuous Improvement Transformation*, DOI 10.1007/978-3-642-35901-9_32, © Springer-Verlag Berlin Heidelberg 2013

the "PDSA Plan" project phase, especially with respect to the critical success factors. From the quality perspective, this includes:

1. *The business needs and expectations (Voice of the Business—V.O.B.).* This is the voice of profit and return on investment. Every "process improvement" project has to enable the enterprise business sustainability and meet the needs of the employees and shareholders.
2. *The customers and the stakeholder's needs and expectations (Voice of the Customer—V.O.C.).* This is the voice calling back at the "improved process" from beyond its outcomes that offer compensation in return for satisfaction of the customers and stakeholders needs and wants. This voice represents the stated and unstated needs, wants, and desires of the customers and stakeholders, referred to as the customers and stakeholders' requirements.
3. *The "improved process" needs and expectations (Voice of the Process—V.O.P.).* The "improved process" must meet the requirements of the customers and stakeholders, and the ability of this process to meet these requirements is called Voice of the Process. It is a construct for examining what the "improved process" is telling about its inputs and outputs and the resources required to transform the inputs into outputs.

The purpose of collecting these data over time is to get sufficient and accurate information to refine and complete improvement of the "improved process" set forth. Most importantly, the purpose is to get accurate and sufficient data to derive complete functional requirements for the "improved process" outcomes. The project's success is directly influenced by the care taken in managing predictions made in the "PDSA Plan" project phase about these V.O.B., V.O.C., and V.O.P. requirements.

The steps undertaken to study each deliverable will vary depending on the type and complexity of the "process improvement" project been undertaken, however its elements can be described here in real details. The key activities required to study each deliverable will be clearly specified within the terms of reference and project plan accordingly. The "Study Deliverables" Process Group includes the following key constituent project management processes:

1. Collect Retrospective Data—V.O.B., V.O.C., & V.O.P.
2. Summarize Overall Data and Display Patterns
3. Analyze Data and Validate Process Performance
4. Develop a Process Control Plan
5. Reinforce a Positive Context of Process Improvement
6. Continuously Monitor New "Improved Process" and Context

32.1 Collect Retrospective Data: V.O.B., V.O.C., & V.O.P.

The first step in studying the project built deliverables is to "Collect Retrospective Data" after the pilots of the developed prototype solution have been concluded. It relates to defining and documenting over the long term, after conclusion of the prototype solution pilots, the "improved process" features and functions needed to achieve the desired results identified in the "PDSA Plan" project phase, especially with respect to the critical success factors.

The project management processes utilized to collect these data have been described during the project scope development in the "PDSA Plan" project phase.

32.2 Summarize Overall Data and Display Patterns

The major purposes of summarizing the collected data and displaying their patterns are to learn from the data and to make final judgments about the "improved process" by answering to the question: "Does the improved process conforms to its quality goals?" As pointed out in the "PDSA Plan" section, the answer to this question is an understanding and a summary of the collected data in some meaningful graphical formats. A well-chosen graphical format conveys an enormous amount of quantitative information that a trained eye can detect quickly and extract salient features. Even for small sets of data, there are many patterns and relationships that are considerably easier to discern in graphical display. The commonly used graphical formats are, but are not limited to:

1. Control Charts
2. Run Charts
3. Scatter Diagrams
4. Frequency Plots
5. Pareto Charts

These graphical formats have been detailed in the "PDSA Plan" section.

32.3 Analyze Data and Validate Process Performance

The retrospective data is collected as basis to make final judgments about the "improved process." However, unless potential signals from the collected data are separated from probable noise from the data collection system, the final judgments taken may be totally inconsistent with the collected data. Thus, the proper use of collected retrospective data requires that the project team uses simple and effective methods of analysis which will properly:

1. Separate potential signals from probable noise;
2. Turn data into information, information into insight, insight into knowledge, and knowledge into wisdom; and
3. Allow learning to take place.

Learning often comes from understanding the themes and patterns in the data. These patterns in data arise in using the different types of data shown in Table 8.1. When dealing with continuous measurements or counts of observations, patterns are often easier to recognize when the data are plotted over time. This maximizes the learning from data. It allows the information to unfold as it happens and eventually display a pattern. The pattern may show improvement or an opportunity for improvement.

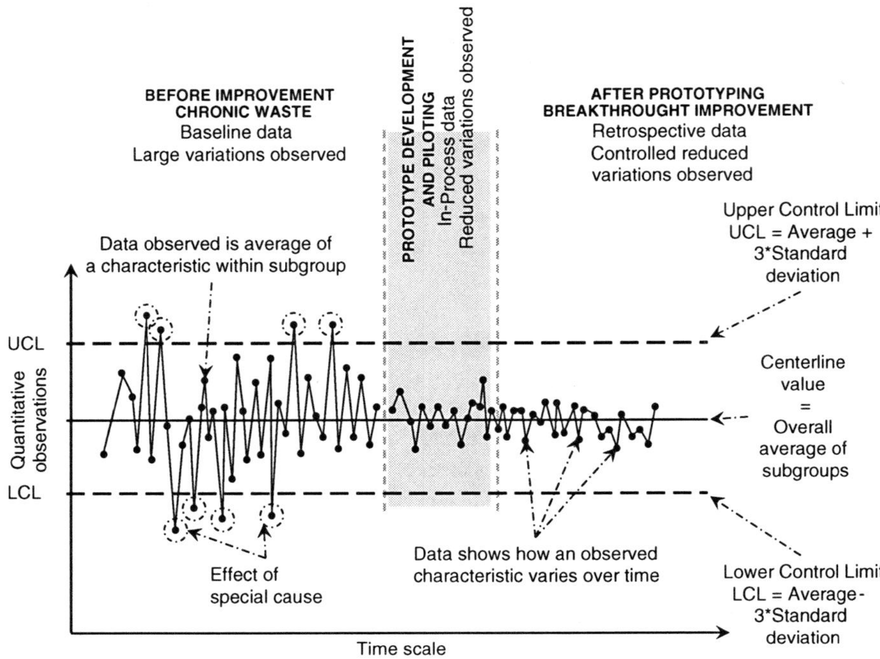

Fig. 32.1 Example of control chart with baseline and retrospective data

Process "Control Charts" describe in the "PDSA Plan" section provide a better approach to the analysis of data. By characterizing all variation as either common, which is predictable, or as assignable, and therefore unpredictable, the process "Control Chart" shifts the emphasis away from the results and toward the behavior of the system that produced the results. This shift in emphasis is a major step on the road of continuous improvement. Thus, using the baseline data collected in the "PDSA Plan" project phase, the in-process data resulting from the pilots conducted in the "PDSA Do" project phase, and the retrospective data collected over time after conclusion of the pilots of the prototype solution, a plot of data over time using a "Control Chart" can be used to see if the change from the original "process to be improved" to the "improved process" results in improvement.

If the data depicts a random pattern within a predictable range, the project team should not infer that a change in performance has occurred. Using a control chart, the summarized data, plotted over time can reveal when the variation in the data no longer follows a predictable pattern. The chart may show an isolated observation or two that are outside the predictable range, or show a new trend. If the random variation in the data is disturbed by some specific circumstance, as shown in Fig. 32.1, improvements can be developed by understanding what these special causes are.

Having collected retrospective data, the project team should determine the new "improved process" rolled throughput yield and establish the rate at which defects occur on a characteristic of the new "improved process" outcomes with respect to the number of "improved process" outcomes inspected.

Under the normality assumption on the observed characteristic of the new "improved process" outcomes, the project team should arrange the collected retrospective data are into subgroups of specific period of time. If the upper and lower specification limits of the process are USL and LSL, the target process mean is T, the estimated expectation of the observed characteristic of the "process to be improved" is $\hat{\mu}$, the estimated variability of the process (expressed as a standard deviation) within a subgroup is $\hat{s}$, and the estimated overall variability of the process (expressed as a overall standard deviation) is $\hat{\sigma}$, then commonly-accepted estimates of process capability indices within subgroups and overall process performance indices are given in Tables 14.5 and 14.6.

At conclusion of the different pilots, the new "improved process" should be in the "Ideal State," toward which every process aspires to. At this stage a point where the prototype solution is sufficiently mature that it can be retrospectively measured, it is desirable that the root-causes and control measures on all specific circumstances that may disturb the "improved process" outcomes be well documented and a Process Control Plan (PCP) developed.

Indeed, it is generally not possible simply to maintain a level of "Ideal State" process performance unless preventive measures are set in place. As a consequence of the second law of thermodynamics, we know that a process will tend to erode no matter what, even if a standard is defined, explained to everyone affected, and posted. This is not because of poor discipline by workers affected, but due to interaction effects and entropy, which says than any organized process naturally tends to decline to a chaotic state if we leave it alone over time as circumstances change. Interaction effects and entropy continually acts upon all processes to cause deterioration and decay, wear and tear, breakdowns and failures. When this happens, traditional processes cease to fit current realities, roles drift out of alignment, relationships become strained, miscommunication occurs, and soon business results suffer.

As Donald J. Wheeler indicates (Wheeler, Two Definitions of Trouble, 2009b), "Entropy is relentless. Because of it every process will naturally and inevitably migrate toward the State of Chaos. The only way this migration can be overcome is by continually repairing the effects of entropy. Of course this means that the effects for a given process must be known before they can be repaired. With such knowledge, the repairs are generally fairly easy to make. On the other hand, it is very difficult to repair something when one is unaware of it. Yet if the effects of entropy are not repaired, it will come to dominate the process and force it inexorably toward the State of Chaos."

Here is what often happens over time in a plant factory once an improved process has been implemented. In every plant factory, small problems naturally occur every day in each production process—the test machine requires a retest, there is some machine downtime, bad parts, a sticky fixture, and so on—and operators must find ways to deal with these problems and still make the required production quantity. Operators only have time to quickly fix or work around the problems, not to dig into, understand, and eliminate assignable cause of variations. Soon extra problem and inventory buffers, many work-around, and even extra people naturally creep into the production process, which, although introduced with good intention, generates even more variables, fluctuation, and problems. Consequently, the production process will tend to erode, decay, wear and tear, and decline to a chaotic state. In many plan factories management has grown accustomed to this situation, and it has become the accepted mode of operating.

32.4 Develop a Process Control Plan

We have defined a process as "a set of logically related discrete elements (tasks, actions, or steps) taken in order to achieve a particular end." Furthermore, most process outcomes (products and services) result from a complex system of interaction among people, equipment, procedures, methods, equipment, materials, and environment. Once all discrete elements of the new "improved process" have been assessed and broken down into their critical elements, a process control plan can be developed.

A Process Control Plan (PCP) is a summary of proactive defect prevention and reactive detection techniques. The primary purpose of the written process control plan is to define the actions that will be taken to impose control over all identified critical elements and to define any associated process direction. The overall intent of the PCP is to control the process outcomes characteristics and the associated process variables to ensure capability (around the identified target or nominal) and stability of the process outcomes over time. The purpose of a written process control plan is to:

1. Encourage a conscious evaluation of each discrete elements of the process;
2. Identify the critical elements which exist within each discrete elements of the process;
3. Define the actions that will be taken to impart control over each identified critical element;
4. Define process direction to consistently assure that special attention is given to each critical element;
5. Facilitate communication throughout the enterprise business and promote employee buy-in;
6. Establish common understanding;
7. Establish the basis for employee training and equipment procurement.

A Process Control Plan (PCP) assures a system is in place to control the risks of the same failure modes as identified in the FMEA developed during the pilots of prototype solution. While PCPs can be developed independently of FMEAs, it is time and cost-effective to link Control Plans directly to FMEAs. A Process Control Plan (PCP) is a natural extension of a FMEA plan, even though it is not considered officially part of a FMEA. The intent of Process Control Plans is to create a structured approach for control of the process and process outcomes characteristics, while focusing the enterprise business on characteristics important to the customer. Typically,

1. A Process Control Plan does assure well thought-out reaction plans are in place in case an out-of-control condition occurs.
2. It also provides a central vehicle for documentation and communication of control methods.
3. Special attention is typically given to potential failures with high RPNs and those characteristics that are critical to the customer.
4. A Control Plan deals with the same information explored in a FMEA plus more. The major additions to the FMEA needed to develop a Control Plan are:
   - Identification of the control factors.
   - The specifications and tolerances.
   - The data collection system.
   - Sample size.
   - Sample frequency.
   - The control method.
   - The reaction plan.

The project team must develop a Process Control Plan (PCP) for use by the process operators to ensure that the new "improved process" will not deteriorate once the improved process is returned to the process owners, but remains permanently effective. In developing a PCP, the project team must address three questions:

1. What has been done to prevent process problems;
2. How is it known when problems occurs; and
3. What will be done when problems in fact do occur? Written descriptions of the systems for controlling parts and processes.

An effective written PCP will define all of the critical elements which have been determined to exist within each component of the new "improved process" (people, equipment, methods, materials, and environment). The plan will also define the actions that will be taken to initiate a level of control that will assure all the critical elements are consistently given special attention. In addition, the plan defines the process direction required to support the desired level of control over all the critical elements.

Like the written quality control plan, an effective written process control plan should be concise, yet thorough. The plan's length is primarily dependent upon how many critical elements have been determined to exist within the new "improved process" and the extent to which the associated process control action and direction is elaborated upon. In summary, an effective written process control plan will:

1. Identify all critical elements of the new "improved process";
2. Define the actions required to control all the identified critical elements;
3. Define all the associated process direction requirements;
4. Define all associated training and equipment requirements.

32.5 Reinforce a Positive Context of Process Improvement

The Context of the new "improved process" is everything that surrounds it, including the social and psychological climate in which it is embedded. Although a "process improvement" project can be an extremely positive and empowering force in achieving the enterprise business intended strategy, its real power can only be realized in a positive context. A positive environment can be truly transformational—both in terms of the internal climate of the system affected by the new "improved process" and externally within the enterprise business as demonstrated by much better results.

The technical and technological aspects of a new "improved process" might improve things temporarily, but the key to the deep and sustainable improvement is the context of the new "improved process." Without it, the technical and the technological aspects might take a leap forward, but the change in form, quality, or state, over time from the original "process to be improved" to the new "improved process" will not be self-preserving. This change will stop when the new "improved process" is deemed "implemented."

The context of a process tends to reflect how the process is perceived by employees and therefore how they respond emotionally to it. Interestingly, even if the process is executed with great technical skill, it can still carry a negative implication. How people respond to a new "improved process" is largely a function of how it is used—that is, what is done with the data that is collected throughout the "process improvement" project lifecycle. The context of a new "improved process" can make the difference between people being energized by the new "improved process" or people just minimally complying with its execution, and even using the new "improved process" for their own personal benefit. That is why the project team must address.

Achieving the full benefits of a new "improved process" requires going beyond simply managing the implementation of the new "improved process" to focusing on maximizing performance and business results under the new model. Those enterprise businesses that proactively reinforce a positive context of process improvement will drive the degree and speed of adoption, limit resistance and opt-outs, create a higher level of operational proficiency, and establish a more effective governance structure—delivering higher overall performance. Thus, the key to progressing toward transformational "process improvement" is to reinforce the context of the new "improved process" in a positive direction.

Reinforcing the context of a new "improved process" is one of the best investments the project team can make, since the context affects all other aspects the system considered. If the positive context created at onset of the "process improvement and management" program is not reinforced, then most people, if they use the new "improved process" at all, will just be "going with the flow" and complying with the business instructions and will very likely also continue using the new "improved process" for their own self-serving purposes.

To reinforce a positive context for the new "improved process" within the system affected, the project manager and its whole project team must break with tradition, keeping in mind that the purpose of a "process improvement and management" initiative within the enterprise business is to provide clearer perception, greater shared and streamlined work knowledge and insight. Creating or reinforcing a positive context means breaking from the employee-as-cog tradition. Encourage employees to be active, think and take ownership of the new "improved process," and enjoy their work. Here are some ways to make this happen:

1. Recognize the difficulty of sustaining the new "improved process" gains;
2. Re-assess people attitudes toward implementation of the new "improved process";
3. Demonstrate visible commitment to execution and adoption of the new "improved process";
4. Keep employees affected by the new "improved process" productively busy;
5. Allocate the required resources to maintain the change in form, quality, or state, over time from the original "process to be improved" to the new "improved process";
6. Reinforce a climate of involvement and appreciation for all those affected by the new "improved process";
7. Maximize employee input to further sustain or improve the new "improved process";
8. Emphasize the importance of learning about and from the new "improved process";
9. Encourage productive social interaction around understanding of the impact of the new "improved process" on the enterprise business intended strategy.

Recognize the difficulty—Reinforcing the context of the new "improved process" from its current baseline stage to its potential maturity stage requires a very significant pattern shift from the way things are currently done in the enterprise business. For others than the project team members, a new "improved process" is not something that most people affected by it want to transit to.

Re-assess people attitudes—As project manager, you should consider re-assessing existing attitudes in the system affected by the new "improved process" in order to gauge how difficult the full-scale implementation will be. This will also help you to determine areas within the system affected that might be more receptive to the change in form, quality, or state, over time from the original "process to be improved" to the new "improved process." It is not important that the entire population of the system affected business be "fit for transformation."

A typical distribution of people attitudes in a positive context is illustrated in Fig. 32.2. Do not try to change those who, through blind ignorance, are clearly resistant to transforming the context of the new "Improved Process." Work to consolidate positive gains on the context and consolidate those visionary managers, leaders and employees who "get it" and who are receptive—people who have "adopted" the transformation.

Demonstrate visible commitment to the new "improved process"—Commitment to the new "improved process" must be truly and authentically valued by those who lead it, or the remaining of the population affected by the new "improved process" will detect the lack of integrity. Therefore, it is important for the project team driving the transformation to become educated in the principles and practices involved in process improvement and management.

Keep employees affected by the new "improved process" productively busy—In a positive context, employees should leave work feeling that they accomplished something worthwhile using the new "improved process." Do not allow them to be

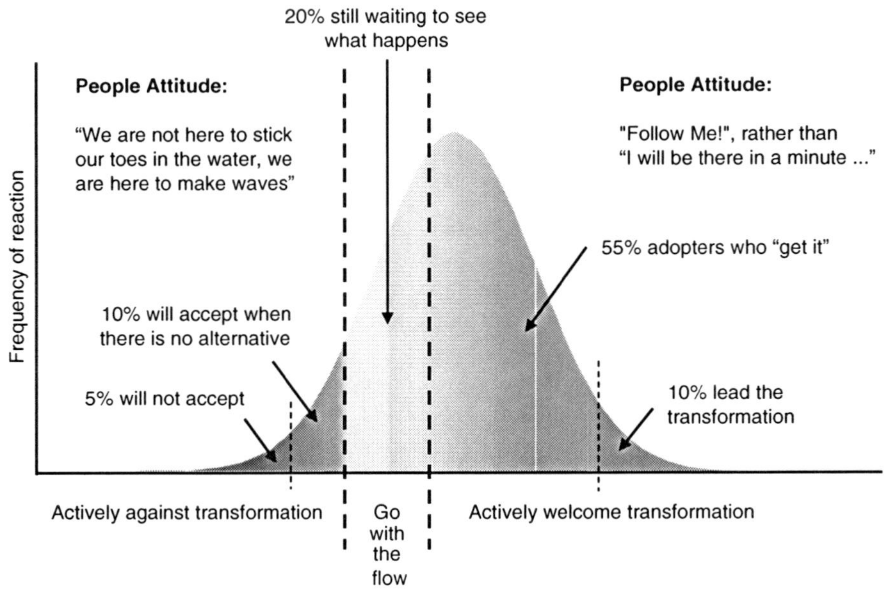

Fig. 32.2 Example of reaction to "Process Improvement" transformation

passive. Instead of letting them wait for assignments, for example, encourage them to use downtime to carry out self-improvement activities or ways to improve further the new "improved process."

Allocate the required resources to maintain the change in form, quality, or state, over time from the original "process to be improved" to the new "improved process"—Although the "process improvement" project has a cost associated with it, if done right, it delivers enormous value to achieving the enterprise intended strategic demands. Don't starve the transformation initiative before it has the opportunity to take root. Allocate the resources, including education and training, necessary for making the transformation from the original "process to be improved" to the new "improved process" a reality.

Reinforce a climate of involvement and appreciation—Most traditional production systems provide a low level of positive recognition. Well thought out expressions of appreciation are powerful drivers of creating and enhancing positive contexts. As the context of the system affected by the new "improved process" progresses and the maturity stage increases, more and more people in the system considered will become involved in initiative (from the lowest organizational level to the highest organizational level) and will begin to experience its positive side. Involvement starts with the "early adopters," but it increases as additional process improvement opportunities are identified, and employees affected by the new "improved process" experience personal involvement in using the improved process to achieve the objectives of other projects and operations work they are assigned to. As the transformation process continues, employees will develop more ownership in process improvement and management as a whole.

Maximize employee input to further sustain or improve the new "improved process"—Employees are a great source of ideas. And they will be committed to those willing to listen to them.

Emphasize the importance of learning about and from the change to the new "improved process"—Learning from the change in form, quality, or state, over time from the original "process to be improved" to the new "improved process" should be considered one of the key outcomes of the "process improvement" project. Nothing is more important in the "PDSA Study" project phase than learning. Learning about and from retrospective data collected over time can make a huge difference in how people relate and respond to the new "improved process" and to each other. As Dean Spitzer pointed out (Spitzer, 2007), "most learning is informational, not transformational. Transformation is about "trans-form-ation," a 'change in form.' Learning aimed at increasing our store of knowledge or existing repertoire of skills is valuable, but it doesn't promote change of form." Learning is how people affected by the new "improved process" acquire knowledge—the "know what," "know why," and "know how" about the new "improved process." The "know what" gives them the basic information about what exists; the "know why" tells them why things happen; and the "know how" tells them "how to do" things (the required skills). People within a given system are products of their learning, or lack of learning. Truly successful people are not only those who possess greater knowledge, but are also the most adept at the process of learning.

Encourage productive social interaction around understanding of the impact of the new "improved process" on the enterprise business in-tended strategy—Interaction enhances communication and cooperation.

32.6 Continuously Monitor "Improved Process" and Context

The "PDSA Study" project phase retains a deep focus on learning throughout its range of activities. It takes on a fresh perspective through the use of a set of systematic observation techniques and activities, focused on ongoing monitoring and control techniques that will help the enterprise business continually collect performance data that can be amassed in order to continuously operate the new "improved process" predictably and on target. As pointed out in previous sections:

1. Operating a process on-target is a necessity simply because, regardless of how large the process capability ratio might be, operating off-target can increase the effective cost of production and use of the process outcomes.
2. Operating a process predictably requires a learning enterprise business; i.e. one where knowledge is both gained and shared. This happens through continuous practice of a way of thinking rather than through implementation of the "right" technique. Without the practice in the way of thinking, simply developing control charts on a process and monitoring and displaying its summarized data over the wall to production will not result in predictable operation.

The control charts developed in the "PDSA Plan" project phase are specifically designed to monitor and assess the process performance and determine whether the process exhibits common cause variation only, or whether, and when, special cause variation is occurring. Two other types of control chart often used in industry are the cumulative sum control chart (often abbreviated as CUSUM chart) and the exponentially weighted moving average (EWMA) control chart.

32.6.1 Cumulative Sum (CUSUM) Control Charts

A cumulative sum control chart is a plot of the cumulative sum of the deviations between each data point, e.g., a sample average, and a reference value T. Thus this type of chart has a memory feature not found in the previous types of charts discussed in the "PDSA Plan" project phase. It is usually used to plot the sample average $\bar{X}$, although any of the other statistics, such as s or p, may be used. It is also often used for individual readings, particularly for chemical processes. Here, we only present the cumulative sum control chart for averages. For a discussion of other types of CUSUM charts, we refer our reader to Wadsworth, Stephens, and Godfrey (1986). For CUSUM charts, the slope of the plotted line is the important aspect considered, whereas for the previous types of charts it is the distance between a plotted point and the centerline.

Cumulative sum control charts, while not as intuitive and simple to operate as Shewhart's charts, have been shown to be more efficient in detecting small shifts (between 0.5 standard deviation and 2.5 standard deviations) in the parameter being studied. For shifts larger than approximately 2.5 standard deviations, Shewhart's-type charts discussed previously in the "PDSA Plan" project phase are just as good or somewhat better and are easier to understand and use.

Cumulative sum control charts, like other control charts, are interpreted by comparing the plotted points to critical limits. However, the critical limits for a CUSUM control chart are neither fixed nor parallel. A mask in the shape of a V is often constructed. It is laid over the chart with its origin over the last plotted point. If any previously plotted point is covered by the mask, it is an indication that the process has shifted.

The following steps may be followed to develop a CUSUM control chart for averages:

1. Obtain an estimate of the standard error of the statistic being plotted; e.g., $\sigma_{\bar{X}}$ associated with the average $\bar{X}$ may be obtained from a range chart or from some other appropriate estimator. If a range chart is used, the estimate is: $\sigma_{\bar{X}} = R/(d_2\sqrt{n}) = A_2\bar{R}/3$.
2. Determine the smallest amount of shift in the mean D for which detection is desired: $\delta = D/\sigma_{\bar{X}}$.
3. Determine the probability level at which decisions are to be made. For limits equivalent to the 3 standard deviations limits, this is equal to: $\alpha = 0.00135$.

Table 32.1 Factors for cumulative sum control chart, $\alpha = 0.00135$

δ	θ	d
0.2	5°43'	330.4
0.4	11°19'	82.6
0.5	14°00'	52.9
0.6	16°42'	36.7
0.8	21°48'	20.6
1.0	26°34'	13.2
1.2	30°58'	9.2
1.3	32°59'	7.8
1.4	35°00'	6.7
1.6	38°40'	5.2
1.8	41°59'	4.1
2.0	45°00'	3.3
2.2	47°44'	2.7
2.4	50°12'	2.3
2.6	52°26'	2.0
2.8	54°28'	1.7
3.0	56°19'	1.5

4. Determine the scale factor k. This is the change in the value of the statistic to be plotted (vertical scale) per unit change in the horizontal scale (sample number). Ewan (1963) recommends that k be a convenient value between $\sigma_{\bar{X}}$ and $2\sigma_{\bar{X}}$, preferably closer to $2\sigma_{\bar{X}}$.
5. Obtain the lead distance d from Table 32.1 using the value of δ obtained in step 2.
6. Obtain the mask angle θ from Table 32.1 by setting $\delta = D/k$ in the table and reading θ from the table. Straight-line interpolation may be used if necessary.
7. Use d and θ to construct the V mask. The V mask is operated by placing it over the last point plotted. If any of the previously plotted points are covered by the mask, a shift has occurred. Points covered by the top of the mask indicate a decrease in the process average, whereas those covered by the bottom of the mask indicate an increase. The first point covered by the mask indicates the approximate time at which the shift occurred. If no previous points are covered by the mask, the process is remaining in control.

8. The sample size for a cumulative sum control chart for averages is usually the same as for the $\bar{X}$ chart. However, Ewan (1963) suggests, for best results, that one uses $n = 2.25\ s^2/D$, where s is an estimate of the process standard deviation.

For some processes it may not be convenient to use a V mask. An alternative tabulation method may be used that is particularly well suited for computer applications. This method is equivalent to the charting method with the mask. The procedure is as follows:

1. Form the CUSUM as $C_1 = \sum_{i=1}^{n} (\bar{X}_i - K_1)$, where $K_1 = T + D/2$, to detect a shift upward.
2. Form the CUSUM as $C_2 = \sum_{i=1}^{n} (\bar{X}_i - K_2)$, where $K_1 = T - D/2$, to detect a shift downward.
3. Tabulate these quantities sequentially with $\bar{X}_i$ ignoring negative values of C_1 and positive values of C_2. That is, reset the upper CUSUM to zero when it is negative and the lower CUSUM to zero when it is positive.
4. Watch the progress of the C_1 and C_2 values. When either value equals or exceeds $D\ d/2$ in absolute value, a signal is produced.

32.6.2 Exponentially Weighted Moving Average Control Charts

The exponentially weighted moving average (EWMA) control chart was first introduced by Roberts (1959) and later by Wortham and Ringer (1971), who proposed it for applications in the process industries as well as for applications in financial and management control systems for which sub-groups are not practical. Like the cumulative sum control charts, it is useful for detecting small shifts in the mean. Single observations are usually used for this type of chart. The single observations may be averages (when the individual readings making up the average are not available), individual readings, ratios, proportions, or similar measurements.

The plotted statistic is the weighted average of the current observation and all previous observations, with the previous average receiving the most weight, that is,

$$Z_t = \lambda Y_t + (1 - \lambda) Z_{t-1}, n = 1, 2, \cdots, n$$

Where Z_0 is the mean of historical data (target), Z_t is the exponentially weighted moving average at the present time t, Z_{t-1} is the exponentially weighted moving average at the immediately preceding time $t - 1$, Y_t is the present observation, $0 \leq \lambda \leq 1$ is the weighting factor for the present observation, n is the number of observations to be monitored including.

The Y_t are assumed to be independent, but the sample statistics Z_t are auto-correlated. However, Wortham and Ringer (1971) demonstrated that for large t, the sample statistic is normally distributed when the Y_t are normally distributed with mean μ and variance σ^2. That is,

$$E(Z_t) = \mu, \qquad Var(Z_t) = \sigma^2\left(\frac{\lambda}{2-\lambda}\right)\left(1-(1-\lambda)^{2t}\right)$$

As time increases, the last term bracketed on the right-hand side converges rapidly to one, and the corresponding expression for the variance becomes

$$Var(Z_t) \approx \sigma^2\left(\frac{\lambda}{2-\lambda}\right)$$

By choosing $\lambda = 2/(t+1)$, the variance approximation becomes the variance of averages of sample size t:

$$Var(Z_t) = \frac{\sigma^2}{t}$$

Under these conditions, the control limits become $\hat{\mu} \pm 3\hat{\sigma}/\sqrt{t}$. For other values of λ, the control limits are $\hat{\mu} \pm 3\hat{\sigma}/\sqrt{\lambda/(2-\lambda)}$.

For the first few observations, the first equation for the variance should be used. If a good estimate of σ is not available, a range chart should be used with $\hat{\sigma}$ estimated by $\bar{R}/d_2$. In the case of individuals, the average moving range can be used as with the control chart for individuals.

Like the cumulative sum control chart, the exponentially weighted moving average control chart is more effective than the $\bar{X}$ chart in detecting small shifts (less than 2.5 standard deviations) in the mean; however, both charts do not perform well than the $\bar{X}$ chart for larger shifts.

For the Shewhart chart control chart, the decision regarding the state of control of the process at any time, t, depends solely on the most recent collected data from the process and, of course, the degree of "trueness" of the estimates of the control limits from historical data. For the exponentially weighted moving average control chart, the decision depends on the exponentially weighted moving average statistic Z_t, which is an exponentially weighted average of all prior data, including the most recent collected data.

32.6.3 Continuously Monitor the People Aspect of the Context

It is vital that people affected by the implementation of the new "improved process" and responsible for monitoring the process performance transform what is measured and monitored from the new "improved process," how it is measured, and what is done with the measurement. Thus, it is vital that they constantly learn about the new "improved process," and improve it context.

Here, the process of learning is similar to the process of converting data into information, knowledge, and wisdom. For the most part, what is taught or learned is data or information collected from operation of the new "improved process."

In formal learning, it is mostly information, since the control charts would have already converted the raw data into informational content. However, in informal learning part of the enterprise business affected by the implementation of the new "improved process" is often confronted with new data, and therefore has to start at the beginning. This can be an advantage, since it gives people affected by the implementation of the new "improved process" the opportunity to perform their own data-into-information conversion, rather than relying on the control charts developed by the project team to do it. And when people affected by the implementation of the new "improved process" convert data into information, it can actually enhance the quality of knowledge we acquire since "they did it themselves." It is also likely to be more meaningful to them.

When people affected by the implementation of the new "improved process" really understand the information, it becomes knowledge. That knowledge can be used together with other knowledge and experience to eventually produce wisdom. That is why people with a certain level of experience within most enterprise businesses tend to be wiser than those who might have a lot of education but lack experience. The better organized an enterprise business internal knowledge is and the better its learning tools are, the more effective and efficient people affected by the implementation of the new "improved process" will be in converting information into wisdom.

Monitor and Control Study Execution 33

This chapter is concerned with the project management process necessary to execute a set of systematic observation techniques and activities focused on collecting, measuring, and disseminating performance information, and assessing collected data and trends to affect process improvement during the "PDSA Study" project phase. Its purpose is to provide an understanding of the project's execution progress during the "PDSA Study" phase so that appropriate corrective actions can be taken when the project's performance deviates significantly from the plan. When actual status deviates significantly from the expected values, corrective actions are taken as appropriate. These actions may require making alteration to the project deliverables, which may include revising the original plan, establishing new agreements, or including additional mitigation activities within the current plan.

As indicated throughout previous sections, monitoring and controlling execution of the "process improvement" project is a very intensive process. It takes place throughout the course of the project. Although the project manager is ultimately responsible to a proper execution of the deliverables building activities, he/she depends on the "eyes and ears" of the project team to make sure that information is captured and acted on. The "Monitor and Control Execution" project management process interacts with every single project management monitoring and control process described in the PDSA Plan process group; namely:

1. Schedule Control Plan;
2. Quality Control Plan;
3. Control Spending Plan;
4. Control Contract Performance;
5. Monitor and Control Risk;

The "Monitor and Control Execution" project management process also builds on the:

1. Project Management Plan
2. Performance Reports
3. Enterprise Environmental Factors
4. Organizational Process Assets

A. van Aartsengel and S. Kurtoglu, *Handbook on Continuous Improvement Transformation*,
DOI 10.1007/978-3-642-35901-9_33, © Springer-Verlag Berlin Heidelberg 2013

33.1 Perform Time Management

As indicated in a previous chapter on the "PDSA Do" project phase, this is the project management process focused on collecting, measuring, and disseminating schedule performance information required to execute the defined schedule control plan and by which time spent by staff undertaking project tasks to build the required deliverables is recorded against the project. Recording the actual time spent by staff on a project has various purposes. It is used to:

1. Calculate the total time spent undertaking each task as well as the total staff cost of undertaking each task in the project;
2. Identify aspects of their relationship with the project; and
3. Categorize each identified customer and stakeholder.

The "Perform Time Management" process during this "PDSA Study" phase of the project is also undertaken through the completion and approval of timesheets.

33.2 Perform Resource Management

As indicated in a previous chapter on the "PDSA Do" project phase, this is the project management process focused on collecting, measuring, and disseminating resource utilization and performance information required to execute the defined resource management plan. It also tracks team member performance, providing feedback, resolving issues, assessing and improving the competencies and interaction of team members, and coordinating changes to enhance project performance. Objectives include:

1. Improve skills of team members in order to increase their ability to complete project activities;
2. Improve feelings of trust and cohesiveness among team members in order to raise productivity through greater teamwork.

33.3 Perform Quality Management

As indicated in a previous chapter on the "PDSA Do" project phase, this is the project management process for performing a set of systematic observation techniques and activities focused on outcomes of the project (i.e., project deliverables and project management processes used to produce the outcomes), to monitor and record results of executing the quality assurance plan in order to:

1. Assess performance of the "process improvement" project and "process to be improved" outcomes; and
2. Recommend necessary alterations to the project objectives and/or "process to be improved" goals.

This project management process focused on collecting, measuring, and disseminating performance information on the quality of the built and studied deliverables and management processes required to build, assured and control the quality of the produced deliverables. It process involves undertaking a variety of reviews to assess and improve the level of quality of project deliverables and processes. More specifically, performing the quality management process involves:

1. Listing the quality targets to be achieved from the quality assurance plan;
2. Identifying the types of quality data to be collected;
3. Implementing quality assurance and quality control techniques;
4. Taking action to enhance the level of deliverable and process quality;
5. Reporting the level of quality attained.

We have indicated in a previous section that the quality management process is performed by the Quality manager and the Quality reviewer during the execution phase of the project. Although quality assurance methods are initiated prior to this "PDSA Study" phase, quality control techniques are implemented during the actual study of each physical deliverable to consolidate the improvement gains obtained at completion of the "PDSA Do" project phase. Without a formal quality management process in place, the basic premise of delivering the project to meet "time, cost and quality" targets may be compromised. The quality management process is terminated only when all of the studied deliverables and management processes have been completed.

At this phase of the project, the quality manager still ensures that the project produces a set of deliverables which attain a specified level of quality as agreed with the customer and reported in the quality assurance plan. The quality manager is responsible for:

1. Reviewing the quality of deliverables produced and management processes undertaken;
2. Ensuring that comprehensive quality targets reported in the quality assurance plan are met for each deliverable;
3. Implementing quality assurance methods to assure the quality of deliverables produced by the project;
4. Implementing quality control techniques to control the quality of the deliverables currently being produced by the project;
5. Recording the level of quality achieved in the quality register;
6. Identifying quality deviations and improvement actions for implementation;
7. Reporting the quality status to the project manager.

With a clear understanding of the quality targets to be achieved, the quality manager executes quality assurance and quality control techniques to assure and control the level of quality of each deliverable constructed. Figure 13.3 describes the processes and procedures to be undertaken to assure and control the quality of deliverables and processes within the project.

33.4 Perform Cost Management

As indicated in a previous chapter on the "PDSA Do" project phase, this is the project management process for performing a set of systematic observation techniques and activities, and by which costs/expenses incurred on the project, are formally identified, approved and paid and recorded in order to:

1. Assess cost performance of the "process improvement" project; and
2. Recommend necessary alterations to the project objectives and/or "process to be improved" goals.

A generic form of the "Control Cost Management" process is shown in Fig. 16.4. As outlined already, the purpose of this project management process is to accurately record the actual costs/ expenses which accumulate during the project life cycle. Cost/expenses incurred are formally documented and recorded through the completion and approval of expense forms.

33.5 Perform Procurement Management

As indicated in a previous chapter on the "PDSA Do" project phase, "Perform Procurement Management" is the project management process for performing a set of systematic observation techniques and activities focused on close monitoring and control of the in-process contracts performance in order to:

1. Ensure compliance and fulfillment of the contract conditions; and
2. Recommend necessary alterations to the contract objectives/goals.

It involves contract administration performed by the project manager (or a qualified designee) after a contract has been awarded to determine how well the portion of the project that is included within the related contract is been implemented and how well the supplier(s) performs to meet the requirements of the contract. It encompasses all dealings between the project manager and the supplier(s) from the time the contract is awarded until the work has been completed and accepted or the contract terminated, payment has been made, and disputes have been resolved.

In performing this process, the focus is on obtaining procurement items, of requisite quality, on time, and within budget. While the legal requirements of the contract are determinative of the proper course of action of the project manager in administering a contract, the exercise of skill and judgment is often required in order to protect effectively the interests of both the enterprise business and the supplier(s).

How well the project manager administers in-process contracts and discusses with suppliers their current performance determines to a large extent how well the portion of the project that is included within the related contract will be implemented and provide value to the enterprise business. By increasing attention to supplier performance on in-process contracts, project managers are reaping a key benefit: better current performance because of the active dialog between the supplier and the project manager.

Monitoring should be commensurate with the criticality of the service or task and the resources available to accomplish the monitoring. A generic form of the "Contract Performance Control" process is shown in Fig. 17.2.

33.6 Perform Communication Management

As indicated in a previous chapter on the "PDSA Do" project phase, "Perform Communications Management" is the project management knowledge area that performs the processes required to ensure timely and appropriate generation, collection, distribution, storage, retrieval, and ultimate disposition of project information as defined in the project communication schedule. The project communication schedule developed during the planning phase describes each communication event, including its purpose, method and frequency as indicated in Tables 18.3 and 18.4. It provides the critical links among people and information that are necessary for successful project communications. Clear project communication therefore ensures that the correct stakeholders have the right information, at the right time, with which to make well-informed decisions.

Various types of formal communication may be undertaken in a project as indicated in Table 18.3. Examples are releasing regular project status or performance reports, communicating project risks, issues and changes, and summarizing project information in weekly newsletters. Regardless of the type of communication to be undertaken, the method for undertaking the communication will always remain the same:

1. Identify the message content, audience, timing and format.
2. Create the message to be sent.
3. Review the message prior to distribution.
4. Communicate the message to the recipients.

These four processes should be applied to any type of formal communication on the project, including the distribution of:

1. Regular project status reports;
2. Results of phase review meetings;
3. Quality review reports documented;
4. Minutes of all project team meetings;
5. Newsletters and other general communication items.

Although the communications process is typically undertaken after the communications plan has been documented, communications will take place during all phases of the project. This process therefore applies to all formal communications undertaken during the life of the project. Without a formal communications management process in place, it will be difficult to ensure that project stakeholders receive the right information at the right time.

33.7 Perform Risk Management

"Perform Risk Management" is the project management process by which risks to the project are formally identified following study of the built project deliverables, quantified and managed during the "PDSA Study" phase of the project. The process entails completing a number of actions to reduce the likelihood of occurrence and the severity of impact of each risk. The risk management process developed during the project planning phase is used to ensure that every risk is formally identified, quantified, monitored, avoided, transferred and/or mitigated.

"Perform Risk Management" has well-established stages that make up the risk management process, as illustrated in Fig. 19.2, although it is presented in a number of different ways and often uses differing terminologies. These stages build into valuable risk management activities, each of which makes an important contribution. In this handbook, the risk management process is taken as a narrow set of activities; the constituent project management processes of which include the following:

1. Identify Project Risks
2. Perform Risk Assessment
3. Develop Risk Response Planning
4. Monitor and Control Risk

These four constituent processes, described in a previous section of the project planning phase, interact with each other and with the project management processes in the PDSA "Process Groups." Each aspect of executing any of these four constituent processes can involve effort from one or more persons, based on the needs of the project. Each aspect occurs at least once in every "process improvement" project and occurs in one or more project phases.

Although the risk management process is undertaken during the "PDSA Do" and "PDSA Study" phases of the project, risks may be identified at any stage of the project life cycle. In theory, any risk identified during the life of the project will need to be formally managed as part of the risk management process. Without a risk management process in place, unforeseen risks may impact the ability of the project to meet its objectives.

33.8 Perform Deliverable Alteration Management

This is the project management process by which alterations or changes to the project scope, built and studied deliverables, timescales or resources are identified, evaluated and approved prior to implementation. The process entails completing a variety of control procedures to ensure that if implemented, the alteration or change will cause minimal impact to the project.

This process is undertaken during the execution and study phases of the "process improvement" project, once the project has been formally defined and planned. In theory, any change to the project during the execution and study phases will need to be formally managed as part of the deliverable alteration or change process. Without a formal deliverable alteration process in place, the ability of the project manager to effectively manage the scope of the project may be compromised. The deliverable alteration management process is terminated only when the execution and study phases of the project are completed.

Figure 29.1 shows the process to be undertaken to initiate, implement and review alterations or changes within the project. Where applicable, alteration roles have also been identified.

33.9 Perform Deliverable Acceptance Management

This is the project management process by which deliverables produced and studied by the project team are reviewed and accepted or rejected by the customer. The process entails completing a variety of review techniques to confirm that the deliverable meets the acceptance criteria outlined in the project management and deliverables plans. It is used to ensure that every deliverable produced by the project is fully complete and compliant to the defined requirements, has been reviewed and approved by the customer.

The acceptance management process, illustrated in Fig. 33.1, is undertaken towards the end of the "PDSA Study" phase of the project, as each studied deliverable is presented to the customer for final sign-off. Depending on the project, one of several approaches may be taken for deliverable acceptance:

1. Each deliverable may be reviewed and presented individually to the customer for sign-off.
2. Sets of deliverables may be reviewed and presented for acceptance at the same time.
3. All project deliverables may be reviewed and presented for acceptance at the same time.

Without a formal acceptance process in place, the customer may not accept the final deliverables produced by the project, thereby compromising the project's overall success. The acceptance process is terminated only when the execution phase is complete. Figure 33.1 describes the processes and procedures required to gain the acceptance of project deliverables by the customer. Where applicable, acceptance roles have also been identified.

Complete Deliverable—Before requesting the formal acceptance of a deliverable by a customer, the deliverable must be completed to a level of quality which is likely to gain customer acceptance. This involves:

1. Undertaking all tasks required to complete the deliverable;
2. Documenting the final deliverable components;
3. Informing the project manager that the deliverable is ready for customer acceptance.

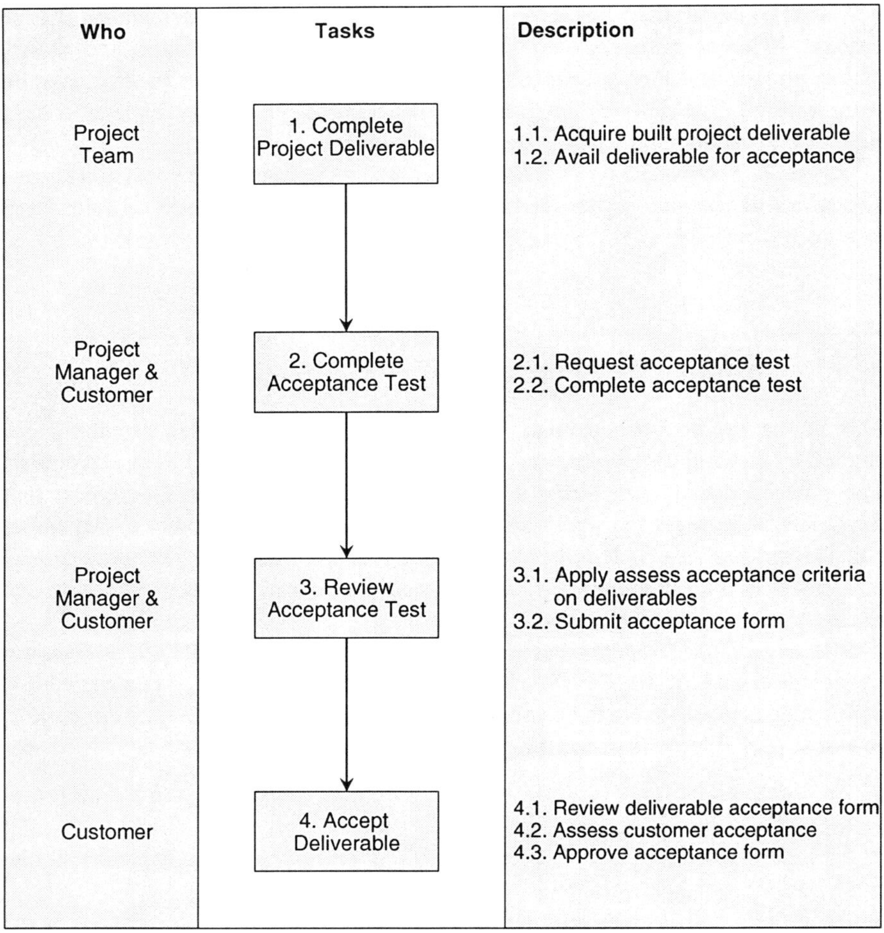

Fig. 33.1 Deliverable acceptance management process

Complete Acceptance Test—The project manager arranges an acceptance test (or review) of the deliverable by the customer to gain agreement that the deliverable matches the acceptance criteria documented in the project plan and is now ready for final sign-off. This involves:

1. Confirming that the review methods outlined in the acceptance plan are still relevant and appropriate. Examples of review methods may include:
   - Physically inspecting the deliverable.
   - Auditing the deliverable by a third party.
   - Analyzing the processes used to create the deliverable.
   - Reviewing the time taken to create the deliverable against the project plan.
   - Reviewing the cost incurred in creating the deliverable against the financial plan.
   - Reviewing the quality of the deliverable against the quality plan.

2. Confirming that the criteria and resources outlined in the acceptance plan are still relevant and appropriate for the review.
3. Scheduling the review with the customer.
4. Undertaking the review with the customer.
5. Documenting the results to present to the customer.

Review Acceptance Test—The acceptance test results are assessed by the customer to determine whether or not they met the criteria specified within the acceptance plan. This involves:

1. Comparing the results against the original acceptance criteria;
2. Determining whether or not those criteria have been met;
3. Initiating further work required to improve the deliverable if required;
4. Completing an acceptance form for deliverable approval.

Accept Deliverable—The deliverable is then finally accepted by the customer. This involves:

1. Reviewing acceptance form to ensure that all final criteria have been met;
2. Obtaining acceptance form approval from the customer;
3. Transferring the deliverable to the customer environment.

This acceptance can be very informal and ceremonial, or it can be very formal, involving extensive acceptance testing against the client's performance specifications.

Ceremonial Acceptance—Ceremonial acceptance is an informal acceptance by the customer. It does not have an accompanying sign-off of completion or acceptance. It simply happens. Two situations fall under the heading of ceremonial acceptance:

1. The first involves deadline dates at which the customer must accept the project as complete, whether or not it meets specification.
2. The second involves a project deliverable requiring little or no checking to see if specifications have been met.

Formal Acceptance—Formal acceptance occurs in those cases in which the customer has written an acceptance procedure. In many cases, especially computer applications development projects, writing an acceptance procedure may be a joint effort by the customer and appropriate members of the project team; it typically is done very early in the life of the project. This acceptance procedure requires that the project team demonstrate compliance with every feature in the customer's performance specification. A checklist can be used and requires a feature-by-feature sign-off based on performance tests. The checklist is written in such a fashion that compliance is either demonstrated by the test or it is not demonstrated by the test. It must not be written in such a way that interpretation is needed to determine whether compliance has been demonstrated. The tests are conducted jointly and administered by the customer and appropriate members of the project team.

33.10 Conduct the Project Retrospective

This is the reflection process performed at the end of each significant milestone of the "PDSA Study" project phase and from which the project team reassembles to look back on what results were actually delivered at the milestone and to what extent the team has met the expectations for the considered milestone time period. The reflection process integrates or links thought and task execution with reflection. As indicated already, it involves thinking about and critically analyzing one's actions with the goal of improving one's professional practice (Schön, The reflective practitioner, 1983; Schön, 1987). Here, engaging in reflective practice requires individuals to assume the perspective of an external observer in order to identify the assumptions and feelings underlying their practice and then to speculate about how these assumptions and feelings have affected the achievement of a significant milestone objective.

When the project team members reflect at the end of a significant milestone, they question the assumptions behind their tacit knowledge revealed in the way they carried out tasks and approached problems and think critically about the thoughts that got them into this fix or this opportunity. The project team members may, in the process, restructure strategies of action, understandings of phenomena, or ways of framing problems. Much of a project team's work is focused on problems that occurred while studying deliverables.

The reflection process should begin when the application of the project team's know-how to build deliverables does not produce the expected milestone results, the activities it conducted and the project management processes used throughout the milestone time period in the "PDSA Study" project phase have failed to meet expectations. As mentioned in a previous section, the project team may decide to ignore the failure or it may respond to it by reflecting in one of two ways:

1. It may reflect "on action" by allowing its members to step away (i.e. assume the perspective of external observers) from the planning process and thinking back on their experience to understand how part of their tacit knowledge, which was revealed in the way they approach problems and carry out tasks required to reach the milestone considered, contributed to an unexpected outcome.
2. Alternatively, the project team may "reflect in the midst of the planning process without interrupting it."

Conducting a project retrospective at the end of a significant milestone during the project's life cycle is the primary means for facilitating learning and continuous innovation in an enterprise business. To be effective, a project retrospective should be facilitated by an experienced, trained, objective facilitator from outside the project team who helps draw people out to share their perspectives, promote effective learning and reflection, and create a positive context for "process improvement" rather than one of finger-pointing, defensiveness, avoidance, or blame.

After the retrospective session, the project manager works with the facilitator and the project management office (PMO) leader to document the results and communicate them to team members, sponsors, and key stakeholders. "Report-out" meetings with senior managers and the PMO may be useful for generating support for the team's improvement actions.

33.11 Perform Phase Review

This is the phase review process performed at the end of the "PDSA Study" phase to ensure that the project has achieved its stated objectives as planned by refining previously provided answers to the three fundamental questions, which form the basis and the preliminary step of the PDSA model:

1. What is intended to be realized or accomplished by the "process improvement" project?
2. How will the realized or accomplished outcome of the "process improvement" project be recognized as is an improvement?
3. What alterations to the system affected by the "process to be improved" can be made based on the realized or accomplished outcome of the "process improvement" project?

A phase review form is completed to formally request approval to proceed to the next phase of a project. The phase review form should describe the status of the:

1. Overall project;
2. Project schedule based on the project plan;
3. Project expenses based on the financial plan;
4. Project staffing based on the resource plan;
5. Project deliverables based on the quality plan;
6. Project risks based on the risk register;
7. Project issues based on the issues register.

The review form should be completed by the project manager and approved by the project sponsor. To obtain approval, the project manager will usually present the current status of the project to the project board for consideration. The project board (chaired by the project sponsor) may decide to cancel the project, undertake further work within the existing project phase or grant approval to begin the next phase of the project. A sample phase review form for the "PDSA Study" project phase is shown in Table 33.1.

33.12 Identify and Document Lessons Learned

That is the final position of the project manager, and the project team, describing for the benefit of "future generations" as well as of next phases of the project just what went well, and what could have been handled perhaps better on the project "PDSA Study" phase. What could have been done better, and should be done differently on the next similar "PDSA Study" project phase?

A lesson learned session focuses on identifying ways of learning that have merit (quality), worth (value), or significance (importance) for the next phase of the "process improvement" project or for future projects within the enterprise business. During the "PDSA Study" project phase, the project team and key

Table 33.1 Phase review form for the "PDSA Study" project phase

PROJECT DETAILS			
Project name: Project manager: Project sponsor:	Report prepared by: Report preparation date: Reporting period:		
Project description: [Summarize the overall project achievements, risks and issues experienced to date.]			
OVERALL STATUS			
Overall status: [Description] Project schedule: [Description] Project expenses: [Description] Project deliverables: [Description] Project risks: [Description] Project issues: [Description] Project changes: [Description]			
REVIEW DETAILS			
Review category	**Review question**	**Answer**	**Variance**
Schedule	Was the phase completed to schedule?	[Y/N]	
Expenses	Was the phase completed within budgeted cost?	[Y/N]	
Deliverables: Deliverable #1 Deliverable #2 ... Deliverable #n	Deliverables: Was the deliverable #1 studied and approved? Was the deliverable #2 studied and approved? ... Was the deliverable #n studied and approved?	[Y/N] [Y/N] ... [Y/N]	
Risks	Are there any outstanding project risks?	[Y/N]	
Issues	Are there any outstanding project issues?	[Y/N]	
Alterations/Changes	Are there any outstanding project alterations/changes?	[Y/N]	
APPROVAL DETAILS			
Supporting documentation: [Reference any supporting documentation used to substantiate the review details above.] **Project sponsor** Signature: Date: This project is approved to proceed to the "PDSA Act" phase.			

stakeholders should identify lessons learned concerning the project management element in which problems arose, how they arose, which positive or negative development was encountered, and what concrete, practical solutions or recommendations was used based on this experience.

The project manager must ask team members, stakeholders, and the project sponsor to help compile the lessons learned document. He/she should them what went well during the development of the project planning and what could have gone better. The following is the information you should include in the lessons learned document:

1. How the project management processes were used throughout the "PDSA Study" project phase and how successful they were in studying deliverables and tracking progress.
2. How well the project plan and project schedule reflected the actual work carried out during the "PDSA Study" phase of the project.
3. How well the alteration/change management process worked and what might have worked better.
4. Why corrective actions were taken and whether they were effective.
5. Causes of performance variances and how they could have been avoided.
6. Outcomes of corrective actions.
7. Risk response plans that have been identified and whether they adequately addressed the risk events.
8. Unplanned risk events that occurred at the "PDSA" project phase.
9. Mistakes that occurred and how they could have been avoided.
10. Team dynamics, including what could have helped the team perform more efficiently.

As indicated at the closure of the "PDSA Plan" and "PDSA Do" phases, lessons learned document should not be limited to only the items on this list above. Anything that worked well, or did not work well, that will help team members perform their next project better or smooth out problems before they get out of hand should be identified and documented here. Lessons learned should include detailed, specific information about behaviors, attitudes, approaches, forms, resources, or protocols that work to the benefit or detriment of projects. They are crafted in such a way that those who read them will have a clear sense of the context of the lesson learned, how and why it was derived, and how, why, and when it is appropriate for use in other projects. Lessons learned at this stage represent both the mistakes made during the "PDSA Study" project phase and the newer "tricks of the trade" identified during a project "PDSA Study" effort.

The content of a lesson learned report should be provided in context, in detail, and with clarity on where and how it may be implemented effectively. Because lessons learned are often maintained in a corporate database, the lesson learned documentation will frequently include searchable keywords appropriate to the project and the lesson.

The process of identifying and documenting lessons learned at this stage of the project lifecycle is particularly useful for projects that failed to pass the phase review, because there are many things that can learn from projects that fail phase reviews that will help prevent next projects from suffering the same fate.

Recording lessons learned information in the organizational process assets is one critical consideration, but equally important is the establishment of protocols to ensure access to the recorded information on a consistent basis. Lessons learned may be captured and logged in depth, but if they are not accessed by project managers and team members within the enterprise business in the future, they do not serve any real function. Access of recorded lessons learned may be encouraged through creative documentation approaches, physical location (hallways and project war rooms), or by including the mandate to access lessons learned as a key component of the performance criteria for project managers and team members.

Conclusion to "PDSA Study" 34

Throughout the previous chapters related to the "PDSA Study" Process Group, we have illustrated and developed the "PDSA Study" constituent processes needed to study the built project deliverables and perform the course of action required to attain the objectives and scope that the project is undertaken to address. The described constituent processes help perform the established project management plan.

The purpose of the "PDSA Study" project phase is to build new knowledge through learning. Without learning and the ongoing communication it entails, the greatest opportunities for using the process improvement project effectively to help making progress on moving the "Process Improvement & Management" initiative from its current maturity stage to "Continuous Improvement" maturity stage are lost.

As the project team builds knowledge about the new "improved process," it will need to calculate whether the change introduced by the new "improved process" will result in improvement under the diverse conditions it will face in the future. This calculation is carried out through continuous and disciplined data collection and monitoring of the new "improved process" performance using control charts. Continuous disciplined data collection and monitoring of the process performance using control charts is a very effective approach as it help to replace habits that most people, especially people who pride themselves on their intuition, often resist. This approach provides a more objective lens for confronting reality—seeing the process performance for what it really is. It help check process operators biases and prevent costly errors in judgment, which otherwise might not have been even detected, especially in the "heat of battle" for producing process outcomes under market or customer pressure.

Continuous disciplined data collection and monitoring of process performance using control charts can help make decisions more objective, disciplined, and less political. But this will only work if done well and with wisdom: it requires a learning enterprise business; i.e. one where knowledge is both gained and shared. In order to attain this capacity, the enterprise business must "internalize" the capability for ongoing and step wise or small, incremental steps improvement. Small, incremental steps let us learn along the way, make adjustments, and discover the path to where we want to be. Since we cannot see very far ahead, we cannot rely

A. van Aartsengel and S. Kurtoglu, *Handbook on Continuous Improvement Transformation*, DOI 10.1007/978-3-642-35901-9_34, © Springer-Verlag Berlin Heidelberg 2013

on up front planning alone. Improvement, adaptation, and even innovation result to a great extent from the accumulation of small steps; each lesson learned helps us recognize the next step and adds to our knowledge and capability.

This happens through continuous practice of a way of thinking rather than through implementation of the "right" technique. Without the practice in the way of thinking, simply developing control charts on a process and monitoring and displaying its summarized data over the wall to production is pointless to enabling the full power of process improvement transformation. This kind of learning requires a context that is open enough to support it. The objective is not just to change a particular instance of learning or improving a process, but to create an ongoing capacity for transforming both learning and improving processes: that is, transformational learning through process improvement.

The key to transformational learning through process improvement is the context associated with the process considered. Is there an environment in the system affected by the process that is conducive to measuring the true process performance? If the context is trending in a positive direction, the project team should be seeing organizational learning increasing as evidence of that. If there is an effort being made to continuously find the right improvement to be made to a process, there will be double-loop learning occurring.

Attaining the optimal environment, as we indicated in previous chapters, requires a specific and intensive set of actions—a transformation process progressing from improving context of transformational learning through process improvement, to improving focus, to improving integration, to improving interactivity—the four aspects of paramount importance to making progress on moving the "Process Improvement & Management" initiative from its current maturity stage to "Continuous Improvement" maturity stage as we have indicated already. Within the "Process Improvement and Management" dimension of "Continuous Improvement" transformation, the factors which contribute to transform the interactivity of "Process Improvement and Management" include the following:

1. Frequent Interactivity
2. Effective and robust dialogue
3. Collaborative learning
4. Appropriate use of technology

Performing the "PDSA Study" constituent processes should include highly interactive and iterative (ongoing) discussions, or dialogues, which are also the most important aspects of learning. As indicated already, these dialogues should be built on the foundation of a positive context, focus, and integration. As was the case with the "PDSA Plan" and "PDSA Do" constituent processes, effective integration and interactivity of the "PDSA Study" constituent processes will also do more than anything else to break down the silos that are also keeping enterprise businesses from realizing "Process Improvement & Management" transformational potential.

Figure 34.1 shows the minimum activities that are part of the "PDSA Study" project phase, in addition to the already listed "PDSA Plan" and "PDSA Do" activities.

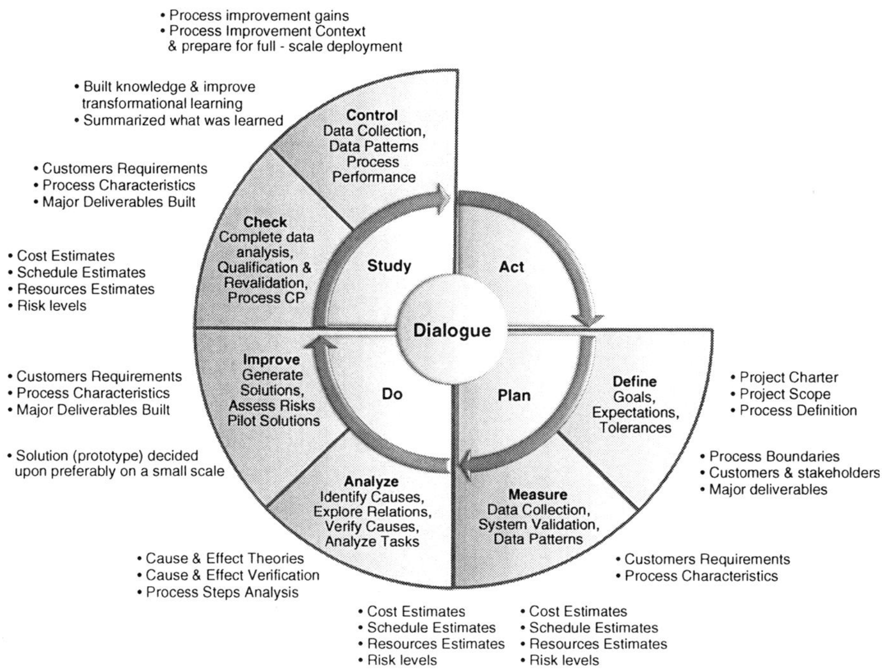

Fig. 34.1 Minimum activities of the "PDSA Study" phase

In this figure we use the "Check" and "Control" nomenclature of the six-sigma literature for convenience and consistency with existing literature. We have placed "Dialogue," which is what enables this continual learning, at the very center of the PDSA Cycle in Fig. 34.1. It is in fact the basic unit of a "process improvement" project work. You cannot plan, execute and learn from a "process improvement" project well without robust dialogue with customers and stakeholders. How people involved in a "process improvement" project talk to each other, talk to customers and stakeholders, absolutely determines how well the "process improvement" project will progress towards its objectives.

As indicated already, the word "dialogue" should be understood in the sense of "sharing collective meaning" and strongly differentiated from "discussion." The word "discussion" comes from the same root word as percussion and concussion and has to do with beating one thing against another. The word "communication" is a more general term meaning "to make something common." So, communication can be done by discussion or dialogue. When information is made common through discussion, it is often two monologues—an attempt to convey your opinion to another person, and nothing more. As we have indicated throughout this handbook, very few people are skilled at dialogue and very few project team members currently have a strong capacity for dialogue.

"PDSA Act" Process Group

35

In the previous chapters, we have characterized a project as a sequence of unique, complex, and connected activities having one goal or purpose and that must be completed by a specific time, within budget, and according to specification. It is a temporary effort undertaken to create a unique product, service, or result. The purpose of a project is to attain its objectives and then terminate. It concludes and must be formally terminated when its specific objectives have been attained. If not, it will limp on, turning an otherwise successful project into a financial failure.

The "PDSA Act" Process Group encompasses the processes needed to act upon the built and studied deliverables based on what was learned from the previous project phase, implement the new "improved process" permanently or stop its implementation, determine what modifications should be made to the system affected, and formally close the "process improvement" project.

Of all the PDSA project phases, the "PDSA Act" phase is the most crucial in the life of a "process improvement" project as it can determine whether, ultimately, the project is a success or failure. This all-important phase, which should ensure that practices needed to sustain the new "improved process" over long term are established and effective and timely completion of the project is performed, is often poorly managed.

The manner this phase is managed can affect how the "process improvement" project is remembered. Often, no one within an enterprise business remembers an effective start-up of a project, but everyone remembers an ineffective close-out as the consequences are felt for a long time. The manner in which the "PDSA Act" project phase is performed will determine how the project will reach conclusion. The way a project concludes and closes out is also influenced by the reason for its termination. Some reasons for termination include, but are not limited to the following:

1. Project objectives have been achieved.
2. It is no longer feasible to achieve the project objectives due to changing market conditions, increasing costs beyond expected bounds, depleted critical resources, lost opportunities, changes in need, lack of feasibility, or change in priorities.

A. van Aartsengel and S. Kurtoglu, *Handbook on Continuous Improvement Transformation*,
DOI 10.1007/978-3-642-35901-9_35, © Springer-Verlag Berlin Heidelberg 2013

3. Simply by default, perhaps due to unsatisfactory project performance, poor quality or workmanship, violation of contract, poor planning and control, bad management, or customer dissatisfaction.

Performing the "PDSA Act" project phase is routine once the project objectives have been achieved and customer's acceptance of the deliverables has been secured. Following customer's acceptance of the deliverables, the project closure activities are undertaken. These activities involve the hand-over of deliverables and documentation to the customer, the termination of supplier contracts, the release of project resource back to the business, and the communication of project closure to all project stakeholders. The constituent project management processes, used to perform these activities during the "PDSA Act" project phase, include the following:

1. Implement "Improved Process" and Install All Deliverables;
2. Complete Project Documentation;
3. Reinforce Mechanisms and Build Capability;
4. Create Standard Practices And Procedures;
5. Release Resources;
6. Settle Contractual Aspects And Final Accounting
7. Conduct Post-Implementation Review;
8. Write Final Report;
9. Celebrate Success and Share the Wealth.

35.1 Implement "Improved Process" and Install All Deliverables

This is the process by which the implementation plan, defined at conclusion of the pilots of the prototype solution, is carried out and the deliverables installed. This commonly occurs in technology related process improvement work. Implementation can involve phases, cutovers, or some other rollout strategy. In other cases, it involves nothing more than flipping a switch. In either case, some event or activity turns things over to the customer. This implementation of the new "improved process" triggers the beginning of a number of close-out activities that mostly relate to documentation, standardization, and report preparation.

35.2 Complete Project Documentation

Documentation, documentation, documentation, it cannot be said enough. If there is one place that pleads to be overlooked it is proper documentation. Record keeping is a mundane and time consuming effort that is most often incomplete or ignored altogether. Projects generate substantial volumes of documentation and it is essential that these are collated and filed for future reference. Certain documents, such as the final cost reports, contracts and claims documents only become available during the close-out and this step in the process ensures that these documents are completed and collated.

The project documentation always seems to be the most difficult part of "process improvement" projects to complete. There is little glamour in doing documentation. That does not diminish its importance, however. There are at least five reasons why the project team needs to do documentation:

1. *Reference for future alterations in deliverables.* Even though the project work is complete, there will be further alterations that warrant follow-up projects. By using the deliverables, the customer will identify improvement opportunities, features to be added, and functions to be modified. The documentation of the project just completed is the foundation for the follow-up projects.
2. *Historical record for estimating duration and cost on future projects, activities, and tasks.* Completed projects are a remarkable source of information for future projects, but only if data and other documentation from them are archived so that it can be retrieved and used. Estimated and actual duration and cost for each activity on completed projects are particularly valuable for estimating these variables on future projects.
3. *Training resource for new project managers.* History is a great teacher, and nowhere is that more significant than on completed projects. Such items as how the Work Breakdown Structure (WBS) architecture was determined, how alteration requests were analyzed and decisions reached, problem identification, analysis and resolution situations, and a variety of other experiences are invaluable lessons for the newly appointed project manager.
4. *Input for further training and development of the project team.* As a reference, project documentation can help the project team deal with situations that arise in the current project. How a similar problem or change request was handled in the past is an excellent example.
5. *Input for performance evaluation by the functional managers of the project team members.* In many organizations, project documentation can be used as input to the performance evaluations of the project manager and team members. Care must be exercised in the use of such information, however. There will be cases where a "process improvement" project was doomed to fail even though the team members' performance may have been exemplary. The reverse is also likely. The "process improvement" project was destined to be a success even though the team members' performance may have been less than expected.

Given all that documentation can do for the enterprise business, to be most effective and useful, the documentation for a given project should include the following parts:

1. Project Overview Statement
2. Project proposal and backup data
3. Original and revised project schedules
4. Minutes of all project team meetings
5. Copies of all status reports
6. Design documents
7. Copies of all alteration notices

8. Copies of all written communications
9. Outstanding issues reports
10. Final report
11. Sample deliverables (if appropriate)
12. Customer acceptance documents
13. Post-implementation audit report

This list is all-encompassing. For a given project, the project manager has to determine what documentation is appropriate. Always refer back to value added considerations. If the document has value, and many will have good value for future "process improvement" projects, then it must be included in the documentation. Note also that the liDesign documentsst contains very little that does not arise naturally in the execution of the project. All that is needed is to appoint someone to care for and feed the project notebook. This task involves collecting the documents at the time of their creation and ensuring that they are in a retrievable form (electronic is a must). Where relevant, original plans and documents as well as the final "as improved" plans must be kept. The documentation can be stored on various media (video, photographic, electronic, paper).

35.3 Reinforce Mechanisms and Build Capability

This is the process by which modifications to be made to the system affected by the implementation of the new "improved process" are determined and realized. Once improvements are implemented, practices need to be established to ensure that the change in form, quality, or state, over time from the original "process to be improved" to the new "improved process" becomes the normal way the business is run. Holding the realized improvement gains usually requires some change in the system to ensure that the change is maintained. Many enterprise businesses make improvements on the job only to discover later that the improved performance has degraded to the old level or some new problem has been encountered. Many times this occurs because mechanisms to reinforce the structure to sustain the change were not established at completion of the "process improvement" project.

Conventional wisdom emphasizes the importance of reinforcing mechanisms and embedding desired changes in structures, processes, systems, target setting, and incentives. In making any changes to an enterprise business structures, processes, systems, and incentives, project managers should pay what might strike them as an unreasonable amount of attention to employees' sense of the fairness of the change process and its intended outcome. Particular care should be taken where changes affect how employees interact with one another (such as head count reductions and talent-management processes) and with customers (sales stimulation programs, call center redesigns, and pricing).

Conventional wisdom also emphasizes the importance of building the capability, skills and talent needed for the desired change in form, quality, or state, over time from the original "process to be improved" to the new "improved process." There are two insights that demand attention from the project manager in order to succeed.

1. Employees are what they think, feel, and believe in. As managers attempt to drive performance by changing the way employees behave, they all too often neglect the context that, in turn, drives behavior.
2. Good intentions are not enough. Good skill-building programs usually take into account that people learn better by doing than by listening.

Some form of training is usually required to implement a new "improved process." If the new "improved process" to be implemented is a simple extension of the work currently being performed, then a one-time discussion of the new "improved process" with the workers affected may be all the training required. Such training could be done on the job or by reviewing the new standards at a meeting.

If the new "improved process" is complex (as when it involves the use of new technology), then extensive, formal classroom training may be required to implement the new "improved process." The type of new "improved process" that is being proposed, who will be asked to implement the new "improved process," and the skill level and work experience of the target group are all considerations in how much training is done. The training required for testing a new "improved process" is minimal, often only requiring a "watch one and do one" approach. Training to support implementation requires a broader, longer-term approach.

We advocate a number of enhancements to these training approaches in order to hardwire day-to-day practice into capability-building processes. First, training should not be a one-off event. Instead, a "field and forum" approach should be taken, in which classroom training is spread over a series of learning forums and fieldwork is assigned in between. Second, we suggest creating fieldwork assignments that link directly to the day tasks of participants affected by the new "improved process," requiring them to put into practice new mind-sets and skills in ways that are hardwired into their responsibilities. These assignments should have quantifiable, outcome-based measures that indicate levels of competence gained and certification that recognizes and rewards the skills attained.

35.4 Create Standard Practices and Procedures

This is the process by which methods of establishing specific recognized policies and practices that act as a model or guidelines for ensuring that critical elements of the new "improved process" are set to perform consistently in the best possible way. The actual documented policies, materials, methods, equipment, and training are usually called "standards" or "best practices."

Enterprise businesses that effectively create standard practices and procedures and use these standards exhibit many of the following conditions:

1. Management requires the use of standards, especially to document improvement efforts.
2. Different employees and shifts use the same standards and expect similar results.

3. Employee training focuses on the documented standards for materials, methods, and equipment. Critical elements and impact on internal and external customers are discussed.
4. Employees document and compare steps, procedures, and results to an easily accessed standard for each when solving problems.
5. Standards are regularly updated and changed on the basis of new knowledge about better methods. To avoid sub-optimization, conditions that get similar results are reviewed for least cost and best results for the overall system.
6. Employees share information with co-workers and managers in ongoing efforts to improve.
7. Variability in the outcomes of processes is reduced, resulting in more predictability.
8. To assist in maintaining improvement, work can be periodically audited to determine whether standard processes are being adhered to.

The following are steps to be used for creating standard practices and procedures:

1. Document the work context of the new "improved process";
2. Collect documents that represent the new "improved process";
3. Compare the documented procedure for executing its critical elements with the actual procedure;
4. Reconcile actual practice within the enterprise business with the documented procedure;
5. Plan to use the documented standard procedure;
6. Use the standard procedure;
7. Check on the use of the standard procedure

35.5 Release Resources and Adjourn Project Team

This is the process by which all resources, equipment, materials, and particularly personnel are released from the project and the project team is adjourned. In addition, all resource managers must be informed that they are relieved of their commitments to the project.

Equipment and materials can simply be returned to stores or suppliers, but people require special attention. Equipment and materials cannot have any further influence on current or future projects. People, however, can have a profound influence on the success of future projects. In fact, project team members may have made significant contributions, or even sacrifices, to the success of the project. If it is not recognized, the project will at best end on a disappointment, at worst it will leave lasting resentment that will roll over into the next project. If the releasing of people is not handled with tact and fairness, a feeling of resentment will form.

Adjourning the project team is the fifth stage of Tuckman's Group Development Model described in the previous chapter. Performing this fifth stage is certainly very relevant to the people in the group and their well-being. As indicated in the previous chapter, teams exist only for a fixed period, and even permanent groups may be disbanded through organizational restructuring. Breaking up a group can be stressful for all concerned and the "Adjourning" stage is important in reaching both group goal and personal conclusions. The break up of the group can be hard for members who like routine or who have developed close working relationships with other group members, particularly if their future roles or even jobs look uncertain.

When adjourned, project team members need to be reassigned, and if this can be organized at an early stage, it will go a long way to removing some uncertainties and lack of motivation that team members often face at conclusion of a project. Team members may be returned to their functional areas or assigned new projects, or both, or may have to be let go. What happens to released personnel can be summed up as one of the following:

1. *Inclusion*—team members are absorbed as part of the project's outcomes into the customer organization.
2. *Integration*—team members are reintegrated into the enterprise businesses and departments from which they were "borrowed."
3. *Extinction*—once the project is closed down, the team members' jobs simply end.

The project manager is responsible for releasing the resources off the project, and is therefore morally obligated to ensure that personnel are released in a fair and proper manner.

35.6 Settle Contractual Aspects and Final Accounting

This is the process by which the project manager ensures that all the contractual aspects are settled and that the final project accounting is done. All contractual commitments to customers, vendors and suppliers must be carefully examined and finalized. This may require reports to be exchanged and final payments to be made and received.

The final accounting of the project must be completed. This will include totaling the costs and revenues, producing the final cost evaluations and reports, paying all accounts, and closing the project's books. Once this is done, no further costs can be incurred against the project, and this is why the closing of the books is done after the resources have been released and contractual commitments settled.

The contractual documentation and final cost and accounting reports are used during the post implementation audit.

35.7 Write Final Report

This is the process by which the project manager writes the project final report summarizing the history of the project and evaluation of the performance. It is one last administrative task the project manager must perform before terminating the project. This is one task that cannot be delegated. It is the project manager and the project manager alone who must write this final report.

The final project report acts as the memory or history of the project. Most of the information about the project is already contained in the project documentation (covered earlier in the first step), or has resulted from earlier steps in the closeout process. Most of the remaining content of the report will be reflective, giving the project manager's honest and candid view of the completed project. It should be written as soon as possible after the project has been completed. It is the file that others can check to study the progress and obstacles of the project. Many formats can be used for a final report, and the following is a summary of suggested contents:

1. *Review:*
   - The initial project objectives in terms of technical performance, time and cost;
   - The soundness of the initial objectives in hindsight;
   - The evolution of the objectives up to the final objectives, and how well the project team performed against them;
   - The reasons for alterations to the objectives, noting which were avoidable and which were not;
   - The activities and relationships of the project team throughout the project life cycle;
   - The interfaces, performance and effectiveness of project management;
   - The relationships among top management, the project team, the functional enterprise business and the customer;
   - The cause and the process of termination;
   - Customer reactions and satisfaction;
   - Expenditures, sources of costs and profitability.
2. *Identify:*
   - Areas where performance was good and note the reasons, organizational benefits, project extensions and marketable innovations;
   - Problems, mistakes, oversights and areas of poor performance, and determine the causes.
3. *Comment on:*
   - *Overall success of the "process improvement" project.* Taking into account all of the measures of success that we considered, can we consider this project to have been a success?
   - *Organization of the "process improvement" project.* Hindsight is always perfect, but now that we are finished with the project, did we organize it in the best way possible? If not, what might that organization have looked like?
   - *Techniques used to get results.* By way of a summary list, what specific things did you do that helped to get the results?
   - *Project strengths and weaknesses.* Again by way of a summary list, what features, practices, and processes did we use that proved to be strengths or weaknesses? Do we have any advice to pass on to future project teams regarding these strengths/weaknesses?
   - *Project team recommendations.* Throughout the life of the project, there will have been a number of insights and suggestions. This is the place to record them for posterity.

The project final report should be made available to senior management and aspiring project managers, to give them the opportunity to apply the experience to future projects.

35.8 Conduct Post-implementation Review

This is the process by which an evaluation of the project's goals and activity achievement as measured against the project plan, budget, time deadlines, quality of deliverables, specifications, and client satisfaction is performed. The purpose of the post-implementation review (PIR) is not only to assess the project's level of success but also to identify lessons learnt and make recommendations for future projects to enhance their likelihood of success. Everyone learns from a project, no matter how big or small. That is inevitable, because every project is different, with its own special risks, new design, and new outcomes and so on. Experiences from one project can strengthen our performance on future projects, even though those projects may be dissimilar in many ways. When a "process improvement" project comes to an end it is important to reflect on all the events and experiences—mistakes as well as successes—so that work on the next and future projects will have an improved experience base.

The post implementation review can be conducted as a formal audit or as a workshop involving the project participants. Whichever method is used, the post-implementation review (PIR) results are recorded in a written document which is retained by the business as the last record of the project.

The PIR is undertaken after the project final report has been approved and all project closure activities completed. Some enterprise businesses wait a number of weeks before undertaking the PIR, to enable the benefits provided by the "process improvement" project to be fully realized by the business. The PIR is typically completed by an independent person who offers an unbiased opinion of the project's level of success. It is presented to the project sponsor/customer for approval and is retained on file within the enterprise business process asset for future projects. The log of the project activities serves as baseline data for this review. In undertaking the PIR, there are six important questions to be answered:

1. Was the project goal achieved?
2. Was the project work done on time, within budget, and according to specification?
3. Was the client satisfied with the project results?
4. Was the customer satisfied with the project results?
5. Was business value realized?
6. What lessons were learned about your project management methodology?
7. What worked? What did not?

Was the project goal achieved?—The project was justified based on a goal to be achieved. It either was or it was not, and an answer to that question must be provided in the review. The question can be asked and answered from two different perspectives.

1. Does the new "improved process" do what the project team predicted it would do?
2. Does the new "improved process" do what the customer required it must do?

The provider may have suggested a solution for which certain results were promised. Did that happen? On the other hand, the requestor may have promised that if the provider would only provide, say, a new or improved system, certain results would occur. Did that happen?

Was the project work done on time, within budget, and according to specification?—Recall from the scope triangle that the constraints on the project were time, cost, and the customer's specification, as well as resource availability and quality. Here the project review team should be concerned with whether the specification was met within the budgeted time and cost constraints.

Was the customer satisfied with the project results?—It is possible that the answers to the first two questions are yes, while the answer to this question is no. How can that happen? Simple; the Conditions of Satisfaction changed, but no one was aware that they had. The project manager did not check with the customer to see if the needs had changed; the customer did not inform the project manager that such changes had occurred.

Was business value realized? (Check success criteria.)—The success criteria were the basis on which the business case for the "process improvement" project was built and were the primary reason why the project was approved. Did the project realize that promised value? When the success criteria measure improvement in profit or market share or other bottom-line parameters, the project manager may not be able to answer this question until some time after the project is closed.

What lessons were learned about this Lean Six Sigma project management methodology?—Enterprise businesses that have or are developing a project management methodology will want to use completed projects to assess how well this methodology is working. Different parts of the methodology may work well for certain types of projects or in certain situations, and these should be noted in the review. These lessons will be valuable in tweaking the methodology or simply noting how to apply the methodology when a given situation arises. This part of the review might also consider how well the team used the methodology, which is related to, yet different from, how well the methodology worked.

What worked? What did not?—The answers to these questions are helpful hints and suggestions for future project managers and teams within the enterprise business. The experiences of past project teams are real "diamonds in the rough;" the project manager will want to pass them on to future teams.

The post-implementation review is seldom done, at completion of most "process improvement" projects. This is unfortunate because it does have great value for all stakeholders. Some of the reasons for skipping the review include these:

1. Managers do not want to know. They reason that the project is done and what difference does it make whether things happened the way we said they would? It is time to move on.
2. Managers do not want to pay the cost. The pressures of the budget (both time and money) are such that they would rather spend resources on the next project than on those already done.

3. It is not a high priority. Other projects are waiting to have work done on them, and completed projects do not rate very high on the priority list.
4. There is too much other billable work to do. Post-implementation reviews are not billable work, and they have billable work on other projects to do.

We cannot stress enough the value in the post-implementation review. There is so much valuable information that can be extracted and used in other projects. Enterprise businesses have such a difficult time deploying and improving their project management process and practice that it would be a shame to pass up the greatest source of information to help that effort.

35.9 Celebrate Success and Share the Wealth

This is the process by which the project celebrates success and share the wealth. Even though the team may have started out as a "herd of cats," the "process improvement" project they have just completed has honed them into a real team. Bonding has taken place, new friendships have formed, and mentor/mentees relationships have been established. The individual team members have grown professionally through their association with one another, and now it is time to move on to the next project. This can be a very traumatic experience for them, and they deserve closure. That is what celebrating success is all about.

The project manager and the senior management team should not pass up an opportunity to show the project team their appreciation. Loyalty, motivation, and commitment by a professional staff are the result of this simple act on the enterprise business part. The important message to convey is that the top leadership of the enterprise business understands the achievements and the frustrations of the employees who have contributed to the "process improvement" project success.

35.9.1 Celebrate Success

Once the improvement is in place and the enterprise business is reaping its benefits, it is motivated to celebrate its success. Success is important to the progress of any endeavor and, therefore, should not be ignored; it should be celebrated often. The tendency is to wait until a project is complete and all the savings have been tallied before we think of celebrating the accomplishments. This method, however, will usually be seen as too little, too late by many of the individuals involved, and in some cases, invariably tends to overlook some of the participants. Each and every person involved did his or her part, each individual will view their contribution as important, and each participant will consider their input as having come about through hard work and added effort, above and beyond their day to day responsibilities. By celebrating success as it happens, no matter how big or how small, everyone involved will not only feel appreciated and important, but motivated and driven to succeed further.

Celebrating success brings optimism and helps improve the context of the system affected by the implementation of the new "improved process." Organizational leaders can make a difference in performance of employees by noticing and celebrating small successes and underplay failures. The leaders can play a pivotal role in allocating the blame for failure to the system affected by the implementation of the new "improved process" and its context, and the leaders can also praise successes to the project team members and employees efforts. In this fashion they can help create more optimistic employees, who are more likely to endeavor to bring about positive context conductive to "Continuous Improvement" transformation.

Acknowledgment of a job well done promotes the spirit we all want to see in our employees and co-workers, it instills pride in their work, and it fosters a sense of worth that culminates in a workforce that looks for problems and willingly brings forth solutions. Participation on a team will no longer be viewed as an added burden to an already heavy workload, but an honor and a responsibility. This ethical mindset will bring more to the bottom line and the future success of the enterprise business than one can imagine.

Celebrating success can be achieved by team events, individual recognition or enterprise business-wide conferences where the enterprise business can look at what has been achieved and recognize that continued success involves ongoing improvements.

Individual recognition is a form of employee motivation in which the enterprise business identifies and thanks employees who have made positive contributions to the enterprise business' success through successful completion of the "process improvement" project. In an enterprise business at the continuous improvement stage of maturity, motivation flows from the employees' pride of workmanship. When employees are enabled by management to do their jobs and produce a product or service of excellent quality, they will be motivated.

The reason recognition systems are important is not that they improve work by providing incentives for achievement. Rather, they make a statement about what is important to the enterprise business. Analyzing the enterprise business' employee recognition system provides a powerful insight into the enterprise business' values in action. These are the values that are actually driving employee behavior. They are not necessarily the same as management's stated values. For example, an enterprise business that claims to value customer satisfaction but recognizes only sales achievements probably does not have customer satisfaction as one of its values in action.

Recognition can be as simple as a commemorative mug, a tee shirt, a pizza party, tickets to a ball game, or something more formal, such as a public recognition. Public recognition is often better for two reasons

1. Some (but not all) people enjoy being recognized in front of their colleagues.
2. Public recognition communicates a message to all employees about the priorities and function of the organization.

The form of recognition can also range from a pat on the back to a small gift to a substantial amount of cash. When substantial cash awards become an established pattern, however, it signals two potential problems:

1. It suggests that several top priorities are competing for the employee's attention, so that a large cash award is required to control the employee's choice.
2. Regular, large cash awards tend to be viewed by the recipients as part of the compensation structure, rather than as a mechanism for recognizing support of key corporate values.

35.9.2 Share the Wealth

As we have indicated in the first chapter, Taylor's 'scientific management' was concerned first and foremost with how a business could survive. Its aims were twofold. Firstly, to improve both the efficiency and the effectiveness of work by eliminating unnecessary actions and activities, improving methods and building in suitable relaxation breaks. Secondly, to share the resulting benefit between employer and employee and so remove the distrust between workers and management which had resulted in 'soldiering', a phenomenon of workers purposely operating well below their capacity, or slow working and restricting output intended by the workers to safeguard employments.

Thus, the project manager and the senior management team should not stop at simply celebrating success of the "process improvement" project, but share the wealth. The old school approach has always been that since the enterprise business footed the bill to complete the projects, purchase the equipment, rearrange the facility, etc. they should reap the rewards. However, sharing the savings will go much farther and pay higher dividends than pocketing the profits. Sharing the wealth should not be seen as distributing the cash, but reinvesting in the enterprise business. Reinvesting in equipment, facilities, personnel, and let not forget, reinvesting in the enterprise business or in the "process improvement" project customers.

As success mounts, the enterprise business will grow, profits will increase, and so too will the workload placed on its employees. The enterprise business has already invested heavily in the training and development of the personnel affected by implementation of the new "improved process"; it does not want to lose them now. In all actuality, the enterprise business should be looking to its employees to take on more of management roll than a laborer attitude, to mentor and train its new hires, and to apply their experience in finding and minimizing defects in the system affected by the new "improved process."

Nothing will stop a person dead in his tracks and send him packing than the thought that his hard work was taken for granted. Hence, as a project manager or senior management team member, make sure the raises you give out are commensurate with the employees' worth: "Promote from within." Make it a given that when a person embraces the responsibilities imposed upon him to improve business processes, that the experience gained has value and that as the enterprise business grows so will its' employees. Why would you take a chance hiring an unknown when a perfectly capable leader is already on the premises? It is important to keep this in mind when setting employee goals and determining what training you will provide. Make a commitment to grow your employees just as you have committed to growing your business.

Spend some of that wealth on improving your equipment. Get that newer, faster, more accurate equipment on the floor; it will increase throughput and productivity, keep you on the leading edge of technology, and open new doors to capabilities and customers you did not have before. Purchase software to reduce the documentation burden, improve the look and conditions of your facility, and hire a higher caliber of employee.

And do not forget about the new "improved process" customers. Passing a significant amount of savings on to the customer will pay back exponentially. By reducing your costs you demonstrate a commitment to improvement and cost control, you also lend credibility to the enterprise business much more than you might think. When everyone else is raising prices and tacking on surcharges, if you can hold, or better yet, reduce prices, who are your customers going to deal with? You will not only retain the customers you already have, but you will attract new ones looking to increase their own bottom line. Additionally, when you inevitably underestimate that one quote, you will stand a better chance of having your customer accept an adjustment. If you can demonstrate time after time that you are reducing your costs and pricing, when the time comes that you need to raise one, your customers will be much more willing to accept it and still feel confident that overall, they are still getting the best value from the market.

35.10 Conclusion to "PDSA Act" Process Group

We cannot stress enough the value of the "PDSA Act" phase of all the PDSA project phases. It is the most crucial in the life of a "process improvement" project as it can determine whether, ultimately, the project is a success or failure. This all-important phase should ensure that practices needed to sustain the new "improved process" over long term are established and effective and timely completion of the project is performed.

The manner in which this phase is managed can affect how the "process improvement" project is remembered. As indicated already, often, no one within an enterprise business remembers an effective start-up of a project, but everyone remembers an ineffective close-out as the consequences are felt for a long time. The manner in which the "PDSA Act" project phase is performed will determine how the project will reach conclusion.

Throughout the previous sections, we have illustrated and developed the "PDSA Act" Process Group processes needed to act upon the built and studied deliverables. This, based on what was learned from the previous project phase, implement the new "improved process" permanently or stop its implementation, determine what

Fig. 35.1 Minimum activities of the "PDSA Act" phase

modifications should be made to the system affected, and formally close the "process improvement" project.

As was the case with the "PDSA Plan," "PDSA Do," and "PDSA Study" constituent processes, effective integration and interactivity of the "PDSA Act" constituent processes will also do more than anything else to help enterprise businesses realize "Process Improvement & Management" transformational potential.

Figure 35.1 shows the minimum activities that are part of the "PDSA Act" project phase, in addition to the already listed "PDSA Plan," "PDSA Do" and "PDSA Study" activities.

In this figure we use the "Standardize" and "Close" nomenclature of the six-sigma literature for convenience and consistency with existing literature. We have placed "Dialogue," which is the basic unit of a "process improvement" project work that enables this continual learning, at the very center of the PDSA Cycle in Fig. 35.1.

Conclusion 36

"Process improvement" projects are vehicles for realizing enterprise business intended strategies; hence, transforming enterprise businesses. They are the means by which enterprise businesses achieve efficient cost structures and more effective operations. They are the means by which enterprise businesses develop new products and execute new business strategies. When "process improvement" projects succeed, they deliver: revenue growth, improved productivity, lower costs, more efficient operations, and higher market valuations. When they fail, they drain critical investment, waste valuable resources, and—directly or indirectly—limit an enterprise business ability to compete.

The chapters of this book have provided the framework and systematic methodology to enterprises business management and to professionals engaged in the "Continuous Improvement" transformation initiative implementation to successfully deliver "process improvement" projects and operations work from end to end.

36.1 Data Collection System: The Fundamental Engine of "Process Improvement"

We have characterized a "data collection system" as consisting of data obtained from a sample, appraisers or people executing the data collection tasks, operational definitions and procedures followed to collect the data, and data collection instruments. The events associated with any one of these constituents are not conveyed to the other constituents; that is, the constituents of a "data collection system" are statistically independent.

As you have realized throughout the chapters of this book, an effective data collection system is the fundamental component crucial to enhance the chance of achieving a "process improvement" project objectives within the "process improvement" project framework developed. Effective management of a "process

A. van Aartsengel and S. Kurtoglu, *Handbook on Continuous Improvement Transformation*, DOI 10.1007/978-3-642-35901-9_36,

improvement" project is based on the foundation of effective data collection system, and almost everything else is based on that. Indeed:

> *"If you cannot collect appropriate data on a process, you cannot understand what that process is doing or your understanding of it is meager. If you cannot understand what the process is doing or your understanding of it is meager, you cannot control it. If you cannot control it, you cannot improve it."*

Furthermore, a "process improvement" project is a conglomeration of many constituent project management processes. Data collected using the data collection system determines what the constituent processes used throughout the course of the project do, and works through these constituent processes to touch every part of the "process improvement" project. An effective data collection system improves decision making and the hallmark of any highly effective "process improvement" project management is making good decisions and making them better, faster, and more consistently throughout the "process improvement" project lifecycle. Unfortunately, few project managers working on "process improvement" have ever formally learned how to make improvement decisions, much less to make data-based decisions. This is why two out of three managers working on "process improvement" projects use "failure-prone" decision-making practices, and also why as many as 50 % of all managerial decisions on "process improvement" projects fail.

One of the major reasons for "failure-prone" decisions is over-reliance on intuition, that is, on an individualistic combination of experience, opinion, mythology, power, politics, and probability, all of which are highly susceptible to bias and personal blind spots. In the absence of data, anyone's opinion is as good as anyone else's—but usually the highest ranking opinion wins! In process improvement initiatives, we would be wise to remember the maxim: *"One accurate collected data set is worth a thousand opinions."* Intuition is good, but by itself, it is not good enough. Among other things, this book also shows to enterprise business executives and managers how the collected data can improve business intuition and significantly increase the "decision making batting average."

Finally, following the writings of Spitzer's in his description of the five phases of "Learning Effectiveness Measurement" methodology (Spitzer, 2005), we should indicate that the data collected throughout a "process improvement" project lifecycle undergoes five stages:

1. Predictive data
2. Baseline data
3. Formative data
4. In-Process data
5. Retrospective data

36.1.1 Predictive Data

Predictive data are collected during the "PDSA Initiate" phase, before the "process improvement" project is selected or planned. They help to make the best "process improvement" investments, target the highest leverage improvement opportunities, and provide crucial data to increase the effectiveness of the "process improvement" project design.

Predictive data is crucial for establishing an expectation, or target, for impact on the system considered. Predictive data should be used actively (as a matter of fact, it should be used pro-actively!) to produce the kind of results we desire. Like a roadmap, it should be used to select and navigate to your destination, not just be used to confirm that you have arrived or not! This requires a transformation of the traditional view of data collected from being predominantly retrospective to being predominantly predictive. Predictive data asks the question: "What alterations should happen to the system considered?" while retrospective data ask the question: "What alterations already happened to the system considered?"

36.1.2 Baseline Data

Baseline data are collected during the "PDSA Plan" phase, before the "process improvement" project plan is performed. They help to identify pre-implementation data and a target value for each process outcome characteristic considered. One of the biggest mistakes made in "process improvement" projects is the failure to collect baseline data. Without baseline data, no before-and-after comparisons can be made, and it is impossible to know if there has been any improvement. Furthermore, without baseline data, credible process improvement targets can not be established. Without such vantage points, enterprise business managers are often designing interventions without knowing how much improvement they want and in what areas.

If relevant data is being continuously tracked throughout the enterprise business, baseline data should be relatively easy to collect. Unfortunately, if data collection is time-consuming, most project managers are often reluctant to invest their scarce resources in collecting appropriate data. This is a serious mistake that has come back to haunt most enterprise business functions that fail to track the effectiveness of their "process improvement" interventions—and, without baseline data, no meaningful data comparisons can be made.

36.1.3 Formative Data

Formative data, which should be collected during the "PDSA Do" phase, makes sure that predictive data is implemented in the "process improvement" intervention design and implementation plan so that maximum effectiveness can be realized.

Figure 36.1, adapted from Spitzer (2005), shows all the places where barriers to achieving business results can occur from performing the "process improvement" project. Formative data involves the review of the "process improvement" plan to

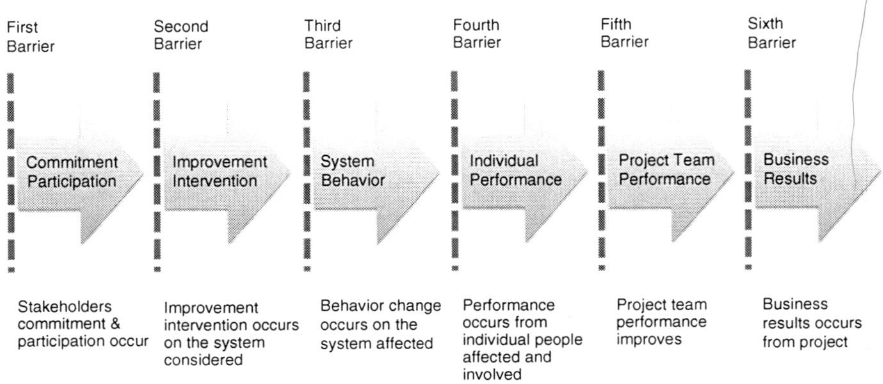

Fig. 36.1 Barriers to achieving business results from a project

ensure that the potential improvement and its implementation plan are powerful enough to achieve the desired results identified in the "Predictive Data," especially with respect to the critical success factors.

In order to achieve a business impact from the "process improvement" project, the "process improvement" project must overcome the many barriers that prevent the potential chain of causality from being realized. Even if the "process improvement" project is "well-planned" in the traditional sense, target people affected by the improvement might not participate in the improvement effort, improvement might not occur, improvement might not result in behavior change needed for "Continuous Improvement" transformation, behavior change might not improve performance from the individual people affected, even if performance from the individual people affected improves it might not impact the project team performance, and the project team performance improvement might not be reflected in the financial results of the enterprise business intended strategy. Real results from the "process improvement" project will be achieved only when these 'barriers to impact' are overcome. This 'chain of impacts,' and all the potential break-points in it, also provide some insight into why disbelief in realizing real transformation from most "process improvement" projects claims is justified from employees "who are in the system considered to make waves."

However, making the right "process improvement" project plan and implementation decisions can promote high return on investment on the improvement interventions with the following generic recommendations:

1. Use formative data to focus on system behavior change, individual performance improvement, team performance improvement, and business results from the "process improvement" project.
2. The objective of the "process improvement" project should no longer be just a matter of producing non-defective outcomes, but disseminating knowledge and developing skills; hence helping the enterprise business transforms.

36.1.4 In-Process Data

In-Process data are also collected during execution of the "PDSA Do" phase to track the effectiveness of the "process improvement" intervention, during piloting and small-scales deployment, and enable corrective actions (if needed). During the piloting of a "process improvement" intervention, data should be collected. Just because a "process improvement" intervention has been carefully planned to meet the pilots and performance requirements does not mean that it will be as effective as anticipated once it is deployed. The frequency of collecting in-process data is dependent on the criticality of the "process improvement" intervention within the enterprise business or within the system considered.

This data should provide timely feedback on how well the "process improvement" intervention is working, on possible problems that might require some corrective action, opportunities to further enhance the "process improvement" intervention, or, in rare instances, discontinue it. It provides early warning signals. No matter how right the prototype solution, what good is a problem that is solved too late? Many people do not recognize a problem until it reaches a crisis level. This is the same situation with a potential improved process. If crucial problems are not identified and addressed early on, the longer-term consequences can be quite severe.

During the collection of in-process data, the project team needs to calibrate the collected data for both their internal and external environments to recognize that change made to the "process to be improved" is an improvement. Hence, answering the fundamental questions, which form the basis and the preliminary step of the PDSA model. Sometimes small but significant changes, which normally might be overlooked, can become visible with the right set of collected data. Good in-process data is a lot less expensive in the long-run than major process overhaul.

In-Process data is high-leverage because it typically requires little effort and can provide extremely valuable, ongoing feedback. Traditional data collection typically occurs too late for such decisions and remedial action to be taken. In order to maximize effectiveness, data collection should occur throughout the piloting lifecycle—from initial conception of the pilots to the end of deployment. Remember: Without data collected, it is impossible to manage anything and the earlier you start collecting data, the more leverage you can get from the data collection system.

36.1.5 Retrospective Data

Retrospective data are collected during the "PDSA Study" phase, after the pilots of the "process improvement" project is concluded and fully implemented. They are 'post-intervention' data and that enable final evaluative decision making. Retrospective data is the final collected data that occurs at the end of the "process improvement" intervention deployment. This does not necessarily mean that the "process improvement" intervention is terminated and taken out of service, but that

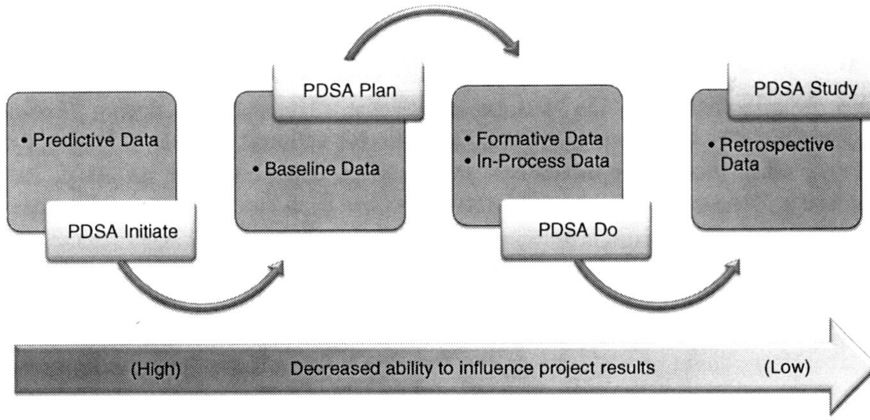

Fig. 36.2 Data ability to influence project results

it has reached a point at which the improvement intervention is sufficiently mature that data can be retrospectively collected. It can also be considered the last in-process data point.

The primary purpose of collecting data at this juncture is to make final judgments about the "process improvement" intervention, including the calculation of returns on investment, if desired. However, "Retrospective Data" is typically too late to inform the most important decisions on the current intervention—although these retrospective data can be used to inform decisions about future "process improvement" projects.

One of the most important distinctions between the five stages of data collected throughout the "process improvement" lifecycle is the relative leverage of the collected data—its ability to influence results. As shown in Fig. 36.2, predictive data has the highest ability to influence the project results, while retrospective data has the least.

The right sets of collected data (predictive data, baseline data, formative data, in-process data, retrospective data) trigger the right decisions and the correct project management constituent processes—because they represent factual information from many sources with each having varying levels of completeness and confidence and from which baselines are established. Similarly, the wrong sets of collected data tend to trigger the wrong activities. Then these wrong activities generate the wrong results—no matter how well-executed the activities are. Most "process improvement" projects do not reach successful conclusions because they do not use effective data collection systems!

Indeed, the goals that a "process improvement" project sets will depend on what data the project team collects. These goals, however, are really nothing more than measurable "target" established on a particular data collection scale. But, the project manager and the project team must first define that scale. The data collection scale can be net profits, customer satisfaction, reduce cost, reduce cycle time, improve productivity, reduce defect rate, decrease potential risk level, stakeholder influence and interest, etc... If the collected data are not reliable, the established project baselines and everything else will be as well. Thus, the data collection system is the engine that drives successful realization of the "process improvement" project.

In order for the full power of data collection system, hence "process improvement" project, to be realized, there must be an optimal environment for effective use of data collected, there must be considerable interaction at each use of the corresponding constituent project management processes leading to new insights about what data to collect, how to collect it, and what are the subsequent right management decisions. Attaining the optimal environment, as we indicated already, requires a specific and intensive set of actions—a transformation process progressing from improving context of data collection activity, to improving focus, to improving integration, to improving interactivity—the four aspects of paramount importance to making progress on "Continuous Improvement" transformation.

Within the "Process Improvement and Management" dimension of "Continuous Improvement" transformation, the factors which contribute to transform the interactivity of "Process Improvement and Management," as shown in our first book, include the following:

1. Frequent Interactivity
2. Effective and robust dialogue
3. Collaborative learning
4. Appropriate use of technology

Thus, performing the PDSA constituent processes should include highly interactive and iterative (ongoing) discussions, or dialogues, which are the most important aspects of a data collection system. These dialogues should be built on the foundation of a positive context.

Regardless of the type of data collected, it should be based on a desire to better understand what is happening—and, at least initially—without judging. Effective data collection should precede any judgment or decision-making although it all-too-rarely does. In order to be credible, improvement of any process should always be based on a solid foundation of data collected.

36.2 Learning and Knowledge

Throughout this book, you have also realized that data collection system is being used to provide real understanding, helpful feedback, and to foster and enable learning and "Continuous Improvement" transformation, rather than just monitor goal achievement. It is a journey from data collection confusion to data collection clarity—but the project team still has a long way to go. Data, which has meaning only within a well defined context, is the raw materials from which information, knowledge, and wisdom can ultimately be created, but it will not take very far until you do something with it. Hence, it is fundamental for the project team to understand the context of the data collected before it begins to perform any associated constituent process management process. It is the background for the collected data that determines how the project team should organize the data, how it should analyze the data, and how it should interpret the results of the analysis. Once the project team ignores the context of the data collected, the "process improvement" project is like a train that has gone off the track, with the inevitable result being a stuttering failure and the train reduced to rubble.

We have shown throughout this book that collecting data is not about reacting to isolated data points, but combining numbers with observations, questions, hypotheses, visualizations, and intuition, and helping everyone understand "the story" behind the data collected that used to be hidden in the abstract numbers. When organized and presented in such a way that its meaning can be recognized by a user, the collected data becomes information. Furthermore, information often describes, defines, or provides perspective, and is created from data by such means as organizing (e.g., sorting, combining), comparing, analyzing, and visualizing. Information is commonly used to support what we already know and to justify decisions; this is okay, but this will not create any new insight, knowledge, or wisdom needed to help move the enterprise business as a system from its current stage of maturity to the desired "Continuous Improvement" stage of maturity.

One of the easiest ways to convert data into information is to add some historical perspective. For example, depicting a relevant trend in numbers on a graph using the baseline data collected in the "PDSA Plan" project phase, the in-process data resulting from the pilots conducted in the "PDSA Do" project phase, and the retrospective data collected over time after conclusion of the pilots of the prototype solution, is information. It is also helpful to have some basis for comparison—a target, a baseline, a benchmark—something that will enable you to add a meaningful context to the data. A valuable rule of thumb for information presentation is to always focus on the meaning, when others might be focused on calculations.

When information is combined internally with other information and personal experience into a form that can be useful it becomes knowledge. Knowledge is personally relevant information you can take action on. Action-oriented insights gleaned from deeper analysis of charts or graphs can become knowledge. "Know-how" is knowledge. For example, "We can attain customer satisfaction from our improved process outcomes forecast by . . ." Personally appreciating the implications of the trade-offs between the collected data in a data collection system is knowledge. The ability to make credible predictions and forecasts requires knowledge.

As additional knowledge and experience is accumulated throughout the course of the "process improvement" project, the base and level of knowledge available on the process studied should increase. When "process improvement" team members achieve real knowledge about some subject of the "process to be improved" being studied, their reaction should reflect the learning satisfaction: "Aha, now we get it!"

Knowledge can be individual or organizational; it can be explicit (documented) or tacit (include one's head or in the enterprise business memory). The more effectively knowledge is managed, the more readily it will grow as knowledge, rather than just as additional data or information. This happens through continuous practice of a way of thinking rather than through collection of data and implementation of the "right" technique or solution. Without the practice in the way of thinking, the way we advocate in our first book, simply developing a new process will not help the enterprise business move to a higher state of maturity.

The process by which a project team (as well as its individual member) gathers and uses new knowledge, with appropriate consideration for the tools, behaviors, and values at all levels, is a learning process. Newly learned knowledge is translated into new goals, procedures, roles, and performance measures. This is the learning process that an enterprise business should develop in order to operate its processes on-target and predictably; hence, decrease the effective cost of production and use of the process outcomes.

The rich understanding and insight that usually develops through a combination of extensive knowledge (knowing) generation, learning and personal experience (doing) over time can be characterized as wisdom. Wisdom is deep and cannot be seen directly, but can be inferred from a track record of consistently good decisions. Wisdom grows through the interplay of existing knowledge, new knowledge (extracted from new information) acquired through study and communication with other knowledgeable people, practical experiences, reflection, etc. In the process improvement arena, there is no such thing as "instant wisdom" on improving processes and it cannot be purchased from a consultant. Every enterprise business (and every person affected by a process being studied) must develop its own wisdom.

A "process improvement" project performed right, in the right context, is a powerful vehicle for realizing the enterprise business intended strategy, developing organizational, as well as individual, and wisdom. Wisdom, when it is acquired, usually looks simple—"Why didn't I realize that before?"—Although the process of acquiring it certainly is not. In fact, very few enterprise businesses are willing to invest much effort in driving learning on data collected from "process improvement" projects from data into wisdom!

By far the most pervasive technique for learning from project experience touted by project management programs and methodologies is the practice of lessons learned; that is, identifying ways of learning that have merit (quality), worth (value), or significance (importance) for the next phase of the "process improvement" project or for future projects within the enterprise business. However, there are three fundamental challenges associated with the lessons learned approach that could render it largely ineffective within enterprise businesses at lower stages of maturity (Julian, 2009).

First, it defers structured learning from experience until the end of a project phase or the end of the whole project, perhaps months or even years after the project began. Project team members can easily forget the problems that arose, having dealt with them and perhaps solved them weeks or months earlier. By the time the lessons-learned session is conducted, the learning has become a distant memory—and that is if collective learning even happened in the first place. Perhaps the most damaging aspect of deferring structured reflection until the end of a project is people's lack of motivation for addressing the real issues. By the time the project is over, nothing can be done to resolve the problems that occurred. Team members may very well decide that addressing difficult conflicts or bringing up past problems is simply not worth it because it would serve only to open old wounds. They may feel that it is better to preserve working relationships among those in their organization than to jeopardize them for a project that is already completed. The desire to maintain harmony may very well outweigh the risks of dredging up the past when nothing can be done to fix the problem. Yes, there is the opportunity to help future teams, but that may not be a compelling enough motivation.

The second fundamental challenge associated with the lessons-learned approach is that it encourages learning from experience only at the project team level. In reality, projects are embedded within a constellation of communities of practice in the enterprise business, getting demands, pressures, support, and guidance from many different sources—almost always from senior managers, from the program managers or the project management office (PMO) (if one exists), and from other functional units within and outside the enterprise business, including customers and key stakeholders. It may be unfair to have project teams be the only source of lessons learned, as this may imply that they are also the primary source of any problems that occurred. Senior managers and project management office (PMO) leaders alike have much to learn from project experience. It could be argued that, for example, since these higher management levels launch and direct a multitude of projects, learning at those levels is even more important for the enterprise business overall health and performance.

It may be no wonder that some project teams consider structured learning from project experience a waste of time. After all, even if they do identify problems that need to be fixed the next time around, it may be the project management office (PMO) and the senior managers who need to make the required alterations, and if they are not part of the learning process, they may not understand the context or have the motivation to carry through with the team's input. They may even be threatened by the prospect of being perceived as part of the problem, choosing instead to focus on other issues that are less threatening. As a result, it may be that neither managers nor teams do anything to fix the problems for the next time around, creating a sense of frustration and futility that undermines future attempts at learning from project experience.

The third challenge with lessons-learned practices is the assumption that people can learn effectively from "lessons" stored on databases. The fundamental dilemma with this assumption is that knowledge can be possessed and therefore can be readily transferred to others in textual form. This view does not take into account the embedded, situated, and tacit nature of knowledge that manifests itself in practice. Some knowledge can be possessed independently of practice while other knowledge is deeply embedded in practice. Furthermore, management's efforts to reuse knowledge from past projects in product or service development could have the unintended consequence of stifling the development of expertise.

Before the reuse strategy was introduced, engineers and technicians developed unique, sometimes redundant designs, which led to "reinventing the wheel." Yet the motivation for learning and collaboration was high, and new engineers were developed through mentoring practices and exploratory learning opportunities.

To overcome these challenges associated with the lessons-learned practice, it is critical to conduct project retrospectives throughout the "process improvement" project life cycle.

Project retrospectives learning expands learning beyond the project team by encouraging reflection at three levels simultaneously: projects, the processes that are common to multiple projects, and the overall project portfolio itself. These levels mirror three types of reflection and include content, process, and premise reflection. The frequency of reflection can be increased by holding regular retrospectives throughout a project lifecycle, not just at the end of a significant milestone. This enables teams to learn from the more recent past, when the memories, emotions, and experiences are still fresh in their minds. The improvements that emerge from these discussions are more robust, realistic, and effective in solving critical challenges.

It is because of the combination of these factors—expanded levels, better quality, and increased frequency of reflection—that teams are more motivated to engage in conscious learning from experience. Rather than going through an exercise aimed at documenting "lessons" for future initiatives, teams are able to identify actions that can solve their immediate problems and improve results at a time when something can still be done. Moreover, as a result of more frequent structured learning, team members become more adept at reflecting collectively in a group format, enabling them to feel more competent and skillful in the art of addressing sensitive issues and communicating in ways that reduce the impact of negative process improvement context.

In addition to enhancing reflective practice in these ways, learning through project retrospective sessions taps the knowledge-brokering role of the project management office (PMO). The PMO leader or program director who oversees multiple projects should find ways to build the improvements into the way work gets done on future "process improvement" projects and programs. Such leaders should diffuse knowledge and maintain connections across multiple communities of practice, including senior management, project teams, and other functional disciplines. By doing so, they should bring learning from retrospectives to the systems level, incorporating it into work routines, systems, methodologies, tools, and templates; hence, helping the enterprise business move forward to the next level.

36.3 Final Admonition

We have indicated in the introductory chapter that the progressive realization of the enterprise full potential by moving from its current maturity stage towards a higher (ultimately "Continuous Improvement") maturity stage, requires a framework and a systematic methodology for studying the constituent elements or processes and systems associated with the eight overarching determining factors. You have now completed all five phases of this methodology.

By now, enterprises business managers and professionals engaged in the "Continuous Improvement" transformation implementation should have gained a detailed understanding of the "Continuous Improvement" transformation methodology by learning the phases, activities and tasks required to undertake a "process improvement" project.

There is no formula or cookbook recipe that can assure success in any complex endeavors. We have described a framework on which to proceed. In every case, the methodology will have to be adapted to the mechanisms, technologies, and the culture of the enterprise business, which the project manager must successfully negotiate to carry out the "process improvement" project. While there is no guarantee of achieving the project objectives successfully, following the project management constituent processes described in this book will put the project manager and his or her team on the right track. Having a track, recognizing what it is, and knowing where it is located is very essential. Certainly, there are hostile forces at work—some created by the barriers which the "process improvement" project itself faces (see Fig. 36.1)—that tend to push the team off course and threaten its well being. Knowing the track will help the project manager and the team proceed, even in the face of those forces.

Thus, to gain the maximum benefit from using the "Continuous Improvement Methodology" described in this book, we recommend enterprises business managers and professionals engaged in the "Continuous Improvement" initiative implementation to consider customizing it to suit their project environment. This can be done by selecting the project management constituent processes in the PDSA life cycle that are most relevant in their project environment. The methodology described is fully scalable, meaning the activities that best suit a particular "process improvement" project needs can be selected separately and the framework from which to deliver projects still remains robust. By selecting the project management constituent processes which are most relevant to a particular project environment, you can use this methodology to undertake any size of "process improvement" project, in any industry.

By adopting the "Continuous Improvement" methodology described in this book as project management methodology for process improvement, you as enterprise business manager or professional engaged in the "Continuous Improvement" initiative implementation will dramatically increase your chances of project success.

We can only hope that the practical tools and techniques we have shared will provide a lasting and valuable store of resources for you to use as you grow in this exciting profession of improving businesses. Whether you are an enterprise business manager or a professional engaged in the "Continuous Improvement" initiative implementation, you should have found value in these pages.

References

Ahuja, H. N., Dozzi, S. P., & AbouRizk, S. M. (1994). *Project management: Techniques in planning and controlling construction projects*. Hoboken, NJ: Wiley.

Badiru, A. B. (1996). *Project management in manufacturing and high technology operations*. New York: Wiley.

Bennett, F. L. (2003). *The management of construction: A project life cycle approach*. Boston: Butterworth-Heinemann.

Bertels, T., & Strong, R. (2003). *Rath & Strong's six sigma leadership handbook*. Hoboken, NJ: Wiley.

Breyfogle, F. W. (2003). *Implementing six sigma: Smarter solutions using statistical methods*. Hoboken, NJ: Wiley.

Breyfogle, F. W., Cupello, J. M., & Meadows, B. (2001). *Managing six sigma: A practical guide to understanding, assessing, and implementing the strategy that yields bottom-line success*. New York: Wiley.

Campanella, J. (1999). *Principles of quality costs: Principles, implementation and use*. Milwaukee, WI: American Society for Quality Control, Quality Costs Committee.

Carmichael, D. G. (2000). *Contracts and international project management*. Boca Raton, FL: Taylor and Francis.

Cendrowski, H., & Mair, W. C. (2009). *Enterprise risk management and COSO: A guide for directors, executives, and practitioners*. Hoboken, NJ: Wiley.

Cockrell, G. W. (2001). *Practical project management: Learning to manage the professional*. Instrument Society of America.

Crawford, J. K. (2006). *Project management maturity model*. Boca Raton, FL: Auerbach Publications.

Curlee, W., & Gordon, R. L. (2010). *Complexity theory and project management*. Hoboken, NJ: Wiley.

Cusumano, M. A. (1985). *The Japanese automobile industry: Technology and management at Nissan and Toyota*. Cambridge, MA: Council on East Asian Studies, Harvard University Press.

Deming, W. E. (1982). *Out of the crisis*. Cambridge, MA: MIT Press.

Deming, W. E. (1994). *The new economics: For industry, government education*. Cambridge, MA: MIT Press.

Dennis, P. (2007). *Lean production simplified: A plain-language guide to the world's most powerful production system* (2nd ed.). New York: Productivity Press.

Doran, G. T. (1981). There's a S.M.A.R.T. way to write management goals and objectives. *Management Review, 70*, 35–36.

Eckes, G. (2002). *The six sigma revolution: How general electric and others turned process into profits*. New York: Wiley.

Efron, B., & Tibshirani, R. (1993). *An introduction to the bootstrap*. New York: Chapman and Hall.

Ewan, W. D. (1963). When and how to use Cu-SUM charts. *Technometrics, 5*(1), 1–22.

Fayol, H. (1949). *General and industrial management*. London: Pitman Publishing Company.

Fleming, Q. W. (2003). *Project procurement management: Contracting, subcontracting, teaming*. Tustin, CA: FMC Press.

A. van Aartsengel and S. Kurtoglu, *Handbook on Continuous Improvement Transformation*, DOI 10.1007/978-3-642-35901-9, © Springer-Verlag Berlin Heidelberg 2013

Gale, B. (1994). *Managing customer value: Creating quality and service that customers can see*. New York: Simon and Schuster.

Greenbaum, T. L. (1998). *The handbook for focus group research*. Thousand Oaks, CA: SAGE.

Gupta, P. (2004). *The six sigma performance handbook: A statistical guide to optimizing results*. New York: McGraw-Hill Professional.

Harry, M. J., & Schroeder, R. (2006). *Six sigma: The breakthrough management strategy revolutionizing the world's top corporations*. Westminster, MD: Currency.

Hill, G. M. (2009). *The complete project management methodology and toolkit*. Hoboken, NJ: CRC Press.

Julian, J. (2009). *Facilitating project performance improvement: A practical guide to multi-level learning*. New York: Amacom.

Kawakita, J. (1977). *A scientific exploration of intellect*. Tokyo: Kodansha.

Kawakita, J. (1986). *The KJ method: Chaos speaks for itself*. Tokyo: Chuo Koron-sha.

Kerzner, H. (2004). *Advanced project management: Best practices on implementation*. Hoboken, NJ: Wiley.

Kerzner, H. (2009). *Project management: A systems approach to planning, scheduling, and controlling*. Hoboken, NJ: Wiley.

Kerzner, H. (2010). *Project management best practices: Achieving global excellence*. Hoboken, NJ: Wiley.

Kliem, R. L., & Anderson, H. B. (2003). *The organizational engineering approach to project management: The revolution in building and managing effective teams*. Boca Raton, FL: St. Lucie Press.

Koichi, S., & Takahiro, F. (2009). *The birth of lean: Conversations with Taiichi Ohno, Eiji Toyoda, and other figures who shaped Toyota management*. Cambridge, MA: Lean Enterprise Institute.

Krafcik, J. F. (1988). Triumph of the lean production system. *Sloan Management Review*, *30*, 41–52.

Langley, G. J., Moen, R., Nolan, K. M., Nolan, T. W., Norman, C. L., & Provost, L. P. (2009). *The improvement guide: A practical approach to enhancing organizational performance*. San Francisco: Wiley.

Lientz, B. P., & Rea, K. P. (2002). *Project management for the 21st century*. San Diego, CA: Academic Press.

Manly, B. F. (1998). *Randomization, bootstrap and Monte Carlo methods in biology*. London: Chapman and Hall.

Mantel, S. J., Meredith, J. R., Shafer, S. M., & Sutton, M. M. (2010). *Project management in practice*. Hoboken, NY: Wiley.

Marco, A. D. (2011). *Project management for facility constructions: A guide for engineers and architects*. New York: Springer.

Maylor, H. (2010). *Project management*. London: Prentice Hall PTR.

Merrill, H. F. (1970). *Classics in management*. New York: American Management Association.

Mezirow, J. (1991). *Transformative dimensions of adult learning*. San Francisco: Jossey-Bass.

Monden, Y. (2011). *Toyota production system: An integrated approach to just-in-time*. Taylor & Francis.

Moore, D. R. (2002). *Project management: Designing effective organisational structures in construction*. Oxford: Blackwell Science.

Morris, P. W., Pinto, J. K., & Söderlund, J. (2011). *The Oxford handbook of project management*. Oxford: Oxford University Press.

Nelson, D. (1992). *A mental revolution: Scientific management since Taylor*. Columbus, OH: Ohio State University Press.

Nicholas, J. M., & Steyn, H. (2012). *Project management for engineering, business and technology*. London: Taylor & Francis.

Ohno, T. (1988). *Toyota production system: Beyond large-scale production*. Portland, OR: Productivity Press.

Pande, P. S., Neuman, R. P., & Cavanagh, R. R. (2000). *The six sigma way: How Ge, Motorola, and other top companies are honing their performance*. New York: McGraw-Hill Professional.

Pande, P. S., Neuman, R. P., & Cavanagh, R. R. (2001). *The six sigma way team fieldbook: An implementation guide for project improvement teams*. New York: McGraw-Hill Professional.

Pearson, K. (1894). Contributions to the mathematical theory of evolution: I. On the dissection of asymmetrical frequency curves. *Philosophical Transactions, 185*, 71–110.

Perez-Wilson, M. (1999). *Six sigma: Understanding the concept, implications and challenges*. Scottsdale, AZ: Advanced Systems Consultants.

Pmi, P. M. (2004). *A guide to the project management body of knowledge: PMBOK guide*. Newtown Square, PA: Project Management Institute.

Pmi, P. M. (2010). *A guide to the project management body of knowledge (PMBOK guide)*. Newtown Square, PA: Project Management Institute.

Pritchard, C. L. (2010). *Risk management: Concepts and guidance*. Arlington, VA: ESI International.

Przekop, P. (2005). *Six sigma for business excellence: A manager's guide to supervising six sigma projects and teams*. New York: McGraw-Hill Professional.

Pyzdek, T., & Keller, P. A. (2009). *The six sigma handbook: A complete guide for green belts, black belts, and managers at all levels*. New York: McGraw-Hill Professional.

Revans, R. W. (1971). *Developing effective managers: A new approach to business education*. New York: Praeger Publishers.

Richardson, G. L. (2010). *Project management theory and practice*. Boca Raton, FL: Auerbach Publications.

Roberts, S. W. (1959). Control chart tests based on geometric moving averages. *Technometrics, 1*(3), 239–250.

Rosenau, M. D., & Githens, G. D. (2005). *Successful project management: A step-by-step approach with practical examples*. New York: Wiley.

Schmidt, T. (2009). *Strategic project management made simple: Practical tools for leaders and teams*. Hoboken, NJ: Wiley.

Schön, D. (1983). *The reflective practitioner*. New York: Basic Books.

Schön, D. (1987). *Educating the reflective practitioner*. San Francisco: Jossey-Bass.

Schwaber, K. (2004). *Agile project management with Scrum*. Redmond, WA: Microsoft Press.

Schwalbe, K. (2010). *Information technology project management*. Boston: Course Technology.

Sherden, W. A. (1994). *Market ownership: The art & science of becoming #1*. New York: American Management Association.

Shewhart, W. A. (1931). *Economic control of quality of manufactured product*. Milwaukee, WI: ASQ Quality Press.

Shewhart, W. A. (1939). *Statistical method from the viewpoint of quality control*. New York: Courier Dover Publications.

Shingo, S. (1986). *Zero quality control: Source inspection and the Poka-Yoke system*. New York: Productivity Press.

Shingo, S., & Dillon, A. P. (1989). *A study of the Toyota production system: From an industrial engineering viewpoint*. Portland, OR: Productivity Press.

Sodhi, M. S., & Sodhi, N. S. (2008). *Six sigma pricing: Improving pricing operations to increase profits*. Upper Saddle River, NJ: FT Press.

Spitzer, D. R. (2005). Learning effectiveness: A new approach for measuring and managing learning to achieve business results. *Advances in Developing Human Resources, 7*(1), 55–70

Spitzer, D. R. (2007). *Transforming performance measurement: Rethinking the way we measure and drive organizational success*. New York: AMACOM: A Division of American Management Association.

Stasiowski, F., & Burstein, D. (1994). *Total quality project management for the design firm: How to improve quality, increase sales, and reduce costs*. New York: Wiley.

Summers, D. C. (2007). *Six sigma: Basic tools and techniques*. Upper Saddle River, NY: Pearson/Prentice Hall.

Taylor, F. W. (1911). *The principles of scientific management*. New York: Harper & Brothers.
Tonchia, S., & Cozzi, F. (2008). *Industrial project management: Planning, design, and construction*. Heidelberg: Springer.
Torkzadeh, D. E., & Gholamreza, A. (2008). *Information systems project management*. Thousand Oaks, CA: SAGE.
Truscott, W. (2003). *Six sigma: Continual improvement for business: A practical guide*. Amsterdam: Butterworth-Heinemann.
Tsutsui, W. M. (2001). *Manufacturing ideology: Scientific management in twentieth-century Japan*. Princeton, NJ: Princeton University Press.
Verzuh, E. (2003). *The portable MBA in project management*. New York: Wiley.
Verzuh, E. (2011). *The fast forward MBA in project management*. New York: Wiley.
Vesely, W. E., Goldberg, F. F., Roberts, N. H., & Haasl, D. F. (1981). *Fault tree handbook (NUREG-0492)*. Washington, DC: US Nuclear Regulatory Commission.
Wadsworth, H. M., Stephens, K. S., & Godfrey, A. B. (1986). *Modern methods for quality control and improvement*. New York: Wiley.
Wang, J. (2010). *Lean manufacturing: Business bottom-lined based*. Boca Raton, FL: CRC Press.
Webb, M., & Gorman, T. (2006). *Sales and marketing the six sigma way*. New York: Kaplan Publishing.
Westland, J. (2007). *The project management life cycle: A complete step-by-step methodology for initiating, planning, executing & closing a project successfully*. London: Kogan Page.
Wheeler, D. J. (2003). Good data, bad data, and process behavior charts. *ASQ Statistics Division Special Publication*. SPC Press.
Wheeler, D. J. (2009a). An honest gauge R&R study. *ASQ/ASA Fall Technical Conference*. SPC Press. (Published in SPC Ink March 2008, Revised January 2009). http://www.spcpress.com/pdf/DJW189.pdf
Wheeler, D. J. (2009b, November). Two definitions of trouble. *Quality Digest Daily*. SPC Press. http://www.spcpress.com/pdf/DJW203.pdf
Wheeler, D. J. (2010a, August). The effective cost of production and use: How to turn capability indexes in dollars. *Quality Digest Daily*. SPC Press. http://www.spcpress.com/pdf/DJW215.pdf
Wheeler, D. J. (2010b, October). What is the zone of eco nomic production? And how can you get there? *Quality Digest Daily*. SPC Press. http://www.spcpress.com/pdf/DJW217.pdf
Womack, J. P., Jones, D. T., & Roos, D. (2007). *The machine that changed the world: The story of lean production—Toyota's secret weapon in the global car wars that is now revolutionizing world industry*. New York: Simon and Schuster.
Wortham, A. W., & Ringer, L. J. (1971). Control via exponential smoothing. *The Logistics Review, 7*(3), 33–40.
Wysocki, R. K. (2004). *Project management process improvement*. Boston, MA: Artech House.
Wysocki, R. K. (2011). *Effective project management: Traditional, agile, extreme*. Indianapolis, IN: Wiley.

Index

A

Activity
  attributes, 145, 153, 173
  list, 145
Actual cost, 289, 291
Adjourning, 187, 612
Affinity, 126, 128
Allocability, 360
Allowability, 360
Alteration request form, 558, 559
Analysis of variance (ANOVA), 114, 122–125
Appraisal costs, 13, 265–268
Assignable causes, 8, 93, 449, 527
Automobile, 10, 47, 89, 129, 184, 216, 217, 231, 232, 260, 283, 284, 345, 408, 418, 487, 542

B

Benchmarking, 134, 338, 343, 408
Bias, 94
Bidding, 46, 301, 307, 337, 339, 340, 350, 385
Binomial distribution, 215, 217
Bivariate data, 470
Block design, 251
Bootstrapping, 121
Brainstorm, 84, 400, 513, 521
Brainstorming, 34, 70, 84, 86, 157, 400, 401, 457, 512, 513, 519–521
Brink of Failure, 243–245
Business case, 45, 47, 180, 274, 301, 616

C

Capability indices, 240
Capital costs, 269, 273
Cartesian coordinates, 223
Cash flow, 262, 271, 276, 301, 343, 394
Central limit theorem, 8, 99, 103, 115, 120, 213
Central tendency, 7, 9–12, 15–17, 24, 221, 222, 231, 233, 242, 243, 268, 536, 537
Chi-square distribution, 119
Cognitive task analysis, 497, 505
Common causes, 7, 93, 153, 527
Communication matrix, 367
Competitive value, 341, 343
Compliance, 26, 29, 39, 135, 176, 201, 230, 270, 336, 341, 343, 352, 353, 356, 359, 368, 394, 487, 554, 592, 597
Confidence, 38, 54, 99–102, 115, 116, 119–121, 189, 195, 223, 275, 276, 314, 321, 398, 439, 440, 442, 476, 531, 565, 628
Conflict, 388
Conformance, 11–13, 15, 16, 38, 176, 200, 201, 231, 241, 242, 265, 266, 268, 296, 297, 358, 359, 426, 486, 536, 537
Consensus, 84, 85, 127, 153, 281, 341, 403, 405, 454, 521, 526
Constraints, 50, 70, 71, 136, 145, 155, 170, 171, 173, 185, 193, 197, 304–306, 309, 312, 321, 342, 367, 382, 389, 406, 439, 486, 492, 493, 496, 497, 508, 509, 533, 548, 616
Content reflection, 431
Contract administration, 281, 282, 351–353
Control charts, 9, 212–217, 223, 225, 575, 576
CoQ. *See* Cost of quality (CoQ)
Corporate, 49, 270, 273, 347, 360, 362, 398, 399, 438, 564, 619
Correctness, 72, 138, 195, 202
Correlation, 104–108, 110, 223, 422, 471–478, 481
Correlation matrix, 421
Cost allocation, 287
Cost-benefit analysis, 507
Cost of quality (CoQ), 12, 15, 16, 268, 486
Cost performance index, 289, 294–296
Cost-plus incentive fee, 317–319

A. van Aartsengel and S. Kurtoglu, *Handbook on Continuous Improvement Transformation*,
DOI 10.1007/978-3-642-35901-9,

Cost variance, 289, 291–294
Covariances, 473, 478, 480, 482
Criticality, 209
Critical path, 159, 161, 163, 165, 167, 169, 171, 295, 296, 365, 377, 405
Critical to Cost (CTC), 131
Critical to Quality (CTQ), 131
Critical to Schedule, 131
Critical-to-X, 131
CTC. *See* Critical to Cost (CTC)
CTQ. *See* Critical to Quality (CTQ)
Culture, 402, 414, 504, 545
Customer value, 67, 80, 535
Cycle time, 491, 492

D
Data collection system, 76, 92, 94, 96, 98, 102–108, 110, 114, 203, 211, 225, 463, 467, 575, 623
Decomposition, 138, 139
Defect, 49, 74, 198, 201, 202, 230, 232, 233, 241, 440, 487, 531, 578, 628
Defect per ubiquitous, 231, 232
Delphi technique, 153, 397
Descriptive statistic, 98, 115, 119
Design of experiments, 468
Detection rating, 209
Diagrams, 42, 44, 45, 48, 147, 149, 150, 156, 157, 159, 160, 163–165, 169, 170, 223, 227, 228, 364, 372, 428, 454, 455, 457, 458, 460, 510, 512, 513, 543, 555
Dialogue, 328, 420, 440–442, 566, 604, 605, 629
Differential cost, 270
Direct cost, 257
Discrete elements, 2, 3, 26, 45, 207, 230, 231, 236, 241, 364, 483, 489–491, 502, 503, 509, 578
Double-loop learning, 527, 529, 530, 604

E
Earliest time event, 161
Earned value, 289–290
Economic value, 9, 221, 233, 291
Effectiveness, 198, 258, 270, 304, 362, 408, 423, 439, 527, 535, 550, 614
Effect size, 104, 121
Efficiency, 57, 139, 143, 179, 197, 198, 251, 294, 295, 343, 348, 362, 418, 530
Effort creep, 72, 73, 195
Estimate at/to completion cost, 289, 291
Ether, 573
Expense costs, 269
Expenses, 258, 261, 264, 268, 269, 272, 275, 281, 298, 321, 336, 361, 435, 437, 552, 553, 562, 563, 592, 599, 600
Experimental study, 206, 250, 465
Expert judgment, 70, 153, 193
Extraneous, 210, 247, 249, 250, 466, 536

F
Facilitated workshop, 70, 193
Facilitator, 84–86, 153, 400, 403–405, 454, 521
Failure costs, 14–15, 265–268
Failure mode and effect analysis (FMEA), 207–209, 466, 538, 578, 579
Feature creep, 72, 73, 195
Firm fixed-price contract, 316
First time yield (FTY), 230, 231
Fishbones, 457
Fisher's distribution, 121, 122, 124
5S, 497–499
Fixed costs, 180, 261–264, 284
Fixed-price
  contracts, 312, 313
  incentive, 313–315, 318, 319
Floats, 164
Fluctuations, 223, 294, 393
FMEA. *See* Failure mode and effect analysis (FMEA)
Focus groups, 80
Frequency plots, 212, 224–227, 575
F-Test, 114, 124, 125
FTY. *See* First time yield (FTY)
Fundamental changes, 261

G
Gantt charts, 157
Gauge R&R, 105, 108
Goal statement, 43, 50, 138
Group decision, 80, 86
Guidelines, 26, 45, 48, 49, 71, 189, 197, 356, 370, 382, 383, 409, 410, 497, 611

H
Hope creep, 72, 195
House of quality, 420
Hypothesis, 96–98, 102, 114, 116–120, 122–124, 206, 465, 468, 475, 476, 488, 492, 493, 496, 507–509, 513

I
Ideal State, 242, 243, 245, 526, 537, 540, 577
Income, 270
Incremental cost, 270, 271, 284
Indirect cost, 257, 258
Infer, 95, 96, 474, 576
Informal learning, 430, 588
Inspection, 26, 71, 195, 223
Integration, 141, 440, 441, 604, 621
Interaction, 88, 135, 207, 253, 326, 440, 469, 470, 483, 485, 488, 489, 497, 503, 505–509, 517, 526, 529, 550, 566, 578, 581, 583, 590, 629
Intra-block analysis, 252

J
Joint Application Development, 84
Just-in-time, 164, 344, 495

K
Kanban, 494, 495
Kano model, 128–130

L
Labor, 49, 131, 151, 179–182, 184, 188, 251, 257–259, 265, 269, 275, 277, 281, 282, 297, 321, 322, 337, 342, 380, 394, 514, 526, 550, 552
Leadership, 84, 183, 187, 388, 400, 545, 617
Legal requirement, 47
Lessons learned, 49, 298, 350, 362, 365, 398, 399, 436, 438, 503, 562, 564, 601, 602
Liability, 68
Likelihood of occurrence, 209, 382, 383, 402, 407, 408, 411–413, 418, 556, 594
Limit of variations, 9–12, 16, 17, 24, 231, 233, 535
Logit, 477

M
Make or Buy analysis, 279–284, 299
Mathematical expectation, 215, 217, 473, 478–480
Maturity, 25, 187, 440, 566, 581, 582, 603, 604, 618, 633
Median, 222
Methodology, 2, 25–27, 44, 202, 280, 282, 319, 352, 398, 399, 514, 615, 616, 623, 634
Milestone, 38, 43, 142, 149, 173, 370, 402, 407
Mixed costs, 263, 264

N
Network diagram, 150, 156, 163
Nominal group technique, 523
Nonconforming/nonconformity, 201, 215–217, 243, 342
Non-controllable costs, 264
Normal distribution, 8, 9, 98, 99, 103, 115, 116, 119, 120, 161, 213, 218, 219, 233
Norming, 187

O
Observation, 80, 550
Observational studies, 468–469
Operational definition, 3, 25–27, 29, 77, 78, 93, 205, 448, 464, 465, 528
Opportunity cost, 268, 270, 273
Ordinary least squares, 480, 481
Organizational process assets, 45, 196, 198, 202, 257, 277, 287, 298, 304, 307, 308, 343, 348, 350, 365, 366, 387, 399, 438, 564, 602
Orthogonality, 479, 480
Overhead costs, 257, 258, 273, 275

P
Pareto, 212, 227–229, 470, 575
Percent spent, 289
Performance, 319
Performing, 187, 441, 566, 604, 608, 612, 629
PERT. *See* Program Evaluation and Review Technique (PERT)
Phase review, 51, 435, 436, 438, 443, 556, 561, 562, 564, 569, 593, 599, 602
Piloting, 443, 523–527, 529–531, 533, 538, 540–544, 565, 569, 573
Plan-Do-Study-Act (PDSA) model, 5, 27–29, 31, 43, 51, 53, 54, 389, 435, 448, 561, 599
Planned value, 278, 284–287, 289–291, 295
Poisson's distribution, 216, 217
Poka-Yoke, 539
Predictable, 93, 94, 107, 134, 196, 212, 239, 242, 243, 245, 388, 449, 454, 526, 537, 576, 583
Premise reflection, 431
Prevention costs, 13, 265

Prioritization matrix, 207, 466, 522, 523
Process
analysis, 70, 193, 194
behavior charts, 212, 448
capability, 229
control plan, 574, 577–579
defect rate, 229
discrete element, 485–489, 491–493, 496, 501–510
future state, 517
map, 510, 513–516
mapping, 512–517
reflection, 431
stability, 240
yield, 229, 230
Productivity, 44, 143, 152, 179, 388, 394, 440, 445, 508, 509, 542, 550, 571, 590, 620, 628
Profits, 135, 264, 287, 313–315, 321, 440, 489, 619, 628
Profound knowledge, 526, 529
Program Evaluation and Review Technique (PERT), 152, 157, 163–165, 169, 397, 406, 407
Project charter, 33, 42, 43, 45, 49, 51, 71, 191, 390
Project retrospective, 432, 561, 598
Prototypes, 80, 524
Prototype solution, 525–527, 529–531, 533–536, 538, 540–542, 544, 545, 565, 569, 573, 574, 576–578, 608, 630
Prototyping, 88, 524
PSDA model. *See* Plan-Do-Study-Act (PDSA) model

Q
Quality
audit, 196
control, 72, 195–202, 258, 283, 309, 388, 454, 551, 552, 579, 591
costs, 12
Questionnaires, 80

R
Random, 7, 8, 73, 90–93, 95, 97, 98, 101–103, 106, 114, 115, 119–121, 124, 125, 153, 196, 213, 215, 216, 218, 219, 234, 240, 244, 250, 251, 352, 379, 382, 384, 469, 470, 477, 526, 527, 576
Randomization, 468
Randomize, 249
Reasonableness, 360
Recurring costs, 265, 335
Reflection process, 429–431, 560, 561, 598
Regression, 153, 224, 477–482
Relationship matrix, 420, 422
Repeatability, 94, 110
Reproducibility, 94, 110
Request for information, 324–326, 350
Request for proposal, 281, 301, 324, 326–337, 341, 346, 415
Resource leveling, 172
Retrospective, 573, 575, 576, 583, 630
Retrospective session, 433, 598
Return on investment, 69, 344, 418, 534, 574
Rework, 11, 12, 16, 152, 170, 193, 201, 202, 212, 230, 231, 486, 487, 489, 511, 516, 536
Risk
categories, 49, 392
constraint, 384
event, 382, 384, 394–396, 407–414, 417–420, 426
response matrix, 420, 422
Rolled throughput yield (RTY), 230, 231, 241, 244, 536, 576
Rolling wave planning, 142
RTY. *See* Rolled throughput yield (RTY)
Run charts, 212, 221–223, 575

S
Sampling, 90
Satisfaction, 11, 24, 33, 69, 76, 129, 130, 132, 136, 319, 344, 345, 354, 355, 358, 440, 512, 534, 574, 614, 618, 628
Satterthwaite's approximation, 117
Scatter diagrams, 212, 223–224, 575
Schedule
baseline, 173, 175–177
performance index, 176, 289, 294–296
variance, 289, 291–294
Scope creep, 72, 73, 298
Scrap, 11, 12, 16, 212, 273, 489
Severity, 209, 210, 382, 417, 556, 594
Shewhart, W.A., 9, 25–27, 233, 530
Sigma, 7–10, 17, 233, 441, 566, 605, 621
Similarity technique, 153
Single-loop learning, 530
Slacks, 163, 164, 172
Slack time, 159, 163, 168, 391
Specifications, 11, 12, 102, 130–136, 170, 183, 189, 200, 201, 234–240, 243, 272, 277, 280, 281, 297, 299, 305, 308–314, 323, 331, 332, 339, 345, 347, 359, 536, 554, 579, 597, 615

Stakeholder registry, 42
Standard costs, 265
Standard deviation, 7, 8, 16, 95, 96, 100, 102, 105, 116, 214, 219, 233, 235, 236, 238, 473, 536, 537, 577
Standardization, 330, 345, 513
Statement of work, 45, 46, 65, 307, 308, 312, 315, 319, 328, 336, 338, 339, 350, 352, 359, 360, 437
State of Total Failure, 244, 245
Statistical inference, 95, 96
Storming, 187
Strategic, 40, 46, 142, 196, 274, 282, 283, 304, 346, 371, 391, 392, 394–396, 398, 422, 582
Student's t-distribution, 116, 117, 120, 475
Student's t-percentile, 116, 117
Student's t-test, 114, 121, 122
Subject matter experts, 70, 193
Sub-optimization, 612
Success
 criteria, 43, 44, 382–384, 394, 402, 409, 410, 412–415, 417, 418, 423–425, 540, 616
 factors, 331, 534, 574, 575
Sunk cost, 271
Surveys, 44, 80, 87
Sustain, 2, 5, 188, 540, 569, 581, 583, 607, 610, 620
System theory, 526

T
Takt time, 491–493
Threshold state, 243, 245, 526
Timesheet, 548, 549
Total cost of ownership, 344
Trust, 84, 550, 590
Tuckman's model, 187, 188

U
Underperformance, 75, 77, 135, 203, 204, 206, 453, 458, 463, 464, 485, 489, 517, 519–522, 524, 534
Unpredictable, 8, 93, 94, 236, 239, 243–245, 449, 468, 527, 576

V
Value-added, 489
Variability, 94, 95, 101, 114, 117, 122–124, 217–219, 222, 234, 236, 251, 311, 406, 482, 537, 577, 612
Variable cost, 259, 260, 263, 264
Variance reports, 375, 377
Vendor, 38, 152, 297, 330, 341, 400, 417, 515
Voice, 67, 69, 78, 80, 192, 193, 205, 206, 238, 534, 574

W
WBS. *See* Work breakdown structure (WBS)
Wisdom, 442, 521, 575, 587, 588, 603, 610
Work breakdown structure (WBS), 46, 48, 50, 137–142, 145–147, 150, 155, 156, 257, 278, 286, 289, 291, 294, 295, 390, 417, 419, 422
Workshops, 80, 84

Z
Zero-defect, 16

CPSIA information can be obtained at www.ICGtesting.com
Printed in the USA
LVOW070507270413

331212LV00005B/199/P

9 783642 359002